T0310399

POWER LINE COMMUNICATIONS

POWER LINE COMMUNICATIONS

PRINCIPLES, STANDARDS AND APPLICATIONS FROM MULTIMEDIA TO SMART GRID

Second Edition

Edited by

Lutz Lampe, Andrea M. Tonello, and Theo G. Swart

This edition first published 2016
© 2016, John Wiley & Sons, Ltd

First Edition published in 2010

Registered office
John Wiley & Sons Ltd, The Atrium, Southern Gate, Chichester, West Sussex, PO19 8SQ, United Kingdom

For details of our global editorial offices, for customer services and for information about how to apply for permission to reuse the copyright material in this book please see our website at www.wiley.com.

The right of the author to be identified as the author of this work has been asserted in accordance with the Copyright, Designs and Patents Act 1988.

All rights reserved. No part of this publication may be reproduced, stored in a retrieval system, or transmitted, in any form or by any means, electronic, mechanical, photocopying, recording or otherwise, except as permitted by the UK Copyright, Designs and Patents Act 1988, without the prior permission of the publisher.

Wiley also publishes its books in a variety of electronic formats. Some content that appears in print may not be available in electronic books.

Designations used by companies to distinguish their products are often claimed as trademarks. All brand names and product names used in this book are trade names, service marks, trademarks or registered trademarks of their respective owners. The publisher is not associated with any product or vendor mentioned in this book.

Limit of Liability/Disclaimer of Warranty: While the publisher and author have used their best efforts in preparing this book, they make no representations or warranties with respect to the accuracy or completeness of the contents of this book and specifically disclaim any implied warranties of merchantability or fitness for a particular purpose. It is sold on the understanding that the publisher is not engaged in rendering professional services and neither the publisher nor the author shall be liable for damages arising herefrom. If professional advice or other expert assistance is required, the services of a competent professional should be sought.

Library of Congress Cataloging-in-Publication Data

Names: Lampe, Lutz, editor. | Tonello, Andrea M., editor. | Swart, Theo G., editor.
Title: Power line communications : principles, standards and applications from multimedia to smart grid /
    [edited by] Lutz Lampe, Andrea M. Tonello, Theo G. Swart.
Description: Second edition | Chichester, UK ; Hoboken, NJ : John Wiley & Sons, 2016. | Includes index. |
    Previously published as: Power line communications : theory and applications for narrowband and broadband
    communications over power lines, 2010.
Identifiers: LCCN 2015048934| ISBN 9781118676714 (cloth) | ISBN 9781118676776 (epub)
Subjects: LCSH: Electric lines–Carrier transmission. | Broadband communication systems.
Classification: LCC TK5103.15 .P695 2016 | DDC 621.382–dc23
LC record available at http://lccn.loc.gov/2015048934

A catalogue record for this book is available from the British Library.

Set in 10/12pt Times by Aptara Inc., New Delhi, India
Printed and bound in Malaysia by Vivar Printing Sdn Bhd

1   2016

Contents

5   Digital Transmission Techniques   261

K. Dostert, M. Girotto, L. Lampe, R. Raheli, D. Rieken, T. G. Swart, A. M.
Tonello, A. J. H. Vinck, and S. Weiss

9 PLC for Smart Grid 509

I. Berganza, G. Bumiller, A. Dabak, R. Lehnert, A. Mengi, and A. Sendin

List of Contributors

Iñigo Berganza, Iberdrola, Section 9.1, 9.4

Gerd Bumiller, Hochschule Ruhr West – University of Applied Sciences, Chapter 7, Section 9.1

Francisco J. Cañete, University of Málaga, Section 2.3.1, 2.3.3, 2.9.2

José A. Cortés, University of Málaga, Section 6.4

Anand Dabak, Texas Instruments, Section 9.3

Salvatore D'Alessandro, University of Udine, Section 6.4

Le Phu Do, Dresden University of Technology, Section 6.2, 6.3

Klaus Dostert, Karlsruhe Institute of Technology, Section 2.3.1, 2.3.2, 5.2.2

Stefano Galli, Assia Inc., Section 2.2, Chapter 8

Mauro Girotto, University of Udine, Section 5.5

George Hallak, Hochschule Ruhr West – University of Applied Sciences, Chapter 7

Holger Hirsch, University of Duisburg-Essen, Section 3.1 to 3.6 (except 3.5.3)

Masaaki Katayama, Nagoya University, Section 2.7

Cornelis J. Kikkert, James Cook University, Chapter 4

Michael Koch, devolo AG, Section 3.1 to 3.6 (except 3.5.3)

Lutz Lampe, University of British Columbia, Editor, Section 2.8, 5.5, 6.5, Chapter 10

Haniph Latchman, University of Florida, Chapter 8

Ralf Lehnert, Dresden University of Technology, Section 6.2, 6.3, 9.1

Martine Lienard, University of Lille, Section 2.9.2

Sina Mashayekhi, University of British Columbia, Section 2.4

Anil Mengi, devolo AG, Section 9.2, 9.3

Dave G. Michelson, University of British Columbia, Section 2.4

Marcel Nassar, The University of Texas at Austin, Section 2.7

Moslem Noori, University of British Columbia, Section 6.5

Fabienne Nouvel, Institut d'Electronique et de Télécommun. de Rennes, Chapter 10

Vladimir Oksman, Lantiq, Chapter 8

Riccardo Pighi, Selta Group, Section 2.5

Antti Pinomaa, Lappeenranta University of Technology, Section 2.9.1

Gautham Prasad, University of Florida, Chapter 8

Riccardo Raheli, University of Parma, Section 5.2

Marco Raugi, University of Pisa, Section 2.9.3

David Rieken, Aclara, Section 5.4

Alberto Sendin, Iberdrola, Section 9.1, 9.4

Theo G. Swart, University of Johannesburg, Editor, Section 5.8

Andrea M. Tonello, University of Klagenfurt, Editor, Section 2.3.3, 2.6, 2.8, 5.3, 5.5, 6.4, 6.5

Mauro Tucci, University of Pisa, Section 2.9.3

Fabio Versolatto, WiTiKee, Section 2.6

A. J. Han Vinck, University of Duisburg-Essen, Section 5.2.1

Stephan Weiss, University of Strathclyde, Section 5.7

Nico Weling, devolo AG, Section 3.5.3

Lawrence W. Yonge III, Qualcomm Atheros, Inc., Chapter 8

Ahmed Zeddam, France Telecom, Orange Labs, Section 3.7

Preface

This book is the second edition of *Power Line Communications: Theory and Applications for Narrowband and Broadband Communications over Power Lines* published in 2010. As for the first edition, it has been our intention to present the most comprehensive coverage of the technical field of power line communications (PLC) that is available in a single publication. The scope of this book is uniquely wide, not only for a book on PLC. Compared to the first edition, the content has been updated and in part restructured. In particular, we have significantly expanded the part dedicated to applications of PLC, which is attributed to the further maturity of PLC technology in terms of consolidated specifications and standards and also reflected in the modification of the subtitle for this edition. Furthermore, recent innovations and changes related to channel characterization, transmission techniques and regulation are included in this edition.

The target audience for the book comprises both newcomers to the exciting field of PLC as well as researchers and practitioners already familiar with PLC. For the former, the book is intended to provide a fairly comprehensive yet readable introduction. For the latter, we expect the book to serve as an authoritative point of reference for information widely dispersed in the literature.

During the writing of this second edition, we involved 42 technical contributors from 29 institutions and 12 countries. Coordination was a huge task, almost more so than for the first edition. The editors would like to express their sincere thanks to all the contributors.

List of Acronyms

AC	Alternating Current
ACF	Autocorrelation Function
ACG	Average Channel Gain
AF	Amplify-and-forward
AM	Amplitude Modulation
AMI	Advanced Metering Infrastructure
AMN	Artificial Mains Network
AMR	Automatic Meter Reading
ARIB	Association of Radio Industries and Businesses
AU	Allocation Unit
AVLN	AV Logical Network
AWGN	Additive White Gaussian Noise
BB	Broadband
BER	Bit Error Ratio
BPL	Broadband Over Power Lines
B-PLC	Broadband PLC
BPRS	Binary Pseudo-random Sequence
BPSK	Binary Phase-shift Keying
BS	Base Station
CA-Msg	Channel Announcement Message
CAN	Controller Area Network
CB-FMT	Cyclic Block Filtered Multitone Modulation
CCDF	Complementary Cumulative Distribution Function
CCo	Central Coordinator
CDCF	Commonly Distributed Coordination Function
CDF	Cumulative Distribution Function
CDMA	Code Division Multiple Access
CE	Conformité Européenne
CEI	Customer-end Inverter
CENELEC	Comité Européenne de Normalisation Electrotechnique
CFP	Contention Free Period
CFR	Channel Frequency Response
CISPR	International Special Committee on Radio Interference
CM	Common Mode *or* Connection Manager

CP	Cyclic Prefix *or* Contention Period
CPE	Customer Premise Equipment
CRC	Cyclic Redundancy Check
CSI	Channel State Information
CSMA	Carrier Sense Multiple Access
CSMA/CA	Carrier Sense Multiple Access with Collision Avoidance
DBPSK	Binary DPSK
DCA	Dynamic Channel Allocation
DCT	Discrete Cosine Transform
DF	Decode-and-Forward
DFT	Discrete Fourier Transform
DLL	Data Link Layer
DM	Differential Mode *or* Domain Master
DPSK	Differential Phase Shift Keying
DQPSK	Quaternary DPSK
DSL	Digital Subscriber Line
DSM	Demand Side Management
DSSS	Direct Sequencing Spread Spectrum
DSTBC	Distributed Space-time Block Codes
DT	Direct Transmission
DWMT	Discrete Wavelet Multitone
EC	European Commission
ECC	Error Correction Code
ECU	Electronic Controlled Unit
EIB	European Installation Bus
EMC	Electromagnetic Compatibility
ETSI	European Telecommunications Standards Institute
EU	European Union
EUT	Equipment Under Test
EV	Electric Vehicle
FB	Filter Bank
FCC	Federal Communications Commission
FD	Frequency Domain
FDMA	Frequency Division Multiple Access
FEC	Forward Error Correction
FFT	Fast Fourier Transform
FH	Frequency Hopping
FIR	Finite Impulse Response
FMT	Filtered Multitone
FSK	Frequency-shift Keying
HDCU	High Data Rate Central Control Unit
HD-PLC	High-definition Power Line Communication
HDR	High Data Rate
HDTV	High Definition Television
HF	High-frequency
HPAV	HomePlug AV

HV	High Voltage, 66 kV and above
ICI	Inter-carrier Interference
IDFT	Inverse DFT
IEC	International Electrotechnical Commission
IFFT	Inverse Fast Fourier Transform
IGBT	Insulated Gate Bipolar Transistors
IH	In-home
IN	Impulse Noise
INL	Interfering Network List
IP	Internet Protocol *or* Integer Programming
IPTV	Internet Protocol Television
ISI	Inter-symbol Interference
ISN	Impedance Stabilization Network
ISP	Inter-system Protocol
ITU	International Telecommunication Union
LAN	Local Area Network
LCL	Longitudinal Conversion Loss
LDCU	Low Data Rate Central Control Unit
LDPC	Low-density Parity-check
LDR	Low Data Rate
LLR	Log-likelihood Ratio
LMS	Least Mean Square
LP	Linear Programming
LPTV	Linear Periodically Time Variant
LTI	Linear Time Invariant
LV	Low Voltage, 110 V to 400 V
LVDC	Low-voltage Direct Current
MAC	Medium Access Control
MAI	Multiple Access Interference
MC	Multicarrier
MDCU	Multiple Data Rate Central Control Unit
MDU	Multi Dwelling Unit
MF	Matched Filter
MIMO	Multiple-input Multiple-output
MLD	Maximum-likelihood Detection
MMSE	Minimum Mean Square Error
MMU	Master Monitoring Unit
MTL	Multi-conductor Transmission Line
MV	Medium Voltage, 7.2 kV to 33 kV
MWR	Multi-way Relaying
NB	Narrowband
OAF	Opportunistic AF
ODF	Opportunistic DF
OFDM	Orthogonal Frequency Division Multiplexing
OFDMA	Orthogonal Frequency Division Multiple Access
OH	Overhead

OOB	Out of Band
OOK	On-off Keying
OPERA	Open PLC European Research Alliance
OQAM	Offset Quadrature Amplitude Modulation
OSI	Open Systems Interconnection
OSTBC	Orthogonal Space-time Block Codes
PAM	Pulse Amplitude Modulation
PDF	Probability Density Function
PHY	Physical
PLC	Power Line Communication
PLCP	Physical-layer Convergence Protocol
PoE	Power over Ethernet
PR	Perfect Reconstruction
PSD	Power Spectral Density
PSK	Phase Shift Keying
PTC	Positive Temperature Coefficient
PVC	Polyvinylchloride
QAM	Quadrature Amplitude Modulation
QC-LDPC	Quasi-cyclic Low-density Parity-check
QoS	Quality of Service
RF	Radio Frequency
RMS-DS	Root-mean-square Delay Spread
ROBO	Robust Modulation
RS	Reed-Solomon
RX	Receiver
SFN	Single Frequency Networking
SINR	Signal-to-noise and Interference Ratio
SISO	Single-input Single-output
SNR	Signal-to-noise Ratio
SST	Spread Spectrum Techniques
STBC	Space-time Block Coding
STFT	Short Time Fourier Transform
SVD	Singular Value Decomposition
TCL	Transverse Conversion Loss
TCTL	Transverse Conversion Transfer Loss
TDM	Time Division Multiplex
TDMA	Time Division Multiple Access
TEM	Transverse Electromagnetic
TF	Time Frame
T-ISN	T-shaped Impedance Stabilization Network
TL	Transmission Line
TS	Time Slot
TWR	Two-way Relaying
TX	Transmitter
TXOP	Transmission Opportunities
UDP	User Datagram Protocol

UPA	Universal Powerline Association
UTP	Unshielded Twisted Pair
UWB	Ultra Wide Band
VDSL	Very High Bit Rate Digital Subscriber Line
VLF	Very Low Frequency
VoIP	Voice Over Internet Protocol
WLAN	Wireless Local Area Network

1

Introduction

L. Lampe, A. M. Tonello, and T. G. Swart

Power line communications (PLC) reuse existing infrastructures (i.e. power lines) whose primary purpose is the delivery of AC (50 Hz or 60 Hz) or DC electric power, for the purpose of data communications. Hence, compared to the electric power 'signal', PLC uses high-frequency signals with frequency components starting from a few hundred Hz up to a few hundred MHz. The plurality of frequency bands used for PLC is related to different applications supported by PLC and their data-rate requirements, the specifics of grid topologies over which PLC is applied, as well as the ability of PLC technology to deal with the harsh communication environment. Before elaborating on this further, we first briefly review the terminology that has been used to describe PLC.

1.1 What is a Name?

Communication over power lines is referred to by different names that are often specific to the considered grid domain and application. The most commonly used terminologies are summarized in the following.

- Carrier-current systems: This term refers to the fact that carrier-modulated data signals are transmitted over power lines. It has often been used to collectively describe relatively narrowband signals with frequencies below 500 kHz. The Code of Federal Regulations, Title 47, Part 15, from the U.S. Federal Communications Commission (FCC) [1] defines carrier-current systems as 'A system, or part of a system, that transmits radio frequency energy by conduction over the electric power lines.'
- Power line carrier: Similar to carrier-current systems, this is an early terminology used for systems that transmit carrier-modulated signals over power lines. A prominent example of its use is the 'Guide to Application and Treatment of Channels for Power-Line Carrier' by the 'AIEE Committee on Carrier Current' of the American Institute of Electrical Engineers

Power Line Communications: Principles, Standards and Applications from Multimedia to Smart Grid, Second Edition. Edited by Lutz Lampe, Andrea M. Tonello, and Theo G. Swart.
© 2016 John Wiley & Sons, Ltd. Published 2016 by John Wiley & Sons, Ltd.

(AIEE) [2], see also [3]. Also due to is earlier use, it typically refers to systems operating at frequencies below 500 kHz.

- Distribution line carrier (DLC): DLC refers to power line communication systems serving applications in the distribution domain. Due to the many line discontinuities and branches in the distribution grid, DLC systems face a more difficult communication environment than power line communication systems operating in the transmission segment of the power grid. DLC usually describes systems using frequencies below 500 kHz.
- Broadband over power lines (BPL): BPL is a more recent terminology that refers to systems operating in the frequency range of about 2 MHz to 30 MHz and beyond, with a signal bandwidth of tens of MHz and with data rates ranging from several Mbps to hundreds of Mbps; hence the term 'broadband'. The application of BPL systems is mainly in the distribution part of the grid, to enable broadband access, as well as for in-home communication. 'BPL' is mostly used in North America. For example, the Subpart G of [1] is entitled 'Access Broadband Over Power Line (Access BPL)'.
- Power line telecommunications (PLT): This term is used similar to BPL, but it is more popular in European countries. For example, the European Telecommunication Standards Institute (ETSI) produced numerous reports and specifications on 'PLT' through its 'ETSI Technical Committee Power-Line Telecommunications (PLT).'

In this book, we understand and use the term 'power line communications (PLC)' as including all of the above, which has been widely accepted by now. For example, the leading scientific conference on the topic is the 'International Symposium on Power Line Communications and Its Applications (ISPLC)' [4], and the IEEE Communications Society has established the 'Technical Committee on Power Line Communications (TC-PLC)' [5]. To differentiate the various PLC technologies, reference [6] introduced a classification of PLC into ultra-narrowband (UNB) PLC, low data rate narrowband (LDR NB) PLC, high data rate narrowband (HDR NB) PLC and broadband (BB) PLC. We will discuss this further in the context of the historical development of PLC in the next section.

1.2 Historical Notes

Figure 1.1 illustrates the evolution of the PLC technology by identifying some early patents, specific application domains and international standards along a timeline.

The origins of PLC can be traced back to the late 1800s and again in the early 1990s. Patents [11] and [12] consider remote meter reading via PLC (see [13]). The first description of remote load management using PLC, or so-called ripple control, is given in [14] (we note that [15] mentions the slightly earlier patent submission [16]). These ripple control systems (RCS) were developed further in the 1930s [17] and at a larger scale in the 1950s [18] to establish unidirectional communication for load management and other control functions in the power distribution grid. RCS use high-power and narrowband PLC signals. The signal frequencies are between 125 Hz and 3 kHz so that signals can pass through the distribution transformers and reach consumers. Before the widespread use of PLC via ripple control in the distribution domain, power line voice communication over medium-voltage and high-voltage transmission lines became popular in the 1920s [19]. These systems operate in the frequency

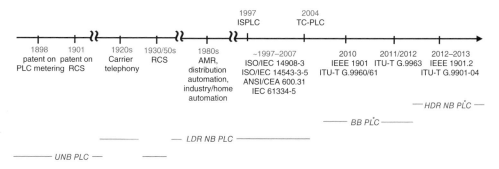

Figure 1.1 Illustration of the evolution of PLC technology. *: BB PLC industry specifications were released in early 2000, e.g. [7]. The first HDR NB PLC systems were presented between 2001 and 2009 [8–10].

range of about 15 kHz to 500 kHz and use signal bandwidths of a few kHz. Later, protective relaying has become a major application for these type of systems [20].

Two-way communication over power lines developed further in the 1980s with PLC systems designed for automatic meter reading (AMR) and automation in the distribution grid, as well as for industry and building automation [6, 18]. The deployment of these systems has been facilitated by the publication of the European Norm EN 50065 'Signalling on low-voltage electrical installations in the frequency range 3 kHz to 148.5 kHz' in 1991.

All of the PLC systems mentioned so far fall under the classes of UNB and LDR NB PLC (see Figure 1.1). The former is defined as operating below 3 kHz and providing data rates of the order of 100 bps, while the latter operate between 3 kHz to 500 kHz with data rates of a few kbps [6]. This changed in the late 1990s, also since electric power utilities considered providing additional consumer services through their lines in the wake of the deregulation of the telecommunication and energy markets in Europe. Broadband PLC systems which use frequency bands between about 1.8 MHz to 250 MHz and provide data rates ranging from several Mbps to several hundred Mbps [6] were developed and deployed for Internet access and in-home multimedia applications. The introduction of BB PLC was accompanied by a surge in research activities, as can be seen from the first ISPLC[1] in 1997, related special issues in journals and magazines [22–26], books [27], and the founding of the IEEE Communications Society TC-PLC in 2004 (see Figure 1.1). Specifications for BB PLC systems were consolidated in the IEEE 1901 [28] and ITU-T G.9960/61 [29, 30] standards in 2010.

A second wave of innovation in the early 2000s shifted the focus back to NB PLC. With the 'Smart Grid' vision taking shape, it was natural for electric power utilities to consider PLC as a means to achieve an efficient and reliable communication infrastructure [6]. While NB PLC solutions have been available to provide, for example, basic AMR services, transmission methods successfully used in BB PLC were adopted for a new class of HDR NB PLC systems supporting between tens of kbps to about 500 kbps [6]. System specifications for HDR NB PLC have been published in the standards ITU-T G.9901-9904 [31–34] and IEEE 1901.2 [35]

---

[1] The International Symposium on Power Line Communications and Its Applications (ISPLC) became IEEE ISPLC in 2006. A data base containing the full papers of the earlier ISPLC conference proceedings is available at [21].

in 2012 and 2013, respectively. Smart grid applications have become one of the main drivers of innovation for both HDR NB and BB PLC, with publications (e.g. [6, 36, 37]) and conference sessions and panels dedicated to this topic.

Modern PLC systems use the latest signal processing techniques including advanced concepts such as multicarrier modulation with adaptive notching and multiple-input multiple-output (MIMO) transmission [38, 39], which have already been adopted in international standards [40] and industry specifications [41] (see Figure 1.1). This goes to show that the PLC technology of today is on a par with what is used for modern wireless communication and communication systems over other wired media. However, since PLC operates over 'live wires', there are distinct features to PLC technology not experienced for other media. An example is the efficient and safe coupling of signals onto and from power lines, which requires special considerations for instance through standards such as [42].

1.3   About the Book

Considering the above-outlined evolution of PLC over the past 100 or so years, as well as the fact that the PLC literature has been widely dispersed, the first edition of the book 'Power Line Communications' [43] was compiled with the objective of being a comprehensive single point of reference for researchers and practitioners either new to the field or already familiar with aspects of PLC. Since its publication in 2010, PLC has experienced important innovations both in terms of technology and its application. For example, new signal processing methods have been introduced to PLC, channel models for PLC have been refined and consolidated, and regulation and standardization have evolved as can be seen from the newly adopted international standards for HDR NB PLC for the smart grid domain. Hence, this second edition provides an updated coverage of the field of PLC. Compared to the first edition, we have reorganized the chapter structure and expanded the coverage of applications of PLC in Chapters 7 to 10. Chapters 2, 3 and 5 have undergone major revisions and include significantly new content, and Chapters 4 and 6 have been rewritten. We thus see this second edition of 'Power Line Communications' as a continuation and extension as well as an in part complementary read of the first edition.

The technical material of this book is organized in Chapters 2 to 10 as follows.

The characterization of power lines and power line networks as transmission media for digital communication signals is one of the very first and fundamental steps towards successful design and implementation of PLC systems. Chapter 2 thus provides an overview of the state of the art in channel characterization and modeling for PLC in different parts of the grid, i.e. the low voltage (LV), medium voltage (MV) and high voltage (HV) sections. Furthermore, it describes the PLC channel in other application domains, namely, the LVDC distribution grid, the in-car domain and the in-ship domain.

Electromagnetic compatibility (EMC) is a cardinal factor when deploying a communication system. Chapter 3 introduces the electromagnetic effects associated with PLC systems, discusses the potential to cause interference to established communication services including other wireline communication systems, and elaborates on EMC regulation.

Chapter 4 discusses the coupling of PLC signals to and from the power line network. It describes the fundamental requirements and principles of coupling networks and covers coupling for LV, MV and HV power lines.

Digital transmission, the core of any communications system, is investigated for PLC systems in Chapter 5. The chapter presents a suite of modulation, coding and detection techniques, which include those used in NB and BB systems mentioned in the previous section. Specifically, the chapter considers multicarrier modulation, current and voltage modulation, ultra wide band modulation, as well as, MIMO algorithms, impulsive noise mitigation techniques, and channel coding schemes.

Chapter 6 addresses medium access control for PLC, considering power lines as a shared medium for which resource allocation strategies are required. This includes cooperative or relay communication methods, which has been a relatively recent topic in PLC.

Different application domains for PLC are covered in Chapters 7 to 10. PLC has been well established for home and industry automation for many years (see Figure 1.1). Chapter 7 reviews pertinent PLC system specifications for different application environments. BB PLC solutions with application to multimedia systems and the latest standards indicated in Figure 1.1 are presented in Chapter 8. Chapter 9 covers the use of PLC for smart grid communications. This includes the standards listed in Figure 1.1, discussion of requirements for different smart grid applications, the regulatory framework for NB PLC, and application and deployment examples. Finally, Chapter 10 provides an overview of the use of PLC for intra-vehicle communication with a focus on automotive applications.

References

1. U.S. Federal Communications Commission (FCC), Code of Federal Regulations, Title 47, Part 15 (47 CFR 15), Sep. 19, 2005.
2. AIEE Committee Report, Guide to application and treatment of channels for power-line carrier, *AIEE Trans. Power App. Syst.*, 73(1), 417–436, Jan. 1954.
3. Power System Communications Committee, Summary of an IEEE guide for power-line carrier applications, *IEEE Trans. Power App. Syst.*, 99(6), 2334–2337, Nov. 1980.
4. IEEE International Symposium on Power Line Communications and its Applications (ISPLC). [Online]. Available: http://www.isplc.org
5. IEEE Communications Society Technical Committee on Power Line Communications. [Online]. Available: http://committees.comsoc.org/plc/
6. S. Galli, A. Scaglione, and Z. Wang, For the grid and through the grid: The role of power line communications in the smart grid, *Proc. IEEE*, 99(6), 998–1027, Jun. 2011.
7. M. K. Lee, R. E. Newman, H. A. Latchman, S. Katar, and L. Yonge, HomePlug 1.0 powerline communication LANs — protocol description and performance results, *Int. J. Commun. Syst.*, 16(5), 447–473, May 2003.
8. G. Bumiller and M. Deinzer, Narrow band power-line chipset for telecommunication and internet application, in *Proc. Int. Symp. Power Line Commun. Applic.*, Malmö, Sweden, Apr. 4–6, 2001, 353–358.
9. I. Berganza, A. Sendin, and J. Arriola, PRIME: Powerline intelligent metering evolution, in *IET CIRED Seminar: SmartGrids for Distribution*, Frankfurt, Germany, Jun. 23–24, 2008, 3–4.
10. K. Razazian, M. Umari, and A. Kamalizad, Error correction mechanism in the new G3-PLC specification for powerline communication, in *Proc. IEEE Int. Symp. Power Line Commun. Applic.*, Rio de Janeiro, Brazil, Mar. 28–31, 2010, 50–55.
11. J. Routin and C. E. L. Brown, Improvements in and relating to electricity meters, British Patent GB 189 724 833, Oct. 1898.
12. C. H. Thordarson, Electric central station recoding mechanism for meters, U.S. Patent US 784 712, Mar. 1905.
13. P. A. Brown, Power line communications — Past present and future, in *Proc. Int. Symp. Power Line Commun. Applic.*, Lancaster, UK, Mar. 30–Apr. 1, 1999, 1–8.
14. C. R. Loubery, Improved method of telegraphing, indicating time, or actuating mechanism electrically, British Patent GB 190 000 138, Jan. 1901.

15. J. Fritz. (2011, Sep.) Rundsteuertechnik. Accessed: March 2015. [Online]. Available: http://www.rundsteuerung. de/

16. C. R. Loubery, Einrichtung zur elektrischen Zeichengebung an die Teilnehmer eines Startstromnetzes, German Patent Nr. 118 717, Mar. 1901.

17. K. Dostert, *Powerline Communications*. Prentice Hall, 2001.

18. D. Dzung, I. Berganza, and A. Sendin, Evolution of powerline communications for smart distribution: From ripple control to OFDM, in *Proc. IEEE Int. Symp. Power Line Commun. Applic.*, Udine, Italy, Apr. 3–6, 2011, 474–478.

19. M. Schwartz, Carrier-wave telephony over power lines: Early history, *IEEE Commun. Mag.*, 47(1), 14–18, Jan. 2009.

20. IEEE guide for power-line carrier applications, IEEE Standards Association, Standard 643-2004, 2005.

21. PLC DocSearch. [Online]. Available: http://www.isplc.org/docsearch/

22. H. A. Latchman and L. W. Yonge (Guest Editors), Power line local area networking, *IEEE Commun. Mag.*, 31(4), 32–33, Apr. 2003.

23. S. Galli, A. Scaglione, and K. Dostert (Guest Editors), Broadband is power: Internet access through the power line network, *IEEE Commun. Mag.*, 31(5), 82–83, May 2003.

24. F. N. Pavlidou, H. A. Latchman, A. J. H. Vinck, and R. E. Newman (Guest Editors), Powerline communications and applications, *Int. J. Commun. Syst.*, 16(5), 357–361, Jun. 2003.

25. E. Biglieri, S. Galli, Y.-W. Lee, H. V. Poor, and A. J. H. Vinck (Guest Editors), Special issue on power line communications, *IEEE J. Sel. Areas Commun.*, 24(7), 1261–1266, Jul. 2006.

26. M. V. Ribeiro, L. Lampe, K. Dostert, and H. Hrasnica (Guest Editors), Special issue on advanced signal processing and computational intelligence techniques for power line communications, *EURASIP J. Adv. Signal Process.*, vol. 2007, article ID 45812, 3 pp.

27. H. Hrasnica, A. Haidine, and R. Lehnert, *Broadband Powerline Communications Networks: Network Design*. John Wiley & Sons Ltd, Chichester, 2004.

28. IEEE standard for broadband over power line networks: Medium access control and physical layer specifications, IEEE Standards Association, IEEE Standard 1901-2010, Sep. 2010. [Online]. Available: http://standards. ieee.org/findstds/standard/1901-2010.html

29. Unified high-speed wire-line based home networking transceivers – system architecture and physical layer specification, ITU-T, Recommendation G.9960, 2011. [Online]. Available: https://www.itu.int/rec/T-REC-G.9960

30. Unified high-speed wire-line based home networking transceivers – data link layer specification, ITU-T, Recommendation G.9961, 2014. [Online]. Available: http://www.itu.int/rec/T-REC-G.9961

31. Narrowband orthogonal frequency division multiplexing power line communication transceivers – power spectral density specification, ITU-T, Recommendation G.9901, Nov. 2012. [Online]. Available: http://www.itu.int/rec/ T-REC-G.9901-201211-I/en

32. Narrowband orthogonal frequency division multiplexing power line communication transceivers for ITU-T G.hnem networks, ITU-T, Recommendation G.9902, Oct. 2012. [Online]. Available: http://www.itu.int/rec/ T-REC-G.9902

33. Narrowband orthogonal frequency division multiplexing power line communication transceivers for G3-PLC networks, ITU-T, Recommendation G.9903, May 2013. [Online]. Available: http://www.itu.int/rec/T-REC-G. 9903

34. Narrowband orthogonal frequency division multiplexing power line communication transceivers for PRIME networks, ITU-T, Recommendation G.9904, Oct. 2012. [Online]. Available: http://www.itu.int/rec/T-REC-G.9904-201210-I/en

35. IEEE standard for low-frequency (less than 500 kHz) narrowband power line communications for smart grid applications, IEEE Standards Association, IEEE Standard 1901.2-2013, Dec. 2013. [Online]. Available: https://standards.ieee.org/findstds/standard/1901.2-2013.html

36. L. Lampe, A. M. Tonello, and D. Shaver (Guest Editors), Power line communications for automation networks and smart grid, *IEEE Commun. Mag.*, 49(12), 26–27, Dec. 2011.

37. J. Anatory, M. V. Ribeiro, A. M. Tonello, and A. Zeddam (Guest Editors), Special issue on power-line communications: Smart grid, transmission, and propagation, *J. Electric. Comput. Eng.*, article ID 948598, 2 pp., 2013.

38. A. Schwager, Powerline communications: Significant technologies to become ready for integration, Ph.D. dissertation, University of Duisburg-Essen, Germany, 2010. [Online]. Available: http://plc.ets.uni-duisburg-essen.de/ Schwager_Andreas_Diss.pdf

39. L. T. Berger, A. Schwager, P. Pagani, and D. M. Schneider, MIMO power line communications, *IEEE Commun. Surveys Tutorials*, 17(1), 106–124, First Quarter 2015.

40. Unified high-speed wire-line based home networking transceivers – multiple input/multiple output specification, ITU-T, Recommendation G.9963, 2011.

41. L. Yonge, J. Abad, K. Afkhamie, L. Guerrieri, S. Katar, H. Lioe, P. Pagani, R. Riva, D. M. Schneider, and A. Schwager, An overview of the HomePlug AV2 technology, *J. Electric. Comput. Eng.*, article ID 892628, 20 pp., 2013.

42. IEEE standard for broadband over power line hardware, IEEE Standards Association, Standard 1675-2008, 2008.

43. H. C. Ferreira, L. Lampe, J. E. Newbury, and T. G. Swart, Eds., *Power Line Communications: Theory and Applications for Narrowband and Broadband Communications over Power Lines*, 1st ed. John Wiley & Sons Ltd, Chichester, 2010.

2

Channel Characterization

F. J. Cañete, K. Dostert, S. Galli, M. Katayama, L. Lampe, M. Lienard,
S. Mashayekhi, D. G. Michelson, M. Nassar, R. Pighi, A. Pinomaa,
M. Raugi, A. M. Tonello, M. Tucci, and F. Versolatto

2.1 Introduction

John David Parsons writes in the preface to the first edition of Mobile Radio Propagation Channel [1]: '*Of all the research activities that have taken place over the years, those involving characterization and modeling of the radio propagation channel are among the most important and fundamental.*' While this view is perhaps not fully unbiased, we need to agree with Parsons that '*the propagation channel is the principal contributor to many problems and limitations that beset [mobile radio systems]*' power line communication (PLC) systems. Reference to mobile radio systems is quite fitting here. Power lines have not been 'designed' for carrying communication signals and, as we shall see in this chapter, wireless and PLC channels share a number of characteristics that affect the design and performance of the communication system.

In the spirit of Parsons' remarks, we start our journey on PLC with the characterization of power lines and power distribution networks as transmission medium for data communication. More specifically, this chapter is intended to provide an overview of the state of the art in channel characterization for PLC. In doing so, we pay particular attention to specific models for the channel transfer function and the characterization of disturbances experienced in different PLC environments.

The treatment is organized in eight parts. Section 2.2 sets the stage with an overview of basic power line topologies and characteristics of power line channels. It is pointed out that nearly all channel models available today fall into three main categories, namely deterministic, empirical and hybrid models, and the advantages and disadvantages of these approaches are discussed. Section 2.3 is focused on the low voltage (LV) PLC channel in both outdoor power distribution grids and in-home networks. Sections 2.4 and 2.5 are dedicated to channel characterization of the medium voltage (MV) and the high voltage (HV) scenarios. When

Power Line Communications: Principles, Standards and Applications from Multimedia to Smart Grid, Second Edition.
Edited by Lutz Lampe, Andrea M. Tonello, and Theo G. Swart.
© 2016 John Wiley & Sons, Ltd. Published 2016 by John Wiley & Sons, Ltd.

multiple wires are available, a multiple-input multiple-output (MIMO) communication channel can be established. The characterization and modeling of MIMO PLC channels is described in Section 2.6. Measurements and mathematical models for the noise experienced in PLC systems are presented in Section 2.7. These models are markedly different and put more structure on the noise than it is seen in the conventional additive white Gaussian noise model, which is the default model used in the communications community. A summary of available reference channel models and tools is reported Section 2.8. This is followed by a description of the main PLC channel characteristics in other scenarios, namely, the low voltage DC distribution grid scenario, the in-car scenario and the in-ship scenario, in Section 2.9.

2.2 Channel Modeling Fundamentals

Among the main technical challenges in power line communications, the power line channel is a very harsh and noisy transmission medium that is very difficult to model [2]–[7]. The power line channel is frequency-selective, time-varying, and is impaired by colored background noise and impulsive noise. Additionally, the structure of the grid differs from country to country and also within a country and the same applies for indoor wiring practices.

Due to the difficulty of modeling the power line transfer function, the first modeling attempts were mostly based on phenomenological considerations or statistical analysis derived from extensive measurement campaigns. More recently, papers attempting deterministic approaches have been appearing, thus indicating that a more basic understanding of the physical propagation of communications signals over power lines is now emerging. It is remarkable that the results of these recent deterministic approaches actually confirm the validity of some of the conjectures formulated during the time that analytical approaches were not deemed feasible, e.g. the multi-path nature of signal propagation along power line cables.

Another important feature of the power line channel is its time varying behavior. The channel transfer function of the power line channel may vary abruptly when the topology changes, i.e. when devices are plugged in or out, and switched on or off. However, the power line transfer function exhibits a time-varying behavior even if the topology of the network and the load (appliances) attached to it do not undergo abrupt changes. In particular, the power line channel exhibits a short-term variation because the high-frequency parameters of electrical devices depend on the instantaneous amplitude of the mains voltage which can translate in periodic variations of the load impedances. In addition, the noise injected into the channel by appliances is also dependent on the instantaneous amplitude of the mains voltage. Therefore, a cyclostationary behavior on the time selectivity of the channel as well as on the noise arises, and the period is typically half the mains period. An example of this behavior unique to the power line channel is shown in Figure 2.1, where the measured time variation of an indoor power line channel transfer function and the noise waveform generated by a halogen light with dimmer are shown. Despite the importance and uniqueness of this behavior, there are few contributions that address this characteristic (see [8]–[13] and references therein) and a contribution that specifically addresses the issue of mapping directly in the discrete time domain the modulated input signal to the output when the time variability of the channel is taken into consideration [14]. In [9], the deterministic and random input-output relation for a Linear Time Varying (LTV) system is found for power line channels both in the time and frequency domain. In [14], the convolution operator is expressed in matrix form and both the Linear Time Invariant (LTI) and the LTV cases are considered; in the LTI case the channel is

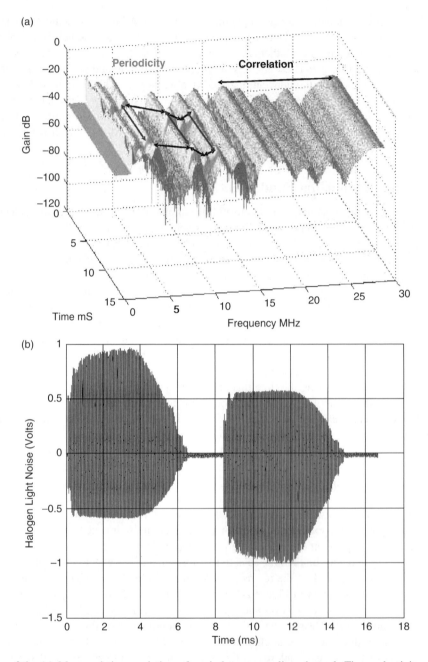

Figure 2.1 (a) Measured time variation of an indoor power line channel. Time selectivity occurs at a period rate, whereas frequency selectivity exhibits correlation at various time instants. (b) Noise waveform created by a halogen light with dimmer over one 60 Hz mains cycle.

modeled by usual Toeplitz matrices, while in the LTV case the channel is modeled by special banded matrices. The important aspect to point out is that traditional channel models based either on multi-path propagation fail to capture time selectivity.

Several groups are pursuing methods to deduce relevant statistical behavior from ensembles of physical models and measurements, e.g. see [15]–[19] and references therein. Other groups are instead following a deterministic approach based on precise channel models, e.g. see [20]–[28], and references therein. Deterministic models require detailed knowledge of the link topology and of the cable models, but do not require measurements. Statistical models can be obtained either by fitting the results of measurement campaigns, e.g. see [29], [30] or by a statistical description of the network topology, e.g. see [25], [31], [32].

Recent results seem to indicate that, if properly modeled, the power line channel transfer function exhibits more determinism than commonly believed. This determinism could be exploited for robust modem design and system optimization. For example, the symmetry of the power line channel (mathematically proven and experimentally validated in [33] and [27]) opens the door to information-theoretic considerations on optimal transmission when the channel is known at the transmitter. For example, treating known intersymbol interference as interference known at the transmitter allows us to use more effectively precoding schemes such as Tomlinson-Harashima or, more in general, dirty paper coding. Moreover, as reported in [27], it is possible to isolate reflections and resonant modes on the basis of specific features of the power line topology. This property of superposition of resonant modes allows us to assess more effectively the similarities (correlation) between the power line transfer functions pertaining to the same home, and this knowledge can be embedded into the adaptive equalizer of a power line modem. It is also important to point out that the hybrid approaches described in [25], [28] and [11] can embed this intrinsic correlation. It is important to be able to model the spatial correlation of channel responses when, for instance, multi-user or cooperative communication protocols have to be analyzed. For instance, in the context of communications with relays, a statistical bottom-up channel model was used in [34], while for the evaluation of physical layer security a top-down statistical channel model was developed in [35].

2.2.1   Brief Review of Indoor/Outdoor Topologies[1]

Indoor and outdoor power line topologies differ greatly country to country, and also within country. The major characteristics of these two environments will be here reviewed, whereas more details will be given in Sections 2.3, 2.4, and 2.5.

2.2.1.1   Low, Medium, and High Voltage Mains Topologies

High voltage (HV) lines bear voltages in the 110–380 kV range and span very large geographical distances. These lines have been used as a communications medium for voice since the 1920s [36] via single-sideband amplitude modulation (power carrier systems). Nowadays, power line communications over HV lines comprises both analog systems (tele-protection) and digital systems (voice and data transmission).

---

[1] Portions of the material in this section are reprinted, with permission, from S. Galli and T. C. Banwell, 'A deterministic frequency-domain model for the indoor power line transfer function,' *IEEE J. Sel. Areas Commun.*, vol. 24, no. 7, Jul. 2006. ©[2006] IEEE.

Typically, Medium Voltage (10–33 kV) and Low Voltage (100–400 V) power distribution lines are used for high-speed PLC communication. Medium-voltage power systems are typically deployed in a loop configuration, but sometimes they can be found deployed as open-loop systems and tree systems with radial arranged lines. Additionally, distribution lines consist of either underground or overhead cables.

In a single-phase configuration, a hot and a return (neutral) wire are fed to the premises main panel. Sometimes a separate ground (earth) wire is also added. This configuration is typical of small residential buildings. Generally, the power company distributes three phases, and only one of these is fed to a house, whereas a neighbor may be served off another phase. In the U.S., voltage ratings[2] are 60 Hz 120 V, allowing a range of 114–126 V (ANSI C84.1). The new harmonized nominal voltage in Europe[3] is 230 V (range, 207–253 V) 50 Hz (formerly, 240 V in the UK, 220 V in the rest of Europe).

The two-phase configuration is not common in Europe, but is typical in the U.S. in the split-phase configuration. In a typical home in the U.S. three cables come into the premises panel from the service. A center-tapped step-down transformer is located on the electrical line pole with the tap grounded and each socket connected across one side of the transformer. Larger devices (electric stoves, central air conditioning units, electric dryers, etc.) are wired across the entire transformer, receiving 240 V. In the U.S., sometimes apartment complexes are even fed with a 120/208 V 'wye' configuration. The transformer is set up in a 'y' configuration with 208 V between any two secondaries, and 120 V between any one secondary and the center-tapped neutral. In some areas, two legs of a 120/208 'y' are fed instead of the usual 120/240 split phase-service.

Three-phase (three hot wires plus one return) configurations are common in Europe, but not in the U.S. The three-wire system that the user sees is typically derived from three-phase distribution, which uses a four-wire or five-wire system. In the five-wire system, there are three hot wires, one neutral wire, and one grounding wire. The common three-wire receptacle uses only one of the three hot wires. In Europe most use 230/400 V, where the 230 V can be found between any of the three phases and neutral and the 400 V can be found between two of the three phases. The phase difference between phases is 120 degrees.

2.2.1.2 Residential and Business Indoor Wiring Topologies

Tree or star configurations are almost universally used. In Europe, both two-wire (ungrounded) and three-wire (grounded) outlets can be found. Interestingly, if two- or three-phase supply is used, separate rooms in the same apartment may be on different phases. The UK has its exceptions and uses special ring configurations; a single cable runs all the way around part of a house interconnecting all of the wall outlets and a typical house will have three or four such rings. Moreover, in the UK the return cable will generally be grounded at the local substation, not locally in the house. There are also some cases, especially in old buildings, where only two wires run around the house (neutral and ground share one wire).

---

[2] The US has a nominal line voltage of 120 Volts (60 Hz) because the original light bulb invented by Thomas Edison ran on 110 volts DC, and that approximate voltage was kept even after converting to AC so that it was not necessary to buy new light bulbs. Many frequencies were used in the 19th Century for various applications, with the most prevalent being the 60 Hz supplied by Westinghouse-designed central stations for incandescent lamps.

[3] Europe's mains voltage is 230 V (50 Hz) because at the beginning of 1900 the German AEG had a virtual monopoly on electrical power systems, and AEG decided to use 50 Hz.

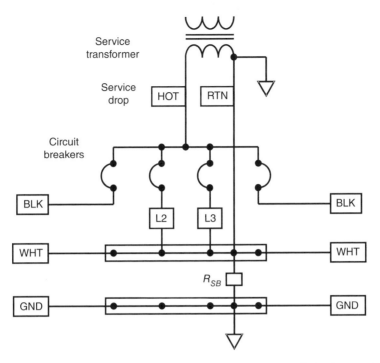

Figure 2.2 Diagram of a typical service panel with one phase, four circuit breakers, and two branch cables shown along with two additional loads. The black (BLK) wires are each fed via separate circuit breakers while the white (WHT) wires connect to the mains transformer return (RTN) via a common terminal block. The safety grounds (GND) are connected to earth ground via a second terminal block. Element R_{SB} represents a low shunt resistance between the ground and return paths, referred to as 'bonding'. [26] © 2005 IEEE.

Power cables used for single-phase indoor wiring are comprised of three or four conductors in addition to the ubiquitous earth ground. These include (in some countries) 'hot' (black), 'return' (white), safety ground and 'runner' (red) wires, all confined by an outer jacket that maintains close conductor spacing.

While the 'white' return wires and safety grounds are isolated throughout all distal network branches, many national and international regulatory bodies today mandate that the return (white) and ground cables be connected together or 'bonded' via a low resistance shunt R_{SB} at the service panel as shown in Figure 2.2. There is substantial mode coupling created by the electrical path through R_{SB}, so that the effects that ground bonding has on the transfer function of the channel are rather substantial [27], [28]. As described in [37], let us consider the simple topology shown in Figure 2.3(a): a two-conductor link, i.e. hot and return wires only, with the far outlet terminated on a matched resistor. This configuration should exhibit a simple attenuation profile which increases with frequency and has no notches. We also consider the case where this topology is composed of three cables (hot, return, and ground), and its return and ground cables are bonded (see dashed line) at the main panel as shown in Figure 2.3(b). The measured transmission responses without bonding (top trace) and with

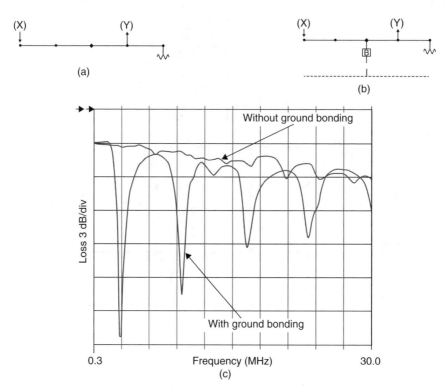

Figure 2.3 (a) Simple topology *without* ground bonding; (b) Same topology as in (a) *with* ground bonding; (c) Measurement of the effects of bonding on the transfer function from X to Y of the topologies in (a) and (b): without bonding (top trace) and with bonding (lower trace). [26] © 2005 IEEE.

bonding (lower trace) are compared in Figure 2.3(c). The upper trace shows a benign 3 dB attenuation over 0–30 MHz in the absence of bonding. The addition of bonding creates significant resonant attenuation at 3.27 and 9.95 MHz, with less pronounced attenuation at 16.89 and 23.27 MHz. This measurement confirms that ground bonding creates significant effects on the channel transfer function. Nevertheless, the modeling of grounding practices has always been neglected and it has been taken into account for the first time by Galli and Banwell [21], [26], [28].

As described above, wiring and grounding practices can be widely different and this makes modem design much more challenging. However, international harmonization has been under-way for the past 20–30 years, and many regulatory bodies, such as the US National Electric Code (NEC), have revised and mandated a harmonized set of practices. In particular, the following practices are now mandatory in most parts of the world:

- Typical outlets have three wires: hot, neutral and ground.
- Classes of appliances (light, heavy duty appliances, outlets, etc.) must be fed by separate circuits.

● Neutral and ground are separate wires within the home, except for the main panel where they are bonded.

Although complex network topologies can indeed exist, the above mentioned regulations greatly simplify the analysis of signal transmission over receptacle circuits.

2.2.2   Some Fundamental Definitions and Properties of Band-limited Channels

Within the bandwidth of the channel W, let us define the channel frequency response $H(f)$ as:

$$H(f) = |H(f)|e^{j\theta(f)} = H_R(f) + jH_I(f) \tag{2.1}$$

where $|H(f)|$ is the amplitude response characteristic and $\theta(f)$ is the phase response characteristic. The group delay of the channel (also called as the 'envelope delay characteristic') is:

$$\tau(f) = -\frac{1}{2\pi}\frac{d\,arg(H(f))}{df} = -\frac{1}{2\pi}\frac{d\theta(f)}{df}$$

A channel is said to be 'ideal' or distortionless if the following conditions are both satisfied:

(a) The channel amplitude response is constant: $|H(f)| = H, \forall f \leq W$.
(b) The channel group delay is constant: $\tau(f) = \tau, \forall f \leq W$. This is also equivalent to saying that the channel phase response $\theta(f)$ is a linear function of frequency.

If only Condition 1 is satisfied, then the channel creates delay distortion. If only Condition 2 is satisfied, then the channel creates amplitude distortion. If both Conditions 1 and 2 are satisfied, then the constant value t of the group delay is also the propagation delay of the channel.

It is interesting to look at the behavior of the group delay near a channel notch. To this regard, it is useful to express the group delay as

$$\tau(f) = -\frac{1}{2\pi}\frac{d\,arg(H(f))}{df} = -\frac{1}{2\pi}\frac{d}{df}\Im[\ln(H(f))] = -\frac{1}{2\pi}\Im\left[\frac{1}{H(f)}\frac{dH(f)}{df}\right], \tag{2.2}$$

where $\Im[z]$ represents the imaginary part of complex number z.

Let us now recall the following property of the continuous Fourier transform:

$$x(t) \leftrightarrow X(f) \Rightarrow t\,x(t) \leftrightarrow \frac{j}{2\pi}\frac{d}{df}X(f)$$

Using the above property, we can now write

$$\tau(f) = -\frac{1}{2\pi} \Im \left[\frac{1}{H(f)} \frac{\mathrm{d}H(f)}{\mathrm{d}f} \right] = \Im \left[\mathrm{j} \frac{\tilde{H}(f)}{H(f)} \right] = \Re \left[\frac{\tilde{H}(f)}{H(f)} \right],$$

where $\Re[z]$ denotes the real part of complex number z, and

$$\tilde{X}(f) = \mathrm{FT}\{t\,x(t)\} = \frac{\mathrm{j}}{2\pi} \frac{\mathrm{d}X(f)}{\mathrm{d}f},$$

where $\mathrm{FT}\{\cdot\}$ denotes the continuous-time Fourier transform operator. From (2.2), it can be seen that the group delay of a channel can become very large at those frequencies where the channel amplitude response vanishes, i.e. in correspondence of channel nulls.

In the next sub-sections we will introduce some important channel metrics. In this regard, let us now define the following sequences:

- *Discrete time impulse response*: $\{h_i = h(t = iT_S),\ i = 0, 1, 2, \dots, N-1\}$, obtained sampling at rate $F_S = 1/T_S$ the continuous time impulse response $h(t)$. The channel memory is $N-1$, and there are N non zero taps.
- *Discrete frequency transfer function*: $\{H_i = |H_i|e^{\mathrm{j}\phi_i},\ i = 0, 1, 2, \dots, N-1\}$, obtained as the N-point discrete Fourier transform (DFT) of the discrete impulse response h_i.

2.2.2.1 Impulse Response Duration

Several definitions of this metric can be found in the literature, and often the channel impulse response duration is mistakenly labeled as 'delay spread' or 'maximum delay spread'. It is generally accepted to define the impulse response duration as the time interval that contains a certain percentage of the total energy of impulse response; typical percentage values are 99%, 99.9% and 99.99%. The truncation of the impulse response is necessary in the cases of measurements as noise contamination is substantial in power lines, and some optimal methods are described in [38]. If the impulse response is generated according to a model, then truncation of the impulse response may be avoided.

2.2.2.2 Average Channel Gain

The power line channel is frequency selective, so average channel gain G is calculated by averaging over frequency:

$$G = \overline{H(f)}^f = \frac{1}{N} \sum_{i=0}^{N-1} |H_i|^2 = T_S^2 \sum_{i=0}^{N-1} |h_i|^2,$$

where the right hand side has been obtained via the Parseval Theorem and T_S is the sampling time.

2.2.2.3   Root Mean Square Delay Spread (RMS-DS)

The RMS-DS is defined as the square root of the second central moment of the power (or energy) delay profile. If $N T_S$ is the duration of the eventually truncated impulse response, the RMS-DS σ_τ can be expressed as follows:

$$\sigma_\tau = \sqrt{\mu_\tau^{(2)} - \mu_\tau^2} = T_S \sigma_0, \qquad (2.3)$$

where the following relationships hold:

$$\mu_\tau = T_S \mu_0, \quad \mu_\tau^{(2)} = T_S^2 \mu_0^{(2)} \qquad (2.4)$$

$$\mu_0 = \frac{\sum_{i=0}^{N-1} i |h_i|^2}{\sum_{i=0}^{N-1} |h_i|^2}, \quad \mu_0^{(2)} = \frac{\sum_{i=0}^{N-1} i^2 |h_i|^2}{\sum_{i=0}^{N-1} |h_i|^2}, \quad \sigma_0 = \sqrt{\mu_0^{(2)} - \mu_0^2}. \qquad (2.5)$$

The RMS-DS is usually much smaller than the impulse response duration, sometimes as much as an order of magnitude, although this is seldom pointed out in the literature. Note that μ_0 and σ_0 are the average delay and the RMS-DS normalized to a unitary sampling time. Quantities μ_0 and σ_0 are sometimes used in the literature (see, for example, [38]), but it is important to remember that they have to be scaled appropriately by T_S as in (2.4).

2.2.3   *Characteristics of the Indoor Channel in the HF and VHF Bands*

A typical indoor power line channel impulse response, frequency transfer function and group delay plot are plotted in Figure 2.4. It is interesting to confirm the theoretical considerations made in the previous sections. For example, it can be seen that: the average group delay is close to the propagation delay; the delay spread is much smaller than the impulse response duration; the group delay has peaks at those frequencies where the amplitude of the transfer function vanishes.

Generally, indoor power line channels in the HF band (2–30 MHz) exhibit the behavior shown in Figure 2.4. The frequency transfer function exhibits high frequency selectivity. Moreover, group delay as well shows high variability with multiple peaks. This means that the power line channel exhibits both amplitude and delay distortion. This can be explained taking into account that bridged taps – which together with ground bonding account for most multi-path generation – have lengths that are comparable with the wavelength λ of HF signals (10 m $\leq \lambda \leq$ 150 m). A bridged tap with a length equal to a quarter of λ causes a pi-shifted reflection to add coherently with the main signal thus producing a notch at the frequency corresponding to wavelength λ. If the bridged tap length is a small multiple of the quarter wavelength, then the pi-shifted reflection will be only slightly attenuated with respect to the main signal and this will cause a dip in the frequency transfer function. Since bridged taps with lengths equal to a small multiple of the quarter wavelength of the frequencies in the HF band are very common for indoor topologies, the indoor power line channel in the HF region is characterized by many frequency notches and dips and, thus, by many group delay peaks as predicted by (2.2).

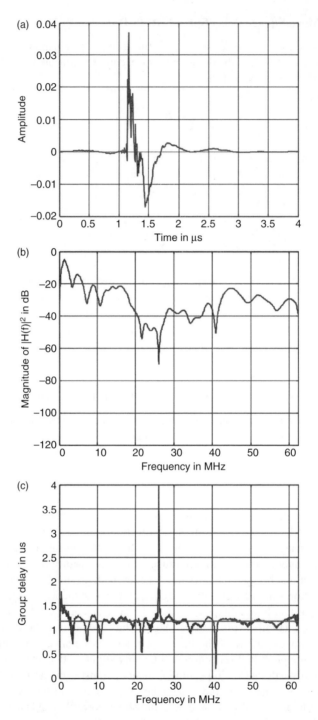

Figure 2.4 Typical measured indoor power line channel: (a) impulse response; (b) frequency transfer function; (c) group delay versus frequency (average group delay is plotted as a bold straight line).

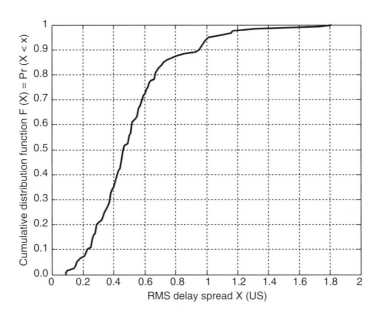

Figure 2.5 Empirical CDF of the RMS-DS of the measured indoor power line channels. For more details, see [43]. © 2009 IEEE.

Several published papers report that typical delay spreads in homes are in the order of few microseconds. For example, papers [39] and [40] report that the measured indoor delay spreads are in the order of 2–3 µs, with some exceptional cases of delay 5 µs. However, it is not really clear how the delay spread has been calculated and if the delay spread reported is actually the RMS-DS defined in (2.3) since the RMS-DS is often confused as the impulse response duration. This confusion on what metric is actually being reported as delay spread is common to most papers on power line channel characteristics. Recently, a detailed methodology, metric definitions and available software scripts have been proposed in [19] to allow a fair comparison of results obtained in different measurement campaigns.

The measurements reported in this section pertain to a set of channels measured by the HomePlug Powerline Alliance [41]. The set contains 120 power line channel impulse responses (both in forward and reverse directions), representing data from six different homes of varying sizes and age in North America. Additional details can be found in [40] and [42].

The impulse response duration on these measured channels was calculated to include up to 99.9% of the impulse response energy. Based on 99.9% energy, RMS-DSs of the channels were calculated and they were found to vary between 0.1 µs and 1.7 µs. The plot of the RMS-DS Cumulative Distribution Function (CDF) is plotted in Figure 2.5 (only the 60 channels in forward direction have been considered). Among the 120 measured responses, the median value was found to be around 0.5 µs and 118 channels exhibited an RMS-DS below 1.31 µs. Two responses only exhibited a much higher RMS-DS: 1.73 µs and 1.81 µs, respectively. This confirms that the RMS-DS of the indoor power line channels is actually much smaller than usually believed. This has an impact on the choice of the parameters of a multicarrier system designed to operate over the power line channel.

The empirical Cumulative Distribution Functions (CDFs) of the channel gain G_{dB} can be easily calculated on the basis of the measured data. The channel gain CDF can also be used to calculate the CDF of the link signal-to-noise power rations (SNRs) available at the receiver. The SNR available at the receiver is a random variable that can be expressed as a scaling of G:

$$\text{SNR} = \frac{P_{\text{TX}}\, G}{P_{\text{N}}},$$

where P_{TX} and P_{N} are the transmit and noise density powers, respectively. Galli reported in [43] for the first time that average and individual channel gains of measured indoor power line links are lognormally distributed. The probability that a link experiences an SNR larger than a certain amount γ can be expressed using the Complementary CDF (CCDF) of the random variable SNR:

$$F_{\text{SNR}}^{(c)}(\gamma) = \text{Prob}\{\text{SNR} > \gamma\} = 1 - \text{Prob}\{\text{SNR} \leq \gamma\}.$$

The resulting empirical CCDF (probability that a link has an SNR higher than γ) is plotted in Figure 2.6 as a dashed curve, assuming a transmit power of $P_{\text{TX}} = -55$ dBm/Hz and a noise floor of $P_{\text{N}} = -120$ dBm/Hz. If the gain G_{dB} is fitted to a normal distribution, then also the SNR is normal and its CCDF is the one shown in Figure 2.6 as a solid curve. The CCDF plotted in Figure 2.6 confirms that SNR in power lines is generally low. For example, the median SNR

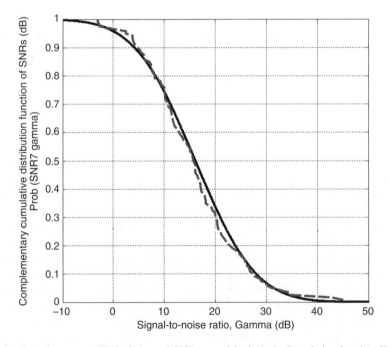

Figure 2.6 Complementary CDF of channel SNRs: empirical (dashed) and simulated (solid) when G is fitted to a lognormal distribution. For more details, see [43]. © 2009 IEEE.

is equal to 15 dB and the probability that the link SNR is above 30 dB is only 8%. Similarly to channel gains, Galli reported for the first time that also the RMS-DSs of measured indoor power line links are lognormally distributed [43].

Although there is today a growing interest in utilizing the VHF part of the spectrum for power line communications, very little data has been published on this matter. The characteristic of the power line channel in the VHF band (30–300 MHz) is very different from the one described above for the HF region, but has not been studied as in depth as for the HF case and very few experimental results have been reported in the literature. In particular, amplitude and delay distortion diminish considerably so that the channel RMS-DS is much smaller in the VHF band compared to the HF band. This can be explained taking into account that indoor bridged taps usually have lengths that are much longer that the wavelength of the VHF frequencies ($1 \text{ m} \leq \lambda \leq 10 \text{ m}$). Therefore, bridged tap lengths are typically a large multiple of the quarter wavelength so that the pi-shifted reflection will be heavily attenuated with respect to the main signal and this will cause just a small dip in the frequency transfer function. Therefore, the indoor power line channel in the VHF region is characterized by much fewer and less pronounced dips than in the HF case. As a consequence, also delay distortion will be small as fewer and less pronounced group delay peaks would be present. On the other hand, attenuation also increases but not as much as one would imagine. A paper to report a quantitative analysis of channel characteristics between 30 MHz and 100 MHz was published in 2006 by Schwager *et al.* [44]. A very interesting result reported there was that the median channel attenuation in the 30–100 MHz range was only 4 dB higher than in the HF band (see Table I of [44]). Another characteristic of the VHF region is that, while channel attenuation increases with frequency, noise decreases. In fact, as shown in [44], noise in the power line channel has a Lorentzian shaped PSD. More recently, the characterization of the in-home PLC channel up to 300 MHz was carried out by Tonello *et al.* in [19]. Herein, for the first time the spatial channel correlation and the relation between the channel attenuation and the physical distance among nodes was studied.

2.2.4 Characteristics of Outdoor Channel (LV and MV)

As for Medium Voltage (MV) channel models, there is not much literature (see for example [45]–[48] and therein references). The MV overhead lines in the US and other countries substantially differ from the underground power cables in Europe. There is still a basic disagreement about how to model MV lines at high frequencies, and this disagreement is based on the underlining model of dissipative TLs above lossy ground. Of interest to this topic are recent results on the modeling of dissipative TLs above lossy ground [49] reporting that previous classical models [50], [51] were not accurate at high frequencies since they did not incorporate ground admittance. More experimental results are needed to confirm which is the appropriate model for MV power line signal propagation. However, if experiments confirm that the most accurate method for modeling signal propagation over MV power cables is the one in [49], then path loss over MV at high frequencies will be lower than currently believed. Generally speaking, MV links are characterized by RMS-DS values around 1 µs or less.

Much more work has been done for the low voltage side of an outdoor power line network. Most approaches use a multi-path or two-conductor TL formalism, although more recently multi-conductor approaches are being suggested [22], [23], and [52]. LV links are characterized by values of RMS-DS larger than the indoor case and also exhibit a more noticeable low pass behavior as the cable length in LV links is often longer than in the indoor case.

More details on the LV and MV PLC scenarios are reported in Sections 2.3 and 2.4.

2.2.5   Characteristics of the Low Frequency Channel and its Impedance

The properties of the power line channel show significant differences for different frequency ranges. This is because the relation between wavelength λ of the injected RF signals and the geometric line length is the essential parameter to determine the possible impact of reflections. Whenever the frequency range is limited to 150 kHz (the CENELEC EN [53] in Europe regulates signaling in the 3 kHz-148.5 kHz range) with a wavelength of $\lambda \approx 1$ km, path length differences of $\lambda/2 \approx 500$ m or multiples must be present to cause deep notches in the frequency response, which is hardly the case. However, due to the lower attenuation at lower frequencies, there may be multiple reflections, so that a kind of 'soft' notches may be observed in some cases. Such effects are usually rather weak though. When moving up to about 500 kHz (available in Asia and the U.S.), the situation is somewhat different as now the wavelength is $\lambda \approx 300$ m. Cables of some hundred meters in length can therefore exhibit strong frequency-selective fading with sharp notches, because the echoes remain relatively strong due to the lower attenuation compared with the higher frequency ranges.

In Section 2.3.2.3, the salient features of power line communication channels at low frequencies in the range 20 kHz–140 kHz will be discussed. Most of the results also apply for the frequency range of up to 500 kHz.

2.2.6   Fundamental Approaches: Deterministic and Empirical Models

There are basically two main approaches to modeling the transfer function of the power line channel: time-domain, and frequency-domain. Although both approaches have been followed to model outdoor (low/medium voltage) and indoor power line links (see Sections 2.2.6.1 and 2.2.6.2), time domain models are generally associated to statistical approaches characterized by averaging over several measurements, whereas frequency domain models are generally associated to deterministic efforts. In the following two sections, we will review these approaches.

2.2.6.1   Time-domain Based Modeling: The Multi-path Model

In the time-domain approach, the power line channel is described as if it were predominately affected by multi-path effects. The multi-path nature of the power line channel arises from the presence of several branches and impedance mismatches that cause multiple reflections. The first published contribution to mention multi-path propagation was authored by Barnes in 1998 [20], although his approach was based on TL theory. According to this model, the transfer function can be expressed as follows [15]–[17]:

$$H(f) = \sum_{i=1}^{N} g_i e^{-j2\pi f \tau_i} e^{-\alpha(f)\ell_i}, \tag{2.6}$$

where g_i is a complex number that depends on the topology of the link, $\alpha(f)$ is the attenuation coefficient which takes into account both skin effect and dielectric loss, τ_i is the delay

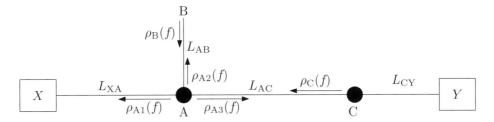

Figure 2.7 Link configuration for the determination of the paths generated in correspondence of a bridged tap located in A and followed by a discontinuity in C. [28] © 2006 IEEE.

associated with the i-th path, ℓ_i are the path lengths, and N is the number of non-negligible paths.

2.2.6.1.1 *Generation of Paths under the Time-domain Model: The Known Topology Case*

The generation of multiple paths in a power line link is due to the fact that each discontinuity or impedance mismatch generates both a reflected and a transmitted signal, so that a part of the signal travels back and forth on the line, bouncing between discontinuities several times before being substantially attenuated enough to be negligible.

Let us now analyze the echoes generated on the power line link shown in Figure 2.7 that has a single bridged tap placed along the line at A followed by a second discontinuity at C. All discontinuities are described by their complex reflection coefficients and we assume that transmitter X and receiver Y are matched to the nominal line impedance.

Let us start analyzing the case when only the bridged tap at A is present, and ignore the discontinuity at C. The distance between A and Y is denoted as L_{AY}. This is a typical configuration analyzed in several published papers [17], [20], [54], [55]. For this configuration, the signal follows a first direct path (X $\rightarrow$ A $\rightarrow$ Y), and a virtually infinite number of secondary paths arising from the signal bouncing between A and B i times, i.e. $i = 1$: X $\rightarrow$ A $\rightarrow$ B $\rightarrow$ A $\rightarrow$ Y, $i = 2$: X $\rightarrow$ A $\rightarrow$ B $\rightarrow$ A $\rightarrow$ B $\rightarrow$ A $\rightarrow$ Y, etc. The complex weights g_i, and the path lengths ℓ_i are given by the following expressions:

$$\text{Direct path } (i = 0): \begin{cases} \ell_0 = L_{XA} + L_{AY} \\ g_0 = 1 + \rho_{A1}, \end{cases}$$

$$\text{Secondary paths } (i > 0): \begin{cases} \ell_i = L_{XA} + 2iL_{AB} + L_{AY} \\ g_i = (1 + \rho_{A1})(1 + \rho_{A2})(\rho_B \rho_{A2})^{i-1} \rho_B. \end{cases}$$

Let us now slightly complicate the situation by adding a generic discontinuity at C following the bridged tap at A (see Figure 2.7). The kind of discontinuity at C does not need to be specified since its behavior is entirely described by the reflection coefficient $\rho_C(f)$. In this case, there are three types of secondary paths arriving at Y:

- Those arising from i bounces between A and B.
- Those arising from j bounces between A and C.
- Those arising from i bounces between A and B *and* j bounces between A and C.

The echo paths and the reflection coefficients pertaining to the above mentioned types of echoes are listed below:

Direct path $(i = 0)$:
$$\begin{cases} \ell_0 = L_{XA} + L_{AC} + L_{CY} \\ g_0 = (1 + \rho_{A1})(1 + \rho_C), \end{cases}$$

Secondary paths of Type 1 $(i > 0)$:
$$\begin{cases} \ell_i = L_{XA} + 2iL_{AB} + L_{AC} + L_{CY} \\ g_i = (1 + \rho_{A1})(1 + \rho_{A2})(\rho_B \rho_{A2})^{i-1}\rho_B(1 + \rho_C), \end{cases}$$

Secondary paths of Type 2 $(j > 0)$:
$$\begin{cases} \ell_i = L_{XA} + (2j + 1)L_{AC} + L_{CY} \\ g_i = (1 + \rho_{A1})(\rho_C \rho_{A3})^j(1 + \rho_C), \end{cases}$$

Secondary paths of Type 3 $(i, j > 0)$:
$$\begin{cases} \ell_i = L_{XA} + 2iL_{AB} + (2j + 1)L_{AC} + L_{CY} \\ g_i = (1 + \rho_{A1})(\rho_B \rho_{A2})^{i-1}\rho_B(1 + \rho_{A2})(\rho_C \rho_{A3})^j\rho_B(1 + \rho_C). \end{cases}$$

As shown above, by adding just one simple discontinuity the number and type of secondary paths drastically increases. The complexity of individually accounting for all of them can soon get out of hand for typical indoor topologies that can be found in the field and that are characterized by many levels of branching. Moreover, the value of N in (2.6) is unknown because it is not possible to know *a priori* how many secondary paths are non-negligible with respect to the direct path. Without a clear lower threshold for g_i, there is a problem of estimating the model order N for the multi-path model.

2.2.6.1.2 *Generation of Paths Under the Time-domain Model: The Unknown Topology Case*

The multi-path model has been proven to describe to some extent signal propagation along power line cables, even if the topology is not known *a priori* but a preliminary measurement of the transfer function is available [15]–[18], [20]. Several examples for both indoor and outdoor channels are given in [16] and [17]. Basically, the model in (2.6) is fitted to the measured channel transfer function. It is worth pointing out that this approach does not allow us to predict the transfer function of the power line link. However, channels could be randomly generated without necessarily having any link to specific topologies by randomizing the parameters in (2.6). This approach to statistical top-down channel modeling was proposed in [56] and a fitting algorithm to determine the parameters for the generation of statistically representative channel responses was described in [30].

2.2.6.2 Frequency-domain Based Modeling: Transmission-Line Models

If a detailed knowledge (topology, loads, cables, etc.) of the power line link is available, then it is possible to pursue a deterministic (bottom-up) approach and arrive at a closed form expression of the transfer function. The first published paper that addressed the problem of the relationship between power line link topology and signal attenuation appeared in only 1998 [57]. Although time domain models such as the multi-path model received more interest in the years following publication of [57], there is today a renewed interest in deterministic modeling based on TL theory. Statistical models can be obtained by developing a random model for the topology and the by applying TL theory to compute the channel transfer function on a

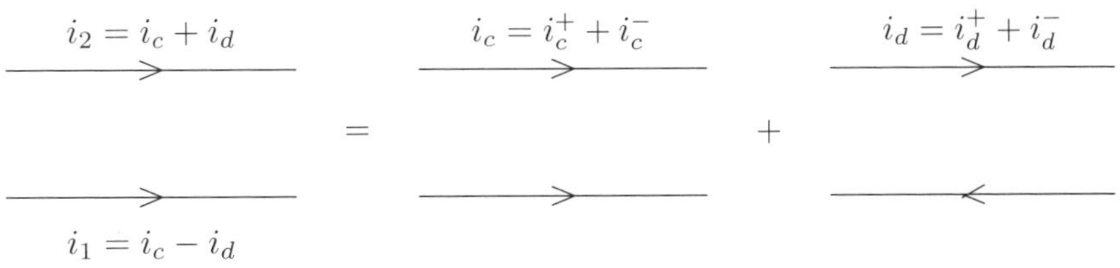

$$i_2 = i_c + i_d \qquad\qquad i_c = i_c^+ + i_c^- \qquad\qquad i_d = i_d^+ + i_d^-$$

$$= \qquad\qquad\qquad +$$

$$i_1 = i_c - i_d$$

Figure 2.8   Differential and common mode currents on a twin wire cable.

given realization of the topology. This was originally done in [25] using a simple and abstract topology model based on a backbone line with a number of randomly placed loads. A realistic statistical description of the power grid geometry in home networks was proposed in [32].

2.2.6.2.1   *Two-conductor Transmission-line Models*

There have been several attempts to model the power line channel as a two-conductor TL (see [24], [25], [54], [55], [57], [58], and references therein) using either transmission matrices, scattering matrices or by simply following a voltage ratio computation approach [32]. Let us consider a single-conductor TL with a ground, the ground being the earth itself or a second conductor. Such a configuration supports four modes of propagation along the TL (TEM approximation[4]), two spatial modes, each with two directions of propagation. The spatial modes are often referred to as differential (or balanced) mode, and common (or longitudinal) mode. This is illustrated in Figure 2.8, where the total currents i_1 and i_2 traveling on the wires are decomposed in common mode currents i_c^+ and i_c^-, and differential mode currents i_d^+ and i_d^-.

The differential mode current is almost always the functional current responsible for carrying the desired data signal along the line. It is possible to excite only a differential propagating mode along a two-conductor TL, e.g. a twisted-pair cable, by differential signaling, i.e. by driving the two conductors with antipodal signals. However, if there are imbalances or asymmetries between the two conductors, common mode components may arise even when driving differentially the two conductors.

The presence of common mode currents on a cable does not in itself pose a threat to the integrity of the differential mode data signals. However, if any mechanisms exist via which energy can be converted from common mode to differential mode, then the common mode current can become a dominant interference signal. This phenomenon is called mode conversion or mode coupling. Therefore, two-conductor TL-based models led to an incomplete circuit representation that was not capable of fully explaining the physics of signal propagation over power line cables. In particular, these analyses neglected three major points:

(a) the presence of a third conductor, which makes the problem one of Multi-conductor Transmission Line (MTL) theory;

---

[4] The Transversal Electromagnetic (TEM) mode is a propagation mode in which both the electric and the magnetic fields are transverse to the axis of the line. Unless the dimensions of the line cross section are a significant fraction of the operating wavelength, the TEM mode is the only mode that can be propagated along a cable. Higher-order modes can exist when the line cross section has dimensions comparable with the considered wavelength and these are analogous to the modes that exist in waveguide systems.

(b) the effects of particular wiring and grounding practices;

(c) estimation of common mode currents related to electromagnetic compatibility.

2.2.6.2.2 Multi-conductor Transmission-line Models

Power line cables used for single-phase power are comprised of three or more conductors, so that the problem of characterizing signal propagation on power cables is a natural problem of MTL theory [59]. MTL analysis involves breaking down a system of N conductors and a ground into N simple TLs, each of these TLs corresponding to the path of a single mode of propagation [59] (see also Section 2.6). On the basis of this analysis, signals at the inputs of an MTL are first broken down into modal components, then sent down the proper modal TLs and, finally, are properly recombined at the output ports. The voltage and current transformation matrices contain the weighting factors that determine the amounts of signal that couple between each of the ports and each modal TL. The manipulation involved in determining modal TL parameters and transformation matrices are called decoupling techniques [60]. A typical application of modal decoupling can be found in SPICE, the well-known program for the simulation of electrical circuits. In fact, SPICE allows the simulation of MTLs via a set of canonical two-conductor TLs [61]. Very often the modes propagating along the cable are not independent and mode coupling often occurs. The effects of mode coupling cannot be described with a two-conductor TL theory approach, so that MTL approaches yield more accurate models.

2.2.6.2.3 The Modeling of Grounded Links

The first approach to indoor channel modeling based on MTL was originally proposed in [21] and then refined in [26]–[28]. This approach can be considered as a natural extension of the two-conductor modeling to include the presence of additional wires, such as the ground wire. The appealing feature of this model is that it is able to describe the power line channel more accurately than the models that ignore the presence of the ground wire. An important result reported in [28] is that it is indeed possible to compute *a priori* and in a deterministic fashion the transfer function of *both* grounded *and* ungrounded power line links by using transmission matrices only. This result is due to the fact that the circuit models for the two dominant propagating modes[5] along the wires are coupled by a modal transformer at the bonding point, and this coupling is amenable to transmission matrix representation as well as the two circuit models. As a consequence, it is then possible to model the indoor power line channel in terms of cascaded conventional 2PNs strongly coupled through a modal transformer. Therefore, once the equivalent 2PN representation is obtained, it is possible to represent the whole power line link by means of transmission matrices only. This allows us to compute *analytically* and *a priori* the transfer function of any indoor power line link, and also allows us to treat with the same formalism (two-conductor TL theory) both grounded and ungrounded topologies and, at the same time, to take into account the most important wiring and grounding practices found in the field.

As an example, let us consider a generic topology of a power line link between two modems located at nodes X and Y as shown in Figure 2.9. If the power line link in Figure 2.9 is ungrounded (no ground bonding at the main panel), then the corresponding topology

---

[5] The first circuit accounts for differential-mode propagation while the second circuit (labeled as the 'companion model') accounts for the excitation and propagation of the pair-mode, which is the second dominant mode and arises prominently with certain grounding practices.

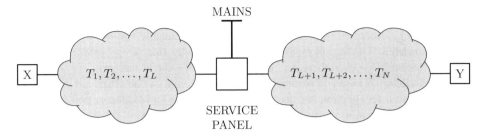

Figure 2.9 Generic indoor power line link between modems located in X and Y. The generic topologies on either side of the service panel are represented by cascaded 2PN. [28] © 2006 IEEE.

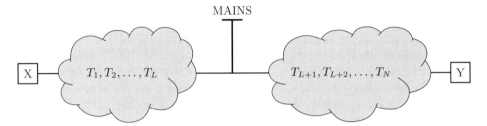

Figure 2.10 The power line link in Figure 2.9 for the *ungrounded* case. [28] © 2006 IEEE.

amenable of simple two-conductor TL theory description via 2PNs is the one depicted in Figure 2.10. If ground bonding is present in the main panel, the mirror topology representing the companion model in [28] must be added to the topology in Figure 2.9. This is shown in Figure 2.11, where it is shown that the companion model appears as a bridged tap located at the main panel.

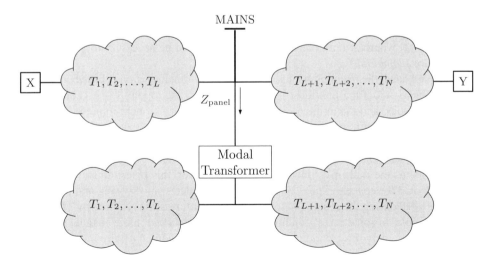

Figure 2.11 The power line link in Figure 2.9 for the *grounded* case. [28] © 2006 IEEE.

2.2.7   Advantages and Disadvantages of Modeling Approaches

The methodology described in Section 2.2.6.1 offers the valuable feature of not requiring a detailed knowledge of the topology of the link, a knowledge that is very seldom available in practical situations. However, there is a price to pay for this useful feature.

This methodology is based on parameters, including N, that can be estimated only *after* the actual channel transfer function has been measured. Lacking knowledge of port impedances, these models do not readily partition into simpler circuits. Resonant effects due to parasitic capacitances and inductances, and the particular wiring and grounding practices cannot be explicitly included in the model, but only 'phenomenologically' observed through the initial measurement. Attenuation parameters can be easily determined from an attenuation measurement using estimation routines like Least-Square-Estimation. However, the determination of path parameters of an impulse response are more complicated also because closely spaced impulses overlay and influence each other.

Finally, especially for the indoor case, there is a high computational cost in estimating the delay, amplitude and phase associated with each of the many paths (it is a time-domain model, therefore all the possible N reflections from the unmatched terminations along the line must be accounted for individually). In fact, as shown earlier, by adding just one simple discontinuity the number and type of secondary paths drastically increases. The complexity of individually accounting for all of them can soon get out of hand for typical indoor topologies that can be found in the field and that are characterized by many levels of branching. Moreover, the value of N in (2.1) is unknown because it is not possible to know *a priori* how many secondary paths are non-negligible with respect to the direct path. As a consequence, there is the additional problem of estimating the model order N. This is a major drawback for modeling indoor links, where the overall short distances make it necessary to include many more signal paths because there are many un-terminated branches and, moreover, the reflected echoes experience much lower attenuation than in the outdoor case given the shorter length of indoor wires. This last problem has often been underestimated, but it constitutes a drawback for any time-domain model.

The major advantage of a frequency-domain model is that its computational complexity is nearly independent of the topology complexity. In fact, a frequency-domain model contains the *composite* of all the signals reflected by the discontinuities (multi-path) over the measured frequency range, whereas in a time-domain model it would be necessary to generate all the different paths *individually*.

On the other hand, the major drawback of a frequency-domain approach based on TL theory is that *everything* about the link must be known *a priori*: the topology, the cable types and their characteristics, and the terminating impedances on every branch: imperfect knowledge about these quantities can impair accuracy of the channel model if they constitute part of the dominant path at a given frequency. As it is often impossible to have such a detailed knowledge of the network, this knowledge can be considered as the price to pay to avoid preliminary measurements, as the multi-path model requires. However, this deterministic methodology could be adopted for characterizing the case of some vehicles since wiring and practices are well documented during the design phase so that all the information required by TL modeling is easily available. In some cases, conditions allow good accuracy in the channel model [62], but in other cases, lack of a ground plane or lack of a fixed geometry (wire movement within the vehicle) can make channel modeling very challenging even if the topology is known [63].

2.2.8   Merging the Deterministic and the Statistical Approaches: Towards a Hybrid Model

The recent growth of contributions that attempt a deterministic approach to the modeling of the transfer function seem to indicate that a better basic understanding of the physics of signal propagation along power lines is currently available. Deterministic approaches allow us to compute deterministically and *a priori* the transfer function of a power line link without the necessity of a preliminary measurement and at the expense of a detailed knowledge of the whole link. Empirically based statistical approaches allow us to express the transfer function in terms of multiple paths at the expense of preliminary measurements when no knowledge about the link is available. It has been recently proposed that combining these two approaches to give rise to hybrid models may indeed offer advantages [28].

In a hybrid model, one would define a set of topologies that can be considered as representative of the majority of topologies that can be found in the field for a specific scenario or by generating randomly a set of statistically relevant transfer functions. Therefore, an important topic of investigation for the research community is the creation of a set of topologies (and practices) that can be considered as representative of the majority of topologies that can be found in the field. In fact, given the wide variability of topologies and practices, it is necessary to devise a set of topologies with their associated transfer functions so that coding and modulation schemes can be tested against them and compared objectively and with repeatability of results, something that only a commonly agreed upon model will allow.

The available deterministic models can also be the basis for the generation of a set of statistically relevant random transfer functions following, for example, the approach described in [25] and in [32] and adapting it to include also grounding as addressed in [28]. For example, on the basis of statistical models, engineering rules, regulatory constraints, it is possible to generate randomly a 'realistic' topology, both for indoor and for outdoor links. This sample topology would represent 'a house', or 'an access link'. For a given topology, one can again randomly generate possible terminating impedances. These variations of the sample topology would represent the variations that can actually be found in the field. These variations are now better understood, and recent models have appeared in the literature (for example, see [9] for the indoor case). Also the effect of these variations on the overall transfer functions are now well understood, and recent contributions suggest methodologies to compute efficiently upper and lower bounds to the variations of the transfer function [10]. Using a deterministic model, the transfer functions of the sample topology with the sample terminating impedances are readily obtained (in the indoor case, this should be done for every pair of power plugs). We can then compute the attainable data rate of all the computed transfer functions, and we can then build a cumulative distribution function with the attainable data rates per home as a function of the percentage of plugs within the home (similarly to radio coverage). It would also be possible to average again over homes or access links, and extract meaningful statistics, e.g. delay spread, attenuation, etc.

The first attempt to define a model for the generation of random topologies has been made by Esmailian *et al.* [25] where the US National Electric Code (NEC) was used to set constraints on the topologies in terms of number of outlets per branch, wire gauges, inter-outlet spacing, etc. A generalization of this approach requires the knowledge of the electric codes of every country. In [64], Tonello *et al.* build on top of the methodology in [25] and particularize topology generation to the European situation. A refined clustered statistical geometry model was then proposed in [32].

Another attempt towards a statistical modeling of the power line channel is reported in [65], where the the goal of the statistical classification was the development of a PL channel generator. Nine classes of channels and their respective transfer functions are defined and it is observed that peak and notch widths, heights, and numbers of a smoothed version of the measured transfer functions are fitted by Rayleigh, triangular, and Gaussian distributions, respectively. However, there is very little physical insight about why such distributions arise and one may wonder if these distributions are just an artifact of the chosen classification procedure. Furthermore, the operation of smoothing of the transfer function applied in [65] is rather questionable since it changes considerably the degree of frequency selectivity of the channel and, thus, the distortion that is introduced.

A top-down random model that makes use of the analytic frequency response in equation (2.6) is described in [56]. Herein, the location of discontinuities are chosen according to a Poisson arrival process while the path gains are modeled as uniform or log-normal random variables. The remaining parameters can be fitted so that the statistics of the average channel gain, delay spread and coherence bandwidth correspond to the target ones obtained by the analysis of measured channels [30].

In another paper, Galli reports several statistical properties of the power line channel that allow the definition of a simple statistical channel model based on two taps and useful for the comparison of modulation and coding schemes [43]. In fact, if the RMS-DS of a channel is less than 20% of the symbol duration time, the dispersion effects of the channel are fully characterized by the RMS-DS alone and are independent of the specific power delay profile as reported for the first time in [66]. Therefore, any channel model with 'realistic' channel gains and RMS-DSs could be used for performing a comparative analysis between communications schemes over PLs. For simplicity, it is then useful to use a two-path, equi-amplitude, and τ-spaced channel to represent the effects of the power line channel distortion. Channel models based on a two-path model are not new as they have been used in wireless TDMA standards (IS-54, IS-136) or for the High Frequency radio channels. These models specify two equal-power independently faded rays spaced by a fixed delay and delays are tabulated to mimic specific link conditions. This simple modeling allowed easy comparison between equalization schemes. The novelty of the model proposed in [43] is that tap amplitude and differential delay are now correlated random variables. Although this model is a simplification of the characteristics of a real power line channel, it allows easy implementation and easy replication of results something that is currently missing in the power line literature.

In the following Sections 2.3 and 2.4, further details on the channel characteristics and models for LV and MV networks will be reported.

2.3   Models for Low Voltage (LV) Channels: Outdoor and Indoor Cases

This section addresses the problem of modeling the low voltage part of the power-grid as a communication medium, but a different treatment is given to the outdoor and indoor cases because the physical features of the power network advice to do so. First of all, a short introduction to the electromagnetic problem of the signal propagation through the electrical cables is provided.

2.3.1   Some Fundamentals of Transmission Line Theory

Before presenting a power line channel model for the access domain, let us briefly review some basics of transmission line theory. The wavelength

$$\lambda = \frac{v_p}{f}, \tag{2.7}$$

corresponding to a signal with the frequency f in relation to the geometrical length ℓ of a line represents a salient figure for the following considerations. The phase velocity v_p in (2.7) is given by

$$v_p = \frac{c_0}{\sqrt{\varepsilon_r \mu_r}},$$

where ε_r is the dielectric constant, μ_r the permeability of the material in use and $c_0 = 3 \times 10^8$ m/s represents the speed of light in a vacuum. A line with length ℓ can be denoted as electrically short, when $\ell \ll \lambda$. If this condition is not fulfilled, wave propagation effects must be taken into account. Also radiation is generally no longer negligible. By strictly keeping symmetry for RF signal injection, it must be ensured that wire-bound propagation remains dominant, so that unintended radiation always remains below allowable levels. In this context symmetry means that forward and reverse currents are 'compensating' at short distance.

The infinitesimally small section of a homogeneous two-wire line depicted in Figure 2.12 can be described as a two-port structure. The following parameters characterize a homogeneous line:

(a) R' is the resistance per length, including losses caused by the skin effect;
(b) L' represents the inductance per length;
(c) C' stands for the capacity per length;
(d) G' is the conductance per length between the two wires, which is mainly caused by dielectric losses of the insulating material between the conductors.

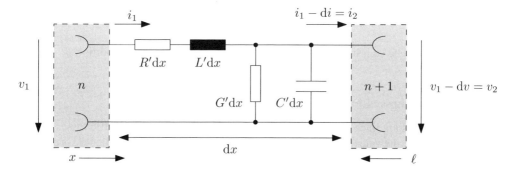

Figure 2.12   Electrical conditions at an infinitesimally small section of a line.

The differential equations

$$-dv = R'\, dx\, i_1(x,t) + L'\, dx\, \frac{di}{dt} \Rightarrow v_1(x,t) = v_2(x,t) + R'\, dx\, i_1(x,t) + L'\, dx\, \frac{di}{dt}$$

and

$$-di = G'\, dx\, v_2(x,t) + C'\, dx\, \frac{dv}{dt} \Rightarrow i_1(x,t) = i_2(x,t) + G'\, dx\, v_2(x,t) + C'\, dx\, \frac{dv}{dt}$$

describe voltages and currents at the two ports in Figure 2.12. With excitation by sinusoidal signals

$$v(t) = \Re\left\{V\, e^{j2\pi f t}\right\} \text{ and } i(t) = \Re\left\{I\, e^{j2\pi f t}\right\},$$

we obtain the well-known solution of the above differential equations in the form

$$V(\ell) = V_2\, \cosh(\gamma\,\ell) + I_2\, Z_0\, \sinh(\gamma\,\ell)$$

and

$$I(\ell) = I_2\, \cosh(\gamma\,\ell) + \frac{V_2}{Z_0}\, \sinh(\gamma\,\ell).$$

In these solutions, we find a quantity denoted as characteristic impedance

$$Z_0 = \sqrt{\frac{R' + j\omega L'}{G' + j\omega C'}}, \tag{2.8}$$

which is determined by the four line parameters (ω denotes the radian frequency). Furthermore, a propagation constant

$$\gamma = \sqrt{(R' + j\omega L')(G' + j\omega C')} = \alpha + j\beta \tag{2.9}$$

appears which can be split into an attenuation portion α and a propagation portion β.

2.3.1.1 Weakly Lossy Lines

RF signal losses along lines result both from the skin effect and from dielectric losses, which are proportional to the value of C'. The losses can be characterized by so-called 'loss angles':

$$\tan \delta_L = \frac{R'}{\omega L'} \quad \text{for the skin effect,} \tag{2.10}$$

and

$$\tan \delta_C = \frac{G'}{\omega C'} \quad \text{for the dielectric losses.} \tag{2.11}$$

Using (2.10) and (2.11) for computing the propagation constant γ, one receives

$$\gamma = j\omega \sqrt{L' C'} \sqrt{1 - j \tan \delta_L - j \tan \delta_C - \tan \delta_L \tan \delta_C}.$$

For weakly lossy lines $R' \ll \omega L'$ and $G' \ll \omega C'$ holds, and thus we get

$$\tan \delta_L \ll 1 \text{ and } \tan \delta_C \ll 1. \tag{2.12}$$

Now the attenuation constant α and the phase constant β can be determined as

$$\alpha = \frac{1}{2} \left(\frac{R'}{Z_0} + G' Z_0 \right) \tag{2.13}$$

and

$$\beta = \omega \sqrt{L' C'}, \tag{2.14}$$

respectively.

Skin Effect: The resistance per length R' is essentially determined by the skin effect at frequencies in the MHz-range. The crucial figure in this context is the so-called penetration depth

$$\delta = \sqrt{\frac{\rho}{\pi \mu f}}$$

for RF currents, whereby ρ is the specific resistance and μ the permeability of the conductor. For a homogeneous line with circular cross-sections of the conductors we get

$$R'(f) = \sqrt{\frac{\rho \mu f}{\pi r^2}} \sim \sqrt{f} \text{ for } f \gg \frac{\rho}{\pi \mu r^2}. \tag{2.15}$$

Dielectric Losses: Dielectric losses essentially contribute to G'. They can be determined by the loss angle δ_C introduced in (2.12):

$$\tan \delta_C = \frac{G'}{\omega C'} \quad \Rightarrow \quad G'(f) = 2\pi f C' \tan \delta_C \sim f, \qquad (2.16)$$

where C' is the capacity per length. For 'fair' insulation materials δ_C is almost constant and relatively small, e.g. 10^{-3}. With the above results, we can rewrite $\alpha(f)$ in the form

$$\alpha(f) = \frac{1}{2} \left(\sqrt{\frac{\rho \mu f}{\pi r^2 Z_0^2}} + 2\pi f C' Z_0 \tan \delta \right) \text{ for } f \gg \frac{\rho}{\pi \mu r^2}. \qquad (2.17)$$

2.3.1.2 Reflections

When a load impedance Z_ℓ is connected to an RF signal generator with the internal impedance Z_i, and $Z_\ell \neq Z_i$, then mismatching occurs. The same also applies for lines when attaching a load impedance $Z_\ell \neq Z_0$. Mismatch is the reason why a part of a forward moving wave is reflected. The reflection factor

$$r = \frac{Z_\ell - Z_0}{Z_\ell + Z_0},$$

defined by impedance conditions, serves as a measure for the strength of a reflection. Another, more illustrative description is given by

$$r = \frac{V_r}{V_f} = \underbrace{\frac{V_{r_0}}{V_{f_0}}}_{r_0 \text{ at the end of the line}} e^{-j2\beta\ell}. \qquad (2.18)$$

In general, the reflection factor r is complex and represents the ratio of the reflected portion V_r and the forward portion V_f. For graphical representation of r within the complex plane usually a polar diagram called a Smith Chart is a quite common tool. According to (2.18), in the Smith Chart the reflection factor r_0, which is present at the line's end, rotates counter-clock-wise with twice the phase velocity when moving e.g. from the load to the generator.

Multiple reflections arise if neither the load's nor the generator's impedance fit to the line's characteristic impedance. First, a wave runs from the generator toward the load. When reaching the point of mismatch at the load, a part of the incident wave is reflected. This portion propagates now in opposite direction, i.e. toward the generator. Arriving there, again a partial reflection occurs according to the degree of mismatch in this place. The reflected portion propagates again into the original direction toward the load. Theoretically these events would repeat infinitely, so that even in a simple line structure the 'impulse response' would be of infinite duration. In practice, however, the magnitudes of the reflection coefficients are always smaller than one, leading to finite impulse responses. Also the line's attenuation will significantly contribute to the decrease of the reflected portions and thus 'shorten' the duration of the impulse response in a 'natural' way. If unmatched branches are connected to a line, further echoes will arise.

2.3.2   Models for Outdoor LV Channels

In this section, we consider power line channel modeling for the low voltage power distribution grid between transformer stations and house connections, the so-called access domain, which has been an important topic of research for over 20 years.

2.3.2.1   Access Network Topologies in Europe, Asia, and USA

The typical structure of the European 50 Hz power distribution grid is shown in Figure 2.13. The long-distance supply is provided on a 110 to 380 kV high voltage level. This way, up to several hundred kilometers are bridged with moderate losses. Then, for further distribution, the medium voltage level follows at 10 to 30 kV, where the used voltage depends on the distance. In rural areas the extension of the medium voltage lines can be up to some ten kilometers, while within towns the radius does normally not exceed 5 km.

For our considerations within this section, both the high and the medium voltage level are not of interest. The focus will be on the low voltage distribution grid, i.e. the 230/400 V level 3-phase supply system as depicted in the lower part of Figure 2.13. Here we find so-called

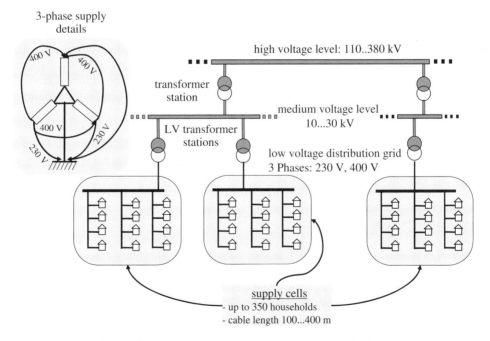

Figure 2.13   The European power supply network structure.

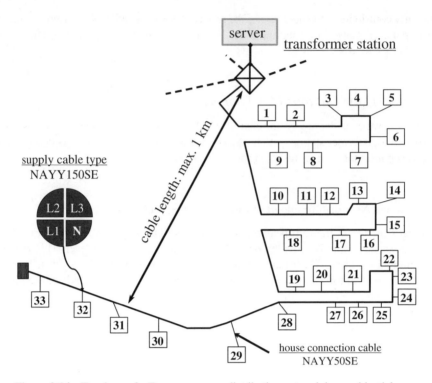

Figure 2.14   Topology of a European power distribution network in a residential area.

supply cells which may include up to 350 households, connected to a single transformer station. Dependent on the load, more than one transformer can be present in a station. Up to ten branches, each of them serving approximately 30 houses, may leave a station. Such a situation is shown in detail in Figure 2.14.

Here we observe that a branch may extend to a maximal length of 1 km, when e.g. one-family houses are the dominant customers, so that the losses along this fairly long line remain moderate. Sub-groups of approximately ten houses are always connected in a rather regular structure with cables usually exhibiting less than 10 m in length. These connection cables have a similar four-sector cross section as the supply cable, but are reduced in diameter. Therefore, they will have a different characteristic impedance, so that each 'stitch line' to a house represents a point of mismatch.

Some typical details with respect to the behavior at high frequencies for the 'last mile' and the 'last meter' environments of European power distribution grids are shown in Figure 2.15. As already mentioned, each connection point of a house is a point of mismatch along the main supply cable. Moreover, at the point where the supply line enters a house extremely low impedance is found, because the numerous distribution lines which run through a house are coming together in this place. Thus, even if there is no load (no power consumption), the characteristic impedances of the lines are permanently in parallel. A typical indoor power line exhibits a characteristic impedance in the range of 40–80 Ω, so that a house connection

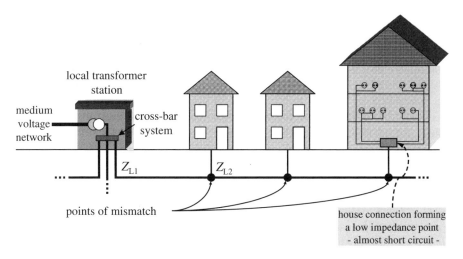

Figure 2.15 Details of the European 'last mile' and 'last meter' environments.

point would show only approx. 5 Ω with 10 lines starting here. In practice, the number will frequently be even larger, so that the RF impedance is often close to a short circuit.

Before we move to the analysis of the RF properties of networks according to Figure 2.14 in different frequency ranges, let us have a look at the power supply structure in Asia and the USA.

As Figure 2.16 indicates, there are significant differences in the supply structures between Europe and these regions, especially with respect to the low voltage distribution grid. While

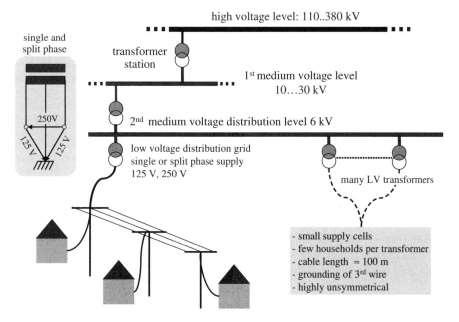

Figure 2.16 Typical power supply system structures in Asia and America.

the high and medium levels (110–380 kV or 10–30 kV, respectively) appear very similar, the second medium voltage level at 6 kV represents a major difference. On this level, a kind of long distance supply is performed, e.g. over extended residential areas, using a high number of rather small transformers, each of which supplies only a few houses with the required low voltage. On the low voltage level we find a further difference in that a split phase structure is used to supply a customer with 125 V or 250 V respectively. So the wiring which runs to a house consists of three lines, one of which is grounded. This leads to a high degree of asymmetry for the transportation of high frequency signals, so that significant problems of electromagnetic compatibility (EMC) may arise. This aspect is especially severe in locations where open wires are used on the low voltage level.

In the access domain it is often distinguished between low frequency/bandwidth, low-speed PLC, and high frequency/bandwidth, high-speed PLC. In Europe the low frequency range is limited to 148.5 kHz[6] and the usage is precisely defined by the norm EN 50065 [53]. In Asia and America similar regulations exist for low frequencies, however, the upper limit is around 480 kHz. During the last 15 years the low frequency range has been widely neglected, because there was a strong demand for high data rates, which definitely could not be provided by the very restricted spectral resources. Thus the spectrum for PLC applications was opened more and more toward frequencies of up to approximately 90 MHz. As we will see in the following, this is not an issue for the access domain where we often have to face a limitation to less than 10 MHz. On the other hand, the low frequency range has recently attracted growing interest for supporting energy-related services, which require moderate date rates of some ten kbits/s, but call for extremely high reliability, i.e. a permanent link to each customer, with no hidden nodes.

2.3.2.2   Echo-based Channel Model

Due to various reflections in power supply networks, a transmitted signal appears usually several times and with different delays at a receiver. This observation suggests the following approach of an echo-based channel model. The fundamental behavior of a channel, where N echoes are effective, can be described by the impulse response

$$h(t) = \sum_{i=0}^{N-1} k_i \, \delta(t - \tau_i), \tag{2.19}$$

where the coefficients τ_i denote the individual delays of an echo, and the coefficients k_i the respective attenuation of an echo. This channel model can be implemented by means of an N-tap finite impulse response (FIR) filter. The first delay τ_0 indicates the 'natural' propagation delay of the main or direct path, and k_0 is the corresponding attenuation. All following taps of the filter are associated with echoes.

---

[6] This rather hard limit has been introduced to protect long wave broadcasting, not because of unintended radiation but more due to the fact that receivers often use the power grid as antenna.

The Fourier transform of (2.19) delivers the complex transfer function

$$H(f) = \sum_{i=1}^{N} k_i \, e^{-j2\pi f \tau_i}.$$

Under real-world conditions the coefficients k_i do not only depend on the cable length, but also on frequency. Evaluating a measurement database of numerous different power line channels has led to the following expression for the attenuation coefficients of echoes:

$$k_i \Rightarrow k(f, \ell_i) = g_i \, e^{-\alpha(f)\,\ell_i}. \tag{2.20}$$

In (2.20) ℓ_i stands for the respective cable length and the g_i denote special weight factors, which include details of the network topology. In fact, g_i can be considered as the product of reflection and transmission coefficients in the course of the path with index i. Summing up the effects of multi-path propagation as well as frequency- and length-dependent attenuation, we eventually receive the complete transfer function

$$H(f) = \sum_{i=0}^{N-1} g_i \, e^{-\alpha(f)\,\ell_i} e^{-j2\pi f \frac{\ell_i}{v_p}}, \tag{2.21}$$

with $\alpha(f)$ denoting the frequency-dependent attenuation coefficient as introduced by (2.13) and (2.17). By using the weakly lossy line model – see also (2.14) – the rightmost portion of (2.21) can be rewritten using the relation

$$\frac{2\pi f}{v_p} = \frac{\omega}{v_p} = \beta = \omega\sqrt{L'C'},$$

so that we finally have

$$H(f) = \sum_{i=0}^{N-1} g_i \, e^{-[\alpha(f)+j\beta]\,\ell_i}. \tag{2.22}$$

Examining and evaluating the attenuation factor $\alpha(f)$ for various types of power lines in detail, the approximation

$$\alpha(f) = \frac{R'}{2Z_0} + \frac{G'Z_0}{2} \approx c_1 \sqrt{f} + c_2 f \approx a_0 + a_1 f^{0.5\dots 1} \tag{2.23}$$

can be derived, which is easier to handle in practice. In the rightmost expression in (2.23) both coefficients a_0 and a_1, as well as the exponent of the frequency generally prove to be constant for a given cable type. As the dimension of $\alpha(f)$ must be m^{-1}, also the coefficients a_0 and a_1 must be specified in m^{-1}, whereby it is already considered in the numerical value of a_1 that the frequency f has to be inserted in Hz.

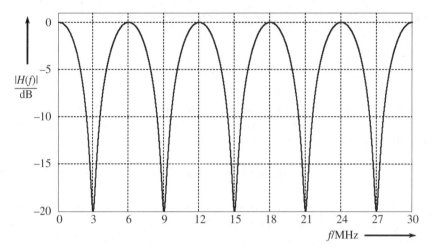

Figure 2.17 Magnitude transfer function for a two-path channel with the delay time difference $T =$ 166.7 ns.

The proposed channel model has been verified in numerous applications, exhibiting good agreement with corresponding measurements [16], [17], [67], [68]. In practice it turns out, that this model should not be used to provide exact agreement with certain individual links, but deliver a description on a statistical basis. This means, that neither the path coefficients g_i nor the length portions ℓ_i do directly correspond to a specific network topology but to a representative channel, whose parameters have been selected from extensive channel measurements.

2.3.2.2.1 *An Illustrative Example*

For illustration of echo conditions at lossless lines let us consider the example in Figure 2.17. Here we have to two paths with a difference of 25 m in length and the path weights 0.55 and 0.45 respectively. The notches appear in such a way that path length difference of 25 m represents an odd multiple of 1/2 of the respective wavelength. In case that the phase velocity is $c_0/2$, we have $\lambda = 50$ m at 3 MHz, so that $\lambda/2 = 25$ m. This explains the first notch at 3 MHz. Similarly, for 9 MHz the wavelength is $\lambda = 16.66$ m, so that $3\lambda/2 = 25$ m, which explains the second notch and so on. Computing the inverse Fourier transform (IFT) of $H(f)$, we receive the following impulse response shown in Figure 2.18. One recognizes 'clean' Dirac impulses with the corresponding path weights 0.55 and 0.45, the location of which is solely determined by the delay over 200 m, or 225 m respectively. This result suggests a channel with no spectral limitation. In practice, we observe such a situation only at wireless links.

When losses occur, a low-pass effect becomes visible, as the attenuation coefficient $\alpha(f)$ will grow with frequency. The impact of this low-pass character will now be closer investigated, using the rightmost approximation in (2.23) with the following parameters:

(a) path weights: $g_0 = 0.55$ and $g_1 = 0.45$;
(b) path lengths: $\ell_0 = 200$ m and $\ell_1 = 225$ m;
(c) attenuation parameters: $a_0 = 0$, $a_1 = 7.8 \times 10^{-10}$ m^{-1}, exponent of f: $\theta = 1$.

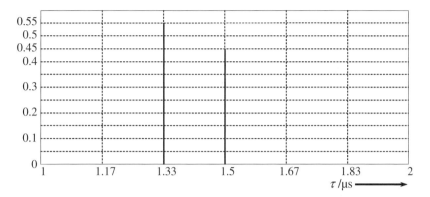

Figure 2.18 Echo delays and amplitudes for the transfer function from Figure 2.17.

With these data we can compute the respective path attenuation:

$$D(f, i) = g_i\, e^{-\alpha(f, \ell_i)} = g_i\, e^{-(a_0 + a_1 \frac{f}{\mathrm{Hz}})\frac{\ell_i}{\mathrm{m}}}.$$ (2.24)

A graphical representation of (2.24) in the usual logarithmic scale is given in Figure 2.19. The shift of the curve of path 1 to higher attenuation is easy to explain by the greater length and its smaller weight.

Combining now both – the pure echoes and the low-pass character – we get

$$H(f) = \sum_{i=0}^{1} g_i\, e^{-(a_0 + a_1 \frac{f}{\mathrm{Hz}})\frac{\ell_i}{\mathrm{m}}} e^{-\mathrm{j}\beta \ell_i}.$$ (2.25)

The graphical evaluation of (2.25) delivers – as Figure 2.20 shows – already a very realistic result for channels to be found on the 230/400 V low voltage level of a power distribution network.

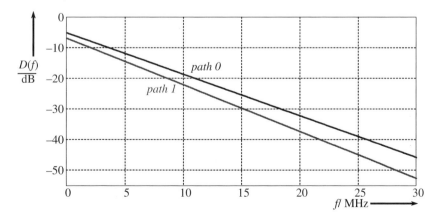

Figure 2.19 Low-pass characteristic of the two-path channel due to skin effect and dielectric losses.

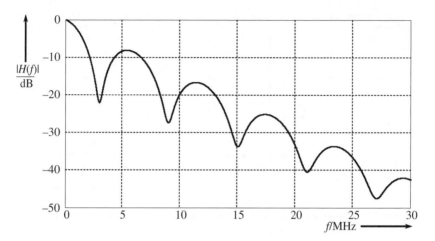

Figure 2.20   Notches and low-pass characteristic are typical features of power line channels.

Again, by means of an IFT, the impulse response can be determined from (2.25). In a normalized representation we receive the result given in Figure 2.21. We observe further contributions besides the two path echoes at 1.33 µs and 1.5 µs. This more realistic impulse response results from the low-pass character of the line. Comparing Figure 2.21 with Figure 2.18 it can be observed that at real power line networks the duration of the impulse response is usually not easy to define. Some kind of threshold must be introduced, so that the echoes not exceeding this threshold can be neglected in communication applications.

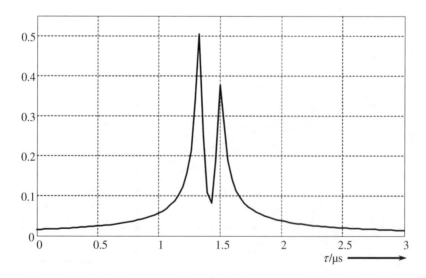

Figure 2.21   Impulse response for a transfer function according to Figure 2.20.

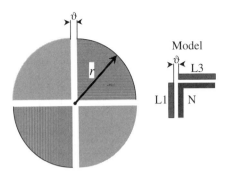

Figure 2.22 Power supply cable and its model.

2.3.2.2.2 *More Realistic Examples*

As a realistic example, we consider a four-sector power supply cable, which is most common for the standard 3-phase supply in the access domain in Europe. We assume an RF signal injection into the two phases L1 and L3 with the neutral conductor N as return. Then the simplified model depicted in Figure 2.22 (which solely needs the distance ϑ between the conductors and the radius r as data) can be used to calculate the line parameters. Then, for the capacity per length we get

$$C' = 2\varepsilon_0\varepsilon_r\frac{r}{\vartheta},$$

and the inductance per length is

$$L' = \mu_0\frac{\vartheta}{2r}.$$

The two further parameters, which are responsible for attenuation, are calculated as introduced with (2.15) and (2.16), i.e. $R' = \sqrt{\frac{\rho\pi f\mu_0}{r^2}}$, and $G' = 2\pi f C'\tan\delta$.

Based on these parameters, the characteristic impedance $Z_0 = \sqrt{\frac{L'}{C'}}$ can be calculated. For a main supply cable we obtain $Z_0 \approx 25\ \Omega$, while for a house connection cable we have $Z_0 \approx 30\ \Omega$. Calculating the expected attenuation of such cables with significant length (1 km), but without branches, leads to the results in Figure 2.23. The higher attenuation for the house connection cable (NAYY50SE) is due to the smaller diameter, so that we have a greater impact of the skin effect. Of course, such a cable would never be used for a distance of 1 km – the usual length is in the range 5–20 m. Attenuation values of approximately 50 dB, as shown in Figure 2.23, would not be critical in the access domain. It must, however, be noted that they only appear for a 'blank' cable without branches, and thus they are not realistic. In practice branching is always present and renders a deterministic analysis and description of a complete supply network generally difficult. Exactly this point will be illustrated in the following, eventually leading us to the definition of reference channels, the selection of which is based on statistical methods.

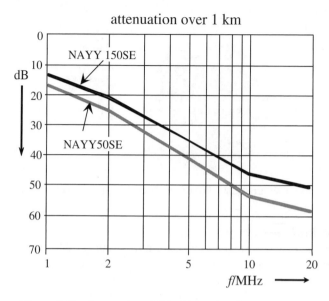

Figure 2.23 Attenuation of 'blank' supply cables over 1 km.

As introduction let us consider a quite simple network with a single branch, the end of which is left open. Assuming that both the generator G and the right end of the line are matched to the characteristic impedance Z_0, we have two points of reflection. In fact, an infinite number of echoes may theoretically occur in this simple structure. But – as already indicated above – due to attenuation, 'distant' echoes will rapidly decrease and can be neglected.

A structure according to Figure 2.24 was built and investigated by measurements. With the measurement results, parameter estimation was executed in order to apply (2.25) as model equation. The estimation procedure led to the following results:

- *Phase velocity*: $v_p = \frac{c}{\sqrt{\epsilon_r}} = 1.5 \times 10^8$ m/s, i.e. $\epsilon_r \approx 4$.
- *Attenuation*: $\alpha(f) = 7.8 \times 10^{-10} \cdot (f/\text{Hz}) \frac{1}{\text{m}}$, i.e. $a_0 = 0$, $a_1 = 7.8 \times 10^{-10}$ m^{-1}, exponent of f: $\theta = 1$.

Furthermore, the list of paths with lengths ℓ_i and weights g_i summarized in Table 2.1 is obtained.

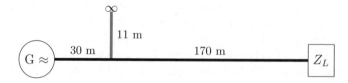

Figure 2.24 Simple network with a single branch.

Table 2.1 Path parameter estimation
results

Path	ℓ_i/m	g_i
1	200	0.64
2	222.4	0.38
3	244.8	−0.15
4	267.5	0.05

The magnitude transfer function $|H(f)|$ plotted in Figure 2.25 clearly shows the notches corresponding to the unmatched 11-m branch. Furthermore, the low-pass character is quite obvious. An interesting detail is the observation that both depth and sharpness of the notches is reduced toward higher frequencies. This is an immediate consequence of attenuation. On one hand the echo amplitudes become smaller, and on the other a kind of 'smearing' occurs with the impulse response as indicated in Figure 2.22. The corresponding (normalized) impulse response is shown in Figure 2.26. We observe that also negative peaks are possible, as indicated by path 3 in Table 2.1. Again we have a time-continuous response and not discrete echo positions. As already mentioned above, this is due to the low-pass character of the line under investigation.

In the next example a rather more realistic network structure is investigated. A sketch of the topology is given in Figure 2.27. The main supply cable has a length of 110 m and 6 branches of 15 m leave the cable for 'real' house connections. Again measurements were performed, and based on the results, model parameter estimation was executed, similarly as outlined above. As the same cable type is used, the results for phase velocity and attenuation are identical with the previous example, i.e. $v_p = 1.5 \times 10^8$ m/s, and $\alpha(f) = 7.8 \times 10^{-10} \cdot (f/\mathrm{Hz})\ \mathrm{m}^{-1}$. The results for the other model parameters are summarized Table 2.2.

As can be seen from the table, now 15 paths are necessary to capture the behavior of the network with sufficient precision. Due to the higher complexity of the topology, it is no longer

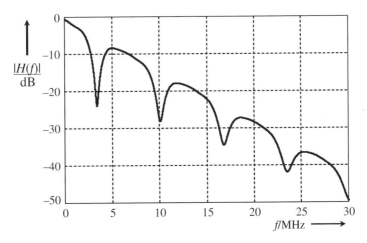

Figure 2.25 Notches and low-pass characteristic for the network from Figure 2.24.

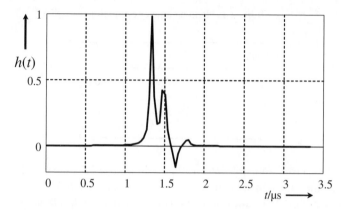

Figure 2.26 Impulse response for the transfer function according to Figure 2.25.

Table 2.2 Path parameter estimation results

Path	ℓ_i/m	g_i
1	90	0.029
2	102	0.043
3	113	0.103
4	143	−0.058
5	148	−0.045
6	200	−0.040
7	260	0.038
8	322	−0.038
9	411	0.071
10	490	−0.035
11	567	0.065
12	740	−0.055
13	960	0.042
14	1130	−0.059
15	1250	0.049

possible to establish a relation between the estimated path lengths and the geometry of the network. The same applies for the coefficients g_i. With this example the necessity of statistical modeling becomes quite obvious. On one hand the detailed topology of power line networks

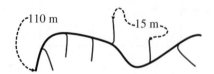

Figure 2.27 Power line section with six branches.

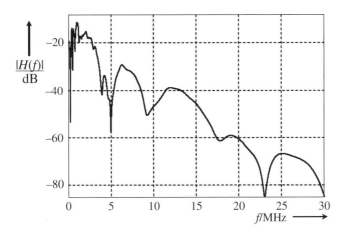

Figure 2.28 Irregular notches and strong low-pass characteristic for the network from Figure 2.27.

is rarely available, and on the other almost no knowledge about the matching of the line ends at the house connection points can be acquired.

The magnitude transfer function with the parameters from Table 2.2 is shown in Figure 2.28. Again notches and the low-pass characteristic are obvious, but the overall shape appears quite irregular. Moreover, in comparison with Figure 2.25, the attenuation is significantly higher, e.g. more than 80 dB at 30 MHz. Note that attenuation values in a range of 80 dB usually rule out the corresponding channels for usage as reliable power line links. In the considered example, a spectrum of up to approximately 20 MHz could still be used.

The corresponding (normalized) impulse response is shown in Figure 2.29. Again we observe also negative peaks, and – in comparison with Figure 2.26 – a significantly increased overall response length. Of course this is due to the higher number of paths. Again we have a time-continuous response without discrete echo positions, which is due to the low-pass character of the network under investigation.

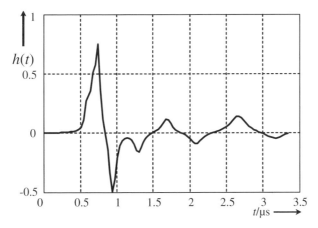

Figure 2.29 Impulse response for the transfer function from Figure 2.28.

2.3.2.3   Differences for the Low Frequency Range 9–500 kHz

Basically the presented power supply cable models, as well as the channel modeling equations, are not limited to certain frequency ranges. Thus, many of the results from the previous section can also be applied to spectra according to EN 50065 [53] or the international extended ranges up to 480 kHz. In detail, however, the network properties exhibit significant differences when moving to very low frequencies. The purpose of this section is to point out these differences and modify the channel modeling accordingly.

As already outlined, the relation between wavelength λ of the injected RF signals and the geometric line length is an essential parameter to determine impacts of reflections. Thus, whenever the frequency range is below 150 kHz – see EN 50965 – the corresponding wavelength on typical power supply cables is $\lambda = 1$ km. As for the appearance of notches path length differences of $\lambda/2$ or multiples must be present, no direct selective fading can occur within the length classes from the previous section. Due to low attenuation, however, there may be multiple reflections, so that a kind of soft notches may be observed in some cases. Such effects are usually rather weak and thus can be neglected.

When moving up to 500 kHz, the situation is different, as now the wavelength is around 300 m. Cables of some hundred meters in length can therefore exhibit selective fading with sharp notches, because the echoes remain relatively strong. In the following the salient features of power supply networks in the access domain will be discussed as an example for communication channels at low frequencies in the range 20 kHz–140 kHz. Most of the results also apply for the range up to 500 kHz.

During the eighties and nineties of the last century hundreds of courses of the transfer function were recorded and evaluated within the scope of extensive studies in various network structures – see [57], [69]–[72]. In this section, we investigate a summary of the characteristics based on Figure 2.30.

For the most important portion of the spectrum specified in EN 50065 Figure 2.30 shows a typical three-dimensional graphical representation of the attenuation over a period of 20 hours. The maximum attenuation values are close to 60 dB. Obviously, both frequency and time have an influence on attenuation, in addition to the path length. One can see frequency-dependent variations up to 30 dB and time-dependent variations of approx. 10 dB. Extreme values measured at certain paths can be even more significant, i.e. attenuations of more than 80 dB can also occur.

At first glance rather stable and remarkably deep notches are around 50–60 kHz. From the above considerations we already know that they cannot be caused by reflections. They must result from resonance effects caused by lumped components at devices and appliances connected to the power lines. For data transmission, the impact of such notches has to be carefully observed, i.e. it is necessary to use modulation schemes which are resistant to various types of frequency-selective attenuation. A link should not fail, even when relatively broad ranges of the transmission spectrum are temporary not usable. It is interesting to note that in Figure 2.30 the attenuation seems to decrease toward higher frequencies, especially in the range from 120–140 kHz. As the cable losses must definitely increase with growing frequency, this strange phenomenon needs a different explanation, which can be given by the behavior of the so-called access impedance. In fact this impedance represents a second major difference between the channel model for high and low frequencies. While access impedance in the high frequency range is dominated by the characteristic impedances of lines, impedances

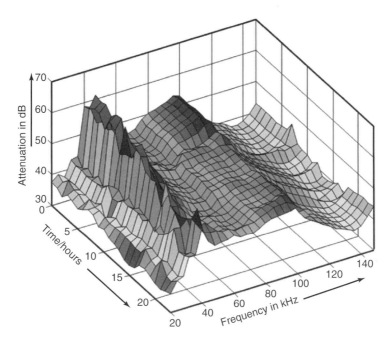

Figure 2.30   Time and frequency dependent attenuation in the low frequency range (EN 50065).

of connected loads play the most important role at low frequencies, i.e. the RF properties of the lines themselves are almost negligible in comparison with the impact of connected devices and appliances.

2.3.2.3.1   The Access Impedance

In the following we present and discuss measurement results of the access impedance Z_A. As an introduction Figure 2.31 gives an overview of the impedance range to be expected for low frequencies. For conciseness, in this preliminary figure no frequencies are specified and no details about the lines under investigation are presented. At first glance the very low magnitudes of both the real and the imaginary portions of Z_A are obvious. A closer look reveals that we always find an inductive character and that the real part's magnitude does not exceed 3 Ω. The general consequences are quite clear: The transmitter of a modem has the task to inject a voltage into the mains that preferably reaches the amplitude limits as specified in according norms, such as EN 50065.

The required transmission power directly depends on the access impedance, in particular on its real portion, as only true power is able to propagate along a line. The smaller the impedance, the more power is required. For the access domain, we expect the most unfavorable conditions at the cross-bar system of a transformer station. Among other factors, the parallel connection of a large number of outgoing trunks causes very low impedances. In Figure 2.32 a representative example of access impedance in a transformer station is shown.

The magnitude is extremely small at the lower end of the frequency range. In fact, with less than 0.5 Ω the injection of significant signal amplitudes causes tremendous effort, so that using higher frequencies is always desirable. This seems possible as we usually find

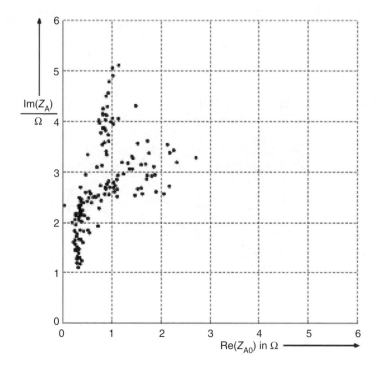

Figure 2.31 Summary of access impedance measurements.

inductive behavior, so that the impedance will grow with frequency. Note, however, that solely considering the value of the magnitude can be misleading, because with pure imaginary impedance no signal propagation would occur. Anyway, even at frequencies around 90 kHz, which is close to the upper limit of the A-band which is envisaged for energy related services and remote meter reading, impedance values around 2 Ω still represent a severe challenge.

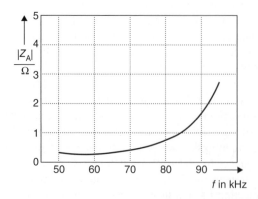

Figure 2.32 Access impedance record in a transformer station.

The transmitter power stages will have to provide at least some 10 Watts in order to establish links.

In summary, it must be stated that also within buildings, especially at house connections points we find very low values which in general do not differ much from the ones in transformer stations. In addition, time-dependence must be taken into account, as in contrast to the other rather stable properties, access impedance at a house connection may strongly vary with load conditions. In consequence, also here sufficiently strong transmitter power stages and 'stiff' coupling will be paramount. Moreover, special care must be taken when multicarrier signaling is envisaged as—due to a certain Crest factor—power peaks must be properly managed.

Therefore, it would generally be preferable to move up to significantly higher frequencies into the range of some hundred kHz or even several MHz. At the moment, this is not possible in a straightforward manner in Europe with regard to EN 50065. Major electrical energy suppliers all over Europe are currently lobbying toward a change of the norm. Achieving a frequency range extension up to 500 kHz would certainly improve the chances for PLC-based energy related services, but it would still not be a significant breakthrough. The benefits to be gained would be that the access impedance problem is clearly mitigated, and that more bandwidth is available. The remaining disadvantages, however, are, that power consumption and operation of appliances still have significant impact on the quality of communication. Moreover, in Europe legal aspects would prevail, because the protection of long-wave broadcasting is an issue.

Therefore, quite novel ideas are currently under investigation, i.e. application and adaptation of cognitive radio principles to the whole spectrum which appears usable in the access domain. From the measuring and modeling results described above, it is quite obvious that a frequency range up to approx. 10 MHz is usable over distances of some hundred meters. In comparison with only 50 kHz or less—being available in the A-band—this represents a tremendous spectral resource, which never must be fully exploited for energy-related services. Moreover, when entering the MHz-range, access impedance—dominated by consumer activity—is no longer an issue. In addition the noise power spectral density caused by operation of appliances, decreases rapidly toward higher frequencies. With respect to modem technology, there is, however, a necessity for completely new approaches to implement cognitive features which will significantly enhance the effort both for hard- and software. There are severe challenges for signal identification due to a variety of possible users and/or interferers. Each of them calls for individual treatment, i.e. either to protect useful services or to cancel noise.

The pending tasks can, however, be perfectly managed by the possibilities of today's micro-electronic components and systems. Also with respect to costs, it can be expected that integrated solutions will be available after a certain learning curve and period of maturation. As only a small fraction of the available bandwidth must be occupied to allow communication for energy services within an abundant range of 10 MHz, high reliability of links can be guaranteed, so that hidden nodes can be excluded.

Novel approaches are of course essential, i.e. some basic research is required for the first steps. One must, however, not start at zero, because the fundamental ideas to use cognitive procedures are already quite common for various wireless LAN solutions. In parallel with the development of devices and systems, the introduction of a novel philosophy with respect to competing use of spectral resources will be necessary in the worldwide regulation and standardization bodies. Basic approaches have indeed already been made, e.g. in ITU and ETSI. It will be paramount to exploit the new possibilities offered by advanced and complex embedded systems. In contrast, during the past, i.e. from the beginning of worldwide broadcasting, strict ruling and agreements were required for exclusive distribution and protection of spectral resources. Such exclusive frequency assignment was necessary due to the limited technical possibilities, i.e. protection against unauthorized users was inevitable. As already mentioned, the situation has already fundamentally changed with the advent of WLAN technology, because here transmitters and receivers are able to analyze not only the status, but also the currently available resources of a transmission channel. This knowledge enables them to perfectly adapt to a variety of scenarios. Communication systems with such capabilities are usually denoted as cognitive radios today. Standardization bodies and regulatory authorities have recognized the benefits of these novel features to fulfill their mandate of managing spectral resources. That is, the protection of services can be guaranteed without exclusive assignment of frequency ranges.

2.3.2.4   Reference Channel Definition for the Access Domain

The preferred usage of the presented model should be in such a way that, from an extended and thus representative channel measurement database, a sophisticated selection of so-called reference channels is made. Usually ten such references can be considered as sufficient to cover all practically relevant power line network situations, i.e. from fair to worst-case behavior.

Setting up a reference channel always starts with an according selection of measurements. In a next step, the salient model parameters g_i, ℓ_i and $\alpha(f)$ are estimated with support of appropriate computer-based mathematical tools. Besides the measurements, this is obviously the most challenging and time-consuming part. When performing this task, it turns out that even at complex channels usually only a small number of echoes must be considered. Often three to five echoes—solely specified by g_i and ℓ_i—are already sufficient, although there may be tens of reflecting locations along the corresponding line. Note that for a certain type of cable, $\alpha(f)$ is usually fixed and must therefore be determined only once. The amazing result that only very few echoes are relevant can be explained by the fact that distant echoes are rapidly decreasing through attenuation. Thus, only the dominant reflections in the proximity of the receiver have to be considered.

To cover numerous real-world power line channels, typical data records have been selected for different classes of length and quality levels. Out of this database, model parameters were determined. Attenuation parameters can rather easily be determined from the corresponding measurements using e.g. 'least squares' estimation methods. However, the acquisition of path parameters to characterize an impulse response causes much more effort, especially for complex network structures with multiple reflections and high attenuation. A considerable amount of experience is needed, because densely adjoining impulses may overlay and influence each other. Thus, sophisticated estimation methods must be found and repetitively executed in order to get a best fitting result.

Regarding a later realization of a channel simulator and emulator respectively, the number of paths must be limited. As already indicated above, the criterion is the size of the smallest echo amplitude which must be taken into account to maintain sufficient precision for the salient properties of a link. From the mentioned extensive measurement database of access channel transfer functions a simplified sketch of length classes as depicted in Figure 2.33 can be extracted. For conciseness the 'fine structure' details (i.e. mainly the notches) have been omitted here. The sketch clearly indicates that the impact of the low-pass character strongly grows with cable length.

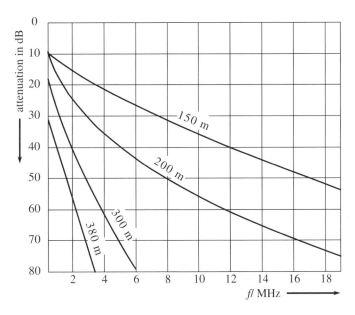

Figure 2.33 Length classes of power supply networks and their attenuation behavior.

The reference channel definition and selection have been performed within the Open PLC European Research Alliance (OPERA) project of the European Union [73]. The value and validity of this selection is emphasized by the fact that the OPERA consortium includes major manufacturers, service providers, and users of PLC technology in Europe. In addition, also partners from Asia and South America have participated in OPERA. As a result of the work within the OPERA consortium, the following length classes have been selected to define reference channels:

- 'short': approx. 150 m;
- 'medium': approx. 250 m;
- 'long': approx. 350 m.

Three levels of quality have been specified for the 150 m and 350 m class, and two levels for the 250 m class. Furthermore, one so-called 'model channel' has been defined. Thus, in total there is a set of nine reference channels covering power line networks in the access domain. The parameters of these reference channels are reported in [73] and the frequency responses shown in Figure 2.34.

2.3.3 Models for LV Indoor Channels

The aim of this section is the characterization of low voltage distribution lines inside homes and small offices as a transmission medium for broadband communications. To this end, the channel properties must be investigated with the help of measurements performed in a wide range of indoor scenarios. Afterward, channel models derived according to the observed

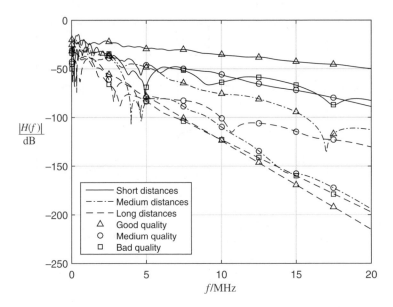

Figure 2.34 Magnitude frequency responses for reference channels from [73].

behavior can be discussed. Here, some measurements are presented both of parameters of the indoor network as a whole system and characteristics of devices that are typically connected to the mains.

The most interesting aspects of indoor PLC channels is their strong selectivity in frequency domain and their time-variation and both of them are addressed here. The frequency selectivity is due to the impedance mismatch problems mentioned in Section 2.3.1. The time-variation is closely related to the mains frequency and leads to a channel behavior that can be modeled by means of a linear periodically time-varying (LPTV) system and an additive noise with cyclostationary and impulsive components. Besides mathematical models, this section also includes practical considerations related to the design of transmission systems.

2.3.3.1 Modeling Principles

Indoor channel modeling probably represents one of the most interesting challenges in PLC systems, from an engineering point of view. There are two reasons for this. Firstly, the topology is less homogeneous in this part of the network, it is more branched, what makes the channel behavior more unpredictable, especially in the medium-high frequency band. The traditional band used for indoor PLC systems extends up to 30 MHz, which constitutes a diffuse frontier from which energy radiation begins to dominate over conduction. The wiring in most countries is a tree-like network deployed from the service panel by means of several branch circuits that reach the outlets in a non-specified manner, as depicted in Figure 2.35. The exact layout of the circuits, the number of sections of wire in each branch circuit and their lengths, are often unknown. Secondly, there are devices with quite diverse characteristics that influence the channel response and introduce disturbances.

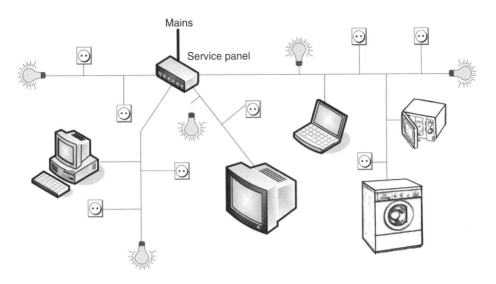

Figure 2.35 Indoor PLC network scheme.

As discussed in Section 2.2, there are different alternatives to tackle the modeling of PLC channels. In the literature, some effort has been made to obtain deterministic models for indoor channels from a bottom-up approach [28]. However, with the objective of assessing the validity of the modeling strategy, testing a particular and well-known network topology, rather than to serve as a general channel characterization. This latter purpose could be fulfilled in other more predictable parts of the power grid, like the outdoor low voltage network [52]. On the contrary, it seems more adequate to consider the physical parameters of the indoor power line networks as an ensemble of random variables, whose statistical parameters are unclear and difficult to estimate.

Other proposed models adopt a top-down strategy, defining a model for the channel behavior directly from measurement campaigns [15], [17], [30], [43], [56], [74]–[76]. In such cases, the models lead to an impulse response characterization by means of a certain number of discrete echoes. These are due to the multi-path effect that appears as the signal propagates through the indoor power line network, encountering an impedance mismatch at each discontinuity (junctions between different sections of cable and terminal loads). The number of significant echoes required to achieve a good model with this strategy may be very large.

The approach described in this section is different. It is based on a bottom-up modeling but not necessarily deterministic, by defining the channel parameters from the physical network features and deriving later a behavioral model as was formerly introduced in [11], [25], and later also employed in [31]. By selecting the physical parameters values in a clever fashion, representative channels or random channels can be obtained [32], [77].

In order to obtain a *structural model* of the indoor channel, the elements of the network must be identified: the transmitter and receiver subsystems, the wiring, the devices connected to them, and the external disturbances. A diagram of the structural model is shown in Figure 2.36. In PLC systems, the transmitter is usually connected to the receiver by using two conductors, the neutral and the line. The connection is established by means of coupling circuits that

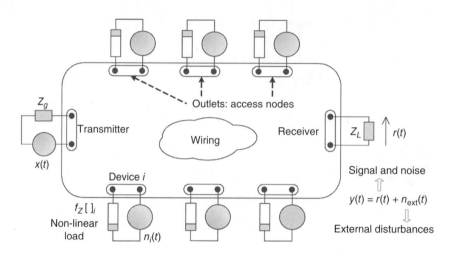

Figure 2.36   Structural model for an indoor power line network.

protect the subsystems from the mains voltage and filter the signals in the desired band. They also try to adapt the subsystem impedance to the one of the power grid, although it is a very difficult task, because the input impedance measured at any outlet is very frequency selective and unpredictable. Values ranging from few to a thousand ohms are common. The coupling circuits effect can be included in the model in two equivalent ways: considering them as part of the channel or including them in the transmitter and receiver subsystems.

The wiring can be modeled as the connection of multiple sections of transmission lines. This is a simplification because the conductors do not exhibit a regular pattern in the transverse plane to the propagation direction (the conductors can be loose inside a tube, blended at corners, irregularly twisted, etc.). The transmitter and receiver (including the coupling filters) can be modeled by their Thèvenin equivalent circuits, characterized by a frequency selective impedance. The devices are the different appliances in the home or small office. Each of them can be considered as a load with a certain impedance (frequency selective and time-varying) and, eventually, a noise source. The access to the general power line network behind the service panel can be modeled as an additional device with a load that usually has a very low impedance. Finally, the external disturbances represent any kind of unwanted signal that is not caused by the devices of the indoor network, among which there are conducted noise, coming through the service panel, and radio-waves, from broadcast services and others.

The characterization of the power line elements determines the final behavioral model. It is reasonable to consider the wiring behavior as linear and time invariant (LTI), but this assumption is not so clear for the devices. On one hand, the devices' working state changes along time, as they are switched on and off. Nevertheless, this kind of transitions occurs at a rate much lower than the typical bit rate of PLC systems and they only have impact on the long-term variation of the channel. On the other hand, many devices' loads have a non linear nature (for instance, due to rectifiers and other similar components) that, under the presence of the mains voltage, leads to a quasi-linear but time-varying behavior, as will be discussed later. Such behavior is responsible for a short-term channel variation.

From the previous argument, different *time scales* must be considered in order to describe the channel time variation. The following classification is proposed:

Invariant scale: At this level, the channel conditions can be assumed as invariant, so an LTI system would be an adequate model. The time intervals at this scale, invariance intervals, must be selected according to the channel coherence time, typically in the order of hundreds of microseconds.

Cyclic scale: At this scale, the channel exhibit periodical variations synchronized with the mains cycle and the appropriate approach is to use an LPTV system to model it. Its natural time unit is the mains period, i.e. 20 ms (50 Hz) or 16.67 ms (60 Hz). However, the mains period can be divided into a series of shorter invariance intervals, at which a snapshot of the channel response is taken. Then, the channel can be represented by a periodical series of such invariant responses that are instantaneous samples of the time-varying response.

Random scale: At this scale, the long-term variation of the channel applies. It is determined by the working state of devices. Since they change randomly in time, it does not have a regular discretization in time. However, the lapse of time between changes is many orders of magnitude above the cyclic scale, since they are associated to the human activity at homes. Each time a transition in the working state of any device occurs, a new realization of the physical parameters of the power line network is obtained and a new LPTV system can be considered (i.e. it can be assumed that the channels remain in their cyclic conditions between transitions).

2.3.3.2   LTI Channel Model

When device loads' non-linearities are neglected, the indoor network can be considered as an LTI system. This assumption has been traditionally adopted in the former literature [17], [78], [79], may be because LTI is the simplest model for a communication channels and because some measurements carried out on these channels seemed to corroborate this initial assumption. However, such measurements where usually obtained with procedures that make impossible to discern the varying behavior. To find it out, it is necessary to synchronize the measurements with the mains period, get estimations and then average them in a clever way; otherwise the averaging not only reduces the noise in the estimations but also deletes the short-term periodical variations.

Nevertheless, the simplicity of the LTI channel model can be preferable when either the channel variation in a certain link is not very important or when the modems to be used on it do not have the capability of adaptation to such variations and only 'see' the averaged channel (time invariant).

2.3.3.2.1   *Some Measurement Results Previous to the Model*
In this section, some results of measurements carried out according to an LTI behavior, both of device parameters and channels, are presented.

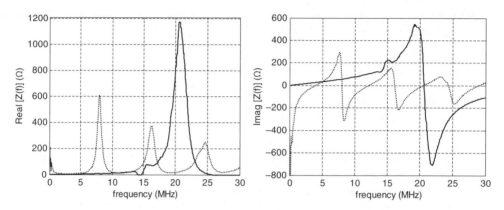

Figure 2.37 Load impedance measurement. Solid line, a computer, dashed line, a vacuum cleaner switched off.

A general model for the devices is a one-port network with two elements: a passive one (a load impedance) and an active one (a noise generator). The latter represents the noise component that some of them introduce to the channel when they are switched on. Both elements can be characterized by measuring their behavior in the desired frequency bandwidth. The huge variety of electrical appliances (even with the same function but different manufacturer) discourages from trying to model the whole network accurately. Nevertheless, after several tests some general features can be summarized. On the one hand, the impedance is usually very selective in frequency and has the shape of a resonant circuit (or some of them with several resonance frequencies), as depicted in Figure 2.37. On the other, the impedance value changes depending on the working state of the appliance.

The values of devices impedance have been measured by means of a network analyzer and a coupling network designed ad hoc. The impedance of some of them does not change considerably when they change their working state, like the computer in Figure 2.37, but this is not the usual case. To give an example of it, the result for an halogen lamp is provided in Figure 2.38. Some more examples of devices measurements and characteristics can be found in [11].

Also, with a similar procedure, channel responses can be measured. In Figure 2.39, three channels registered at the same apartment, but selecting different outlets for the transmitter and receiver location, are presented. Channel A corresponds to a link of about 30 m between outlets of different branch circuits, i.e. the main signal path passes through the main panel, which causes more dispersion of the signal energy and, thus, more attenuation and distortion. Channel B was measured between outlets in the same branch circuit and the transmission distance was about 25 m. The channel response exhibits a lower linear distortion than the previous one. Channel C corresponds also to a link in which the main path is restricted to the same branch circuit but with the shortest distance, approximately 15 m. These results indicate that the link distance is not the only, nor even the most important, factor in the amount of signal degradation.

2.3.3.2.2 *Channel Response Modeling*
A classical model, shown in Figure 2.40, comprises an LTI system, which can be modeled with its impulse response or frequency response, and an additive noise, which can be partially

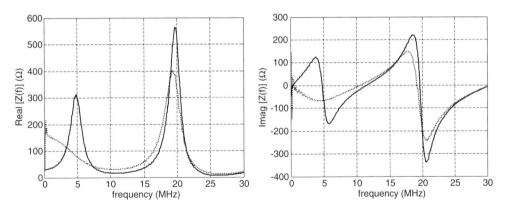

Figure 2.38 Load impedance measurement. Solid line, halogen lamp switched off, dashed line, halogen lamp switched on.

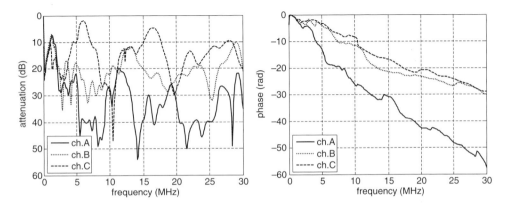

Figure 2.39 Channel frequency response measurements.

modeled with a certain power spectral density (corresponding to its stationary component) and some additional impulsive components (see Section 2.7).

The channel response can be derived from the structural model. Once a certain network structure is considered, by means of transmission line models the system can be seen as the concatenation of several two-port networks from which the channel frequency response can be estimated. The most common configuration of the wiring comprises three conductors:

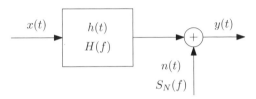

Figure 2.40 LTI channel model.

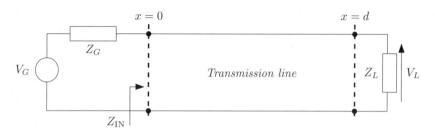

Figure 2.41 Loaded transmission line.

line (or hot), neutral and ground. As mentioned, PLC systems usually employ differential transmission between the first two of them and the electromagnetic field concentrates in the space between both conductors. A reasonable simplification is to neglect the influence of the unused third conductor and only consider a two-conductor transmission line analysis to calculate the channel response. However, sometimes neutral and ground wires are bonded at the service panel. This creates mode coupling with the third conductor that requires a multi-conductor transmission line analysis [28]. This is also the case in more recent PLC transmission systems in which MIMO (Multiple-Input Multiple-Output) strategies are being implemented, but this problem will be conveniently covered in Section 2.6, hence, in the following analysis, only the two-conductor approximation is focused.

The transmission line theory is a classical field in electromagnetism [80], and some fundamentals have already been presented in Section 2.3.1, thus no more details will be given here.

The most simple configuration of an indoor power grid could be a loaded transmission line, as depicted in Figure 2.41. It serves to model a transmitter generator, a section of wire and a receiver impedance. The selection of the adequate kind of transmission line presents some particular problems. The closest structure to the actual layout depends on the installation practices in each country, but the two parallel wires may be the most common. However, the separation between conductors is not constant (when the wires are loose in a tube) and the dielectric material in between is not homogeneous (there are both the PVC insulation and the air) what obliges to assume some approximated values for the physical parameters [81].

A transmission line can be seen as a two-port network. A convenient way to characterize its behavior is by means of the set of transmission parameters, also known as ABCD parameters, which are described in Figure 2.42. ABCD parameters allow to calculate the global parameters of the concatenation of several two-port networks by simple multiplication of the parameters of each of them. This approach has been used in other environments like Digital Subscriber Lines (DSL) [82] and allows unveiling some interesting properties of the PLC channel like the symmetry [33].

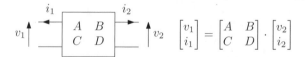

Figure 2.42 Matrix of transmission, or ABCD, parameters.

The relation between ABCD parameters and the secondary parameters of a transmission line, with characteristic impedance Z_0 and propagation constant γ defined in (2.8) and (2.9) respectively, is

$$A = D = \cosh(\gamma d)$$

$$B = Z_0 \sinh(\gamma d)$$

$$C = Z_0^{-1} \sinh(\gamma d) = B Z_0^{-2}.$$

These parameters are frequency selective due to the fact that Z_0 and γ depend on frequency. The matrix of global ABCD parameters for the loaded transmission line results in

$$\begin{bmatrix} V_G \\ I_G \end{bmatrix} = \begin{bmatrix} 1 & Z_G \\ 0 & 1 \end{bmatrix} \begin{bmatrix} A & B \\ C & D \end{bmatrix} \begin{bmatrix} V_L \\ I_L \end{bmatrix} = \begin{bmatrix} A' & B' \\ C' & D' \end{bmatrix} \begin{bmatrix} V_L \\ I_L \end{bmatrix}$$

and, considering that $V_L = I_L Z_L$, the response of the system at the frequency of the excitation can be obtained as

$$H(f) = \frac{V_L}{V_G} = \frac{1}{A' + B'/Z_L}. \tag{2.26}$$

When sections of transmission line are connected in parallel, the property of loads translation permits representing any of them as an equivalent load at the connection point. Using this property, the input impedance Z_{IN} in Figure 2.41 can be expressed as

$$Z_{\mathrm{IN}} = Z_0 \frac{Z_L \cosh \gamma d + Z_0 \sinh \gamma d}{Z_0 \cosh \gamma d + Z_L \sinh \gamma d} = \frac{A Z_L + B}{C Z_L + D}. \tag{2.27}$$

By means of matrix manipulation it is possible to obtain the frequency response of any system composed by multiple sections of transmission lines and loads in any arrangement. Hence, the *procedure to get the channel response* can be summarized in the following steps:

(a) The power line network is described in terms of sections of transmission lines and loads. The result is a tree-like structure, with intermediate nodes between sections and terminal nodes at the outlets.
(b) The input and output ports are defined. They are the points at which transmitter and receiver are placed and the main path is the direct connection between them. Any outlet can be considered one of these ports.
(c) The remaining set of transmission lines out of the main path become branches hanging from it, and they can be collapsed into an equivalent impedance by using the property of load translation in (2.27).
(d) The transmitter and receiver subsystems can be modeled as one-port networks with a load of frequency selective impedance that includes the effect of the coupling circuits.
(e) The ABCD matrix of the global structure is calculated.

(f) Finally, the frequency response $H(f)$ of the LTI system that models the channel can be obtained with (2.26).

Alternatively, a scalar method known as voltage-ratio approach can be applied to determine the channel frequency response as proposed in [32]. Since it is a scalar version of the ABCD method, its complexity is lower.

2.3.3.3 LPTV Channel Model

In this section, a more accurate model for the indoor PLC channel response is provided. It is based on the fact that many appliances do not behave as linear time invariant devices. In the first instance, an experiment illustrates the need for a better characterization of the devices' behavior. Actually, some of their electrical parameters, like the instantaneous value of the impedance, present a high correlation with the instantaneous voltage of the mains. Due to this effect, a good model for the whole channel is a linear periodically time-varying (LPTV) system.

2.3.3.3.1 Empirical Basis: Tests with Time-varying Devices

It has been observed that the dependency on the mains voltage of the impedance value of most appliances can be classified into two categories. The first one is a commuted behavior with changes between two states (probably due to rectifiers, silicon-controlled rectifiers or similar components), usually one of high impedance and the other with lower value. These transitions commonly occur twice every mains period, which suggests its relation to the absolute value of mains voltage. The second class is a more continuous variation in time, with a softer shape, usually harmonically related to the mains frequency.

The following experiment supports the previous argument. The impedance value variation of a device is measured with a network analyzer. The selected device is a common compact fluorescent lamp (a low-energy light). It corresponds to the first type of variant behavior, commuted (which may be due to the rectifiers inside the electronic ballast). As the high impedance state presents a value that exceeds the dynamic range of the analyzer, the impedance of the parallel of the mains network and the lamp was registered. The result of the real part is shown on the left side of Figure 2.43. The surprising ripple in the graph is due to the fact that the analyzer sweeps the frequency range in a time longer than the mains period. Actually, there are 38 mains periods in a sweep time and, hence, there are 76 commutations in the curve. When the lamp is in its high impedance state, the measure exhibits essentially the input impedance of the mains. At the low impedance state, the measure is the parallel combination of both impedance values and the lowest one dominates the overall value. On the right side of Figure 2.43, the real part of the input impedance of the mains, without connecting the lamp, is graphed. As observed, it describes the envelope of the previous curve.

The time variation of the device can be better observed in the Figure 2.44. The graph on the left shows the evolution of the real part of the impedance in the time and frequency plane. The white color represents a high impedance state, while the dark zone represents the state of lower impedance. The curve on the right has been included to appreciate the frequency selectivity in the dark zone. It has been obtained by measuring the lamp switched on but triggering the analyzer with the mains signal, so that every point in frequency is obtained at the same instant

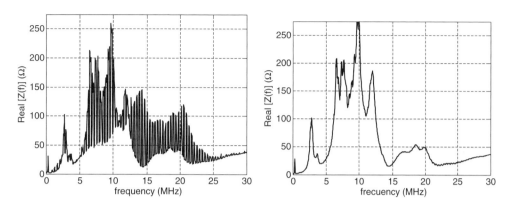

Figure 2.43 Device impedance test. On the left, compact fluorescent lamp in parallel with mains; on the right, mains impedance alone.

of the mains period. The value of the lamp in the high-impedance state can be considered as an open circuit.

Some additional measurements results of devices with a time-varying behavior are given. The graph in Figure 2.45 corresponds to a device with a commuted behavior. There are two different states of impedance clearly visible with a notable difference in the frequency selectivity. On the contrary, in Figure 2.46 a device with a continuous behavior is shown. It can be observed that the impedance is like the one of a resonant circuit but the resonance frequency changes synchronously with the mains signal.

More recently, the characterization of the impedances of typical appliances has been done in [13]. It has been found that the time variant behavior of the load impedances is more pronounced at low frequencies and it is significantly reduced by the presence of an EMI filter installed in the appliance.

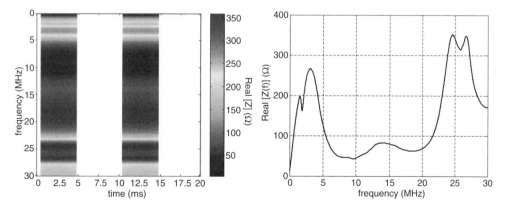

Figure 2.44 Estimation of compact fluorescent lamp impedance value. On the left, time-frequency variation; on the right, detail of the frequency selectivity.

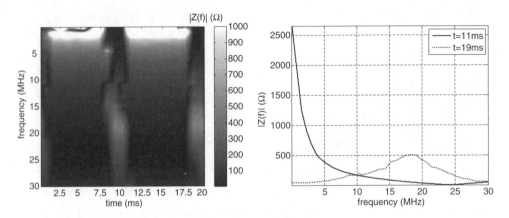

Figure 2.45   Absolute value of the measured impedance of an electric shaver.

2.3.3.3.2   *Theoretical Basis for the Time-varying Response*

A PLC channel can be considered as a non-linear system (NLS) because it contains some non-linear devices. An adequate description for a NLS is through a Volterra series, but in this case there are conditions that simplify the analysis [9]. At the system input, the superposition of two signals can be assumed: the communication signal, of low level and high frequency content, and the mains signal, of very large level and very low frequency, as depicted in Figure 2.47. Infinite terms created by the system nonlinearity appear at the output which are combinations of products of powers of the system response, the mains signal and the excitation signal. However, many of them can be neglected. At the receiver, and also at the transmitter, there are high pass filters that prevent the mains voltage from entering the communication equipment. Hence, all the terms in the Volterra series that do not contain the communication signal are filtered due to their low frequency content. Also, all the terms that contain the communication signal more than once can be obviated, due to their very low level. The resultant simplified

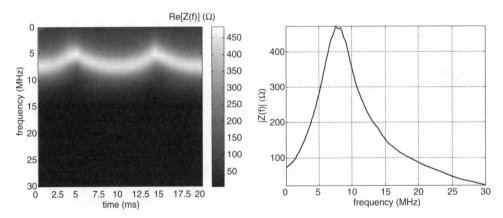

Figure 2.46   Real value of the measured impedance of an electric coffee machine. On the right, variation of the impedance with frequency at time = 10 ms.

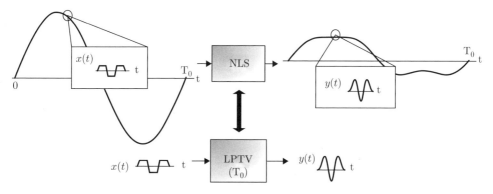

Figure 2.47   Diagram of the origin of the LPTV behavior.

superposition integral corresponds to a quasi-linear but periodically time-varying system as shown in Figure 2.48.

A linear time-varying (LTV) system can be described by its input output relation

$$y(t) = \int_{-\infty}^{+\infty} h(t, u)x(u)du,$$

where $h(t, u)$ is the impulse response of the system, which represents the response at the instant t to an impulse applied at the instant u. When the response is periodic in time, it is an LPTV system, that is

$$h(t, t - \tau) = h(t - nT_0, t - nT_0 - \tau),$$

being T_0 the fundamental period (and n any integer). The frequency response is obtained after applying the Fourier transform, to the impulse response, in the τ variable,

$$H(t, f) = \int_{-\infty}^{+\infty} h(t, t - \tau)e^{-j2\pi f \tau}d\tau.$$

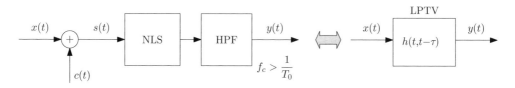

Figure 2.48   Time varying channel model.

It is also periodic in t with period T_0 and, hence, it can be expressed by means of a Fourier series, whose coefficients are

$$H^\alpha(f) = \frac{1}{T_0} \int_{-T_0/2}^{T_0/2} H(t,f) e^{-j2\pi\alpha t/T_0} dt. \tag{2.28}$$

It can be demonstrated [9] that the relation between the input signal $x(t)$ and the output $y(t)$ of the LPTV system in the frequency domain is

$$Y(f) = \sum_{\alpha=-\infty}^{+\infty} H^\alpha \left(f - \frac{\alpha}{T_0}\right) X \left(f - \frac{\alpha}{T_0}\right).$$

Nevertheless, the LPTV model for PLC channels admits a further simplification. The channel variation is quite slow, that is, the channel coherence time (the time during which the channel properties can be considered invariant) is several orders of magnitude above the duration of the impulse response. The former is in the region of hundreds of microseconds (μs) [9] while the latter is only of some μs [83]. Also, in practice, input signals used by current modems are shorter than the channel coherence time (common values are around tens of μs). For this reason, it is possible to make a locally invariant approximation of the channel response and to represent the LPTV system as a collection of successive LTI states that appear periodically.

This idea can be expressed by the following rationale. Let $x_\sigma(t)$ be an input signal of short duration, compared to the channel coherence time, and located around an interval $t \in [\sigma - \Delta t, \sigma + \Delta t]$ for a small Δt that is denoted by $t \approx \sigma$. Then, the output of the system can be expressed as

$$y_\sigma(t) = \int_{-\infty}^{+\infty} h(t, t - \tau) x_\sigma(t - \tau) d\tau \simeq \int_{-\infty}^{+\infty} h_\sigma(\tau) x_\sigma(t - \tau) d\tau.$$

Since the impulse response $h(t, t - \tau)$ does not change notably in $t \approx \sigma$, it has been substituted by $h_\sigma(\tau) = h(t, t - \tau)_{t=\sigma}$, which is the response of the LTI system measured at that interval of time. Moreover, as the input is of short duration, the output $y_\sigma(t)$ will be also of short duration and located in $t \approx \sigma$. If the Fourier transform is applied to the last expression, the result is

$$Y_\sigma(f) \simeq H(t,f)_{t=\sigma} X_\sigma(f).$$

This simple formula is similar to the one of LTI filtering and indicates that the output signal spectrum depends on both the input spectrum and the frequency response of the LPTV system $H(t,f)$, sampled in the interval $t \approx \sigma$.

2.3.3.3.3 Channel Time-varying Response Modeling
Under the assumption of slow variation, $H(t,f)$ can be estimated from snapshots of the LTI channel response at different instants of the mains cycle. The model must manage data sampled both in the the frequency and the time axis, with an adequate time resolution along the mains period. The procedure to synthesize the response may be the same presented in the LTI

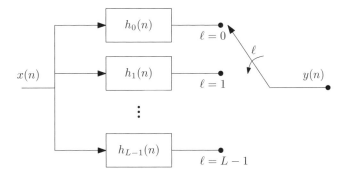

Figure 2.49   Zero-order hold interpolation of the LTI channel states.

channel model section (that is, based on a network structure and by means of a transmission line analysis) but including loads with a time-varying impedance.

This way, a collection of LTI states of the channel response $H_\ell(k)$ are obtained. The number of invariance states, indexed by $\ell = 0, 1, 2, \ldots, L - 1$, are distributed in the mains period T_0, i.e. $T_0 = L\,T_\ell$ (where T_ℓ denotes the invariance interval duration). From $H_\ell(k)$, by means of the Fourier inverse transform, the impulse response of L FIR filters $h_\ell(n)$ can be calculated. These states must be interpolated later to get the final time-varying channel response at the desired system sampling rate, according to a sampling period denoted as T_s. Hence, an invariance state is interpolated by a factor M that satisfies $T_\ell = M\,T_s$ or, equivalently, $T_0 = L\,M\,T_s$.

Depending on the time resolution of the available data, the interpolation method may be as simple as a zero-order hold structure (in Figure 2.49). (However, if the resolution is scarce, a linear interpolation may be preferable.)

The resultant channel time-varying impulse response is

$$y(n) = \sum_i h(n, n - i)x(n - i).$$

This response will be periodic in the discrete-time index n with a period of LM samples, and the index i must cover the effective duration of $h(n, n - i)$.

A channel emulator can be developed based on this model and implemented on a FPGA development board, as described in [84][7].

2.3.3.3.4   Measurements of Actual Channel Responses
In order to characterize the response of actual channels a measurement set-up that takes into consideration the mains instantaneous voltage has been defined. It comprises a digital board for signal generation at the transmitter side and a data acquisition board at the receiver. To establish the link, both cards are connected to the power grid at the selected outlets by means of a coupling circuit (essentially a passband filter plus a transformer and a transient suppressor). At the transmitter, this circuit is a more restrictive filter than the reconstruction filter of the

---

[7] In this prototype not only a channel time-varying filter is included, but also disturbances generators for narrow-band interference, colored background noise and several kind of impulsive components.

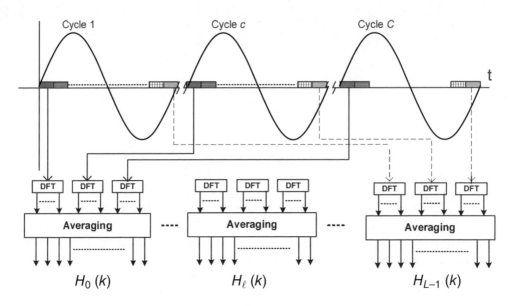

Figure 2.50 Diagram of the measurement procedure for LPTV channels.

board. At the receiver, the coupling circuit acts as an anti-aliasing filter for the acquisition board. The sounding signal is a set of N sinusoids with constant amplitude and phase and distributed uniformly between 0 and 25 MHz. However, the presented measurements only reach 20 MHz due to the passband of the coupling circuits. Since $N = 512$, the spectral resolution is 48 kHz. The received signal is recorded during C mains cycles (hundreds of them) and processed later.

The *processing algorithm* is represented in Figure 2.50 and can be described as follows:

(a) *Sounding signal*: N tones harmonically related between 0 and the maximum frequency.
(b) *Received signal*: Tones received with a periodical amplitude variation (due to channel filtering) and with noise.
(c) *Arrangement in time*: Decomposition of the captured signal in a series of intervals of invariance (compensating for the mains jitter)

$$x_{\ell,c}(n) = x(LT_\ell c + 2N\ell + n),$$

with $0 \leq n \leq 2N - 1$, $0 \leq \ell \leq L - 1$ and $0 \leq c \leq C - 1$. As well as calculation of the DFT

$$X_{\ell,c}(k) = \frac{1}{2N} \sum_{n=0}^{2N-1} w(n)x_{\ell,c}(n)e^{-j\frac{2\pi kn}{2N}},$$

where $w(n)$ is the employed window.

(d) *Averaging*: It reduces the noise in the estimate but is synchronized with mains cycle to unveil periodical variations.

$$X_\ell(k) = \frac{1}{C} \sum_{c=0}^{C-1} X_{\ell,c}(k).$$

(e) *Response estimation*: Estimation of frequency response in every interval of invariance.

$$H_\ell(k) = X_\ell(k)/S(k),$$

where $S(k)$ is the transmitted signal DFT.

This methodology has been employed to obtain the response of tens of channels at different indoor scenarios, in an apartment, a detached house and offices and laboratories in a university building. Some illustrative results are included here. The sampling frequency employed was $1/T_s = 50$ MHz, hence the time resolution T_ℓ is 20.48 μs and the number of invariant states in a mains period is $L = 976$. In Figure 2.51, on the left, the estimation of the amplitude response measured in a link at the apartment is presented. It is plotted in the plane of frequency and time, for the band of interest and the mains period ($T_0 = 20$ ms in Europe) respectively. The periodical variation of the channel in the band around 5 MHz is clear, with a period of half mains period. The channel variations are often frequency selective, that is, usually there are bands in which the shape of the variation is different from other bands. Moreover, it is not strange to have bands in which the channel is invariant while in others it is strongly time-varying. To appreciate this aspect better, represented on the graph on the right is the evolution of the channel amplitude along the mains period at some frequencies, i.e. $|H(t,f)|$ for three particular values of f. As seen, the response exhibits quite different variation profiles at the selected frequencies, with significant excursions in all cases. On the contrary, the channel is approximately invariant in the higher frequency band. This channel response shape must be

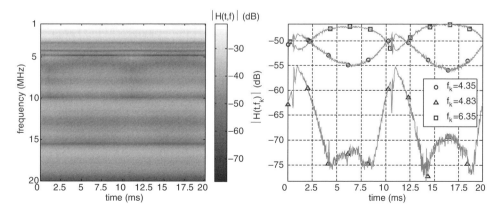

Figure 2.51 Amplitude response of an actual channel. On the right, variation along the mains cycle of the response at several frequencies (in MHz).

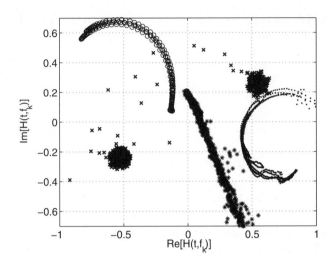

Figure 2.52 Evolution along mains cycle of the complex amplitude of four channel responses at different frequencies.

caused by devices of the second class, but many other responses with a commuted behavior, created by devices of the first class, have been measured as well [9].

Regarding the phase response, there are also remarkable cyclical variations. In Figure 2.52, measurements of four different channels have been plotted together. The graph represents the evolution of the channel response in the complex plane at certain frequencies, estimated at each of the L states of the mains period. For clarity, the plotted response values are relative, i.e. they have been normalized by the maximum value in each case. Sometimes there are frequencies at which the channel exhibits an invariant behavior in the amplitude while the phase commutes drastically and vice versa, in other channels both the amplitude and phase changes. It is worth noting the strong effect that such kind of variations in the complex plane have on a digital communication transmission. Unless a correct time-varying equalization is employed, the error rate will be drastically increased. These results lead to conclude that if the time variations are disregarded, transmission systems will lose part of the channel capacity. Some of the current PLC systems incorporate processing techniques to cope with the time-varying nature of the channel. Otherwise, the system exhibits less performance than expected because of this fact.

Apart from these examples, some statistical values extracted from the accomplished measurements are now summarized. Two parameters have been defined to study the magnitude of the time variation. The first one is the maximum excursion of the amplitude along the mains cycle, measured in dB,

$$\hat{H}(f) = 20 \log \frac{\max_t |H(t,f)|}{\min_t |H(t,f)|}.$$

The second one is the maximum excursion of the phase along the mains cycle, measured in radians,

$$\sphericalangle H(f) = \max_{p,q} \left[\sphericalangle (H(p,f)H^*(q,f)) \right].$$

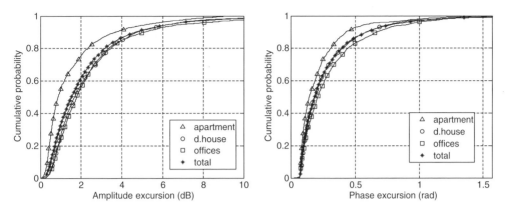

Figure 2.53   Estimation of the CDF for the amplitude excursion and the phase excursion along the mains cycle.

To compute the time variation, the channel response measured at each frequency has been processed independently, although those with notches have been removed from the analysis. Hence the results would represent the characteristics of all the valid carriers in a multicarrier transmission system approach.

The time variation is very frequency selective, there are bands highly variant, whereas, a 35% of the analyzed frequencies can be considered almost invariant. The criterion to consider a frequency as LTI has been that the channel response variations, in amplitude and phase, are small enough to guarantee a good symbol detection without any equalization in a 64-QAM modulation transmission. The remaining frequencies are considered LPTV and for this group an estimation of the cumulative distribution function (CDF) of both parameters is shown in Figure 2.53. (For clarity, the curves have been truncated, but there are values exceeding 17 dB in the amplitude excursion and π radians in the phase excursion.) It is observed that, in approximately 30% of the variant frequencies, the excursion in the amplitude is higher than 2.6 dB and the phase excursion is over 0.3 radians. Such channel variations would cause symbol detection errors in not too dense constellations, e.g. a 16-QAM modulation, unless time-varying equalization is applied.

In order to assess the rate of the channel time variation an adequate parameter is the *Doppler spread*, which measures the spectral broadening that the channel causes to an input sinusoid. In this case, the spectral broadening appears as harmonics of the mains frequency, so the *Doppler spread* can be defined as the frequency of the largest coefficient in the Fourier series in (2.28) for which $H^\alpha(f)$ is 40 dB below the maximum value, $H^0(f)$. It has been calculated that in 50% of the frequencies the Doppler is equal or greater than 100 Hz, while in 10% of the cases it exceeds 400 Hz [9]. The channel *coherence time*, which is inversely related to the Doppler spread, is no smaller than 600 μs in the worst case.

An important parameter in multicarrier systems is the cyclic prefix length, that is set according to the channel *delay spread*. Fortunately, it has been tested that in PLC channels this parameter exhibits little variation in time. The mean value of the delay spread, for the LTI average of the channels, is in the range from 0.3 μs to 0.65 μs. It is inversely related to the channel *coherence bandwidth* that, using a measure of the 90% of space-frequency correlation, can be estimated between 150 kHz and 250 kHz [83].

2.3.3.4   Reference Channels Definition for In-home

As has been discussed in this section, the deterministic characterization of a certain power line network is a hard task. Alternatively, or complementary, the adoption of a structural modeling approach with a set of reference channels can be very helpful to benchmark indoor PLC transmission systems.

2.3.3.4.1   Setting Parameters in a Structural Modeling Approach

When testing and benchmarking PLC modems, it is not necessarily important to replicate a multitude of particular power line networks, but rather the consideration of a relatively small number of sample models that represent the expected behavior of typical networks is sufficient. One way to achieve this is to generate randomly network topologies, with well chosen distribution parameters [31], [32], [81], [85], and then solve for the channel responses by means of the procedure described in Section 2.3.3.2.2. The following aspects are important:

(a) *Parameters of cables*: Estimation of cables characteristics can be obtained from manufacturers data. For example, typical cable diameters are 1.5, 2.5, 4, 6, and 10 mm^2, and the insulation material is usually PVC or similar.

(b) *Topology layout*: The parameters that need to be defined include number of cable sections and their lengths and relative position. In Table 2.3, some reasonable mean values are provided for three indoor network scenarios of different sizes: the number of electrical circuits, the number of outlets per circuit and section lengths. From them, different parameter ensembles can be obtained with random number generators leading to random network realizations.

(c) *Device characteristics and working state*: These characteristics can be taken from a database created from measurements. But, a more straightforward possibility is to create synthetic impedance functions, and also select randomly among them. In the next section some examples of these functions are given.

2.3.3.4.2   Reference Channels Generator

The degrees of freedom for the above-described structural modeling can be further limited to reach a set of reference channels. Below, a proposal by Cañete *et al.* is explained that follows these principles and that can be called a *Simplified Bottom-up approach*. It is based on the experience from analyzing hundreds of actual channels and that was previously presented in [86]. The proposal consists of three main ideas:

Table 2.3   Mean values to generate random topologies

Scenario Type (area, m^2)	Number of Circuits	Number of Outlets	Section Length
Small (60)	5	5	4
Medium (100)	7	6	6
Large (200)	10	7	10

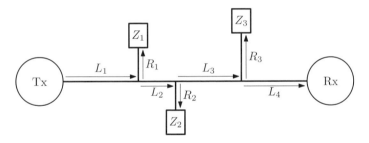

Figure 2.54 Simplified network topology for reference channel model.

(a) *Topology*: Let us define a simple power line network topology as shown in Figure 2.54. It comprises only seven cable sections and five outlets: there is a main path, in which the transmitter and receiver are located, and three stubs in parallel with the corresponding load impedances, which represent devices at each outlet. The layout parameters to be set are the seven lengths involved, denoted as L_k ($k \in \{1, 2, 3, 4\}$) and R_n ($n \in \{1, 2, 3\}$). Their values can be fixed manually or generated with a random number generator with a uniform distribution and will determine to a great extent the channel delay spread. Intervals of values in the order of tens of meters lead to reasonable results.

(b) *Transmission line parameters*: It is necessary to adopt the transmission line parameters: R, L, G and C (per unit length) or γ and Z_0. In Table 2.4 some example values are given, they have been estimated according to cable characteristics from manufacturers [81]. The type of cable (depending on the cross-section area) can be selected manually or at random. (The equivalent permittivity ε_{eq} is estimated including the effect of non-homogeneous dielectric between parallel wires, i.e. PVC and air. An overestimation factor for the losses ℓ is included in the G parameter to compensate for the very simplified network topology, so that the final channel attenuation is closer to the one in actual channels. An heuristic value of $\ell = 5$ has been proven to offer good results.)

(c) *Loads*: A reduced set of synthetic impedance functions is used as loads. Three groups of such impedances can be defined: constant values, frequency selective functions and time-varying functions. Each load in the topology can be selected among these types of impedances.

Table 2.4 Characteristics of actual indoor power network cables

Cable	H07V-U	H07V-U	H07V-R	H07V-R	H07V-R
Section [mm²]	1.5	2.5	4	6	10
ε_{eq}	1.45	1.52	1.56	1.73	2
C [pF/m]	15	17.5	20	25	33
L [µH/m]	1.08	0.96	0.87	0.78	0.68
R_1	1.2×10^{-4}	9.34×10^{-5}	7.55×10^{-5}	6.25×10^{-5}	4.98×10^{-5}
G_1	30.9	34.7	38.4	42.5	49.3
Z_0 (Ω)	270	234	209	178	143

$R = R_1 \sqrt{f}$ (Ω/m) and $G = 2\pi f \cdot \ell \cdot G_1 \cdot 10^{-14}$ (S/m).

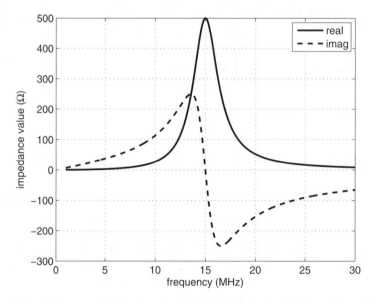

Figure 2.55   Frequency selective impedance for $R = 500\ \Omega$, $Q = 5$ and $\omega_0/2\pi = 15$ MHz.

(i) *Constant values*: A reasonable set would be $\{5, 50, 150, 1000, \infty\}\ \Omega$. They correspond respectively to the cases of: low impedance, RF standard impedance, similar to cable characteristic impedance Z_0, high impedance, and open circuit.

(ii) *Frequency selective functions*: They can be defined as the impedance of a parallel RLC resonant circuit that contains three parameters: R, resistance at resonance; ω_0, resonance angular frequency; and Q, quality factor (determines selectivity),

$$Z(\omega) = \frac{R}{1 + jQ(\frac{\omega}{\omega_0} - \frac{\omega_0}{\omega})}.$$

Reasonable values for these parameters are: $R \in \{200, 1800\}\ \Omega$; $Q \in \{5, 25\}$; and $\omega_0/2\pi \in \{2, 28\}$ MHz. See Figure 2.55 for an example.

(iii) *Time-varying functions*: There are two types of devices impedance behavior in time (see Section 2.3.3.3.1) and both can be modeled with simple mathematical functions as described in Figure 2.56.

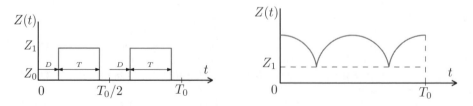

Figure 2.56   Time-varying functions for impedance model, with commuted (left) and continuous (right) behavior synchronized with mains cycle, T_0.

- *Commuted behavior*: It is modeled as two impedance states with abrupt periodical transitions with a period of half mains cycle. For each state a constant value or a frequency selective function can be selected. The parameters that describe the time variation are: T, the state duration, and D, the delay with respect to the mains voltage zero-crossing. The values for these parameters can be set manually or at random, with T uniformly distributed between 2 and 8 ms. Then, D value is in the interval $[0, T_0/2 - T]$.
- *Continuous behavior*: Its mathematical model corresponds to

$$Z(\omega, t) = Z_1(\omega) + Z_2(\omega) \left| \sin\left(\frac{2\pi}{T_0} t + \phi \right) \right|, \quad 0 \le t \le T_0,$$

where a half mains cycle periodicity is assumed (due to the rectified sinusoid), and constant or frequency selective impedance values can be selected for Z_1 and Z_2. A phase term ϕ can be included to reference the variation with respect to the mains voltage zero-crossing. Again, a uniform distribution is a reasonable choice for ϕ.

The presented channel generator is analyzed in [77] and its validity is tested over measured channels. The generator can be downloaded from [87], with a user guide and additional information, for instance regarding PLC noise models.

2.3.3.4.3   *Channel Generator Based on the Statistical Description of the Topology Geometry*

The channel generator described in the previous section is statistically representative when a point-to-point link has to be simulated since it is based on an abstract description of the network topology. When it is required to model a multiple access channel, it is important to generate channel responses that capture the shared network characteristics resulting in signals that propagate through the same wiring infrastructure. This results in a spatial correlation between channel responses of pair of nodes belonging to the same network [19] that cannot be neglected when assessing the performance of media access techniques and cooperative schemes [34].

To more realistically implement a channel generator, the bottom-up approach was applied on in-home networks obtained from a statistical model of the topology geometry in [32]. Herein, from the observation of European norms and practices, it has been shown that the topology can be partitioned in area elements (called clusters) that contain all the outlets connected to a derivation box (as shown in Figure 2.57). Derivation boxes are then connected to the main panel. The clusters have a rectangular shape with a variable dimension ratio, but the same area on average. Then, an efficient method, referred to as voltage-ratio method, to compute the channel transfer function between any pair of outlets belonging to a topology realization can be used. The statistics of the topology geometry and the wiring paths is also described in [32].

This reference channel model has been used to infer the behavior of the in-home PLC channel as a function of the topology geometry characteristics in [31]. An extension for modeling MIMO channels has then been described in [88], [89] (see also Section 2.6).

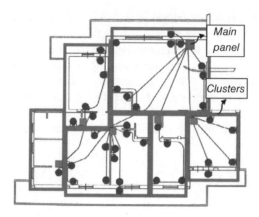

Figure 2.57 Topology geometry model: clusters, outlets (circles), derivation boxes (squares) and main panel.

2.3.3.4.4 Other Alternatives for Channel Generators

Apart from the bottom-up approach, also behavioral modeling can be exploited to construct directly indoor PLC channel generators from a *Top-down approach*. Two remarkable efforts following this strategy can be mentioned. The first one was initially proposed by Zimmermann *et al.* in [17] for the outdoor scenario. It consists in a multi-path model for the channel frequency response with a limited number of paths. Afterward, it was adapted to indoor PLC channels by Tonello in [56], by defining statistical distributions for its parameters values. A generator based on this model is available for download in [90]. The second one was proposed by Galli firstly in [91] and lately extended in [74]. It is based on a model for the channel impulse response with L-taps, and the amplitudes and delays of the coefficients are selected according to statistical distributions (but after forcing a correlation between the channel attenuation and the RMS-DS).

The performance of the three reference channel models mentioned in this section have been analyzed and compared in [92], [93]. All of them exhibit a reasonable fitness to the behavior observed in measured indoor PLC channels.

A significantly improved version of the top-down channel generator in [56] has been developed in [30] to statistically fit channels measured in European home sites.

2.4   Models for Medium Voltage (MV) Channels

Medium voltage (MV) transmission lines (nominally between 1 and 35 kV, as specified in ANSI/IEEE 1585-2002 [94] and IEEE Std 1623-2004 [95] are used to connect the distribution substations that terminate the High Voltage (HV) transmission network (110 kV and above) to the MV/LV distribution transformers that drive the Low Voltage (LV) local distribution network (600 V and below). The range of voltages used for this purpose represent a compromise between the need to reduce transmission losses by reducing the current carried on the line and the need to ensure safe operation in public environments by reducing the line voltage.

For decades, the monitoring and control capabilities of the electrical power grid focused on the generation and transmission networks and extended only as far as the distribution substations. A key objective of grid modernization is to extend monitoring and control through the distribution network to the customer premises. Recent interest in Power Line Communications over the MV distribution network (MV PLC) has developed largely in response to: (i) the need to supply connectivity to devices used for Distribution Automation (DA), Fault Location, Isolation and Service Restoration (FLISR), and Demand Response (DR) and by the Advanced Metering Infrastructure (AMI) and (ii) the potential for MV PLC to supply such connectivity in a reliable and cost effective manner.

Grid architects have expressed particular interest in using MV PLC to enable: (i) telecontrol or monitoring of intelligent electronic devices (IEDs) attached to the distribution network, including protective relaying devices, capacitor bank switches, recloser controllers, voltage regulators, etc. (ii) telecontrol of smart meters and backhaul of smart metering data, (iii) exchange of demand side management and similar messages, and (iv) provision of Internet connectivity to downstream customers. Developers of MV PLC equipment must overcome significant propagation impairments introduced by the MV channel. Channel models play a key role in quantifying these impairments and presenting them in a form useful in simulation and design.

In this section, we review progress in characterizing MV channels and developing channel models that will meet the needs of MV PLC developers both now and into the future. In Section 2.4.1, we consider the nature of MV transmission lines and the distribution network, including transmission line construction, attached devices and network topology. In Section 2.4.2, we review the types of MV PLC channel models that have been proposed, including those realized using the so-called bottom up and top down approaches; forward looking efforts, such as channel models that support emerging MIMO PLC technology, and characteristics of noise on MV lines.

2.4.1   The Medium Voltage Scenario

In virtually all respects, the physical attributes of MV distribution networks and transmission lines are intermediate between those of their high voltage and low voltage counterparts. A variety of standards and design practices contribute to the diversity of MV operating voltages, distribution substation designs and MV network layouts and topologies that are employed by various utilities around the world. Here we summarize the essential characteristics that are shared by the vast majority of MV distribution transmission lines and distribution networks in use today.

2.4.1.1   Distribution Substations

The heart of the MV distribution network is the distribution substation that accepts power from one or more high voltage transmission lines and steps it down to an intermediate voltage that is suitable for local distribution via one or more MV feeders and distribution lines. These feeders and distribution lines power the MV/LV distribution transformers that are located near customer premises and which step the voltage down to the even lower values, typically less than 600 volts, that are suitable for consumption by end users.

Distribution substations perform important secondary functions. First, they provide a mechanism for isolating faults in either the transmission or distribution systems and thereby increase the overall reliability of the power system. Second, they often serve as points of voltage regulation (although additional voltage regulators may be installed at various points along longer feeders). Finally, they often serve as an endpoint and gateway for MV PLC networks that allow monitoring and control of devices attached to the distribution network.

The simplest distribution substations consist of little more than switchgear and a step-down transformer but may be realized in different forms. Some distribution substations are based upon indoor equipment installed inside either: (i) a purpose built room or space within an existing building or (ii) a dedicated enclosure, possibly prefabricated, in an outdoor environment. Alternatively, some distribution substations are based upon outdoor equipment that has been mounted either upon a ground pedestal or atop a pole. The most complicated substations are typically found in the central business districts of large cities and include various features designed to enhance reliability and make it easier to reconfigure their interconnections to the grid when faults occur.

2.4.1.2   Network Layout and Topologies

The layout of the MV feeders and lines within a service area is ultimately dictated by the distribution and density of customers and the preferred locations for the distribution transformers that serve these customers. Branch lines are attached to the main feeder as required to achieve the desired coverage. MV lines are typically several kilometers in length but are composed of many shorter segments that are joined or spliced together, often at points where branch lines are attached.

The topology of an MV network is a description of the internal connections and alternative transmission paths within the network. The choice of network topology is driven mainly by the reliability requirements and density of customers within the service area and opportunities to increase reliability through increased redundancy.

Three possible MV distribution network topologies include: A single-line service realized through connection to an MV radial network, a ring-main service realized through connection to an MV loop network and a duplicate supply service realized using parallel MV feeders. Their attributes are summarized in Table 2.5 and Figure 2.58.

The simplest way to realize an MV distribution network is to implement a tree-like radial topology in which each distribution transformer is connected to its substation by a single path. Switches, reclosers, voltage regulators, capacitor banks and related IEDs are deployed throughout the network in accordance with local design policy in order to permit isolation of faults and to ensure power quality. A radial distribution network of this type combines simplicity with ease of operation and the lowest possible installation costs but offers no redundancy or options to quickly restore service to affected customers in the case of a fault or line break. More sophisticated network topologies can help to improve redundancy and thereby reliability.

While limited reliability is often accepted in rural environments that present low load densities (e.g. often defined as any region in which the load is less than 1 MVA/km^2), they are not acceptable in urban areas that present high load densities (e.g. greater than 5 MVA/km^2) and where high reliability of the electricity supply is expected. Open loop layouts provide

Table 2.5 Three possible connections to an MV network

	Service		
Characteristic	Single-line	Ring-main	Duplicate Supply
Activity	Any	Any	High tech, sensitive office, health-care
Topology	Radial	Loop	Parallel/Independent Feeders
Service area	Single building	Single or several buildings	Single or several buildings
Service reliability	Minimal	Standard	Enhanced
Power demand	≤ 1250 kVA	Any	Any
Setting	Isolated site	Low density urban area	High density urban area

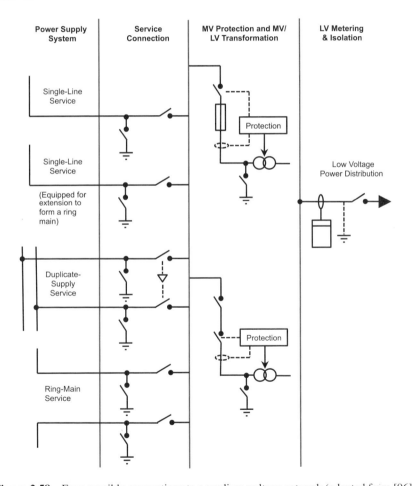

Figure 2.58 Four possible connections to a medium voltage network (adapted from [96]).

each distribution transformer (or branch feeding several distribution transformers) along the loop with two possible paths to its substation. Each transformer or branch line is connected to the loop by a pair of MV switches. During normal operation, all of these switches are closed except one so that the system functions as a pair of radial arms. When a fault occurs and a portion of the loop becomes isolated, it is a simple matter to close the appropriate switch so that service is immediately restored to the isolated branch and affected transformers.

The ring main topology is a variant of the open loop topology in which the loop is fed by more than one feeder. Switches are used to isolate portions of the ring that are driven by different feeders. If one feeder is not available due to fault or maintenance, the appropriate switches can be opened or closed as required to ensure that the remaining feeders energize all portions of the loop. The number of feeders that must be connected to the ring main electrical power distribution system increases with the maximum load presented by the system, the total length of the loop and the required voltage regulation.

When reliability requirements are very high or the density of customers can justify the additional cost, parallel or independent feeders may be used to increase redundancy and, therefore, reliability. Variants of the above schemes, including lattice type/closed loop topologies and simplified lattice type/open loop topologies are employed in dense urban areas in order to: (i) increase both network reliability and flexibility of and (ii) make it easier to modify and upgrade the MV distribution network.

2.4.1.3 Overhead Lines and Underground Cables

MV feeders and branch lines may be implemented using overhead lines or underground cables. Overhead lines are far less expensive to deploy but are also far more susceptible to damage due to weather and external factors. They are widely deployed in rural settings and frequently found in suburban settings. Underground cables are more expensive to deploy, but once buried in a trench or run through conduit are relatively unaffected by weather and external factors. They are widely deployed in suburban and particularly in urban areas.

The proportion of the length of overhead lines and underground cables on MV networks varies greatly between countries. European countries tend to have a high proportion of underground cable installations. North American countries, which tend to have larger and more extensive rural deployments, tend to have a high proportion of overhead line installations.

2.4.1.4 Overhead Lines

MV overhead lines are typically deployed atop utility poles at approximately 10–15 m above ground level. In suburban areas, poles are normally spaced 30–50 m apart but the separation increases to 100–130 m in rural settings. Most utility poles only carry a single MV feeder, also referred to as a primary circuit, although more than one primary circuit may be carried in exceptional cases. The MV feeder usually shares the pole with LV distribution wiring and communications circuits, including television and cable television that are mounted in pre-defined zones below the uppermost MV zone.

In North America, the feeder comprises three phases and a neutral. In Europe, only the three phases are distributed. The three phases are usually separated horizontally and mounted on a 2.5–3-meter cross arm mounted at the top of the pole. Porcelain or polymer insulators support the conductors. Such insulators are broadly classified as either pin type, which support the conductor above the structure, or suspension type, where the conductor hangs below the structure. The neutral wire may be installed either below or above the cross arm depending upon local practice and shielding priorities. Armless construction involves use of fiberglass insulator standoffs or post insulators to carry the feeder. Occasionally, and especially when no secondary or communications circuits are carried, the three phases may be vertically separated and mounted on the side of the pole itself.

Although copper was popular in the past, conductors in overhead lines are now manufactured from lighter and cheaper aluminum. Aluminum-conductor steel-reinforced (ACSR) cable is a composite material that combines the strength of a steel core with the high conductivity of aluminum. All-aluminum-alloy conductor (AAAC) is composed of a high strength Aluminum-Magnesium-Silicon Alloy that offers better strength to weight ratio and offer improved electrical properties, excellent sag-tension characteristics, and superior corrosion resistance when compared with ACSR. Aluminum-conductor composite-core (ACCC) uses a carbon and glass fiber core that offers a coefficient of thermal expansion about 1/10 of that of steel and allows creation of conductors with greater current capacity but lower thermal sag than ACSR. Cross-sectional views of conventional ACSR and modern ACCC conductors are compared in Figure 2.59. The impact of different construction materials on PLC is rarely if ever considered.

Although manufacture and installation of overhead lines are far less costly than manufacture and installation of underground cables, overhead lines are vulnerable. Wind and ice may cause conductors to touch and cause temporary short-circuits. Insulators may break or shatter or be compromised by dirt or other foreign matter. Because many of these fault conditions are only temporary, circuit breakers with an automatic reclosing facility or reclosers are often used to protect MV overhead lines.

Power line communications over overhead lines requires devices that allow broadband signals to be coupled into the lines. Although both capacitive and inductive approaches are possible, capacitive couplers are by far the most common. A typical installation is shown in Figure 2.60.

Figure 2.59 Conventional ACSR and modern ACCC conductors used in overhead transmission lines. (Photo by Dave Bryant, licensed under CC BY-SA 3.0 Unported license.)

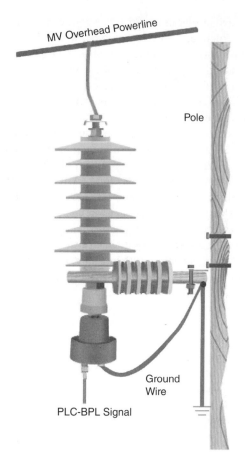

Figure 2.60 A typical capacitive coupler for injecting PLC-BPL signals into medium voltage overhead lines. © 2014 Arteche.

2.4.1.5 Underground Cables

Underground MV cables are typically deployed in trenches and conduits below city streets (along the municipal right of way) and within buildings. An MV cable circuit consists of three single core cables or one three-core cable with terminations at each end that allow it to be connected to the substation equipment or distribution transformer as appropriate. When the length of the MV circuit exceeds the length of a single cable, joints are used to connect individual cables. MV cables can be laid directly into the ground, pulled into ducts laid in the ground, or run in tunnels or above ground on cable trays or ladders. They may be laid flat (side-by-side) or in trefoil (three cores laid in a triangular formation). When multiple single cores are laid, care must be taken to ensure that the phases are connected in a way that ensures that a balanced current distribution is achieved. Although underground cable networks suffer from fewer faults than overhead networks, faults that do occur are invariably permanent and take longer to locate and resolve.

The components of an MV cable include the conductor, the conductor screen, insulation, the insulation screen, the metallic sheath and the anti-corrosion sheath. The conductor is rated by its effective cross sectional area; for MV cables, this normally falls in the range from 35 to 1000 mm^2. The conductor may be solid, formed from several layers of concentric spiral-wound wires, or consist of sector shaped conductors that help make the cable flexible while minimizing its overall diameter for a given current carrying capacity. Aluminum conductors are usually used in cables used to cover long distances where their low cost and weight are most useful. Copper cables are used in short links in substations or industrial installations where their high conductivity and small size for a given power handling capacity are most useful. The insulation layer is a critical component of high power cables. For many years, paper-insulated lead covered (PILC), separate lead screened (SL), and oil filled (OF) cables were widely used. In recent years, cross-linkable polyethylene (XPLE) insulation has replaced previously used materials.

Power line communications over underground cables requires devices that allow broadband signals to be coupled into the cables. Although both capacitive and inductive approaches are possible, inductive couplers are by far the most common. A typical installation is shown in Figure 2.61.

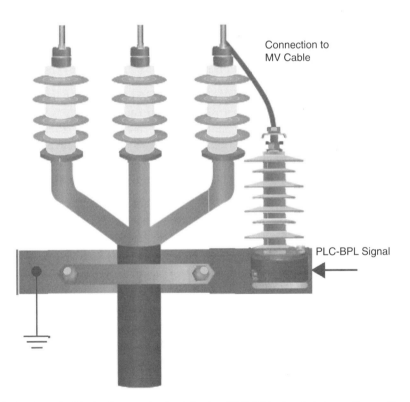

Figure 2.61   A typical inductive coupler for injecting PLC-BPL signals into medium voltage underground cables. © 2014 Arteche.

2.4.2   Medium Voltage Channel Models

MV power lines were not designed for data transmission and often present a challenging environment for communication. Attenuation, frequency selectivity, delay spread and temporal and spatial variability of the transmission response between two points on the MV distribution network are determined by: (i) the cross-sectional geometry and length of the feeder, (ii) the nature of splices, joints and couplers that connect feeders, (iii) the length and location of the branch lines attached to feeders, (iv) the nature and location of the transformers, capacitor banks, and other devices that are attached to the distribution network and (v) loads attached to the MV network are switched on or off, or as the MV network or network-attached IEDs are reconfigured.

Just as the physical attributes of MV distribution networks and transmission lines are intermediate between those of their high voltage and low voltage counterparts, the channel impairments encountered on MV lines are also mostly intermediate between those encountered in HV and LV scenarios. Assessment of these impairments and alternative techniques for mitigating them requires channel models that capture our knowledge and understanding of these impairments in a form useful in design and simulation.

Development of MV channel models is complicated by the diverse set of standards and design practices upon which both MV distribution networks and MV PLC around the world are based. As a result, the vast majority of current MV channel models correspond to specific cases and scenarios. While some have suggested that development of a universal MV channel model is a worthy goal, doing so would likely be an extremely difficult proposition. In general, both the top-down and the bottom-up channel modeling approaches can be followed, as it is done for in-home channel modeling (see also Section 2.3.3). The former requires a measurement-based or parametric method that quantifies the statistical relationship between a body of measured channel characteristics in the time and/or frequency domains and relevant parameters of the corresponding physical scenario. The latter generally refers to a theory-based or physical approach that seeks to predict channel characteristics strictly from detailed knowledge of the scenario geometry. Both approaches may be used to characterize segments or functional blocks. The results can then be used to construct signal flow graph representations of much more complex networks that are composed of a multiplicity of segments or functional blocks.

Within each approach, one can also distinguish between deterministic and stochastic models. The former are more applicable to specific scenarios while the latter are intended to capture channel behavior over a range of circumstances.

2.4.3   Measurement-based Characterization of MV Channels

Measurement-based characterization of MV channels takes place at two levels: the component level and the network level. The effort begins with characterization of the transmission lines that form the heart of the channel. Experience has shown that overhead lines attenuate signals less (and incur less capital cost) compared to underground cables. They are, however, more susceptible to interference and damage due to external factors. The obvious inference is that underground cables attenuate signals more (and incur greater capital cost) compared to overhead lines. They are, however, less susceptible to interference and damage due to external factors.

Component-level characterization of the PLC channel generally takes place in a test lab and may involve the use of a vector network analyzer or similar stimulus-response test set to characterize both the input impedance (or reflection response) and transmission response of the component as a function of frequency. The frequency range and resolution over which the response of the component should be characterized depends upon the range of frequencies that are occupied by the PLC communication system of interest.

While component-level characterization is conceptually simple, practical difficulties abound. First, standardized couplers that allow the stimulus signal to be applied to the component and the response signal to be retrieved are not generally available. Second, standardized techniques for removing the effects of the couplers from the measured response of the component, i.e. for calibrating the couplers and de-embedding the actual response, are not generally available. Third, standardized data formats for sharing the results in a way useful in Electronic Design Automation (EDA) software are not generally available. Progress in any of these fronts would likely be of considerable benefit to the PLC industry.

Network-level characterization of the PLC channel is generally conducted using either a test bed assembled from operational and possibly prototype components or within an operational MV distribution network, e.g. [48], [97]–[100]. The results may be used to verify both the input impedance (or reflection response) of an input port or the response of a transmission path between ports as a function of frequency as predicted by simulation or calculation. An experimental system for characterizing the PLC channel between two MV substations using a stimulus-response test setup is shown in Figure 2.62. In this example, most of the major

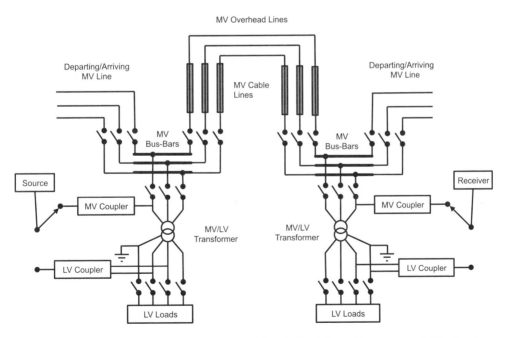

Figure 2.62   An experimental system for characterizing the PLC channel between two MV substations using a stimulus-response test setup (adapted from Cataliotti *et al.* 2013 [108]).

components are represented with the exception of branch lines and attached devices such as capacitor banks. The signal can be injected and received either through direct coupling into the MV network or via an LV network.

The transmission characteristics of an MV-PLC network may change as switches are opened and closed and as loads are switched on and off. Dynamic loading effects have been extensively studied on LV (low voltage) and indoor PLC networks, e.g. [101], [102], but relatively little effort has been devoted to the corresponding phenomenon on MV-PLC networks. Additional study is likely indicated, particularly in networks with relatively simple topologies. Both simulation and measurement show that in a complex power network with several parallel-connected branches, time variation of the transfer function due to on-off switching of loads is strongly reduced [103], [104].

2.4.4   Theory-based Characterization of MV Channels

2.4.4.1   Overhead Lines

Characterization of a single wire transmission line over a lossy earth was first considered in the early days of telegraphy and telephony. In 1926, Carson [50] presented the earliest solution for this problem. He calculated the distribution parameters of a quasi-TEM mode based upon assumptions that restrict the solution to very low frequencies and/or perfectly conducting earth. In 1956, Kikuchi [51] derived an exact modal equation for very thin wires above the earth. Through use of quasi-static and asymptotic expansion of the exact modal equation, his formulation permits investigation of the transition from quasi-TEM to surface wave propagation.

In 1972, Wait [105] extended Kikuchi's work and obtained an exact numerical solution for a single wire over ground. Later authors, including D'Amore [106] and Amirshahi and Kavehrad [47] extended these results to the multi-conductor case and accounted for lossy ground return. When combined with typical network topologies and appropriate noise models, such as those defined by Lazaropoulos [107], models of this type permit accurate prediction of the channel response and capacity of PLC systems based upon overhead lines. Distinguishing character-istics of such responses include the rate at which the channel frequency response decays with both frequency and distance and the number and depth of spectral nulls in the response.

2.4.4.2   Underground Cables

Characterization of a coaxial transmission line with lossy dielectric was first considered in the early days of telegraphy and coincides with the emergence of electrical engineering as a distinct discipline. The multiconductor nature of MV distribution cables gives rise to many of their most important features and has been an important research topic since the 1930s. While a two-conductor transmission line supports one forward- and one backward traveling wave, a multiconductor transmission line (MTL) with $n + 1$ conductors may support n pairs of forward- and backward-traveling waves or modes, each with a distinct propagation constant. In underground cables, n modes may be supported, including the common mode (CM) that propagates via the n conductors and returns via the shield and the $n - 1$ differential modes (DM) that propagate and return via the n conductors.

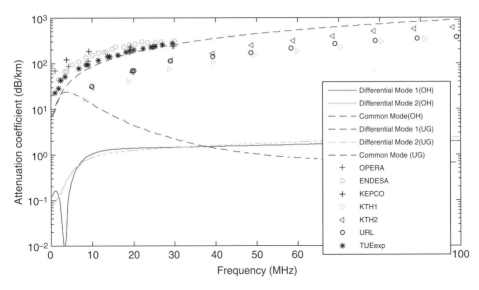

Figure 2.63 Comparing frequency spectra of attenuation coefficients of BPL power distribution networks for overhead and underground MV lines (adapted from Lazaropoulos and Cottis 2010 [107]).

An exact evaluation of MTL traveling-wave solutions generally requires a hybrid-mode full-wave analysis. For underground cables, much of the challenge lies in capturing the relatively complex cross sectional geometry of the cable and the dielectric structures used to provide both insulation and mechanical support. However, quasistatic approximations based upon TEM modes yield acceptable results if the cable's transverse dimensions are far smaller than a wavelength and the longitudinal field components, which are normally associated with lossy or longitudinally varying dielectric, are very small. Through a matrix approach, the standard transmission line analysis for two conductors can be extended to the multiconductor case. The n modes satisfy a set of $2n$-coupled first-order partial differential equations that relate the line voltages to the line currents. Predicted values of the attenuation constants of typical underground cables and overhead transmission lines are plotted as a function of frequency and compared to measured values in Figure 2.63.

2.4.4.3   MIMO PLC in MV Networks

The application of MIMO techniques to PLC is a fairly recent development. As in wireless implementations, it seeks to establish multiple transmission paths between the transmitter and receiver in order to increase both capacity and reliability [89], [109]. Various inductive MIMO PLC coupling schemes have been proposed. Only the delta-style and T-style schemes are recommended, however. Several PLC standards have been extended to incorporate provision for supporting MIMO operating modes in the future. Enhanced performance comes at the expense of higher cost, however. The commercial success of these schemes will depend on how designers trade off between enhanced performance and greater cost and complexity. Accurate MIMO PLC channel characterization is required in order to assess the achievable performance.

The vast majority of measurement campaigns have been conducted in LV environments, e.g. [110]. MV-specific campaigns will be required to assess the performance of MIMO PLC in MV environments.

2.4.5   Noise and Interference

Noise and interference impose fundamental limits on the performance of MV PLC systems. They may arise from: (i) processes involved in the functioning of the electrical power grid (endogenous noise), (ii) processes external to the electrical power grid (exogenous noise) and (iii) interfering signals from other PLC devices. Non-Gaussian noise in MV-PLC networks may be as much as 50 dB above thermal noise and is considered to be the main source of errors in PLC communications [111].

The noise in MV-PLC networks falls into three categories: (i) generalized or colored background noise, (ii) periodic impulsive noise and (iii) asynchronous impulsive noise. The generalized or colored background noise is characterized by narrowband interference (mostly below 1 kHz) that arises from devices such as switching power supplies that is superimposed upon an exponentially decaying power spectral density that becomes white at higher frequencies. The periodic impulsive noise is characterized by very short pulses with amplitudes up to 2 kV that arise from devices such as switching regulators and motors that are attached to the electrical power system. The asynchronous impulsive noise is mainly due to switching transients associated with on-off switching of loads connected to the network. Various studies have sought to model the impulse durations and inter-arrival times of the various forms of impulsive noise, e.g. [112]. Other works have focused on the instantaneous amplitude statistics of the impulsive noise.

2.5   Models for Outdoor Channels: High Voltage Case

High voltage (HV) transmission lines (110 kV and above) are used to transmit energy from generating power plants to a remote electrical substation or to interconnect two electrical substations. Most HV transmission lines usually adopt high voltage three-phase alternating current (AC) solutions, although high voltage direct-current (HVDC) technology can be used to increase the transmission efficiency over very long distances (hundreds of kilometers). In this section we provide an introduction to the main aspects related to power line communication over HV lines, starting from a picture of the common scenario where HV lines are involved and ending to some analytic relations that can be used to approximate the communication channel behavior.

2.5.1   High Voltage Scenario

The HV PLC telecommunication networks have been originally created to support an analog telephone direct communication between substations and control centers [113]. Later, with the implementation of Remote Terminal Unit (RTU) in substations for SCADA (Supervisory Control and Data Acquisition) applications, the analog data channel has been replaced by digital modems able to create multiplexed channels with data rate up to 64 kbit/s or higher.

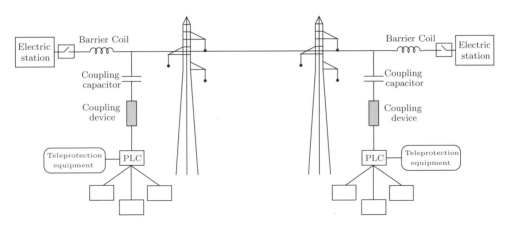

Figure 2.64   Example of an HV transmission system network.

The available carrier-frequency band for HV PLC communications spans from 40 to 500 kHz, even if in some country it is possible to transmit at higher frequency (up to 1 MHz). The quality of data transmission is directly dependent on the line and coupling attenuation, the signal-to-noise ratio and line parameters that can vary during adverse weather conditions. Using line traps and coupling devices the PLC system has become a part of the power line infrastructure covering basic requirements for telecontrol and teleprotection applications. The behavior of HV power line communication channel cannot be fully understood without a deep comprehension of all the devices involved in the application: in the next few pages the HV scenario with each of its characteristic component will be presented.

Figure 2.64 shows a general HV PLC equipment configuration. Electric utilities use a variety of different HV overhead conductors and transmission towers and poles to transmit energy and data. The conductor of high voltage cables can be made of copper or aluminum and is either round stranded of single wires or additionally segmented in order to reduce the current losses: the structural elements are surrounded by an insulated layer and an external protective jacket. The current-carrying capacity along with the presented resistance of HV lines depends on the conductors sizes, which can range from 10 mm^2 up to 750 mm^2. The individual strands are often preshaped to provide a smoother overall circumference. The insulation may consists of *cross-linked polyethylene*, also called XLPE, or *polyvinylchloride* (PVC) which exhibit a reasonable flexibility, can tolerate operating temperatures up to 120°C and serve as an anti-corrosion layer. The insulation of the cable must not deteriorate due to the HV stress, ozone produced by electric discharge in air or tracking. A conductive metallic shield (called a Hochstadter shield in honor of its inventor) connected to earth ground often surrounds the insulated HV conductors in order to equalize the dielectric stress on the cable insulation.

The transmission towers are steel lattice structures used to support the overhead HV power line [114]: they can be designed to carry three (or multiples of three) conductors with one or two ground wires, also called 'guard' wires, placed on top to intercept lightning and divert it to ground. The transmission towers are insulated from the HV power lines by either glass or porcelain discs or by composite insulators made by silicon rubber materials assembled in strings or long rods whose lengths are function of the line voltage and environmental conditions. Tower heights can vary considerably according to the design span and the nature of the ground. Conductors may be arranged in one plane or by the use of several cross-arms may be arranged in a roughly symmetrical, triangulated pattern to balance the impedances of all three phases. Modern practice is to build overhead lines with phase conductors symmetrically disposed to each other: this solution reduces the electrostatic and electromagnetic unbalance that may arise from an asymmetrical configuration and prevents the use of transposition towers (the common solution used to eliminate the electromagnetic imbalance). It must be pointed out that the line configuration and conductor spacing are influenced by the operating voltage along with many other factors including the conductor sag, the nature of the terrain, the type of insulators, the span length and the external climatic conditions.

Tower structures can be classified considering the way in which they support or arrange the line phase conductors: *suspension* towers support the conductors vertically using suspension insulators hanging down from the tower or two insulator strings making a 'V' shape. In either case, several insulator are used in parallel to increase the mechanical strength of the structure. A *transposition* tower is a transmission tower that changes the relative physical positions of the HV phase conductors in order to balance the electrical impedances between phases. A *dead-end* tower uses horizontal strain insulator to support the HV conductors and can be used when the circuit changes to a buried cable, when the transmission line changes its direction by more than a few degrees or as branch pylon. In recent years many utilities prefer the use of steel tube towers in place of lattice steel poles for new power line and tower replacement because of their durability and ease of manufacturing and installation.

As illustrated in Figure 2.64 the connection of conveyed wave line terminals to HV lines requires the use of the following components:

- wave (or line) traps with their tuning devices;
- coupling capacitors;
- coupling circuits (also known as coupling devices).

Line traps consist of a cylindrical coil with inductance values ranging from 0.2 mH to 0.5 mH, as shown in the left part of Figure 2.65: they confine the conveyed wave signals to a particular sections of the HV line, preventing them from being sent in unwanted directions and enabling a reuse of frequency bands in the power network. The wave traps connected in series to the HV line conductors may be assembled 'free-hanging' or 'supported' as shown in Figure 2.66 and have to be designed to support the maximum carried currents, i.e. every type of short circuit currents and must present a negligible impedance at network frequency (50 Hz or 60 Hz) to avoid interfering with the normal carried power whereas a sufficiently high impedance in the frequency band dedicated to the PLC data transmission. Moreover, line traps allow the connection attenuation and impedance features in all network configurations to remain unaltered, including earthing of phase from both side of the section, and protect communication apparatus against atmospheric discharges and any surges during operation. Wave traps consists of a single or multiple conductor, wound in a cylindrical shape, with a section that depends on the tolerable current entity and inductance value: if there are multiple cylindrical coils, they are usually arranged concentrically and electrically connected in parallel. The conductor is made of aluminum and has a rectangular section; the short side of the section is arranged parallel to the inductor's vertical axis in order to make the structure more mechanically robust. The spirals constituting the inductor are evenly separated at several points by fiberglass spacers soaked in epoxy resin, resistant to atmospheric agents, in order

Figure 2.65   Standard line trap and tuning device used in HV PLC communications.

to guarantee insulation and mechanical consistency: the distance is measured to optimize cooling and high frequency electrical features. These passive circuit elements are significantly stressed, both mechanically and electrically, are subject to lightning and stress due to short circuits in the line which cause intense transient phenomena: standards ANSI C93.3 and IEC 60353 describe the requirement for the realization of line traps.

Along with wave traps, a tuning device, shown in the right part of Figure 2.65, consisting of an RLC circuit set up in parallel to the main inductor provides a programmable anti-resonant circuit for an extended band of frequencies, making the attenuation and impedance, typical of

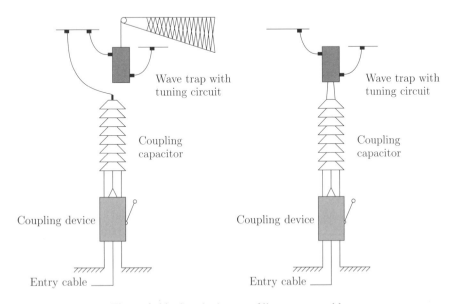

Figure 2.66   Standard types of line trap assembly.

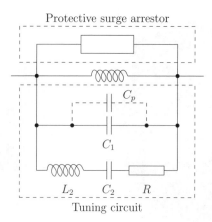

Figure 2.67   Standard wire diagram of a line trap with broadband tuning device.

the high voltage lines, insensitive to the status of the components of the electric substation. The circuit that constitutes the tuning device, shown in Figure 2.67, can be often decomposed into three main parts: a series-tuned arm consisting on a programmable resistance R, a capacitor C_2 and a multi socket inductance L_2 with a connection to the circuit that depends on the frequency band used by the communication equipment, a parallel-tuned arm containing a programmable capacitor C_1 based on the selected frequency band and a trimmer capacitor C_p that can be used to adjust the value of C_1. The capacitor C_1 in parallel with the line trap builds a resonant circuit at frequency f_1 while capacitor C_2 is rated so that the entire circuit arrangement will block also the frequency f_2: frequency f_1 is the higher, f_2 is the lower of the two carrier frequencies used by the communication terminal. All line traps are usually equipped also with a protective surge arrestors whose nominal operating voltage ranges from 3 kV to 6 kV: the surge arrestor, connected in parallel to the main inductor and the tuning device, keeps any surges during operation under the limits of the coil's isolation levels.

Coupling capacitor is the passive component used to connect the communication equipment to the HV line: its value can vary from 2000 pF to 10 000 pF. In outdoor switchyards coupling capacitors for low operating voltages are suspension-mounted, while pedestal-mounted versions are preferred for higher voltages. Power transmission networks have the neutral solidly grounded when the voltages are higher than 150 kV, so that the coupling capacitor interposed between one conductor and ground may be rated for a voltage lower than the nominal power line voltage: the charging current of the coupling capacitor increases with the operating voltage and with the coupling capacitance. Standard IEC 60358 describes the requirements for coupling capacitors.

Finally, an HV coupling device (also known as coupling circuit) guarantees the signal connection between the PLC communication equipment and the HV line for the entire range of frequency between 24 kHz and 500 kHz (or 1000 kHz). As Figure 2.68 shows, it generally consists of wiring between the power line carrier equipment and the coupling device, one or two coupling devices, one or two coupling capacitors, one or two HV line phases and optionally may be equipped by a high-frequency hybrid. The function of variable tuning coupling is performed by a variable tuning circuit inside the coupling device, while the isolation and

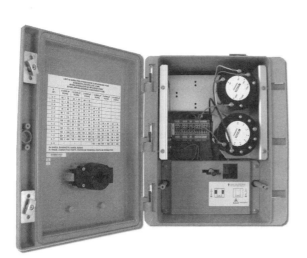

Figure 2.68   Standard coupling device.

impedance adapter transformer that usually equips the coupling device allows the impedance adaptation between the power line and the connection line along with a galvanic isolation between the primary and secondary terminals of the coupling device. Moreover, in addition to the above-mentioned functions, the coupling device can perform transit functions (bypass) allowing an intermediate power station located within the transmission path to be bypassed. It is common practice to include inside the coupling device also a hybrid device which allows dual insertion of PLC terminals with close transmission bands and signal group-insert in a bypass configuration. The insertion loss due to the coupling device should be less than 2 dB while the return loss should be greater than 12 dB. The insertion loss measurements are applicable on the single coupling device as well as on the entire connections: in the former case, the measured losses can be entirely attributed to the coupling transformer and the filter, whereas in the latter case the measured losses due to the HV line are included. Standard IEC 60481 describes the requirements for coupling devices.

Carrier communication terminals may be connected to two conductors of an HV power line through different kind of coupling schemes, as depicted in Figure 2.69:

- *phase-to-ground*: the signal injection is carried out between one conductor and the earth. Given its simplicity and low-cost, the single-phase coupling scheme is mainly used for short connections to not-so-high voltage (typically 110 kV): on the other hand, its very high attenuation usually tends to put the PLC terminals out of service in the case of a fault in the ground of the transmission phase;

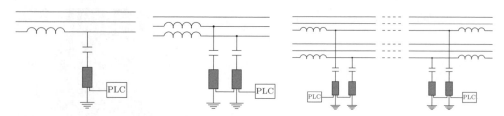

Figure 2.69   Common adopted coupling schemes: phase-to-ground, phase-to-phase and inter-system.

- *phase-to-phase*: the signal injection is carried out between any two phase conductors belonging to the same three-phase system (or circuit). Such a coupling scheme is usually adopted on high voltage lines (220 kV and above) for connections that cover long distances, or in situations where teleprotection systems are equipped in addition to PLC services;
- *inter-system* (or *circuit-to-circuit*): the signal injection is carried out between any two phase conductors belonging to different circuits but on the same tower.

The phase-to-phase coupling schemes are usually preferred with respect to phase-to-ground schemes because of their lower attenuation and higher robustness against noise and disturbances and because they present a high stability in terms of attenuation and characteristic impedance against atmospheric phenomena. Moreover, they afford a higher measure of dependability in the event of a broken conductor: a fault on one of the coupled phases does not interrupt the service since transmission can be conveyed to the other phase (phase-ground coupling). Inter-system coupling offers the additional advantage that one of the two circuits can be grounded without interposition of line traps and the carrier communication system will still remain operative: however, inter-system coupling schemes suffer by a higher attenuation with respect to the one offered by the phase-to-phase connection.

2.5.2   HV Channel Model

Having defined all the devices and apparatus involved in PLC communications over HV lines, it is now possible to introduce a general model for HV channels.

It is common practice to adopt a top-down approach and define a model for the HV channel starting from measurement campaigns. However, the presence of multiple reflections along the line caused by impedance discontinuities, non-ideal couplings and different non-homogeneous loads, along with the frequency dependent attenuation provided by all the physical devices involved in the communication cause a transmitted wave to propagate along several different paths: deep narrowband notches, spread over the whole frequency range, occur in the channel transfer function. Hence, the receiver observes a superposition of multiple attenuated replicas of the transmit signal: this phenomenon is referred as *multi-path propagation*. Moreover, due to different lengths of the echo paths and the time-varying behavior of the loads along the HV line, the individual multi-path components may experience a different time shift: even if the HV cable itself is not physically dispersive the overall PLC channel may be considered *time-dispersive*.

Taking into account all these aspects, one can conclude that the HV channel can be modeled as a linear *slowly* time-variant filter and described by the impulse response

$$h_{HV}(t, \tau) = \sum_{i=0}^{L-1} \alpha_i(t)\delta(t - \tau_i(t)) \quad (2.29)$$

where the set of coefficients $\{\alpha_i(t)\}_{i=0}^{L-1}$ represents the complex multi-path echoes amplitudes with their associated propagation delay $\{\tau_i(t)\}_{i=0}^{L-1}$. Since the HV time-dispersive channel has a multiplicative effect on the transmitted signal in the frequency domain, it can also be considered *frequency-selective* in the sense that different frequencies are differently attenuated. An additive noise term $n(t)$ can also be introduced to represent the additive noise experienced by PLC equipments: noise in HV links may be roughly modeled as a sum of a stationary component and some additional impulse components (see Section 2.5.3 and Section 2.5.4).

In-field measurements are usually carried out during the installation phase of an HV PLC link in order to analyze the line attenuation, line impedance, line reflection and noise levels: such an approach is used to characterize, over a short time interval of observation, the line behavior and make a first attempt in the link budget design as well as communication equipment set-up. An example of the HV channel behavior is presented in Figure 2.70, where the channel attenuation as a function of frequency and time is shown for a 132 kV line of 11.5 km length. Even if Figure 2.70 represents just a single channel time-frequency response, it describes well the typical impairments PLC equipments have to cope with.

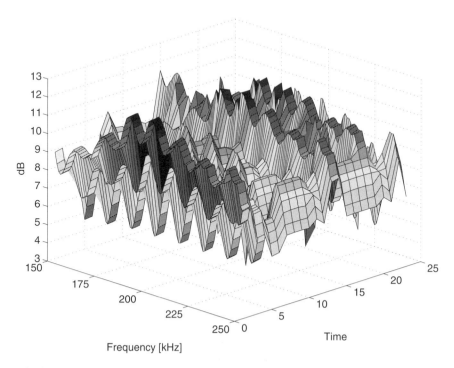

Figure 2.70 Line attenuation versus frequency and time for a 132 kV line of 11.5 km length. Acquisitions every 30 min.

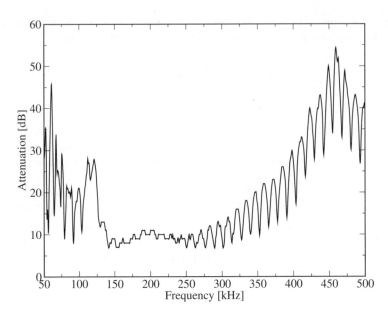

Figure 2.71 Line attenuation for a 51.3 km 380 kV line.

Figure 2.71 shows the attenuation as a function of frequency for a 380 kV line of length 51.3 km: it is interesting to note that the coupling circuit used and the line trap tuned over the frequency range 168–240 kHz have a low-pass effect on the line attenuation, while for the remaining frequencies the attenuation presents deep notches. Since this is a typical effect, one can assume that, under steady-state conditions of the link, it is possible to derive a few approximated formulas to describe, at least in terms of the link attenuation, the behavior of the HV channel. The next section will detail this approach.

2.5.2.1 Attenuation in HV Links

HV electric lines can rarely be modeled as symmetric and homogeneous lines because of asymmetries along the signal path such as different terminating conditions: moreover, the spatial disposition of the conductors with respect to the earth or guard wires can vary considerably along the path. However, starting from experimental data and using the modal analysis theory applied to an homogeneous line it is possible to derive some general considerations that can be used as guidelines for the evaluation of the attenuation in HV links.

The total attenuation in HV communication links is mainly a function of the frequency and the length of the line. It can be decomposed into the sum of the different contributions produced by each part of the link plus the attenuation related to the non-ideal impedance matching between the transmitter (and the receiver) and the line: as a consequence, it is possible to write the attenuation of a complete transmission section as

$$A(f) = a_\ell(f) + [2a_{cp}(f) + 2a_{st}] + a_{cbl}(f) \cdot \ell_{cbl} \tag{2.30}$$

where $a_\ell(f)$ is the attenuation of the line of length ℓ, $a_{cp}(f)$ is the attenuation introduced by coupling circuits, a_{st} is the loss introduced by the power network station, $a_{cbl}(f)$ is the attenuation of high-frequency (HF) entry cable (which depends on production method and frequency) and ℓ_{cbl} is the length of the HF entry cable.

The line attenuation $a_\ell(f)$ as well as the line characteristic impedance is a function of

- line configuration type (single triangular circuit or single plane circuit) and coupling scheme;
- distance between conductors and the earth as well as between conductors and guard wires;
- material and sections of the conductors;
- presence of one or two guard wires;
- number and schemes of transpositions;
- relative permittivity, permeability and conductivity of the earth above the conductors;
- geometric configuration and position of the coupling conductors with respect to other conductors (if present), to the earth and to the guard wires.

The line attenuation $a_\ell(f)$ is primarily a function of frequency, the adopted coupling scheme and the position of the coupled conductors with respect to other conductors not interested in data communications, to the terrain and to the guard wires. The overall line attenuation can be viewed as a sum of the conduction loss, the radiation loss and the induction loss. Radiation and induction losses depend on the ratio d/h where d and h are, respectively, the separation between conductors as far as phase-to-phase or inter-system coupling schemes are adopted or a mean distance between the conductor used by the transmission and the other two nearest conductors when phase-to-ground coupling scheme is used, and the height h is the mean ground clearance at the towers. The ratio d/h can experience a high degree of variability along the line either as a function of the catenary profile or the terrain profile between two successive towers. However, under the assumptions that no transpositions are presented along the line, assuming that d/h may be less than 0.6 (good coupling conditions), and assuming a medium value condition for the earth resistivity, it is possible to express the line attenuation $a_\ell(f)$ for phase-to-phase and inter-system coupling schemes as

$$a_\ell(f) = \left(a_1 \frac{\sqrt{f}}{d} + a_2 f \right) \ell \quad \text{[dB]} \tag{2.31}$$

where d represents the external diameter of the conductor (expressed in millimeters), f the frequency in kHz and ℓ the length of the line. The coefficient a_1 takes into account the conductor material ($a_1 = 0.055$ for copper conductors, $a_1 = 0.07$ for aluminum conductors) while a_2 depends on the fraction d/h as shown in Figure 2.72. The first and second terms of (2.31) represent, respectively, the conduction loss and the combination of the radiation and induction losses.

For the phase-to-ground coupling scheme, (2.31) can be modified as

$$a_\ell(f) = \left(a_1 \frac{\sqrt{f}}{d} + a_2 f \right) \ell + a_3 \quad \text{[dB]} \tag{2.32}$$

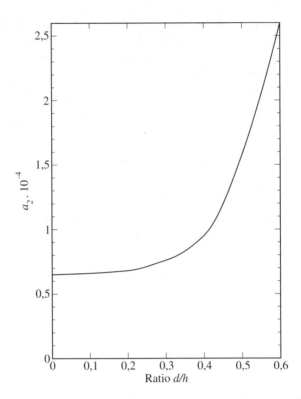

Figure 2.72   Values of coefficient a_2 as a function of the ratio d/h.

where a_3 is an additional term in which the magnitude is a function of the input impedance presented by the power station to the non-coupled conductors. It has been determined by field tests for the two extreme cases of short-circuit (grounding) and open circuit (disconnection) and it was found that the line attenuation a_3 for phase-to-ground coupling scheme can be about 2.2 dB and 5.7 dB, respectively. For transmission over several line sections, these additional terms must be added for each section. Note that normally no line traps are provided for the non-coupled conductors: as a consequence, the additional attenuation term a_3 depends on the conductor-to-ground impedance of the two conductors entering the station and, consequently, on their switching condition.

Equations (2.31) and (2.32) represent a reasonably approximation of the real line attenuation $a_\ell(f)$ as far as the validity of the hypothesis under which they are derived holds. Otherwise, only on-site measurements can be carried out in order to estimate $a_\ell(f)$. As an example, if all experimental data are gathered together to determine the line attenuation $a_\ell(f)$, planning values as shown in Figure 2.73 can be obtained.

Finally, it is worth highlighting that (2.31) and (2.32) do not take into account other limiting effects experienced in a real world scenario. To cite just few of them, it must be remembered that line attenuation is not affected appreciably by rain, but a serious increase in loss may occur when the phase conductors are thickly coated with hoar-frost or ice: attenuation $a_\ell(f)$ of up to three times the fair weather value can been experienced.

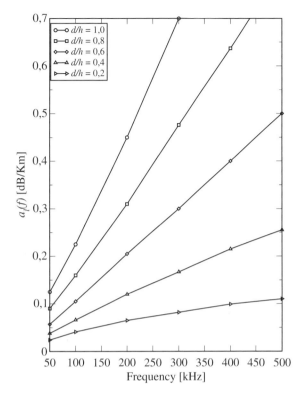

Figure 2.73 Planning values for the line attenuation $a_\ell(f)$ at operating voltages between 110 kV and 400 kV as a function of the ratio d/h.

The attenuation of coupling circuits $a_{cp}(f)$ depends on the non-perfect matching between the impedance of the communication equipment and the line impedance. It is common practice to use broadband coupling circuits so rated that the mismatch loss introduced will not exceed 1.3 dB: considering also small losses in the coupling capacitor and in the coils of the filter circuit, the calculated values should be incremented by a total of approximately 0.4 dB. As a consequence, the attenuation of coupling devices $a_{cp}(f)$ can vary from 0.5 dB to 2.0 dB as a function of the bandwidth used by the communication equipment and the central frequency as well. Figure 2.74 shows a characteristic example of the attenuation $a_{cp}(f)$ introduced by broadband coupling circuits.

The attenuation a_{st} introduced by the power network station is the result of different contributions and can be expressed as

$$a_{st} = a_{wt} + a_{cc} \quad [\text{dB}] \tag{2.33}$$

where a_{wt} is the shunt loss given by the wave traps and a_{cc} is the attenuation due to coupling capacitors. Wave traps are used to prevent the HF energy from traveling away into the connected power network station: their blocking efficiency may be limited and a certain amount of shunt loss must be expected. The impedance presented by the wave traps is practically insignificant

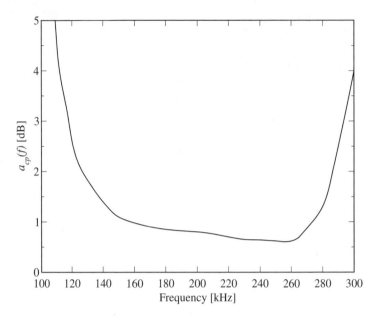

Figure 2.74 Insertion loss $a_{cp}(f)$ of a typical broadband coupling circuit.

as far as the power frequency is concerned, so that the voltage drop caused by the operating current can be neglected. For the carrier frequency current, however, the impedance should be higher than or at least equal to the characteristic impedance of the line. If the power network station at the end of the line is disconnected and the line is not grounded, the station does not introduce any further attenuation in the transmission circuit. If, on the other hand, the line is grounded or the station is connected, the wave trap, or station and wave traps, are arranged in parallel with the coupling circuits: this condition produces an increased attenuation which should be kept as low as possible by an adequate rating of the trap impedance. To obtain a sufficiently low shunt loss, the reactance should not be particularly high: to trap one or two frequency bands within the working range, traps of a low inductance (0.2 mH) are used and combined with a capacitor to form a resonant circuit. Since a switching station represents normally a capacitance, the inductive reactance both of a resonated and a non-resonated wave trap may be partially, if not fully, compensated by the capacitive reactance of the station: as a consequence, a substantial increase in the shunt loss must be expected if the short circuit or ground is removed and the line is connected to the station. Generally speaking, the attenuation introduced by the wave traps can vary from 0.5 dB to 2.5 dB.

Starting from the tangent of the loss angle of the coupling capacitor $\tan(\delta)$, i.e. the ratio between the equivalent series resistance and the capacitive reactance of the coupling capacitor (at specified sinusoidal alternating voltage and frequency), and taking into account the capacity value of the coupling capacitor, it is possible to express the coupling attenuation as

$$a_{cc} = 10 \log_{10} \left(\frac{R_s + Z_0}{Z_0} \right) \quad [\text{dB}] \tag{2.34}$$

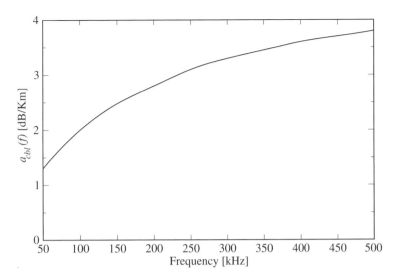

Figure 2.75   Attenuation of high-frequency entry cable $a_{cbl}(f)$ usually used in HV power line carrier.

where Z_0 is the characteristic impedance of the power line and $R_s = \tan(\delta)/\omega C$ is the equivalent series resistance of the coupling capacitor: the attenuation introduced by the coupling capacitors ranges from 0.1 dB to 0.3 dB.

Finally, the attenuation of high-frequency (HF) entry cable $a_{cbl}(f)$ can be easily derived from transmission line theory: the entry cable can be treated as an homogeneous transmission line whose unitary attenuation and length are known. It is common practice to use a high frequency cable of preferred characteristic impedance of 75 Ω with a value of $a_{cbl}(f)$ ranging from 2 to 4 dB/km at HV power line transmission frequency of interest, as shown in Figure 2.75.

2.5.3   Noise in High Voltage Lines

We conclude this section with a brief discussion of the noise in HV lines.

Figure 2.76 shows a noise profile measured in a 11.5 km length 132 kV line: the adopted coupling scheme is of phase-to-ground type with a line trap of 200 mH and a coupling capacitor of 4000 pF. The three-phase system is of single-line single-circuit type. Figure 2.76 is representative of the noise characteristics often encountered in HV lines: the noise is heavily colored, time-varying and almost non-Gaussian. Even if the average noise level decreases as the frequency increases, it is evident how deep notches are spaced out by high noise spikes in a non-regular and time-varying mode. It is common practice [115] to decompose the overall measured noise as a mixture of a colored background quasi-Gaussian noise, whose power spectral density is relatively lower and decreases with frequency plus impulse noise spikes that may depend on the loads and unbalance of the line or atmospheric discharges or also on the coupling effect between other PLC equipments with the considered line under measurement. As to the background noise affecting HV lines, it is usually generated by discharges across insulators and line fittings, and by discharges on the lines proper. As to impulse noise, several

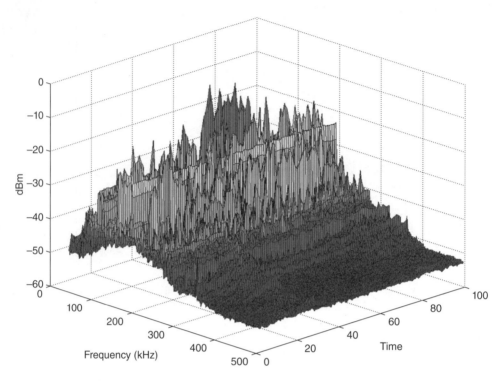

Figure 2.76   Noise levels, as a function of time and frequency, usually encountered in HV power lines. Acquisitions every 30 min.

models have been presented in literature to represent it, the most accurate are the Middleton [116], [117] and the mixed Bernoulli-Gauss [118] models.

High noise levels arise also from lightning strikes and system fault inception or clearance. Although they are of short duration, lasting only a few milliseconds at the most, they may cause overloading of power line receiving equipment.

Finally, another main noise source in HV (but also in MV) lines is corona noise. The impact of corona noise on PLC communications has been reported in several articles [119], [120]. Corona is a common phenomena associated with all energized transmission lines and is heard as a crackling or hissing sound: under certain conditions, the localized electric field near conductors can be sufficiently concentrated to ionize the air close to the conductors, resulting in a partial discharge of electrical energy which, in addition to the noise level that results, can also damage system components over time. Corona noise becomes more relevant at higher voltages (345 kV and above) even if power flow does not affect the amount of corona produced by a transmission line: during wet or foul and humid weather conditions, the conductors will produce the greatest amount of noise and a humming sound may be heard in the immediate vicinity of the line. Since corona noise depends on the conductor electric field gradient, lines carrying more than 220 kV are made up of bundled conductors to reduce the electric field strength around the wires, i.e. large diameter conductors are usually used given their lower electric field gradients at the surface and, hence, lower corona than smaller conductors: as a consequence, corona noise level on extra high voltage lines increases only slowly compared with 22 kV lines. Section 2.5.4 presents a practical model for corona noise.

Another phenomenon, appearing in addition to the independent corona noise voltage, is *corona modulation*. The periodic variations of corona noise cause the transmitted carrier signal to be modulated in the rhythm of the power frequency: corona modulation is likely to affect double-sideband transmission where the carrier has a greater amplitude than the sideband so that its noise spectrum will also have a greater amplitude.

Finally, broadcasting transmitters or other radio transmitters can contribute to generate interference signal along the HV lines. If interference is caused by radio transmitters, a lower noise level may be expected for phase-to-phase coupling than for phase-to-ground coupling: this is also true in those cases where the interferences are PLC carrier equipment operating at the same frequency over other line sections.

2.5.4   Corona Noise[8]

The HV and MV powerline channel may consist of one or more conductors, depending on the considered coupling scheme, i.e. phase-to-ground or phase-to-phase. Corona noise is a common noise source for MV and HV transmission lines, since it is permanent and its intensity depends on (i) the service voltage, (ii) the geometric configuration of the power line, i.e. the locations of the conductors in relation to each other, (iii) the type and diameter of conductors involved in the line, (iv) the atmospheric conditions [121], (v) the elevation of the line above sea level and (vi) the condition of the conductor and hardware.

Corona noise is caused by partial discharges on insulators and in air surrounding electrical conductors of power lines [122]. When HV power lines are in operation, the voltage originates a strong electric field in the vicinity of the conductor. This electric field accelerates free electrons present in the air nearby conductors: these electrons collide with molecules of the air, generating a free electron and positive ion couple. This process continues forming an avalanche phenomenon called 'corona discharge'. The motion of positive and negative charges induces a current both in the conductors and ground [123].

The induced current appears like a train of current pulses, with random pulse amplitude variations and random interarrival intervals. The injected current due to corona noise on one conductor can be modeled by a current source [122], [123]: according to the Shockley-Ramo theorem [121], a corona discharge induces current in all conductors, i.e. each conductor of the power line channel is connected to the ground by a current source. Irregularities, such as nicks and scrapes of the conductor surface or sharp edges on a suspension device, concentrate the electric field at these locations and thus increase the electric field gradient and the resulting corona at the spots.

A few corona noise models are present in the literature [123]–[126]: here the model proposed in [124], [125] will be considered. Corona noise, as a random signal, is characterized equivalently through its autocorrelation function or its power spectrum. To this purpose, the corona noise spectrum is generated by a method that takes into account the generation

---

[8] Portions of the material in this section are reprinted, with permission, from R. Pighi and R. Raheli, 'Linear predictive detection for power line communications impaired by colored noise,' in *Proc. IEEE Int. Symp. Power Line Commun. Applic.*, Orlando, USA, Mar. 27–29, 2006, pp. 337–342. © [2006] IEEE.

Table 2.6   Values of the digital filter coefficients $\{v_i\}_{i=1}^{4}$ for various service
voltages (lateral phase-to-ground coupling). [119] © 2006 IEEE

Voltage [kV]	v_1	v_2	v_3	v_4
225	−1.225	1.052	−0.603	0.217
380	−1.298	1.109	−0.625	0.210
750	−1.302	1.041	−0.611	0.207
1050	−1.292	1.080	−0.647	0.224

phenomena of corona currents injected in the conductors and the propagation along the line
[127], [128]. This spectrum is utilized to synthesize an autoregressive (AR) digital filter [129],
whose output is described by the expression

$$n_k = \sum_{i=1}^{N} v_i n_{k-i} + w_k \tag{2.35}$$

where $\{w_k\}$ is a sequence of independent zero-mean Gaussian random variables and $\{v_i\}_{i=1}^{N}$
is the set of coefficients modeling the corona noise process. The synthesis of the digital filter
essentially calls for the identification of the coefficients $\{v_i\}_{i=1}^{N}$ and can be done using a
procedure based on the maximum entropy method proposed in [130] or on the minimization
of the difference between estimated and measured power spectra.

Table 2.6 shows, for $N = 4$, a complete set of coefficients modeling the corona noise for
different voltage lines with carrier couplings of lateral phase-to-ground type while Table 2.7 the
same set of coefficients is shown assuming a central phase-to-ground coupling configuration.
Note that, as already outlined, (2.35) defines a corona power spectrum whose frequency
components are over the entire frequency domain, i.e. its bandwidth is generally greater than
that used by the transmission system. In Figure 2.77, the corona noise power spectrum obtained
with the model presented in (2.35) with coefficients shown in Table 2.6 is also presented in
terms of the power frequency response $|V(f)|^2$ of the AR digital filter.

2.6   MIMO Channels

Multiple-input multiple-output (MIMO) communications exploit the spatial diversity to
improve coverage [132], or rather, to improve data rate with respect to conventional SISO

Table 2.7   Values of the digital filter coefficients $\{v_i\}_{i=1}^{4}$ for various service
voltages (central phase-to-ground coupling) [131]

Voltage [kV]	v_1	v_2	v_3	v_4
225	−1.235	1.110	−0.669	0.252
380	−1.219	1.112	−0.658	0.250
750	−1.212	1.103	−0.661	0.252
1050	−1.185	1.079	−0.649	0.238

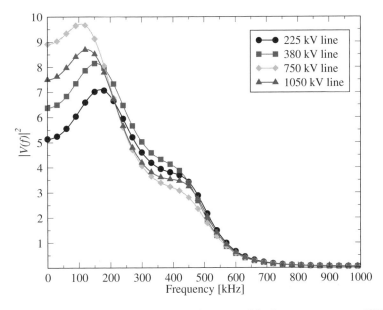

Figure 2.77 Corona noise power spectrum, shown in terms of the frequency response $V(f)$ of the AR filter as defined in (2.35). [119] © 2006 IEEE.

solutions. In wireless, MIMO techniques are enabled by the use of multiple antennas at both the transmitter and the receiver side [133], and wireless MIMO schemes were recently standardized. For instance, they are adopted by the standards IEEE 802.11n [134] and 3GPP LTE [135].

MIMO can be also applied to wireline communications. According to the Kirchhoff's laws, with N conductors, $N - 1$ circuits are available and thus MIMO communications can be established. Power delivery networks deploy three or more conductors. In the high voltage distribution grid, the power is transferred according to a three phase system in a Δ-style configuration that deploys three cables. Therefore, two different circuits are present.

In the medium and low voltage distribution grid, different configurations may exist. Namely, the power is transferred according to a three phase system and a Δ-style or Y-style configuration, depending on the country. Δ-style is the standard in Europe and it is more common worldwide. Y-style is deployed in North and Latin America. The Δ-style configuration is equivalent to that of high voltage lines and the number of conductors is three. In the Y-style configuration, a fourth conductor is present, namely, the neutral, and the number of available circuits is three.

Similarly, the configuration of the outdoor low voltage distribution grid varies in different countries. In North America, the power is distributed by means of two phase wires and a neutral conductor. The voltage between the phase wire and the neutral and between the phase wires is 120 and 240 V, respectively.

In Europe, the power is distributed according to a three phase system in a Y-style configuration that comprise the neutral, and not all loads (houses) are fed by all the three phases. In the in-home scenario, conventional SISO PLC signals over the phase wire (P) and the neutral wire (N). For safety reasons, a third conductor is present, namely, the protective earth (E). Thus, MIMO is still possible.

So far, most of the efforts toward MIMO PLC has been put on the in-home scenario, where MIMO technology is expected to provide an effective increase of the PLC data rate, matching the ever increasing requirements of the new high-speed multimedia applications. Hence, in the following sections, the focus will be mostly on in-home MIMO PLC.

2.6.1   Grounding Methods

The protective earth wire ensures that, at the mains frequency, all the metal surfaces that are exposed to the human touch have the same electrical potential of the surface of the Earth. In the presence of an insulation fault, namely, a short circuit, the short-circuit currents flow through the E wire toward the main panel, where the circuit breaker recognizes their presence and shuts down the power supply [136]. In the higher frequency range, the low impedance connection between the E wire and the physical earth is no longer present because of the inductivity of the grounding wires.

A detailed analysis on the worldwide deployment of the E wire is presented in [136]. Basically, the use of the E wire is growing and it is becoming mandatory in a increasing number of countries. Furthermore, it has been estimated that the E wire is already present in more than 90% of the installations of some countries, as Australia, China and UK [110].

According to the grounding rules, the E and N wires can be short-circuited at the customer location. US grounding methods are reported in the Article 250 of the NEC [137]. European grounding methods are reported in the IEC norm 60364-1 [138]. Basically, NEC imposes short-circuiting the wires E and N at the customer location, and the grounding arrangement is TN-C-S. Indeed, in Europe, TN-C-S is not mandatory. For instance, in Italy, N and E are not short-circuited in the main panel of the customer location (CEI 64/8), and the installations are compliant with the grounding arrangement TT, that is also adopted in Japan. For further details on grounding arrangements, we refer to [138].

2.6.2   MIMO PLC Principles

The use of the E wire for MIMO PLC is quite recent. Formerly, MIMO PLC was proposed for multiple phase installations [139], where multiple channels were supposed to be established between two uncoupled pairs of wires that did not include the E conductor. In such a configuration, all the phases are supposed to be available at the transmitter and the receiver end. Recently, the use of MIMO PLC has been extended to networks where only one phase is available at each termination outlet (though different outlets can be fed by different phases) in conjunction with the N and E wire. In this Section, we limit the results to this latter case, and, in Figure 2.78, we show all MIMO modes that are available in the presence of the wires P, N and E. In general, the wires are coupled. We refer to a mode as the signal that can be observed on a termination port. Furthermore, according to the literature [136], we refer to the modes of Figures 2.78(a), 2.78(b) and 2.78(c) as delta (Δ), star (Y), and CM, respectively. From now

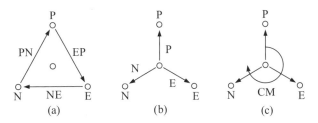

Figure 2.78 PLC MIMO modes. From left to right, Δ, *Y* and CM. The center denotes the reference and the arrows indicate the mode direction (the opposite of the current flow). CM accounts for the signal that flows with the same intensity and direction on the three wires.

on, Δ and *Y* will refer only to the MIMO modes and no longer to the configuration of the three-phase power distribution system.

The Δ modes involve two conductors each. Namely, the signal is sent or received symmetrically between pairs of wires. There are three Δ modes, and they are denoted by PN, NE and EP. The mode PN is the signal that is observed between the wire P and the wire N. Similarly, modes NE and EP are between wires N and E, and E and P, respectively. At the transmitter, the three Δ modes cannot be deployed together because Kirchhoff's laws impose their algebraic sum to be zero. Therefore, the third mode is a linear combination of the other two. At the receiver, the three Δ modes can be exploited together to set up a 2 × 3 MIMO communication system. In this respect, we note that all the three modes exhibit some differences (diversity) due to parasitic components of the probes and the network [140]. At the transmitter side, Δ modes are the standard because they allow for a purely differential signal injection that, in turn, allows for the lowest radiated emission.

At the receiver, the use of the *Y* modes was firstly proposed in [140]. Basically, the *Y* modes are the signals received between one conductor and the reference. The reference is the physical earth (not the E wire). In this respect, we point out that the algebraic difference of two star modes yields to a Δ mode. The star modes are three and we denote them with the identifier of the conductor to which they are associated, namely, P, N or E. The star modes are suited to be deployed in combination with the CM.

The CM accounts for the signal received with the same intensity and direction by all the three conductors with respect to the reference. Even if the injection of the PLC signal is purely differential, between couples of wires, the asymmetries of the PLC network lead to the partial conversion of the differential mode currents into common mode components. Common mode currents are responsible for most of the radiated emissions, according to the Biot-Savart rule [140]. Therefore, the use of the CM at the transmitter side is deprecated. To enable the reception of the star modes and to enforce the CM, the use of a large metal plate is required at the receiver. Basically, the metal plate acts as a ground plane that establishes a capacitive coupling path toward the physical earth [136]. Note that high-definition TVs are equipped with large metal backplanes that can be exploited to this scope. Furthermore, the dimension of the ground plane is related to the wavelength of the transmitted signal. The higher the frequency, the lower the dimension. It follows that the use of the reference plane is not suited for narrowband PLC.

2.6.3   Experimental Measurement Results

MIMO PLC was characterized experimentally from the results of the measurement campaigns reported in [136], [141], [142]. The campaigns adopted different setups, and the differences have to be taken into account when comparing the results. The details on MIMO couplers are provided in Section 2.6.3.1. The main findings on the statistics of the channel response are reported in Section 2.6.3.2.

2.6.3.1   Equipment: MIMO Couplers

Couplers are required to protect the equipment from the mains and to enable the signal injection/reception in the higher frequency band. They should exhibit low attenuation in the range of frequencies of interest and, further, they also should provide some protection against the high impulsive noise spikes that may occur in the power delivery network. Simple MIMO couplers for Δ-style modes can be designed as the combination of three conventional SISO couplers, each between one possible couple of wires. The design of conventional SISO couplers is detailed in Chapter 4.

The reception of both Δ and Y modes require a more complex design [110]. In Figure 2.79, we report the block diagram of the delta-star coupler. Basically, the coupler consists of four parts, i.e. the protection circuitry, the CM block, the Y-mode ports, and the Δ-mode ports. The protection circuitry is the front-end stage. It consists of high-pass filters that remove the mains providing a low attenuation in the higher frequency range, and gas discharge tubes, varistors and protection diodes that cope with voltage surges and noise spikes. The aim of CM block is twofold. To measure the CM components, when the switch is open, and to block the CM currents, when the switch is closed. In the latter case, the CM currents are short-circuited through the switch. The CM transformer is magnetically coupled [136] and the CM signal is measured at the CM port. The CM port is connected to the balun. The baluns are

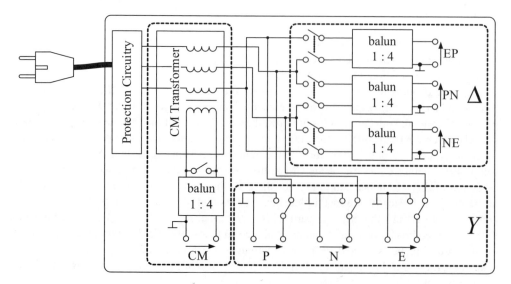

Figure 2.79   MIMO delta-star coupler schematics [110].

low-loss Guanella transformers with transform ratio 1:4 and they enable a better matching between the network, and the input impedance of the equipment, namely, 50-Ω. The baluns are deployed also for the reception of the Y modes. In detail, Y and Δ mode ports are connected to the center of the coupler, after the CM transformer. Y-modes are measured between the reference plane and one of the three wires. Δ-modes are measured between couples of wires.

An optional T-style coupler can be connected to the Y-mode ports to enable the parallel transmission between wires P and N with E being the reference, and the emulation of the SISO mode. In this respect, we note that mode PN is not equivalent to the conventional SISO because it assumes the remainder Δ ports to be closed into a 50-Ω load. In SISO, the remainder of the ports are left open.

The reference is obtained with a large metal plane. To enable stable measurements in the band 1–100 MHz, the area of the plane should be at least 1 m^2. In this respect, it should be noted that the area increases as the frequency decreases. Therefore, Y-modes are not suited for narrowband MIMO PLC.

2.6.3.2   Statistical Channel Characterization

Besides the preliminary results in [29], MIMO PLC channels have been characterized from experimental basis according to the results of three different measurement campaigns. The first campaign was presented in [141] and it was focused on the 3×3 MIMO configuration where Δ modes were deployed both at the transmitter and receiver side. In this respect, note that only two out of three transmitting modes can be used at the same time due to the Kirchhoff's laws. The measurement campaign assessed five sites in France for a total amount of 42 links. The measurements were performed in the frequency domain with a vector network analyzer (VNA). In this respect, one combination of transmitting-receiving mode was measured at a time. Thus, nine measurements were needed to obtain the full 3×3 channel matrix. The targeted frequency range was 2–150 MHz, and broadband Δ-mode couplers were developed and deployed. According to the standard Δ mode configuration, the ports of the coupler that were not interested by the transmission were closed into a 50 Ω load. The channel frequency response is defined as the scattering parameter s_{21}, and the acquisitions are divided between same-circuit and different-circuit channels. The first class refers to the channels that are defined between outlets that are fed by the same circuit breaker. The second class groups the remainder channels whose signal path includes the main panel. The aim of the classification is to evaluate the effect of the circuit breakers on the channel response. We refer to this measurement campaign as Campaign (A).

A second measurement campaign was performed by the HomePlug technical working group, it focused on the North American scenario, and the results were presented in [142]. Basically, it collected 96 MIMO PLC channels in five houses and it targeted the Δ-Δ configuration, i.e. Δ modes were deployed both at the transmitter and the receiver side. The number of transmitting modes was two, and the number of receiving modes was three. Measurements were performed in the frequency domain, according to a channel sounding method that deployed OFDM-modulated and orthogonally-coded signals. The channel frequency response was obtained as the ratio between the received frequency sample and the OFDM transmitted symbol. The estimation was averaged over 10 subsequent transmissions, and it targeted the band 0–100 MHz. Furthermore, the frequencies above 88 MHz were notched in order to remove

Table 2.8   Distribution of assessed sites per country

Country	No. of sites	No. of links
Belgium	5	60
France	7	86
Italy	2	8
Germany	13	121
Spain	5	30
UK	4	48

the interference due to broadcast FM radios. In the following, we refer to this measurement campaign as Campaign (B).

The third measurement campaign is the most extensive. It was performed by the special task force 410 (STF-410) of the European Telecommunication Standards Institute (ETSI) and it was presented in [110], [143]. The campaign assessed a total amount of 36 sites in seven countries across Europe. For each site, MIMO PLC channel responses were measured between up to 4 pairs of outlets. The transmitting and receiving outlets were selected as the most probable for a high speed communication, e.g. where an access point or a TV was supposed to be connected. The total number of assessed links is 353. In Table 2.8, it is reported the number of sites and measured links per country. The targeted configuration is Δ-Y, namely, Δ at the transmitter and Y plus CM at the receiver. A 3×4 MIMO system was set up. Measurements were aimed at obtaining the scattering parameters in the frequency domain, according to the measurement procedure that is detailed in [136], and that was followed by all members of STF-410. Basically, the scattering parameters were measured with a VNA in combination with two coaxial cables, delta-star couplers and large metal plates. The metal plates were used to enforce the physical earth path. The campaign addressed the band 1–100 MHz and the channel frequency response was defined as the scattering parameter s_{21}.

A statistical analysis of the experimental data was performed for all the campaigns, and the results were reported in [141]–[143]. In the following, we review and compare the main findings concerning the statistics of the CFR, the attenuation as a function of the topology and the country, the ACG, and the capacity. Firstly, we focus on the CFR. When the reception modes are compliant with the Δ configuration, the dB-version of the CFR shows a normal distribution regardless the frequency [142]. When Y-modes are deployed, the CFR is comprised between -5 and -100 dB and the PN feeding mode turns out to be the worst [143]. In frequency, the CFR decreases linearly, with a slope that has been quantified in 0.2 dB/MHz [142], [143]. In this respect, it is interesting to note that similar results were obtained by both [143] and [142] though they adopted different measurement procedures and they defined differently the MIMO channel responses. Furthermore, in [142], the attenuation was computed on a normalized version of the measured channel response.

In Figure 2.80, we show the mean profile of the CFR of the Δ-Y configuration. The following evidences can be pointed out. First, the PN transmitting mode yields the most attenuated channels (on average). Second, as intuition suggests, the best reception modes are those associated with the wires where transmission takes place. Finally, the CM mode is the most attenuated, on average. However, CM provides benefit in the worst cases as it is pointed out from the observations on the values in Table 2.9. In Table 2.9, we report the 20th, 50th and

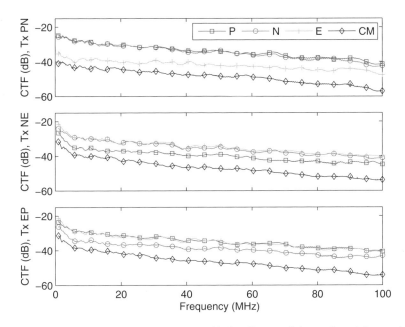

Figure 2.80 Mean channel frequency response profile for all transmitting and receiving modes in Δ-Y configuration.

80th percentiles of the distribution of the channel response of all modes of the Δ-Y MIMO configuration at frequencies 10, 40 and 70 MHz. It can be noted that, first, the reception on the conductor that is not involved in the transmission is the most attenuated. This result is consistent with Figure 2.80. Second, the CM mode is less spread for all the considered

Table 2.9 Channel frequency response quantiles at three frequencies and for all transmitting and receiving modes of the Δ-Y configuration

		$f = 10$ MHz			$f = 50$ MHz			$f = 70$ MHz		
		20th	50th	80th	20th	50th	80th	20th	50th	80th
PN	P	−55.7	−41.3	−24.2	−64.0	−50.2	−35.8	−69.7	−54.7	−37.8
	N	−53.7	−42.6	−24.9	−65.6	−50.9	−34.5	−69.3	−55.7	−38.7
	E	−61.1	−49.1	−37.4	−67.0	−52.7	−40.4	−70.7	−58.6	−44.0
	CM	−56.1	−48.5	−41.0	−62.4	−53.5	−44.0	−67.9	−58.0	−48.3
NE	P	−54.9	−44.8	−32.2	−65.2	−52.9	−38.6	−69.5	−57.5	−42.3
	N	−54.4	−43.5	−28.3	−65.0	−50.8	−36.0	−68.2	−56.0	−41.3
	E	−56.9	−43.4	−28.6	−65.7	−51.7	−36.8	−69.6	−57.9	−44.3
	CM	−52.5	−45.4	−37.5	−62.4	−52.3	−42.9	−67.0	−57.4	−47.5
EP	P	−53.6	−42.5	−26.9	−63.0	−48.9	−37.7	−66.8	−55.9	−40.0
	N	−53.7	−43.6	−31.3	−64.6	−50.4	−38.2	−67.4	−56.0	−42.1
	E	−55.5	−42.4	−29.1	−63.9	−49.7	−36.2	−68.1	−57.9	−42.9
	CM	−50.6	−43.0	−36.7	−61.2	−51.9	−42.1	−65.4	−56.6	−48.2

frequencies and regardless the transmission mode. Namely, the 20th and 80th quantiles are closer to the value of the 50th quantile. Third, the worst configuration is PN-E, that exhibits the highest attenuation at all frequencies. In a dual manner, the best configuration is PN-P. Fourth, the quantiles exhibit a frequency decreasing behavior that matches the profile in Figure 2.80. At high frequencies, the EP transmission mode ensures the lowest attenuation in the worst cases, namely, the 20th quantile. Indeed, the PN transmission turns out to be the best choice for low attenuated channels, namely, the 80th quantile.

The attenuation is a function of the country and the size of the topology [144]. In Germany, where the outlets of different rooms are fed by different phases, the attenuation is, on average, larger than that experienced in other countries. Furthermore, from the robust regression of the experimental measures, the slope of the mean attenuation versus the size of the location turns out to be -0.1240 dB/m^2. With reference to Δ-Δ configuration, further intuitive conclusions are the following. The same-circuit channels are, on average, less attenuated than different-circuit channels. In addition, the same-circuit channels defined between the same mode at both the transmitter and receiver side are less attenuated than the others [141].

Now, we compute the ACG as defined in Section 2.2.2, and, in Figure 2.81, we report the CDF of the ACG for all feeding and receiving modes of a Δ-Y configuration. We note the followings. First, in terms of ACG, P and N are equivalent. In detail, the distribution of the ACG of the mode P overlaps that of mode N, regardless the transmission mode. Second, when the transmission mode is PN, the ACG of mode E is 2 dB less than that of the P and N modes. Indeed, when the transmission mode is NE, the distribution of the ACG of mode E is close to

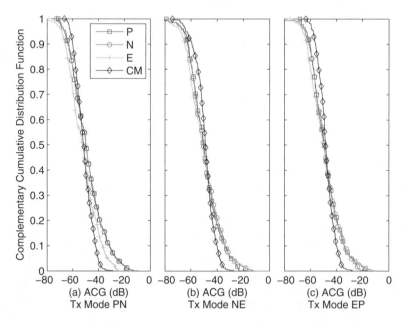

Figure 2.81 Complementary cumulative distribution function of the ACG for all transmitting and receiving modes of the Δ-Y configuration.

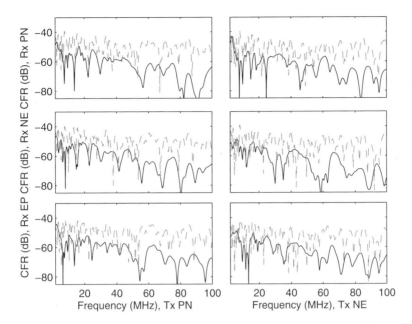

Figure 2.82 Comparison between the MIMO channel responses generated with the frequency domain (blue continuous line) and the time-domain (red dotted line) top-down approaches. The channels generated in the frequency domain belong to a different-circuit channel class as defined in [141].

that of modes P and N. Third, the ACG of the CM is less spread than that of other receiver modes. In detail, the ACG of CM is almost confined between −60 and −30 dB. Indeed, the ACG of the other modes spans between less than −60 dB and up to −10 dB. Fourth, the ACG of CM behaves similarly regardless the transmission mode. It follows that the CM reception offers the lowest attenuation in worst channels. Thus, it can improve coverage. Furthermore, in the former analysis, the effect of the couplers is not removed. Since the couplers attenuate the CM mode 4 dB more than the other modes, the actual CM channel is even less attenuated. Finally, the ACG is negatively related to the RMS-DS and the slope is that found for the SISO channels [142]. This result has been obtained from the robust regression of the ACG and the RMS-DS of all MIMO channel entries in Δ-Δ configuration, regardless the transmission and the reception mode.

The MIMO channels are correlated, thought there is no clear indication about the degree of correlation. The correlation has been studied through the singular value decomposition (SVD) analysis [142], [143], [145] and the study of the Pearson correlation coefficient [141]. The SVD enables computing the eigen-spread. The eigen-spread is defined as $\Lambda(f) = \max\{\lambda_1(f), \lambda_2(f)\} / \min\{\lambda_1(f), \lambda_2(f)\}$, where $\lambda_1(f)$ and $\lambda_2(f)$ denote the singular values. In this respect, we note that the maximum number of nonzero singular values is two, as the maximum number of transmitters allowed by the Kirchhoff's laws. The larger Λ, the higher correlation. From the experimental evidence, $\Lambda(f)$ can be modeled as frequency-independent [144]–[146], regardless of the receiver configuration. Furthermore, when the configuration is Δ-Δ, the eigen-spread exhibits a Rayleigh distribution [146] and the channels seem to be highly correlated. Indeed, when the configuration is Δ-Y, the largest correlation is experienced for the 2×2 MIMO, and the number of highly correlated channels is small. The analysis of the Pearson correlation coefficient turns out the following evidence. First, the distribution of the correlation coefficient of all possible combination of channels at all frequencies is uniform. Second, channels H_{11}^{Δ} and H_{22}^{Δ} are strongly correlated. Third, the pair of channels H_{11}^{Δ}-H_{33}^{Δ} and H_{22}^{Δ}-H_{33}^{Δ} exhibit a similar correlation.

Preliminary studies of the MIMO channel capacity reveal that the higher the number of receiving ports, the higher the capacity regardless the configuration. The general expression of the MIMO channel capacity is [147]

$$C = \int_B \log_2 \left(\det(\mathbf{I}_{N_R} + \mathbf{N}_c^{-1}(f)\mathbf{H}(f)\mathbf{Q}(f)\mathbf{H}^H(f)) \right) df, \qquad (2.36)$$

where B is the signaling band, $\mathbf{I}_{N_R}$ is an $N_R \times N_R$ identity matrix, $\mathbf{N}_c^{-1}(f)$ is the inverse of the noise covariance matrix at frequency f, $\mathbf{H}(f)$ is the MIMO channel matrix at frequency f and $\mathbf{Q}(f)$ is the covariance matrix of the transmitted signal. The performance improvement provided by the use of 2×3 instead of 2×2 MIMO in Δ-Δ configuration under space-correlated noise was reported in [146]. Indeed, in [144], it was shown that 2×4 is the best Δ-Y configuration.

Some other statistics can be extracted from the study of the MIMO channels as independent SISO channel entries. In detail, the RMS DS is smaller than 150 ns in 95% of the cases and the coherence bandwidth at level $\rho = 0.9$ is 1.5 MHz [141].

A time-domain analysis was presented in [142]. The channel impulse response is obtained by means of inverse Fourier transform of the measured channels and they are aligned (shifted) in time in order to exhibit the peak at a given time instant, namely, n_{peak}. The study of the amplitude statistics is at the basis of the top-down time-domain model of Section 2.6.4.1.

2.6.4   Modeling and Generation of MIMO PLC Channels

MIMO PLC channel modeling has been addressed following both a top-down and a bottom-up approach. Concerning the top-down approach, both time-domain and frequency domain channel generation algorithms were presented. Indeed, the bottom-up MIMO channel generation was limited to the frequency-domain. In this section, we review the main results. To this aim, we introduce the notation $H_{\kappa\ell}^{\Delta}(f)$ and $H_{\kappa\ell}^{\star}(f)$. Basically, $H_{\kappa\ell}^{\Delta}(f)$ denotes the CFR between the receiver mode κ and the transmitter mode ℓ when Δ modes are deployed at both the transmitter and the receiver end. Furthermore, $\kappa = 1, 2, 3$ refer to modes PN, NE and EP, respectively. Indeed, $H_{\kappa\ell}^{\star}(f)$ denotes the channel frequency response when Y modes are considered at the receiver side. In this case, $\kappa = 1, 2, 3, 4$ refer to modes P, N, E, and CM, respectively.

2.6.4.1   Top-down Modeling Approach

Top-down MIMO PLC models can be derived both in the frequency domain and in the time domain. A frequency-domain top-down statistical MIMO channel model was proposed in

[141]. It is based on the multi-path propagation model that was firstly presented in [17], extended in statistical terms in [56] and fitted to the experimental data in [30]. Basically, the multi-path propagation model describes the frequency response of the channel as the sum of multiple path components. The p-th path component is characterized by the path gain g_p and the path length d_p. The number of paths is N_p. The other parameters of the model are the attenuation A and the coefficients a_0, a_1 and k that describe the attenuation of the power line cables. In [30], the parameters g_p, N_p and d_p are made random, according to the statistics that enables the best fitting with measurements, while a_0, a_1 and k are left constant, and their value is chosen in order to ensure the best fit with measurements. In [141], the idea of using random parameter values has been exploited and extended to characterize the 3×3 MIMO system defined between all possible Δ modes. In detail, the Δ-Δ CFR is given by

$$H^\Delta(f) = A \sum_{p=1}^{N} g_p e^{-j\phi_p} e^{-j\frac{2\pi d_p}{v}f} e^{-(a_0+a_1 f^k)d_p}, \tag{2.37}$$

where ϕ_p is a random phase contribution, and v is the speed of light in the cable and it is equal to $2/3 \, c$, namely, the speed of light in the vacuum. Furthermore, in (2.37), the random phase ϕ_p models the spatial correlation. The model fits the MIMO channels that were measured during the campaign (A) of Section 2.6.3.2, and it distinguishes between same-circuit and different-circuit channels. The former are the channels defined between outlets that are fed by the same circuit breaker. The latter are the channels defined between outlets that are fed by different circuit breakers. In the second case, the backbone, namely, the shorter signal path between the transmitter and the receiver outlet, includes the main panel. The MIMO channels can be obtained as follows.

(a) Generate the channel response $H_{11}^\Delta(f)$ according to (2.37), with $\phi_p = 0 \, \forall p$. The statistics of the random parameters is reported in Table 2.10. Parameters ΔA and a_0 are obtained as the

Table 2.10 Statistics of the random parameters of the frequency-domain top-down model

Model Parameter		Distribution	Statistics	
			Stat. parameter	Value
A	Same circuit	Uniform	Interval	$[0.005, 025]$
	Diff. circuit	Exponential	Mean	0.00238
ΔA	Same circuit	Exponential	Mean	0.3659
		—	Offset	0.45
a_0		Exponential	Mean	0.00827
		—	Offset	0.005
d_p		Uniform	Interval	$[0, L_{max}]$
g_p		Uniform	Interval	$[-1, 1]$
K		Gaussian	Mean	1.01748
			Std. Deviation	0.01955
N		Poisson	Mean	ΛL_{max}

sum of a random variable with the statistics specified in the Table plus the offset reported in the table. The values of the remainder (constant) parameters are $a_1 = 4 \times 10^{-10}$, $\Lambda = 0.2$ and $L_{max} = 800$ m. In this respect, note that, differently from [56], the attenuation factor A is made random in order to provide the best matching with the experimental results.

(b) Generate the channel response $H_{33}^{\Delta}(f)$[9]. To this aim, introduce a random phase contribution ϕ_p to all paths in (2.37), yet keeping all the values of the remainder parameters of the model equal to that used to generate $H_{11}^{\Delta}(f)$. Let ϕ_p be uniformly distributed in $[-\Delta\phi, \Delta\phi]$, where $\Delta\phi = \pi$ and $\Delta\phi = \pi/2$ for different-circuit and same-circuit channels, respectively.

(c) Generate the channel response $H_{22}^{\Delta}(f)$. To this aim, start from $H_{33}^{\Delta}(f)$ and apply a further random phase contribution $\hat{\phi}_p$ to all paths. Let $\hat{\phi}_p$ be uniformly distributed in $[-\Delta\hat{\phi}, \Delta\hat{\phi}]$, with $\Delta\hat{\phi}$ equal to $\pi/2$ and $\pi/4$ for different-circuit and same-circuit channels, respectively.

(d) Generate all the remainder channels as $H_{33}^{\Delta}(f)$, but letting ϕ_p be uniformly distributed between $[-\Delta\phi, \Delta\phi]$, with $\Delta\phi$ equal to half of that used for $H_{33}^{\Delta}(f)$. Furthermore, for the same-circuit channels, multiply the channel response by ΔA, where ΔA is an exponential random variable. The statistics of ΔA is detailed in Table 2.10.

A time-domain top-down statistical MIMO channel model was presented in [142]. Later, in [145], the model was refined with the new findings on the spatial correlation of the MIMO channels. Basically, the model is based on the experimental results that were reported in [142] for the 2×3 MIMO configuration with Δ modes at both the transmitter and the receiver end. Measurements were performed in the frequency domain, with sampling frequency of 200 MHz, and the impulse response of the measured channels was obtained by means of inverse Fourier transform (IFT). For further details on the experimental results, we refer to the Campaign (B) of Section 2.6.3.2. The experimental evidences suggest to model the 2×3 MIMO channels as six independent SISO channel impulse responses, and to introduce a statistical correlation between the generated channels assuming that the correlation acts at the termination ends of the propagation path.

The generation of the SISO channel impulse responses is accomplished according to the following algorithm [142].

(a) Let the sampling frequency be that of the measurement campaign, namely, 200 MHz, and deal with the discrete-time version of the channel impulse response (CIR).

(b) Limit the CIR to 2000 samples, namely, 10 μs. In this respect, it is shown that the assumption enables preserving 99% of the channel energy [142].

(c) Divide the CIR into three time intervals, as a function of the position of the peak of the absolute value of the channel impulse response, namely, n_{peak}, and let n_{peak} be equal to 300. The first $n_{peak} - 2$ samples are random values according to a Weibull distribution with a random sign flip. The shape parameter is 3/4. The scale parameter is a function of the time instant n, namely, $\lambda(n) = f(n)$, where

$$f(n) = e^{\sum_{i=0}^{5} C_i n^i} \tag{2.38}$$

---

[9] Strictly, [141] deals with the opposite of $H_{33}^{\Delta}(f)$, namely the CFR defined between modes PE.

Table 2.11 Constant coefficient values of the model in (2.38) for the scale parameter λ and the standard deviation σ

	Scale param.	Std. deviation
C_0	-6.83	1.10
C_1	9.50×10^{-3}	-1.59×10^{-2}
C_2	-2.25×10^{-4}	1.84×10^{-5}
C_3	2.07×10^{-6}	-1.37×10^{-8}
C_4	-6.98×10^{-9}	5.62×10^{-12}
C_5	9.15×10^{-12}	9.22×10^{-16}

where C_i are the constant coefficients in the second column of Table 2.11. The samples in the interval $[n_{\text{peak}} - 1, n_{\text{peak}} + 1]$ are normally distributed with mean and standard deviation equal to the ones of Table 2.12. Finally, the samples of the last time interval, namely, $n > n_{\text{peak}} + 1$, are normally distributed with zero mean and standard deviation σ. The latter can be modeled as (2.38), where the constant coefficients are equal to the ones of the third column of Table 2.11. The peak of the generated impulse response is located at the time instant n_{peak}. Therefore, samples larger than $h(n_{\text{peak}})$ must be removed.

(d) Introduce the correlation in time. In the second interval, around the peak, there is no experimental correlation between the samples. Indeed, the correlation is significant for the samples with low standard deviation, namely, that of the first and the third interval. Therefore, within these intervals, decimate the generated samples with a decimation factor that is variable in order to obtain a suitable set of time instants [142]. In [142], it is not specified the criteria for the selection of the decimation factor. In this respect, a possible choice is the following. In the first interval, select the time instants $\chi = \{\hat{n}_0, \hat{n}_1, \ldots, \hat{n}_l\}$ where $\hat{n}_i = \Delta_i + \hat{n}_{i-1}$, $\hat{n}_0 = 0$ and $\Delta_i = 1/\left(20 \cdot \lambda(\hat{n}_{i-1})\right)$. Similarly, in the third interval, choose the samples $\chi = \{\tilde{n}_0, \tilde{n}_1, \ldots, \tilde{n}_l\}$ where $\tilde{n}_i = \Delta_i + \tilde{n}_{i-1}$, $\tilde{n}_0 = n_{\text{peak}} + 2$ and $\Delta_i = 1/\left(200 \cdot \sigma^2(\tilde{n}_{i-1})\right)$.

(e) Remove the samples that are larger than the peak, located at n_{peak}, and interpolate the remainder samples using a linear interpolation function. Then, compute the delay spread of the generated channel, and adjust the gain. To this aim, we can exploit the inverse relation between the delay spread and the average channel gain that was pointed out in [43]. The six resultant independent channel impulse responses can be arranged in the

Table 2.12 Mean and standard deviation of the normal distribution that fits the channel impulse response in the intermediate time interval

	Time instant		
	$n_{\text{peak}} - 1$	n_{peak}	$n_{\text{peak}} + 1$
Mean	1.45×10^{-1}	2.56×10^{-1}	1.47×10^{-1}
Std. deviation	6.96×10^{-2}	8.16×10^{-2}	7.49×10^{-2}

matrix $\hat{\mathbf{h}}^{\Delta}(n)$, where $\hat{h}^{\Delta}_{\kappa\ell}(n)$ is the impulse response between the transmitter mode $\kappa = 1, 2$, and the receiver mode $\ell = 1, 2, 3$ at the discrete-time instant n. Strictly, $\kappa = 1, 2, 3$ refer to the Δ modes PN, NE and EP.

(f) Introduce the spatial correlation to obtain the MIMO channel impulse response matrix that reads

$$\mathbf{h}^{\Delta}(n) = \left(\mathbf{r}^{1/2}\right)^H \hat{\mathbf{h}}^{\Delta}(n)\mathbf{t}^{1/2} \tag{2.39}$$

where $\mathbf{r}$ and $\mathbf{t}$ are the correlation coefficient matrices that model the spatial correlation introduced at both the termination ends, and $\{\cdot\}^H$ denotes the Hermitian operator. In detail, $\mathbf{r}$ and $\mathbf{t}$ are 3×3 and 2×2 matrices with elements $r_{\kappa\ell} = \rho_r^{|\kappa-\ell|}$ and $t_{\kappa\ell} = \rho_t^{|\kappa-\ell|}$, respectively, where $\rho_r = 0.6$, and $\rho_t = 0.4$, according to the experimental evidence in [145]. A frequency-domain version of the approach was also reported in [145].

2.6.4.2 Bottom-up Modeling Approach

The bottom-up approach is based on the TL theory and it is the most accurate method to compute the channel transfer function under certain conditions. It provides close matching between the channel response, the underlying propagation phenomena and the topology. The main drawbacks are the following. First, it requires a perfect knowledge of all the details about the topology, e.g. the impedance of the loads, the characteristics of the cables, the description of the network, etc. Second, it is computationally-expensive. Third, it can be applied only under the TEM or quasi-TEM mode assumption. The first two points were addressed and solved by the introduction of a topology-generation algorithm and an efficient approach to compute the channel response [89]. The latter point cannot be overcome. Basically, the TEM assumption holds true when the transversal dimension of the overall cable structure is relatively small if compared to the transmission signal wavelength. In the range of frequencies of a PLC transmission, namely, below 100 MHz, the TEM assumption is validated when the signal propagates through the conductors that are either enclosed in the same plastic cap or nearby placed inside small raceways. Additional transmission modes that consider the physical earth as a propagation path do not fulfill the TEM assumption. It follows that they cannot be modeled with a bottom-up approach.

With three wires, two different circuits sharing the same return conductor can be defined. The per-unit-length (p.u.l.) model is shown in Figure 2.83. The elements r_i, l_k, g_k and c_k

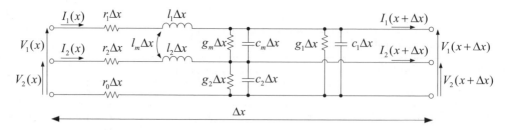

Figure 2.83 Per-unit-length equivalent model of the three-conductor transmission line.

with $i \in \{0, 1, 2\}$ and $k \in \{1, m, 2\}$ denote the p.u.l. resistance, inductance, conductance and capacitance. The p.u.l. inductance l_m, capacitance c_m and conductance g_m take into account the mutual interactions between the conductors. Coupling effects provide interactions between them, therefore by transmitting and receiving on both the circuits, a 2×2 MIMO system is defined. According to the TL theory, the signal propagation can be modeled applying the multiconductor extension of the telegraph equations. In this respect, the current and the voltage at the generic coordinate x of a uniform transmission line read

$$\mathbf{I}(x) = \mathbf{T}\left(e^{-\Gamma x}\mathbf{I}_m^+ + e^{\Gamma x}\mathbf{I}_m^-\right), \tag{2.40}$$

$$\mathbf{V}(x) = \mathbf{Y}^{-1}\mathbf{T}\Gamma\left(e^{-\Gamma x}\mathbf{I}_m^+ - e^{\Gamma x}\mathbf{I}_m^-\right), \tag{2.41}$$

respectively, where $\mathbf{V} = [V_1, V_2]^T$ is the voltage vector, $\mathbf{I} = [I_1, I_2]^T$ is the current vector, $\mathbf{I}_m^+$ and $\mathbf{I}_m^-$ are current vectors whose coefficients are determined by the boundary conditions, $\mathbf{T}$ and Λ are the eigenvector and the eigenvalue matrix of $\mathbf{YZ}$, $\Gamma = \mathrm{diag}\{\gamma_1, \gamma_2\}$ such that $\Gamma\Gamma = \Lambda$, $\mathbf{Y} = \mathbf{G} + j2\pi f\mathbf{C}$, $\mathbf{Z} = \mathbf{R} + j2\pi f\mathbf{L}$ and $\{\cdot\}^T$ denotes the transposition. We neglect the frequency-dependence for notation simplicity. Furthermore,

$$\mathbf{R} = \begin{bmatrix} r_1 + r_0 & r_0 \\ r_0 & r_2 + r_0 \end{bmatrix}, \quad \mathbf{L} = \begin{bmatrix} l_1 & l_m \\ l_m & l_2 \end{bmatrix},$$

$$\mathbf{C} = \begin{bmatrix} c_1 + c_m & -c_m \\ -c_m & c_2 + c_m \end{bmatrix}, \quad \mathbf{G} = \begin{bmatrix} g_1 + g_m & -g_m \\ -g_m & g_2 + g_m \end{bmatrix},$$

are the p.u.l. parameter matrices of the resistance, inductance, capacitance and conductance, respectively. For the symmetric cables, $r_1 = r_2 = r_0 = r$, $l_1 = l_2 = 2l_m = \mu_0/\pi \log(d/r_w)$, $\mathbf{LG} = \mu_0 \sigma_d \mathbf{U}$, σ_d is the conductivity of the dielectric, $\mathbf{U}$ is the 2×2 identity matrix, $\mathbf{LC} = \mu_0 \varepsilon_0 \varepsilon_r \mathbf{U}$, $\varepsilon_0 = 8.859 \times 10^{-12}$ F/m, and $\varepsilon_r(f) = 1.661 \times 10^6/f + 2.9701$ for typical PVC-insulated cables. In [148], it was shown that r varies with frequency, and it depends on the skin depth and on the structure of the conductor, e.g. stranded or solid [89]. Finally, the characteristic impedance matrix and the reflection coefficient matrix can be defined as

$$\mathbf{Z}_C = \mathbf{Y}^{-1}\mathbf{T}\Gamma\mathbf{T}^{-1}, \tag{2.42}$$

$$\rho_{L_l} = \mathbf{T}^{-1}\mathbf{Y}_C\left(\mathbf{Y}_L + \mathbf{Y}_C\right)^{-1}\left(\mathbf{Y}_L - \mathbf{Y}_C\right)\mathbf{Z}_C\mathbf{T}; \tag{2.43}$$

respectively, where $\mathbf{Y}_C = \mathbf{Z}_C^{-1}$, and $\mathbf{Y}_L$ is the load admittance matrix. Relations (2.40)–(2.41) are the starting point for the computation of the channel response in multiconductor complex networks. In this respect, several approaches can be followed. The most complete (and complex) is based on the multiconductor extension of the chain-parameter matrix method [148]. Alternative solutions were presented in [149] and in [89]. The former is the MIMO extension of the two-conductor TL theory channel simulator presented in [150] which, in turn, exploits the method of the modal expression for the electrical quantities. The latter is based on the multiconductor extension of the voltage ratio approach (VRA) that was presented firstly in [151].

According to the VRA, the complex network is modeled as the cascade of elementary units and the channel frequency response is computed as the product of the transfer function

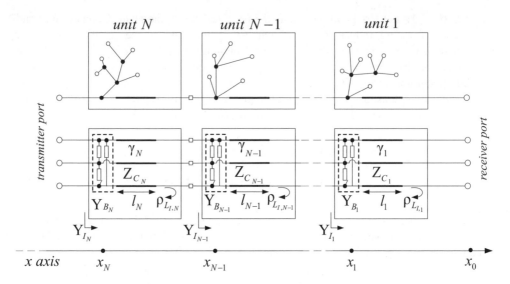

Figure 2.84 On the top, unifilar description of a topology remapped in units. On the bottom, equivalent representation of the units in the admittance matrix terms. [89] © 2011 IEEE.

of the units. The unit description of the generic network is depicted in Figure 2.84. Thick and thin lines represent physical wires and zero-length connections, respectively. Each unit $b = 1, \ldots, N_u$ contains an homogeneous piece of backbone line and the equivalent admittance of the branch connected to the node n_b. $\mathbf{Z}_{C_b}, \mathbf{\Gamma}_b, l_b$ and $\boldsymbol{\rho}_{L_{I,b}}$ denote the characteristic impedance matrix, the propagation constant matrix, the length of the piece of backbone line, and the load reflection matrix of the unit b, respectively. Furthermore, $\mathbf{Y}_{L_b}$ and $\mathbf{Y}_{I_b}$ denote the load and the input admittance matrix of the unit b, respectively. The first is the receiver admittance matrix when $b = 1$, otherwise it is the input admittance matrix of the unit $b - 1$. The latter is the sum of the load admittance matrix $\mathbf{Y}_{L_b}$ carried back at the input port of the unit b, and the branch admittance matrix $\mathbf{Y}_{B_b}$. Now, the CFR of the unit b is given by

$$\mathbf{H}_b = \mathbf{Z}_{C_b} \mathbf{T}_b \left(\mathbf{U} - \boldsymbol{\rho}_{L,b}\right) \left(\mathbf{e}^{\mathbf{\Gamma}_b l_b} - \mathbf{e}^{-\mathbf{\Gamma}_b l_b} \boldsymbol{\rho}_{L,b}\right)^{-1} \mathbf{T}_b^{-1} \mathbf{Z}_{C_b}^{-1}, \tag{2.44}$$

and the overall CFR reads

$$\mathbf{H}(f) = \prod_{b=1}^{N_u} \mathbf{H}_b(f), \tag{2.45}$$

namely, the product of the unit channel frequency responses as a function of the frequency. In [89], the method has been validated experimentally by comparing the simulated and the measured results for two test networks. Either symmetrical or ribbon cables were considered. In the second case, a model refinement has been proposed to account for the surface charge densities [152]. Furthermore, the method has been deployed in combination with the random TGA that was presented in [32] to obtain a MIMO PLC random channel generator [88]. The TGA enables for the random generation of single-phase PLC in-home networks and it is based

Table 2.13 Parameter set for the bottom-up generator

Parameter		Statistics		
		Distribution	Stat. parameter	Value
A_f	[m^2]	Uniform	Interval	[100, 300]
A_c	[m^2]	Uniform	Interval	[15, 45]
Λ_o	[outlets/m^2]	Constant	—	0.5
p_v		Constant	—	0.3

on the observation that in-home networks are made of two levels of connections. At a first layer, the outlets are connected in groups to special nodes, i.e. the derivation boxes. Then, at a second layer, the derivation boxes are connected together with dedicated cables. The outlets fed by the same derivation box are nearby placed and the derivation boxes are almost regularly spaced inside the topology. Therefore, the topology can be split into area elements each of which contains all the outlets connected to a derivation box and the derivation box itself. These area elements are the clusters. The clusters can be modeled as square-shaped elements with area A_c that is constant for a specific topology realization and uniformly distributed in a realistic interval. It follows that the number of clusters for a given topology of area A_f is $[A_f/A_c]$. The topology area A_f and the number of outlets in each cluster can be modeled as a uniformly distributed random variable, and a Poisson variable with intensity $\Lambda_o A_c$, respectively. Clusters are interconnected via derivation boxes and the in-home network is connected to the energy supplier network via the main panel. The main panel is a derivation box. The interconnections are modeled as multiconductor transmission lines, and the wires are assumed to be nearby placed and enclosed in a uniform dielectric cap. The conductor inter-distance is further supposed to be constant. Finally the effect of loads is also considered. A set of 50 loads was obtained from measurements, and the probability that no loads are connected to a plug is p_v. Thus, the probability that a load extracted from the measured set is connected to a plug is given by $(1 - p_v)/50$. The values of the parameters for the generation of random topologies are reported in Table 2.13.

2.6.5 Beyond the Channel Frequency Response

Experimental measurement campaigns addressed the MIMO line impedance, the noise and the EMC related aspects. These results provide a significant further deep insight on MIMO PLC, thought they have to be intended as preliminary results. In the following sections, the main findings are briefly reviewed.

2.6.5.1 Line Impedance

The line impedance is the load that is seen by the transmitter. The line impedance can be obtained from the scattering parameter s_{11} as $Z_i(f) = Z_0 \left(1 + s_{11}(f)\right) \left(1 - s_{11}(f)\right)$, where Z_0 is the reference impedance. The knowledge of the line impedance is fundamental to quantify the amount of injected power that is reflected toward the transmitter due to a non-perfect matching.

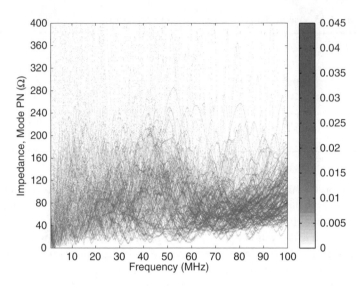

Figure 2.85 Probability density function of the line impedance experienced at the transmitter port of the mode PN.

The line impedance varies in frequency, time, and location. Some preliminary results about the statistics of the SISO line impedance were reported in [153]. Concerning MIMO, the statistics of the line impedance was studied in [144] for modes Δ, Y, and for other configurations obtained using the T-style coupler or considering Δ modes where not all ports are connected to a 50-Ω load. The analysis is based on the results of an experimental measurement campaign performed by the STF-410, during which 35 locations were assessed in Germany, Spain, France, Belgium, Italy and UK for a total amount of 146 measures. The measures were performed according to the setup of the measurement campaign (C) of Section 2.6.3.2. Due to the presence of the baluns (see Figure 2.79), the reference impedance is $Z_0 = 200\ \Omega$. The PDF of the absolute value of the line impedance of the Δ mode PN is reported in Figure 2.85. It has been computed as the histogram of the measured samples. The PDF of the line impedance of the remainder Δ modes is close to that of Figure 2.85. Therefore, we limit the study to Figure 2.85 without loss of generality.

The figure shows the following evidence. First, the line impedance is less spread in the higher frequency range. This observation is consistent with the results in [153] for the SISO case. Second, the line impedance in absolute value exhibits a frequency increasing behavior. Third, from the analysis of the real and the imaginary component of the line impedance, it can be shown that a) the PDF of the real part exhibits a concave behavior in frequency and b) the mean reactance increases with frequency. Both the former findings are consistent with the SISO case. Furthermore, the frequency increasing trend of the PDF in Figure 2.85 is determined by the reactive component.

The value of the line impedance of the three Δ modes obtained by averaging the measures at all locations and frequency samples is very close. Namely, it reads 102.91, 105.25 and 104.15 Ω for modes PN, NE and EP, respectively. Furthermore, the statistics of the line impedance does not vary as a function of the country, except for UK [144]. In UK, the line impedance is smaller because the network is ring-shaped and it can be modeled as the parallel of two branches. Differently, in the rest of Europe, the network has a tree-like structure that can be modeled as a single branch. The median value of the line impedance measured in all countries except to UK reads 86.86, 89.73 and 88.27 Ω for modes PN, EP and NE, respectively. Indeed, the median value of the line impedance in UK reads 77.39, 81.96 and 78.18 Ω for modes PN, EP and NE, respectively.

2.6.5.2   EMC Related Aspects

The PLC signal produces radiated emissions that may interfere with radio systems. The main contribution to the radiated emissions is due to the CM components, as described in Chapter 3. In SISO PLC, the CM components arise from the mode conversion of the differentially injected signal. Basically, the signal injected between the wires P and N, namely, Δ mode PN, is converted into CM components due to the asymmetries of the power delivery network. Similar effects take place for MIMO signals. To quantify the radiated emissions of a MIMO transmission, field measurements were carried out and the results are reported in [110], [154].

Measurements were performed in the frequency domain, with a VNA, broadband Δ-Y couplers, coaxial cables, a biconical antenna for the measure of the electric field and a ring antenna for the measure of the magnetic field. The coaxial cables were used to connect the antenna and the coupler to the instrument, and they are surrounded by ferrites to limit the disturbances. The antenna was placed at distances of 3 m and 10 m from the measurement site. Results were provided in terms of coupling factor and interference to broadcast radios. The coupling factor is the intensity of the radiated field when a 0 dBm signal is injected into the network. For further details, we refer to Chapter 3. The interference to broadcast radios is quantified subjectively, as the power of the injected signal that can be detected as a disturbance while listening to a broadcast radio. To measure the interference, a random noise signal is FM-modulated and injected. At the receiver, a commercial radio is deployed.

Differential modes yield to similar levels of radiated emissions. Strictly, the magnitude of the coupling factor is similar when Δ modes PN, NE or EP are deployed. Therefore, Δ modes are equivalent in terms of radiated emissions. Similar results are obtained for other differential modes, i.e. when not all the transmitting ports are closed into a 50-Ω load. For these modes, the coupling factor is close to that obtained with a conventional Δ mode transmission. Indeed, the CM signal injection yields to an increase of the coupling factor that can be quantified between 5 to 15 dB. As expected, the common mode signal injection must be avoided. Furthermore, the coupling factor of the CM is a function of the frequency, and it is larger at higher frequencies.

The experimental evidence on the interference to broadcast radios confirms the previous results. Basically, for a given level of disturbance, the power of the CM signal is 10 dB lower than that of differentially injected signals.

2.6.5.3   MIMO Background Noise

Similarly to SISO, the MIMO PLC background noise can be modeled as additive colored Gaussian noise. Furthermore, the noise experienced by different modes is correlated and the statistics is a function of the receiver configuration. A characterization of the MIMO noise experienced by Δ modes was reported in [155] and [146], from experimental basis. In [155], the focus is on the PSD profile. The PSD of the noise was obtained from the time-domain measurements by means of the Welch's periodogram. The target frequency range is between 2 and 150 MHz, and it is shown that the noise PSD can be modeled as follows

$$P_W(f) = 10 \log_{10} \left(\frac{1}{f^\alpha} + 10^\beta \right) \text{ [dBm/Hz]}, \tag{2.46}$$

where α and β are uniformly distributed random variables. In detail, $\alpha \sim \mathcal{U}(1.86, 2.2)$, $\mathcal{U}(1.75, 2.1)$ and $\mathcal{U}(1.76, 2.1)$ for modes PN, NE and EP, respectively, and $\beta \sim \mathcal{U}(-16.1, -15)$, regardless the considered mode. The noise experienced by mode PN turns out to be the lowest. In [146], the focus is on the correlation between the noise experienced by different modes. The analysis is performed in the frequency domain, and it is shown that the noise of the modes PE and NE is highly correlated.

The noise experienced by Y modes was studied in [144] and [156]. The analysis is based on the results of the experimental campaign that was performed by STF-410 across Europe, during which 31 sites were assessed in Germany, Spain, France, Belgium and UK. Measurements were performed in the time domain, with a digital storage oscilloscope, a Δ-Y coupler and a set of low noise amplifiers that enabled measuring low noise levels with sufficient accuracy. The main findings are the following.

(a) The noise is comprised between -80 and -160 dBm/Hz and, from the comparison of the results in [156] and [146], it seems that Y modes experience a lower noise with respect to Δ modes.
(b) Modes P, N and E experience similar noise levels. Some minor deviations can be found for the tails of the noise amplitude CDF. Strictly, at the 90th percentile, mode E experiences the largest noise. Indeed, at the 10-th percentile, mode P experiences the largest noise.
(c) Mode CM is the noisiest [156]. Namely, the noise of the CM mode is 5 dB larger than that of the other Y modes, and the gap is even larger if we consider the attenuation introduced by the coupler. Basically, it seems that the CM mode is more susceptible to the radiated noise, especially in the lower frequency range, below 40 MHz [144].
(d) The dimension of the measurement site does not impact on the statistics of the noise. Thus, the noise depends only on the devices that are connected close to the measurement outlet [144]. Furthermore, no variation of the noise was observed between different countries [144].

Finally, from the analysis of the time-domain correlation between the noise experienced by different modes, it was pointed out that the FM and AM broadcast radio interferers tend to increase the level of correlation because they are received with almost the same intensity by all modes.

2.7   Noise and Interference

Power line communication networks, initially designed for power transfer, present a hostile environment for communication systems. Power line noise, resulting from various devices connected to the grid as shown in Figure 2.86, is a significant factor that contributes to

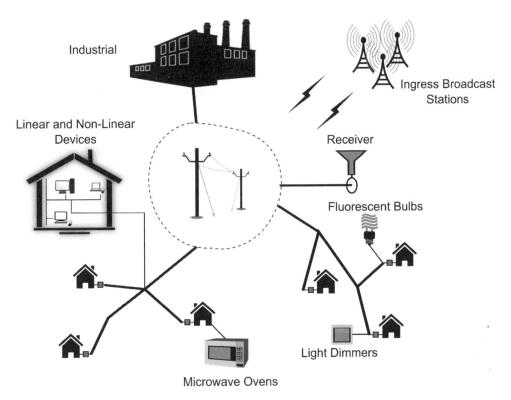

Figure 2.86 Various sources of noise in power line communication networks that interfere with the communication signal at a given receiver. [157] © 2012 IEEE.

this hostile environment. This noise deviates significantly from the additive white Gaussian (AWGN) assumption typically used to design and analyze communication systems and is characterized by the presence of strong, time-varying non-white, and often non-Gaussian noise. This section describes the tools necessary to analyze PLC noise in multiple scenarios and summarizes different properties of PLC noise based on measurements performed in outdoor narrowband and indoor broadband PLC systems. Our treatment includes statistical noise models and their possible applications in communication system design.

2.7.1 PLC Noise Analysis

In order to understand the effect of PLC noise on the performance of communication systems, it is important to analyze its structure in both time and frequency domains. The structure of the noise in these domains typically determines the design of both single and multicarrier communication systems. In this section, we look at the properties exhibited by PLC noise traces in three domains: temporal, spectral, and spectro-temporal. The analysis of the noise structure in these domains allows for choosing an appropriate model that suits the communication system under consideration.

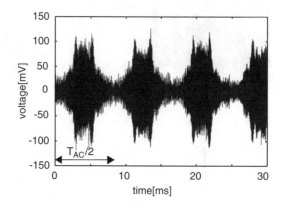

Figure 2.87   Noise waveform by an inverter driven fluorescent lamp (30 W). [159] © 2006 IEEE.

2.7.1.1   PLC Noise in Time Domain

Experimental studies based on time-domain traces of PLC noise show that it can be roughly categorized into several classes as described in the following [68], [112].

2.7.1.1.1   Continuous Noise

- *Time-Invariant Continuous Noise* has a constant envelope for a long period of time (at least more than a few cycles of mains AC voltage). This noise is also referred to as *Background Noise* and may include the thermal noise caused by front-end amplifiers of receivers.
- *Time-Variant Continuous Noise* has an envelope that changes synchronously to the mains absolute voltage. Thus the noise period is half the mains AC cycle duration, $T_{AC}/2$. In narrowband PLC systems, this noise often dominates the system performance [157]. A typical source for this type of noise is an appliance with an oscillator whose power supply is a rectified but not smoothed voltage. Induction heaters and inverter-driven fluorescent lamps are examples of such appliances. Figure 2.87 shows an example of a noise waveform of this class. In addition to the behaviors of appliances, channel characteristics between the noise sources and a receiver may vary synchronously to the mains voltage, see section 2.3.3.3, [158]. This is another cause of the fluctuation of noise waveforms synchronous to the mains voltage.

2.7.1.1.2   Impulsive Noise

PLC systems often encounter abrupt noise with large amplitudes with short (typically a few micro-seconds to milli-seconds) durations, which can be classified as follows.

- *Cyclic impulsive noise synchronous to AC mains* is a class of noise waveforms composed of a train of impulses with the frequency of AC mains or twice of it. A typical cause of the noise of this class is a silicon controlled rectifier or thyristor based light dimmer. This device controls brightness of a light by switching AC current based on its phase, and thus switching noise (impulse) occurs synchronously to the mains voltage. An appliance with a brush motor is another source of noise of this class. In this case, switching at brushes of a

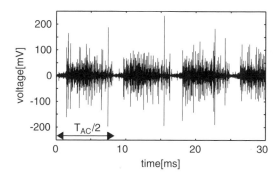

Figure 2.88   Noise waveform by a vacuum cleaner with brush motor. [159] © 2006 IEEE.

motor is more frequent. And since noise amplitude depends on the AC (absolute) voltage, the impulses show the periodicity of the mains frequency as shown in Figure 2.88. Noise from some electronic circuits are also classified in this class as shown in Figure 2.89.

- *Cyclic impulsive noise asynchronous to AC mains* is a class of noise waveforms composed of a train of impulses with a frequency much higher than that of mains AC. The typical cause of this class of noise is a switching regulator.
- *Isolated impulsive noise* is composed of impulses that occur at random timing, often with long (more than seconds) interval. This noise is caused when a wall switch or a thermostat in heaters/foot-warmers, for example, makes/breaks the mains AC current [160].

2.7.1.2   PLC Noise in Frequency Domain

Figures 2.90 and 2.91 show the amplitude distributions of two different subbands: the lower subband noise is approximately Gaussian, while noise tends to be more impulsive in the

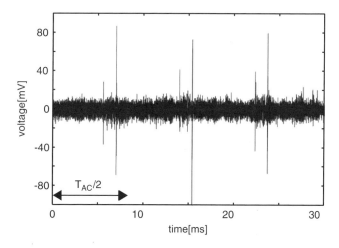

Figure 2.89   Noise waveform by a CRT TV. [159] © 2006 IEEE.

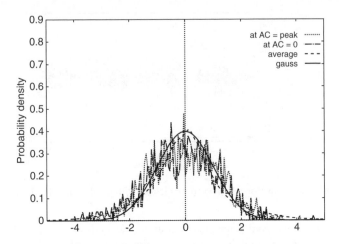

Figure 2.90   Noise amplitude distribution in 0–2 MHz band.

higher subband. This shows that PLC noise is frequency-band dependent exhibiting a non-white statistics.

2.7.1.2.1   *Colored Noise*

PLC noise is colored and it has larger power in lower frequency region. This is because the propagation between each noise source and a receiver has larger attenuation in higher frequency. In addition, many noise sources have power concentration in lower frequency range. Especially in narrowband PLC systems using the kHz band, noise power decreases approximately exponentially with frequency. In broadband PLC in the shortwave band, this tendency still stands, but noise spectrum is more complicated. This is because of frequency notches caused by multi-path propagation in PLC channel and narrowband-noise from outside.

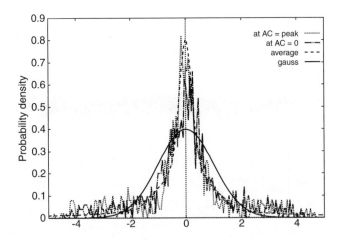

Figure 2.91   Noise amplitude distribution in the 3–3.5 MHz band.

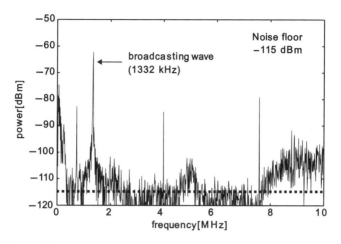

Figure 2.92 PLC noise spectrum at a house near a broadcasting station.

2.7.1.2.2 Narrowband Noise

Broadband PLC systems share the frequency bands with broadcasting and wireless communication systems, and the radio signals of these system may contaminate PLC channels as narrowband noise. Thus this noise can be regarded as 'interference from wireless to PLC' and also called *tone jammer*. Figures 2.92 and 2.93 show an example of PLC noise waveform taken at a house located about 5 km from a 100 kW AM broadcasting station [161]. It is observed that the radio signal is the dominant factor of PLC noise and the noise envelope is almost identical to the modulating audio signal.

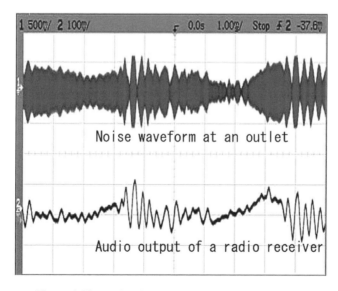

Figure 2.93 PLC noise waveform by a broadcast signal.

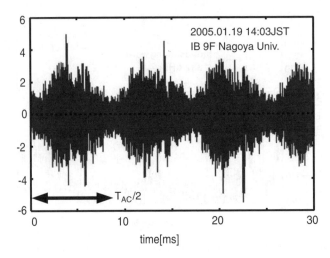

Figure 2.94   Snap-shot of a measured noise waveform (normalized amplitude). [159] © 2006 IEEE.

2.7.1.3   PLC Noise in the Spectro-temporal Domain[10]

The previous section showed that PLC noise is spectrally colored with various narrowband interferers. However, it did not reveal whether the PLC noise had *non-stationary* features; i.e. its spectral structure varied with time. A commonly used technique for non-stationary signal analysis is the *Short time Fourier Transform* (STFT) [162]. The STFT divides the signal of interest into multiple, possibly overlapping, sections and computes the Fourier transform of each section, revealing how the signal's frequency content changes with time. The STFT of a given signal $x[n]$ is given by

$$X[m, \omega] = \sum_n x[n]w[n - m]e^{-j\omega n} \tag{2.47}$$

where $w[n]$ is a window function. The spectrogram (magnitude of the STFT) of a noise trace collected at a low voltage site is given in Figure 2.95 (the STFT was computed using a Hamming window). This noise exhibits strong cyclostationary features in time and frequency domain with period $T = T_{AC}/2 \approx 8.3$ ms. In addition, there is a higher concentration of noise power in the lower frequency band with broadband impulses occurring every T and some weaker narrowband interference.

---

[10]Portions of the material in this section are reprinted, with permission, from M. Nassar, A. Dabak, I. H. Kim, T. Pande, and B. L. Evans, 'Cyclostationary noise modeling in narrowband powerline communication for smart grid applications,' in *Proc. IEEE Int. Conf. Acoustics, Speech and Sig. Proc.*, Kyoto, Japan, Mar. 25–20, 2012, pp. 3089–3092, © 2012 IEEE, and K. F. Nieman, J. Lin, M. Nassar, K. Waheed, and B. L. Evans, 'Cyclic spectral analysis of power line noise in the 3–200 kHz band,' in *Proc. IEEE Int. Symp. Power Line Commun. Applic.*, Johannesburg, South Africa, Mar. 24–27, 2013, pp. 315–320, © 2013 IEEE.

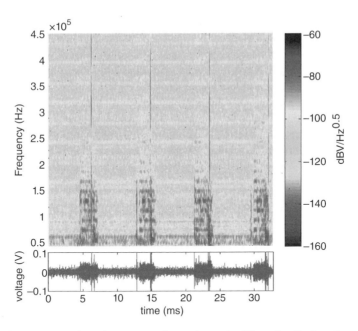

Figure 2.95 Spectrogram of a noise trace at a low voltage site. The noise displays the cyclostationary features both in time and frequency. [163] © 2012 IEEE.

Cyclostationary signals are a special class of non-stationary random processes that have a periodic *instantaneous auto-correlation function*. The *symmetric instantaneous auto-correlation function* of a discrete-time random process Y is defined as

$$R_{YY}[n, l] = \mathbb{E}\left\{ Y\left[n + \frac{l}{2}\right] Y^*\left[n - \frac{l}{2}\right] \right\} \tag{2.48}$$

where $\mathbb{E}\{\cdot\}$ represents the expectation operator. As a result, the auto-correlation $R_{YY}[n, l]$ of a cyclostationary random process Y satisfies the following relationship [164], [165]

$$R_{YY}[n, l] = R_{YY}[n + kN, l], \quad \forall k \in \mathbb{Z} \tag{2.49}$$

where N is the period of the cyclostationary process. Using the 2-D Fourier transform, the cyclic auto-correlation function can be transformed to yield a 2-D function of cyclic frequency α and frequency f:

$$S(\alpha, f) = \sum_{n=-\infty}^{\infty} \sum_{l=-\infty}^{\infty} R_{YY}[n, l] e^{-j2\pi\alpha n} e^{-j2\pi f l}. \tag{2.50}$$

Note that since $R_{YY}[n, l]$ is time-periodic (in time index n), it admits a Fourier series representation in the time domain and has a discrete cyclic spectra in α at multiples of the fundamental frequency F_S/N where F_S is the sampling frequency.

This $S(\alpha, f)$ describes the behavior of the cyclostationary signal on two different scales: a macro-scale characterized by α that describes the cyclic behavior of the signal (determined by the time-domain of (2.48)), and a micro-scale characterized by f that describes the correlation of the noise samples within one cycle (determined by the lag-domain of (2.48)). In contrast, a stationary signal would be completely characterized by the micro-scale frequency f, and a deterministic periodic signal would be completely determined by the macro-scale cyclic frequency α. Figure 2.96 illustrates the application of spectro-temporal analysis to analyzing PLC noise samples collected in a residential complex. In particular, Figure 2.96(a) shows the time domain captured noise trace and its corresponding STFT. The STFT is computed using a 512-point FFT with a 256-length Hamming window with an overlap of 170 samples. The signal's structure repeats every 8.3 ms (half of AC main cycle). Furthermore, this figure reveals a large number of noise sources with various spectral and temporal properties. In particular, we see pronounced energy in the noise at frequencies 4, 45, 58–70, 75, 80–85, 90–120, and 133–140 kHz. The cyclic spectral density in Figure 2.96(b) reveals frequency peaks at a distance of about 120 Hz corresponding to the fundamental cycle period of 8.3 ms (half of AC main cycle). These peaks are a direct consequence of the autocorrelation being periodic in the time domain resulting in the discrete cyclic frequency spectrum as discussed previously. The cyclic frequencies have been binned 1 through 150 (indexing by zero) corresponding to cyclic frequencies of 25 Hz to 3750 Hz with a 25 Hz step size. Additional applications of cyclic spectral analysis to PLC noise can be found in [166].

2.7.1.4   Overall Noise Waveform

When a PLC receiver is located in close proximity to a noise source and no other noise emitting appliance is around the receiver, the noise source dominates the received noise waveform. In this case the noise waveform may be classified in one of the above mentioned classes. But in a general environment, many appliances connected to the power line network emit noise and thus PLC noise is the superposition of noise waveforms from the different classes. An example for an overall noise waveform is shown in Figure 2.94 [159].

2.7.2   Statistical-physical Modeling of PLC Noise[11]

Statistical models of PLC noise and interference are important for communication system design and communication performance. These models attempt to capture the characteristics of the noise at the receiver that are important for communication performance such as noise sample distribution and temporal sample dependance. The two main modeling approaches are *statistical-physical* and *empirical* models. In this section, we introduce statistical-physical modeling of PLC noise. Empirical models are discussed in the following section.

Given a statistical model of signal propagation and a statistical model of interference emissions across the power grid, statistical-physical modeling captures the interaction between the two models to derive the statistics of the interference at the receiver. While challenging

---

[11]Portions of the material in this section are reprinted, with permission, from M. Nassar, K. Gulati, Y. Mortazavi, and B. L. Evans, 'Statistical modeling of asynchronous impulsive noise in powerline communication networks,' in *Proceedings IEEE Global Telecommunications Conference*, Houston, USA, Dec. 5–9, 2011, pp. 1–6. © 2011 IEEE.

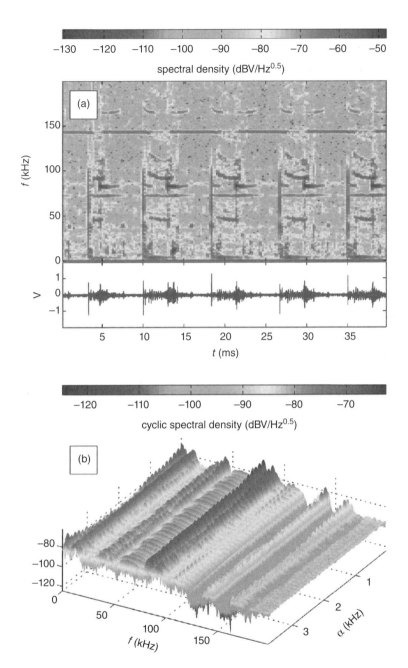

Figure 2.96 Analysis of noise samples collected in the living room of an apartment: (a) its normalized spectrogram and noise samples, and (b) its cyclic power spectrum. A large number of narrowband and impulsive noise sources are observed, many of which are cyclical with an 8.3 ms period. [166] © 2013 IEEE.

for general PLC noise, the following section presents the derivation of the popular Gaussian mixture and Middleton's class-A models in the context of PLC networks. Both models are statistical-physical models for asynchronous emissions typically encountered in broadband PLC networks and dense communication networks [116], [117], [167].

2.7.2.1   Gaussian Mixture and Middleton's Class-A: Model Description

First order statistics (such as sample pdf) are commonly utilized to evaluate the performance of communication systems. The Gaussian mixture and Middleton's class-A models are such first order statistical models of impulsive noise.

A random variable $\mathbf{X}$ has a Gaussian mixture (GM) distribution with centered components if its probability density function (pdf) is a weighted summation of different zero-mean Gaussian distributions

$$p(x) = \sum_{k=0}^{K} \pi_k \cdot \mathcal{N}(x; 0, \gamma_k), \tag{2.51}$$

where $\mathcal{N}(x; 0, \gamma_k)$ denotes a Gaussian pdf with zero mean and variance γ_k, and π_k is the mixing probability of the kth Gaussian component.

A special case of the GM model is the *Middleton class-A* (MCA) model [116]. The MCA model further parametrizes the GM model using two parameters, an impulsive index $A \in [10^{-2}, 1]$ and power ratio $\Omega \in [10^{-6}, 1]$, as follows:

$$\pi_k = e^{-A}\frac{A^k}{k!}, \quad \gamma_k = \sigma^2\frac{k/A + \Omega}{1 + \Omega} = \sigma_i^2\frac{k}{A} + \sigma_g^2$$

where σ_i^2, σ_g^2 are the impulsive and background powers respectively, and $\sigma^2 = \sigma_i^2 + \sigma_g^2$ is the total noise variance. On one hand, the impulsive index A controls the impulsiveness of the noise: smaller A corresponds to more impulsive noise; and, as $A \to \infty$, the noise becomes Gaussian. On the other hand, the power ratio $\Omega = \sigma_g^2/\sigma_i^2$ represents the ratio of the background to impulsive noise powers. As a result, the MCA pdf is given as

$$p(x) = \sum_{k=0}^{K} e^{-A}\frac{A^k}{k!} \cdot \mathcal{N}\left(x; 0, \sigma^2\frac{k/A + \Omega}{1 + \Omega}\right). \tag{2.52}$$

In practice, only the first few significant terms are retained. This model can be extended for modeling baseband noise by replacing the real Gaussian pdf $\mathcal{N}(x; 0, \gamma_k)$ by the circularly symmetric complex Gaussian pdf $\mathcal{CN}(x; 0, \gamma_k)$ (a circularly symmetric complex Gaussian distribution models the in-phase and out-phase components as independent Gaussian variables with equal variance). In such a case, it can be shown that the marginals of the in-phase and quadrature components are still GM or MCA [168]. While uncorrelated, the in-phase and quadrature components are no longer independent as in the AWGN case. In fact, one of these components having a large value implies that the other component is also likely to be large since both were generated by the same Gaussian component of the mixture model.

As described, both the Gaussian mixture and Middleton class-A models only characterize the first order statistics; thus, they do not describe the power spectrum or the autocorrelation of

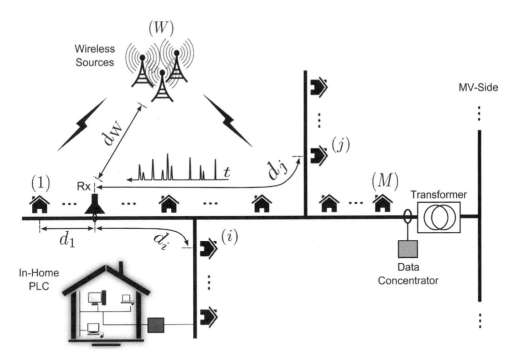

Figure 2.97   A system model for a low voltage power line communications network with interference sources. Each interferer emits asynchronous impulsive noise emission at a distance d_m from the receiver for $m = 1, \ldots, M$. 111 © 2011 IEEE.

the noise. As a result, independent sampling of these models will have limited application in PLC systems with the exception of broadband PLC noise modeling. Extending these models to capture higher order statistics is an important area for future PLC research.

2.7.2.2   Gaussian Mixture and Middleton's Class-A: Model Derivation

Let us consider a power-distribution PLC network in which a randomly located receiver receives a signal of interest in the presence of interfering emissions. A typical system model for a low voltage PLC network is given in Figure 2.97: M interferers, made up of residential and industrial sources and possibly wireless transmissions, emit interference that impinges on the given receiver.

The interference observed by the receiver at a reference time $t = 0$ due to emissions that arrived within a time interval of duration T from the reference time is given by

$$\mathbf{I}(T) = \sum_{m=1}^{M} \mathbf{I}_m(T) \tag{2.53}$$

where $\mathbf{I}_m(T)$ is the interference resulting from interference source m (see Figure 2.98). Here, we assumed that the interference is stationary: an assumption that holds on time scales comparable to the signaling time and interference impulse durations [112]. As a result, the instantaneous interference that accounts for all emissions until time t is given by $\Psi = \lim_{T \to \infty} \mathbf{I}(T)$. The goal

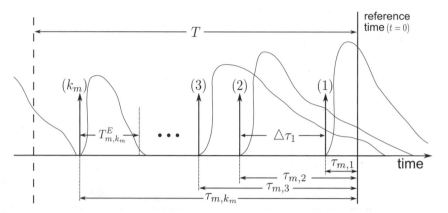

Figure 2.98 Superposition of impulses generated by source m: vertical arrows are illustrations indicating arrivals, k_m is the number of arrivals within time duration T, and $t = 0$ is the reference time. [111] © 2011 IEEE.

is to find the first-order statistics of this total interference Ψ by calculating the characteristic function of $\mathbf{I}_m(T)$ for each interferer m.

We start by defining the statistical model of the interference emitted by a single interferer. Experimental studies reported in [112] have demonstrated that the inter-arrival time and the duration of asynchronous impulses in broadband PLC systems follow an exponential distribution. Exponential inter-arrival times are well-captured by Poisson point processes; thus, for each source m, we assume that the process $\Lambda_m = \{\tau_{m,i} : i \in \mathbb{N}\}$ representing the impulse arrival times with respect to the reference time is a time Poisson point process with rate λ_m. The impulse duration $\mathbf{T}^E_{m,i}$ ranges between 10 µs to 1 ms with a distribution that is loosely exponential and a typical value of hundreds of µs [112] (the exact distribution of the impulse duration is not important since the derivation depends only on its first moment $\mathbb{E}\{\mathbf{T}^E_{m,i}\}$). The impulse response of the PLC channel has a delay spread τ_h between 1 µs to around 4 µs [17], [169]. Since $\mathbf{T}^E_{m,i} \gg \tau_h$, the response of channel to the impulsive emission will have only one resolvable component and the channel will be a flat fading channel.

2.7.2.3 Resulting Statistical Models

Using the statistical models described in Section 2.7.2.2.1, [111] showed the interference at the receiver due to source m, $\mathbf{I}_m(T)$, follows a Middleton Class-A distribution with parameters given by

$$A_m = \lambda_m \mu_m = \lambda_m \mathbb{E}\{\mathbf{T}^E_m\} \tag{2.54}$$

$$\Omega_m = \frac{A_m \times \mathbb{E}\{\mathbf{h}^2_m \mathbf{B}^2_m\}}{2} = \frac{A_m \gamma(d_m) \mathbb{E}\{\mathbf{g}^2_m \mathbf{B}^2_m\}}{2} \tag{2.55}$$

where $T^E_{\max}$ is the maximum impulse duration, A_m is the overlap index that indicates the amount of impulsiveness of the interference originating from source m, and Ω_m is its mean intensity.

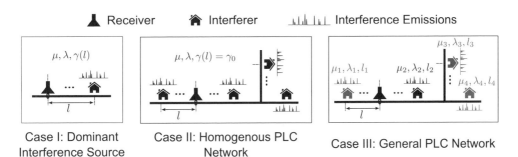

Figure 2.99 Interference scenarios in PLC networks. Each interferer emits a random sequence of emissions onto the power line, which add up at the receiver. Each interferer is described statistically by a mean number of emission events μ, mean duration between emission events λ, and the path loss to the receiver γ. [170] © 2013 IEEE.

Using this result, the interference statistics for typical PLC network scenarios given in Figure 2.99 is given in Table 2.14 [111].

2.7.3 Empirical Modeling of PLC Noise

Empirical models begin with noise data collected in the field to propose models that match the characteristics exhibited by that data. These models are particularly useful for accurate prediction of communication performance under particular scenarios and receiver design.

2.7.3.1 Time Domain Approach for Impulsive Noise

Modeling the impulsive noise in time domain requires a statistical characterization of three parameters: (i) pulse amplitude, (ii) pulse width, and (iii) inter-arrival time. These parameters have been experimentally investigated in [112], [171]–[174]. A popular approach to capture impulsive noise in time domain is to model its marginal distribution. Prior work fits noise data to different statistical models such as Middleton's Class A [175, 176], Gaussian mixture [176], Nakagami-m [177], and Rayleigh [171] distributions. These distribution explicitly model the

Table 2.14 Statistical-physical models of interference in PLC networks categorized by network types. The different scenarios with their corresponding parameters are described in Figure 2.99 where M is the number of interferers. The numerical values are based on the experimental results in [112]

Scenario	Statistical Model	Example Values
Dominant Interferer:	Middleton Class A: $A = \lambda\mu$,	$A = 1.11 \times 10^{-5}$
Rural Area, Industrial Area	$\Omega = A\gamma E\left[h^2 B^2\right]/2$	$\Omega = 131.82$
Homogeneous PLC Network:	Middleton Class A: $A = M\lambda\mu$,	$A = 9.41 \times 10^{-5}$
Urban Area, Residential Buildings	$\Omega = A\gamma E\left[h^2 B^2\right]/2M$	$\Omega = 1380.4$
General PLC Network:	Gaussian Mixture:	$\pi = [0.99\ 0.01]$
Dense Urban Area, Commercial	π_k and γ_k given in [111]	$\gamma = [0.99\ 4.12]$

pulse amplitude distribution. However, they assume the noise samples are independent and identically distributed (i.i.d.) and, as a result, lack the flexibility to capture the pulse width and the inter-arrival times. In [112], a partitioned Markov chain with multiple states have been proposed to capture both the pulse width and the inter-arrival times for wideband PLC systems.

2.7.3.2 Frequency Domain Approach

In order to generate PLC noise waveforms with a non-white power spectral density (PSD), frequency domain approaches dividing the noise bandwidth into several sub-bands and assigning a set of PDFs to each subband have been used. As a set of PDFs, [69] uses the sum of two Rayleigh distributions, while [177] employs the same concept with a Nakagami-m distribution. In [178], noise voltages are sampled at each sub-band to define a PDF by a histogram of noise voltages.

This class of noise models well represents noise features in the frequency domain, and it can be used especially for the design of the systems with multi-tone modulation schemes. On the other hand, it is necessary to modify the model to represent time-variation or non-stationary features of noise.

2.7.3.3 Periodic and Cyclostationary Noise Model[12]

Section 2.7.1.3 and various studies targeted specifically for narrowband PLC (NB-PLC) have shown that periodic noise and cyclostationary noise are the dominant components of the additive noise in these systems [157], [159], [179]. A typical time-domain noise trace and a spectrogram of noise samples collected in the field at medium and low voltage sites with some possible sources is given in Figure 2.95. This trace, in addition to field data presented in the IEEE P1901.2 contribution [180], exhibit cyclostationarity in both the time and frequency domain.

In [159], the authors propose to use a *cyclostationary* Gaussian model to characterize this aspect of the additive noise in NB-PLC systems. This model captures the temporal characteristics of the noise by a Gaussian process with a time-varying variance that is periodic with a period equal to half the AC cycle; i.e.

$$n[k] \sim \mathcal{N}\left(0, \sigma^2[k]\right) \text{ with } \sigma^2[k] = \sigma^2[k + mT] \tag{2.56}$$

where the notation $n[k]$ refers to the sample of the discrete time process n at time index k. In (2.56), $T = 0.5 T_{AC} \times F_S$, T_{AC} is the duration of the AC cycle, F_S is the sampling frequency, and $m \in \mathbf{Z}$. Spectral coloring is introduced by passing this noise through a linear time-invariant shaping filter $h_s[k]$ with an exponentially decaying spectral density (see Figure 2.100). As a result, this model decouples the frequency and temporal shaping. Since the main

---

[12]Portions of the material in this section are reprinted, with permission, from M. Nassar, A. Dabak, I. H. Kim, T. Pande, and B. L. Evans, 'Cyclostationary noise modeling in narrowband powerline communication for smart grid applications,' in *Proc. IEEE Int. Conf. Acoustics, Speech and Sig. Proc.*, Kyoto, Japan, Mar. 25–20, 2012, pp. 3089–3092. © 2012 IEEE.

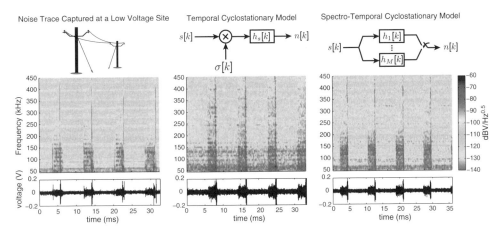

Figure 2.100 The two cyclostationary noise models for NB-PLC: (i) the temporal cyclostationary model [159] shapes an AWGN signal $s[k]$ by the temporal function $\sigma[k]$ and filters it through a single shaping filter $h_s[k]$, (ii) the spectro-temporal cyclostationary model [157] filters an AWGN signal $s[k]$ by a filter bank of different shaping filters $h_i[k]$ each corresponding to a stationary region of the noise period. Since the temporal cyclostationary model is constrained to a single spectral shape, it overestimates the noise power spectral density in the 50–200 kHz region by as much as 50 dBV/Hz. The spectro-temporal model provides a closer fit to the measured noise spectrogram. [157] © 2012 IEEE.

cyclostationarity captured is temporal, we refer to this model as the *temporal cyclostationary model*. In its characterization, this model requires a parametric form to characterize $\sigma^2[k]$ which might lead in some scenarios to a data-intensive parameter estimation procedure.

If the noise trace exhibits a time varying spectral content (see Figures 2.95 and 2.96), the temporal cyclostationary model might not be enough to produce an accurate presentation of the noise trace. In order to capture the cyclostationarity in both the time and frequency domain, the authors in [180] and [157] propose a cyclostationary model, which we refer to as *spectro-temporal cyclostationary model*. This model divides each noise period into M regions during which the noise is assumed stationary (see Figure 2.95). Each region is characterized by a spectral shape, a corresponding shaping filter, and the set $\mathcal{R}_i$ of time indices corresponding to each region. Using this model, the PLC noise is modeled as the convolution of an AWGN signal $s[k]$ with a linear periodically time-varying system $h[k, \tau]$ given by

$$n[k] = \sum_{\tau} h[k, \tau] s[k - \tau] = \sum_{i=1}^{M} \mathbf{1}_{k \bmod(T) \in \mathcal{R}_i} \sum_{\tau} h_i[\tau] s[k - \tau] \qquad (2.57)$$

where $\mathbf{1}_{\mathcal{A}}$ is the indicator function, $\bmod(T)$ represents the T modulo operation, and $h[k, \tau] = \sum_{i=1}^{M} h_i[\tau] \mathbf{1}_{k \bmod(T) \in \mathcal{R}_i}$. This can be implemented using a filter bank as shown in Figure 2.100. The linear time-invariant filters $h_i[k]$ correspond to spectral shaping filters for each region of the spectrogram in Figure 2.95 and can be estimated using spectrum estimation techniques [157]. A comparison between the temporal cyclostationary model and the spectro-temporal cyclostationary model is discussed in Figure 2.100.

In addition to the cyclostationary noise, [179] shows that in the *very low frequency* (VLF), below 10 kHz, there is an additional wide-sense periodic noise component. Presently, the VLF range is extensively used by AMR providers. This periodic structure is exploited in [179] to filter out this component and increase the post-processing SNR of these systems.

2.7.4 PLC Noise Features for Adaptive Coding Modulation and Demodulation

In the design and the analysis of conventional communication systems, stationary additive white Gaussian noise (AWGN) is often used as a model of noise. As follows from the discussion above, such a model cannot be applied for PLC systems.

The non-Gaussian, especially impulsive, noise is often believed to be the cause of low reliability of PLC. However, the fact that the Gaussian distribution has the largest entropy for a given power implies that the communications under non-Gaussian noise may achieve better performance than under AWGN, if the detection takes the noise statistics into account [181].

Since the noise is non-white, it is possible to introduce multicarrier modulation schemes with adaptive allocation of power, modulation index, and coding gain for each carrier, which has been done in wireless communication systems for frequency selective fading channels. PLC noise is non-uniform not only in frequency domain but also in time domain. And when the noise is dominated by cyclostationary components, estimation of future noise statistics and adaptive transmission in time domain based on the estimates is also possible.

It has been mentioned in Section 2.7.2.1 that in-phase and quadrature components of narrowband impulsive noise whose PDF is described by Middleton's Class-A model are not independent. This fact implies that a measurement of one of these components provides information on the other noise component. This information can be used as the side information in signal detection as proposed in [182].

Figures 2.101 and 2.102 show examples of noise waveforms in different frequency bands measured at the same time for the duration of three AC voltage cycles. As shown in these figures, the instantaneous noise powers in different frequency bands are dependent. Under such environments, the instantaneous noise powers in a frequency subband can be estimated by observing the noise in other frequency subbands not used for signal transmission. For example,

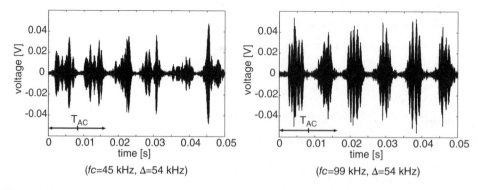

Figure 2.101   Power line noise waveforms in different frequency bands (lower frequency region). [183] © 2005 IEICE.

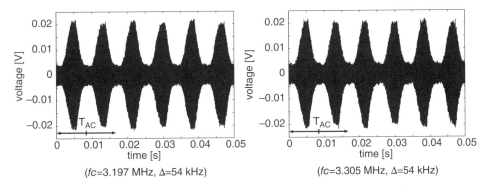

Figure 2.102 Power line noise waveforms in different frequency bands (higher frequency region). [183] © 2005 IEICE.

[183] proposes an orthogonal frequency division multiplexing (OFDM) receiver which uses the estimated noise statistics as side information in decoding process.

In many wireless systems, thermal noise dominates noise statistics, and the noise waveforms at different transceivers are independent. In PLC systems, however, the dominant factors of noise sources are electrical appliances connected to a power line network, and it can be expected that the noise waveforms at different outlets are correlated. In [161], [184], it is shown that instantaneous noise voltages at two different outlets have high correlation as shown in Figure 2.103 if they are connected to the same live conductor. In addition, it is reported that the instantaneous noise powers and cyclic averaged noise powers have large correlations even for a pair of outlets connected to the different live conductors. These correlations of noise waveforms at different locations in a power line network can be used to improve the performance of PLC systems. For example, a transmitter can allocate power, modulation, and

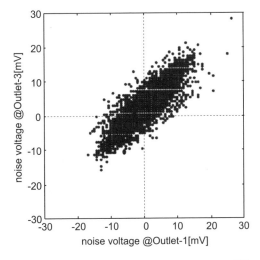

Figure 2.103 Scattering diagram of instantaneous noise voltages at two different outlets. [161] © 2007 IEICE.

codes estimating the noise statistics of its receiver without noise-state information feed-back
from the receiver. This scheme is especially suitable for one-way and multi-cast communication
systems. Multicarrier modulation is discussed in Chapter 5, while adaptive resource allocation
algorithms are discussed in Chapter 6.

2.8 Reference Channel Models and Software

The previous sections have presented characterization and modeling approaches for PLC over
AC power lines. The majority of the studies and developments have been concerned with the
LV part of the grid, which is not surprising as this is the PLC channel environment that is most
difficult to characterize.

Table 2.15 summarizes the modeling approaches for signal propagation in the LV domain.
We categorize the models into

- top-down: empirical modeling approaches which try to capture the propagation phenomena
  like signal echoes;
- bottom-up: modeling based on properties of electric signal propagation over power lines;
- deterministic: models comprise one or a small set of channel realizations (often through
  parameter sets for top-down approaches or given topology, cable types, etc. for bottom-up
  models);
- stochastic: models include probability distributions to produce random channel realizations.

We note that the majority of the top-down models have been developed for broadband PLC,
which provides the more diverse scenario in terms of path loss and frequency selectivity that
is experienced for different networks and links within a network. Due to their derivation from
physical signal propagation properties, bottom-up approaches are generally applicable to a
wide range of frequencies where TL theory can be applied. Furthermore, while much of the

Table 2.15 Overview of channel models for signal propagation over power lines

Domain	Channel		Approach	Type	Model	Reference
Access	SISO	LTI	top-down	deterministic	CFR	[73], Section 2.3.2.4
						[185, App. D]
LV/MV	SISO	LTI	bottom-up	deterministic	CFR	[185, App. D]
Indoor	SISO	LTI	top-down	stochastic	CFR	[65]
					CFR	[30], [56], [186]
					CIR	[43], [74], [91]
					CFR	[187]
			bottom-up	stochastic	CFR	[24], [25]
					CFR	[31], [32], [151], Section 2.3.3.4
		LPTV	top-down	stochastic	CFR	[77], Section 2.3.3.4
			bottom-up	deterministic	CFR/CIR	[85], Section 2.3.3.4
	MIMO	LTI	top-down	stochastic	CFR	[141], Section 2.6.4
					CIR/CFR	[142], [145], Section 2.6.4
			bottom-up	stochastic	CFR	[88], [89], Section 2.6.4

Table 2.16   Available software implementations of channel models

Channel		Approach	Type	Software	Link	Based on
SISO	LTI	top-down	stochastic	MATLAB	[90]	[30], [56], [186]
	LPTV	top-down	stochastic	MATLAB	[87]	[77]
	LPTV	bottom-up	deterministic	ns-3	[188]	[32], [77]
	LTI	bottom-up	stochastic	MATLAB	[189]	[31], [32], [77], [151]
MIMO	LTI	bottom-up	deterministic	MATLAB	[190]	[89]

results reported in Table 2.15 apply to the indoor (or in-home) scenario, the deterministic bottom-up modeling approaches can be extended to outdoor links.

While many of the references provided in Table 2.15 allow the reader to reproduce the results and generate CIRs or CFRs, software implementations have been made available for some of the presented methods. These are summarized in Table 2.16. The source codes available at [90] and [87] are self-contained MATLAB modules that produces CFRs according to the top-down models presented in [30], [56], [186] and [77], respectively. The ns-3 suite of software modules available and described in [188] uses the bottom-up approach from [32] together with the modeling of time-varying impedances, including periodically time-varying impedances according to [77], and it allows the integration of a physical PLC channel model into PLC network simulations. Reference [189] presents a MATLAB software that implements the stochastic bottom-up approach from [31], [32], [151] and uses the time-invariant frequency-selective impedance model from [77]. Finally, [190] describes a MIMO bottom-up software implementation that includes a module for the numerical computation of per-unit-length parameters and allows for varying numbers of conductors at different network nodes. Its computation of CFRs is based on the method introduced in [89] where the analytical computation of the per-unit-length cable parameters is also addressed. One of the target applications in [190] is PLC in vehicles, as discussed further in Section 2.9.2, but it is equally applicable to MIMO PLC in indoor and outdoor environments and can be extended to stochastic modeling as done in [89].

2.9   Channels in Other Scenarios

In this section, other relevant application scenarios of PLC are considered: PLC in the low voltage DC distribution grid and in the vehicular (car and ship) power grids. The main channel and noise characteristics are described.

2.9.1   Low Voltage Direct Current (LVDC) Distribution System

The low voltage DC (LVDC) electricity distribution system concept presents a novel approach to low voltage distribution grids by applying DC in the power supply and power electronics for conversions between AC to DC and back to AC. Originally, the LVDC system was designed for rural areas, where a single long medium-voltage (MV) grid branch typically supplies an MV/LV (20/1/0.4 kV) transformer substation of a small area with only a few end-users. In this case, the MV line has to be brought close to the end-users, because the transmission capacity of the low voltage AC grid is limited. Similarly, the size of the geographical area and the number of customers in each low voltage AC network are limited.

The LVDC system provides advantages over traditional AC distribution systems. Because of the use of DC in the low voltage distribution grid, the power transmission capability is improved compared with traditional AC low voltage grids. In addition, both the reliability of electricity distribution and the quality of network service received by the electricity end-users can be enhanced by the LVDC [191], [192].

Besides the advantages of the LVDC for the power distribution, an LVDC system equipped with data communication infrastructure provides smart grid functionalities. In this context, application of PLC turns out interesting. Hence, in the next sections, the focus is on the main characteristics of the LVDC system with the application of PLC, including the PLC channel characteristics.

2.9.1.1   Structure and Characteristics of an LVDC Distribution System

In general, the main objectives in evolving electricity distribution systems are the cost efficiency and reliability of electricity distribution. In the LVDC system, a single MV overhead line, or a part of it, and low voltage AC distribution grids are replaced with a low voltage DC distribution system implemented with underground cables. As the existing AC distribution grids are aging and they are also sensitive to adverse weather conditions, there is public discussion of replacing old overhead lines by underground cables in several countries, including Finland.

After the MV/LV (20/1 kV) transformer, low voltage AC is rectified to DC and supplied to the customers by a low voltage DC grid. In addition, the rectified DC is smoothed according to the standard on low voltage electrical installations [IEC60364]; the ripple of DC voltage is in the range of 10% of rated DC voltage [191], [192]. Each customer is equipped with a customer-end inverter (CEI), which converts the DC back to AC voltage suitable for the customers. The structure of the LVDC system and the main differences compared with the traditional MV/LV AC system are illustrated in Figure 2.104.

The main differences between the LVDC system and the traditional low voltage AC systems are:

(a) The geographical area may be larger and the distances in the LVDC grid longer.
(b) Parts of the MV grid can be replaced with the LVDC system, and thus, the length and number of medium-voltage AC lines can be decreased.
(c) The number of MV/LV (20/0.4 kV) and LV/LV (1/0.4 kV) transformers is decreased.
(d) The number of power-electronic devices, i.e. inverters, is increased.

There are basically two approaches for implementing an LVDC system; a unipolar system with 1500 VDC and 0 V, and a bipolar one with ±750 VDC and 0 V voltage levels (Figure 2.105), of which the bipolar LVDC system is more feasible [191], [192]. According to the directive [LVD 73/23/EEC], the maximum allowed voltage 1500 VDC, which is rated as low voltage DC [193], can be used for power delivery with the same low voltage underground cables as in traditional AC systems. These cables have a voltage rating of 900 VDC against earth and 1500 VDC between two conductors. By adopting a bipolar LVDC system, low voltage underground cables of this kind can be used.

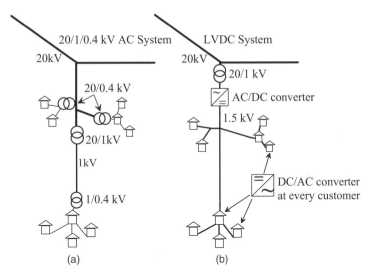

Figure 2.104 Differences between the traditional AC and LVDC systems that include a medium- and low voltage distribution grid [191].

The LVDC distribution system concept has been under study at Lappeenranta University of Technology (LUT) in Finland since 2006. One of the key outcomes of the project is that the LVDC distribution system with the application of power electronic converters is a viable technical alternative for the electricity distribution grids in the future [191]. The first LVDC laboratory prototype system implemented with a one-phase inverter was built in 2008 [194]. The customer-end inverter was upgraded into a three-phase inverter with improved EMI filtering in 2011 [195].

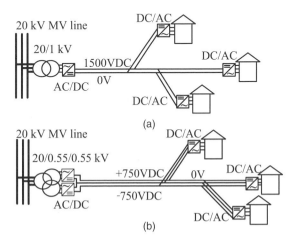

Figure 2.105 Basic structure of a unipolar (1500 VDC, 0) and bipolar (±750 VDC, 0) LVDC system with power electronic converters used for power conversions (adapted from Pinomaa *et al.* 2011 [198]).

Two underground low voltage cable types have been applied to the bipolar DC grid in the research project; an AMCMK cable comprising three phase conductors (L1, L2, L3) with a concentric conductor, protective earth (PE) (3 × 16+10 mm²), and an AXMK cable with four conductors (4 × 16 mm²). The rectifier is implemented with half-controlled diode-thyristor bridges. Every customer in the grid is equipped with a CEI implemented with insulated gate bipolar transistors (IGBTs), which converts the DC back to AC (230 VAC / 50 Hz).

Normally, the low voltage underground cables are delivered in cable reels of 500 m. The DC grid between the rectifier and the customer-end inverters consists of 500 m long cable segments, and the cables are joined in overground cable connection cabinets, where the DC conductors and the neutral (0 V) conductor of the cables are joined to the corresponding DC and neutral conductors between two cable reels. The case is somewhat different with AXMK cables; there are two conductors reserved for ±750 VDC and two neutral (N) conductors (0 V). Thus, the N conductors of every AXMK cable are short circuited, and these loops are joined together between two 500 m cables.

The first bipolar LVDC field pilot installation including four customers with loads (the number of CEIs in the grid being three) was implemented and has been in use since June 2012 [196]. The low voltage cable used in the field pilot is AMCMK 3 × 95+21 mm². Both DC poles and 0 V are delivered to all three CEIs in the grid, but each CEI is supplied either from +DC and N, or N and −DC. The structure of the bipolar LVDC field pilot system with its dimensions, and the PLC network is depicted in Figure 2.106 [197].

The LVDC system provides benefits in electricity distribution compared with traditional AC distribution systems; with the customer-end inverter, the quality of the AC low voltage supply for the customers can be improved, and by replacing sections of the MV overhead branch lines with underground low voltage DC cables, the number of possible faults in the MV grid and thereby the costs caused by outages are decreased [199], [200].

In the future, transition to DC distribution could be possible also inside customer premises; most of the customers' present appliances and small-scale electricity generation units applied in new smart grids operate with DC. The main advantage of the LVDC system when considering the transition to smart grids is that no synchronization is required for distributed electricity generation units and electrical vehicles (EVs), the number of which has been increasing on

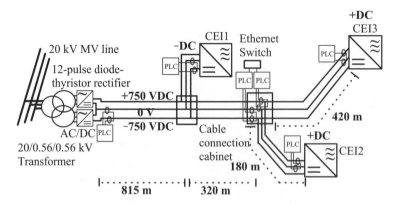

Figure 2.106 Structure of the bipolar LVDC field pilot system with the proposed PLC network implemented in an actual distribution grid (adapted from Pinomaa *et al.* 2013 [197]).

the grids over the past few years. Furthermore, the applied converter technology provides new opportunities to grid management and electricity markets, thus promoting the advancement of smart grids.

In order to add and efficiently apply small-scale electricity generation units or EVs to the DC grid, new functionalities, such as control and monitoring of the statuses of the distributed electricity generation units have to be included in the system. These functionalities, again, require a communication network infrastructure on the DC power grid. Ubiquitous communication plays a key role in smart grids and also in the proposed LVDC distribution systems. The LVDC concept is considered as a smart grid platform, meaning that different smart grid applications, such as grid control (data and power flow) and protection applications, are integrated into the LVDC system; further, data transmission between nodes in the DC grid is required. In this respect, the use of PLC appears interesting. PLC has been commonly used in traditional AC distribution grids, and proposed as a communication solution for novel smart grids.

2.9.1.2 PLC in the LVDC Distribution System

The LVDC grid structure and dimensions (1–5 km between the rectifier and the CEIs), which are larger than in traditional low voltage AC grids, bring challenges to the PLC applied in the DC grid [198], [200]. The PLC signaling range and repetition intervals in the LVDC grid have to be known to provide reliable communications in the system. In addition, it has to be noted that power electronic devices (the rectifier and the CEIs) in the system generate interferences and voltage and current harmonics to the channel.

Commonly, narrowband (NB) PLC techniques, such as PRIME and G3-PLC operating at a low-frequency band from 9 kHz to 500 kHz, are considered appropriate for smart grid applications in traditional AC systems [201]. The NB PLC (i.e. PRIME in Europe and G3-PLC) provides low data rates with long data transmission distances between PLC modems. However, the total noise power in distribution grids tends to rise as more and more power electronic devices are continuously connected to the grid. This is also the main challenge to the LVDC PLC; the noise power including high-power impulsive noise components generated by the switching devices of the CEI IGBTs is significantly higher in the frequency band from kilohertz to megahertz compared with traditional AC grids. Therefore, and based on the LVDC PLC studies, the HF band appears more suitable for PLC transmission in the LVDC system. One applicable commercial PLC technology for the system is the HomePlug protocol. HomePlug 1.0 is considered herein appropriate for the LVDC system as it includes a synchronization method that does not require the AC mains signal [202].

2.9.1.3 PLC Channel Characteristics in the LVDC System

Figure 2.107(a) illustrates a possible PLC configuration (cf. [198] and [200]) for the bipolar LVDC system built in the laboratory setup. The bipolar LVDC laboratory setup consists of a 35 kVA double-tier transformer (400/562/562 V) and a half-controlled diode-thyristor rectifier (6-pulse thyristor bridges for both poles). A 198-meter-long AXMK cable is used for DC power delivery and is connected between the rectifier and the three-phase CEI, which is implemented with an IGBT unit [195], [197]. The CEI converts the DC back to AC, which is also smoothed

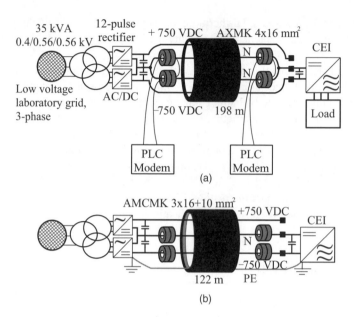

Figure 2.107 PLC concept implemented into a bipolar LVDC laboratory system between a rectifier and a three-phase CEI. The AXMK cable used for a DC link in (a) and the AMCMK in (b) (adapted from Pinomaa *et al.* 2013 [197]).

with ferrite cores (each phase is smoothed individually). The inverter is connected to resistive loads, which can be changed with fixed load steps.

Because of the arrangement and coupling method of the N conductors in AXMK cables, inductive coupling is considered for the proposed PLC configuration with the two options below [198]:

(a) PLC modems are coupled differentially between the N conductors or between the +750 and −750 VDC; the cable conductor structure is symmetrical for the communications.
(b) Coupling between the N and ±750 VDC conductors; the cable conductor structure is asymmetric.

Figure 2.108 illustrates an inductive coupling implemented with commercial ferrite rings (Ascom Powerline type IC-R-27-200) in the inverter end coupled differentially between the N and −750 VDC conductors of the AMCMK power cable. The two N conductors with inductive couplers, namely the NN coupling, provide the following advantages:

(a) The communication channel can be divided into segments without branches in the NN coupling, while the channel between the 0 V and ±750 contains branches having a negative effect on the channel response.
(b) Short-circuited NN loops provide a low-impedance channel for current, and are thus ideal for inductive coupling. In addition, non-galvanic coupling is achieved; this is an installation

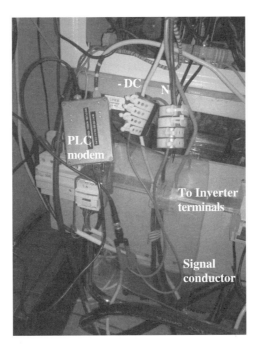

Figure 2.108 Inductive couplers implemented with commercial ferrite rings coupled differentially between the N and −750 VDC conductors of the AMCMK cable at the inverter end in the LVDC laboratory setup. The signal conductor connected to the PLC modem is coupled through the ferrites.

advantage, and inductive couplers are not prone to overvoltages that may occur on the grid.

(c) The common-mode interferences in the NN conductors, generated by power-electronic devices in the channel ends are canceled out by each other with the transformer couplings formed by the inductive couplers.

(d) The loads and load impedance changes caused by the CEIs do not affect the impedance of the communication channel in the NN couplings.

Capacitive couplers cannot be used in the NN coupling case. However, the NN coupling cannot be used in all cases. For example, in AMCMK cables (Figure 2.107(b)) there are only three line conductors, and a cable of this kind does not provide a better communication channel. Thus, the cable characteristics and channel properties are studied with inductive couplers for the both coupling cases available. The frequency band studied is from 100 kHz to 30 MHz, which covers the HF band applied in the concept.

2.9.1.3.1 *High-frequency Characteristics of Low Voltage Power Cables*

Signal voltage attenuation in low voltage cables is an important parameter when power cables are used as the medium for PLC. It defines the maximum signaling range, when the network topology, loads, the data transmission power spectrum density (PSD), and the noise PSD with the required signal-to-noise ratio (SNR) at the receiver are known.

Cable parameters, including the signal voltage attenuation coefficient α, can be estimated from the distributed cable parameters, which can be derived from the two-conductor transmission line model. In the two-conductor cable model, the crosstalk with the conductors not used for signaling is neglected. According to the proposed PLC concept, the application of the two-conductor model is used as a first approximation. The signal voltage attenuation coefficient can also be estimated by measuring the input impedances of the cable, with the other cable end first short circuited and then open circuited. The unused conductors of the cable are left open at the both cable ends. The input impedances of the cables are measured with an HP4149A impedance analyzer with an impedance probe HP4194A. The linear frequency sweep is supplied to the cable, and 401 measurement points including the impedance absolute value and its phase in the frequency band between 100 kHz and 30 MHz are measured. From the measured input impedance, the cable parameters for the AXMK and AMCMK cables are estimated. The estimated signal voltage attenuation coefficient α for the AXMK cable between the N conductors, and between the N and −750 VDC conductors, and for the AMCMK cable between N and −750 VDC conductors is illustrated in Figure 2.109.

According to [203], with polyvinyl-chloride (PVC) insulated MCMK low voltage cables, the attenuation coefficient at the 20 MHz frequency is approximately 100 dB/km, which is multiple times more than the attenuation in the AXMK cable. The difference in attenuation between these two cables is primarily caused by dielectric losses in the insulation material. The dissipation factor $\tan \delta$ of the cross-linked polyethylene (PEX) insulation material is lower than in PVC; at 1 MHz, $\tan \delta$ is around 0.09–1 for PVC and for polyethylene, $\tan \delta$ is >0.0005. Furthermore, in polyethylene, the dielectric constant ε_r (2.25 for polyethylene), and the dissipation factor remain relatively constant as a function of frequency [203].

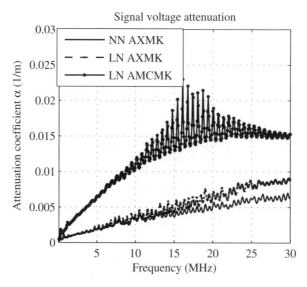

Figure 2.109 Measured signal voltage attenuation coefficients α between the N conductors (NN), and the L and N conductors (LN) of the AXMK cable, and between the LN conductors of the AMCMK cable as a function of frequency.

The size of the cable conductor cross-section is larger in actual LVDC grids. However, the ratio of the conductor size to the insulation material thickness remains unchanged; low voltage cables are scaled versions of each other, and thus, the effect on the cable distributed parameters is slight (l and c remain approximately unchanged). The resistance of the cable depends on the conductor diameter, and in cables with larger cross-sections, the longitudinal r of the cable distributed parameter is smaller (transverse r is dependent on the insulation material). Thus, signal attenuation in an actual grid is lower compared with the case in the studied cable, and the estimated signal-voltage attenuation can be used for AXMK and AMCMK cables with larger cross-sections.

2.9.1.3.2 *Channel Gain*

For the frequency response of the channel, the S-parameters are measured from the terminals of the inductive coupling interfaces at the cable ends at the rectifier and the CEI. The scattering parameters S_{12} and S_{21} indicate the power loss in the studied channel from the input (rectifier) to the output (CEI), and vice versa. The S-parameter measurements are carried out with an Agilent 3495A network analyzer with the Agilent 87511A S-parameter test equipment. Linear frequency sweep is measured between 100 kHz and 30 MHz.

As can be seen in Figure 2.108, the channel gain is higher in the NN coupling case compared with the LN coupling in the AXMK cable case. Even if the AMCMK cable is over 70 meters shorter than the AXMK cable, the channel gain is poorer compared with the 198-meter-long AXMK cable. This is due to the higher signal attenuation coefficient in the AMCMK cable as seen in Figure 2.110.

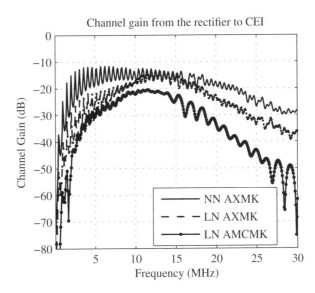

Figure 2.110 Channel gain measured from the terminals of the inductive couplers (ferrite rings) with small-signal diodes used for overvoltage transient protection in the NN and LN couplings in the AXMK cable (198 m), and the LN coupling in the AMCMK cable (122 m). The frequency band observed is 100 kHz–30 MHz.

Based on the measurements carried out in the laboratory environment, estimations of the channel characteristics can be made for the actual LVDC field pilot system, where the distances between the rectifier and the CEIs are longer. In the laboratory, an overoptimistic channel gain is achieved in the LN coupling case. The reason for this is that in this setup there is only one CEI on the grid, and no branches, unlike in actual grids. The branches would have a major effect on the channel gain in the LN coupling case; they would generate deep notches for some frequencies, which is unfavorable for communication purposes. Furthermore, the time-varying impedance of the customer loads causes time variations to the channel response.

2.9.1.3.3 *Channel Noise*

Along with the channel gain (including signal attenuation in the power cable), noise is an essential factor affecting the PLC channel characteristics. The noise in the LVDC PLC channel is mainly impulsive. The impulsive noise is generated by the diode-thyristor rectifier bridge at one end, and by the IGBTs at the CEIs at the other ends of the branched DC grid.

To analyze the characteristics of the LVDC PLC channel noise, noise measurements were carried out in the LVDC field pilot system. The LVDC field pilot grid is wider and includes branches, unlike the simple laboratory setup (Figure 2.106). Noise samples of 20 ms were measured from the terminals of inductive couplers connected differentially to the end of the DC grid at the rectifier and at the other ends of the grid at the CEIs in different load conditions. Noise measurements were carried out with a Rohde & Schwarz RTO 1014 oscilloscope. The measured noise samples of 20 ms (one 50 Hz AC cycle) are divided into segments of 10 μs, which includes the duration of single HomePlug 1.0 symbol 8.4 μs [202]. A periodogram was calculated for each time-domain noise sample segment to detect and analyze the variation in the noise PSDs in the time domain. The noise sample measured from the terminals of the inductive couplers at the CEI in the load condition in the LVDC field pilot is illustrated in Figure 2.111. Variation in the noise PSD in the time domain estimated from the measured noise sample is depicted in Figure 2.112. The analysis shows the variation in the noise PSDs with the appearance of impulses in the time domain. The noise impulses with these variations have a direct effect on the performance of the PLC applied in the channel.

According to [112], the noise can be divided into five categories; the colored background noise floor is approximately at −115 and −110 dBm/Hz at the rectifier and the CEI, respectively, in the LVDC field pilot in the nominal load condition [197]. In addition, two impulsive noise sources are detected in the measured noise samples. The diode-thyristor bridges of the rectifier together generate a 150 Hz voltage and current harmonics, and thus, noise impulses with a repetition frequency of 150 Hz to the N conductor of the DC cable from both ±750 VDC poles (periodic impulsive noise synchronous to the mains frequency of 50 Hz). Each CEI presents a noise source by generating high impulsive noise to the LVDC PLC channel. The switching frequency f_s of the IGBT module in normal mode is 16 kHz, and these switchings are seen in the noise samples measured from the terminals of the inductive couplers (periodic impulsive noise asynchronous to the mains frequency). These impulses in the time domain are illustrated in Figure 2.113.

Moreover, the CEI generates 100 Hz harmonics to the DC grid (periodic impulsive noise synchronous to the mains frequency 50 Hz), and also other impulsive noise components to the DC grid. The CEI mixes and sums the impulses generated and received from the rectifier and other CEIs with its own switching frequency and harmonics.

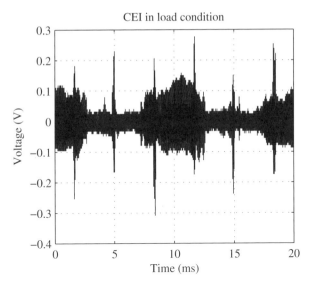

Figure 2.111 Noise sample of 20 ms measured from the terminals of ferrite rings coupled differentially between the −750 VDC (L) and neutral (N) conductors of the AMCMK cable at the CEI in the load condition.

Furthermore, the performance of the PLC modems (HomePlug 1.0 modified for the application) was tested in the LVDC field pilot to observe and analyze the common actions and effects of long communication distances and noise sources at the channel ends on the PLC. The results published in [197] show that a CEI as an impulsive noise source in the channel ends has a major effect on the HF band PLC performance.

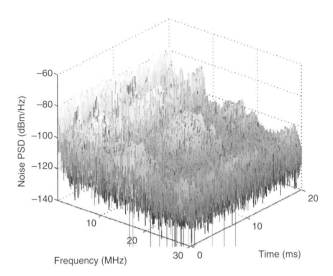

Figure 2.112 Variation in the noise PSD in the time domain in the CEI in the load condition calculated from the 20 ms noise sample.

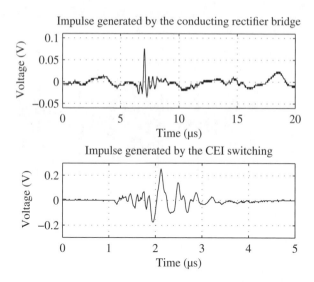

Figure 2.113 Noise voltage waveforms measured from the inductive couplers coupled between the −DC and N conductors in the load condition. A single impulse generated by the conducting rectifier bridge, and impulse generated by the CEI IGBT switchings (f_s = 16 kHz). An 800 W load connected to the −DC pole.

2.9.2 In-car Power Line Communication Channels

New generation of vehicles needs a wide set of sensors and electronic equipment to ensure high reliability and safety condition for the driver. Protocols for automotive applications using dedicated buses have been proposed in the past and applied in most of actual vehicles. In the near future, the need of high data rate for both multimedia application and control command communication will become undeniable. Communication between an Electronic Controlled Unit (ECU) and electronic equipment could be possible, without adding additional buses, by using the power distribution grid as a communication medium. The main advantages of this solution are numerous in terms of reliability since it reduces the amount of splicing, simplifying and lightening the cable bundles as well. However, it is difficult to transpose results obtained on the in-building power grid directly to vehicles since the topologies and the geometric characteristics of the in-vehicle cable bundles are quite different from the in-building power grid. Moreover the activation of actuators, such as antilock braking systems (ABS) and engine controls, can produce impulse noise and fast changes of the load impedances, leading to a highly frequency and time variable channel. Knowing and modeling the channel is therefore essential to optimize the transmission scheme. After describing the configuration of an automotive wiring, this section gives an overview of the broad-band channel characteristics and of the noise and interference properties.

2.9.2.1 Configuration of an Automotive Wiring Harness

Let us consider, for example, a communication between a Front Left Light (FLL) and a Rear Left Light (RLL) on the in-vehicle cable bundle (Figure 2.114). The medium between these

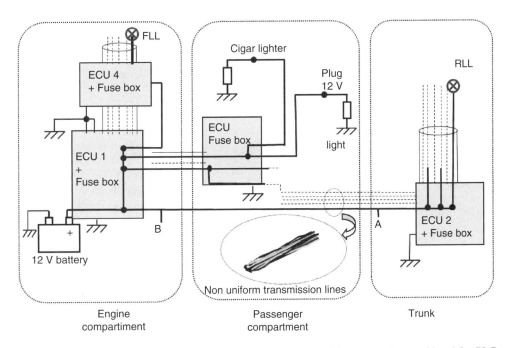

Figure 2.114 Conceptual power line harness diagram of a car with some nodes considered for PLC (adapted from Carrion *et al.* 2006 [204]).

two access points is a series of bundles of multiconductor lines. These conductors are power supply lines and other wires transmitting either analog or digital signals (like communication buses). The number of conductors inside a bundle can be up to several dozens. Therefore, coupling between lines inside a bundle must be taken into account into the propagation model. The impedances of hundred electric and electronic equipment found in cars, including but not limited to light bulbs, sensors and motors (in electric cars), take complex values whose amplitudes range from 1 Ω to 1 kΩ. These impedances are often unknown and, in any case, frequency dependent. This could be a serious problem for modeling the propagation along the cabling but considerations on the harness itself allow simplifying the approach. Indeed, the impedances are connected to wires of unspecified length, often on the order of the wavelength, and all wires are coupled together within the various bundles. Therefore, introducing in the model impedances of real but randomly chosen values, instead of their complex actual values, would not change the statistical results.

To investigate the properties of the automotive harness, we have chosen terminal points (transmitting or receiving points) belonging to the DC network supplying each ECU and on-board computers. Indeed, since there are about 20 to 70 ECUs distributed in the car, from the front to the rear of the vehicle, this network presents many branches. In the following, a connection is labeled 'direct link' if the shortest path between the modems does not pass through the battery and 'indirect link' in all other cases. This distinction is based on the fact that, in the PLC frequency range, the battery acts as very low impedance. An indirect link connection is thus highly penalized by the presence of the battery which partially short-circuits the transmission line. It must be mentioned that, in this section, cars under consideration are combustion engine vehicles. For electric vehicles, results described in [205] consider a bundle of short length due to the small size of the car. Even if the cable architecture in the electric car is different from the one deployed in the 'combustion engine car', some of the common problems are: the strong frequency selectivity of the channel and the presence of impulsive noise, for instance the one generated by the DC/DC converters.

2.9.2.2   Channel Transfer Function

In order to determine the expected channel statistical properties between a transmitter and a receiver connected to an in-car electrical grid, a large number of measurements must be examined. They have to be carried out on different vehicles and during different conditions of the vehicle, for instance when it is stopped, it is moving, when various electrical devices are intentionally activated, etc. Some measurement results of these kind have been reported in the literature [204], [206]–[208]. Since the number of tests is limited by the available number of access points on the car power line, the amount of data can be increased by using a theoretical propagation model, based on electromagnetic concepts, to simulate the tree-shaped bundle configuration behavior [209] and therefore to be able to infer the channel characteristics.

In this section, a brief review of the most relevant features of in-car channels is provided by presenting some measurement examples and also statistics of behavioral parameters. The common way to characterize a system/channel transfer function is to measure the scattering parameters by means of a network analyzer and to derive the transfer function or insertion losses. However, this procedure is essentially adequate for LTI channels, a condition not always satisfied in the case of in-car channels because the loads connected to the power grid are changing over time (as long as on-board devices are turned on/off, e.g. the lights). Nevertheless, the rate of these changes can be considered very low compared to the expected bit rate of communication systems and the channel can be assumed as steady between successive changes. The frequency range selected for the measurements can be extended up to tens of MHz, to explore the channel capability of supporting broadband communications.

In Figure 2.115, some channels measured in the first car, corresponding to the diagram in Figure 2.114, and in another car are presented. In particular, the amplitude response, or channel gain, of two different links are shown. Both cars have combustion engines, more details can be found in [204], [206]. Regarding the location of the transmitter and receiver in the car harness, two classes of channels can be identified: those with a direct path between both terminations, without passing through the battery, and those with an indirect path, passing through the battery. The latter ones usually present longer transmission distance with more branches and terminations, what has a remarkable influence on the channel behavior; they exhibit higher attenuation and selectivity in frequency (or time dispersion). This fact can be observed in the plots of Figure 2.115. A level of attenuation increase of about 20 dB is encountered when the link is indirect. Although the shape of the channel gain varies considerably among measurements of different cars, and even among different channels of the same one.

Another source of time-variation in the channel response can be the presence of devices in the network like the alternator, which is engaged to the engine revolutions and hence is continuously changing. This situation makes more difficult the channel response estimation. To face this problem, in [206], measurements are carried out by setting the engine to a certain regime, approximately constant. Three engine states are defined to explore its effect: 'off', with ignition switch off (only those electronic systems connected directly to the battery are

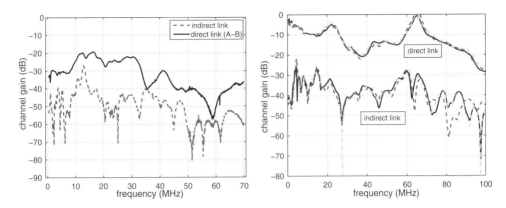

Figure 2.115 Channel amplitude response measured over two different kind of links: direct and indirect. On the left, for the first car in Figure 2.114 (adapted from Carrion *et al.* 2006 [204]). On the right, measurement on a second car and for three different engine states: off (solid line), idle (dotted-dashed line) and 2000 rpm (dashed line) (adapted from Vallejo-Mora *et al.* 2010 [206]).

powered); 'idle', with ignition switch off and the engine idling (approximately with 750 rpm -revolutions per minute-, and all electronic systems are powered); '2000 rpm', when engine is running at 2000 rpm (the gear is in neutral position and all electronic systems are powered). During the measurements, no electronic devices were turned on to isolate their influence in the channel estimation. The observed behavior is that the higher the engine activity the worse the channel conditions (and also the more noise level ingresses the receiver).

To highlight the grade of variation that the channel response can reach on the same car Figure 2.116 is included here. The estimated responses for 12 different links measured on

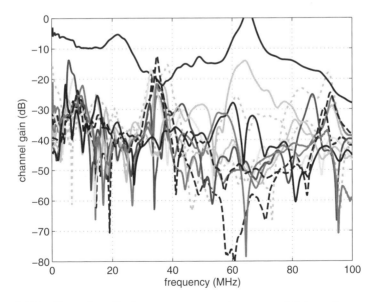

Figure 2.116 Channel amplitude response measured over several links on the second car.

Table 2.17 Statistical characterization of measured channels response.

Engine State	Off	Idle	2000 rpm
Mean B_C (kHz)	527.06	568.60	482.92
Std B_C (kHz)	235.58	233.58	183.84
Mean $\overline{|H|}$ (dB)	−37.93	−38.21	−39.15
Std $\overline{|H|}$ (dB)	8.38	8.79	9.94

the second car are superimposed (all links are under a 2000 rpm engine state). Only one of them correspond to a direct link, the upper curve with low attenuation, while the remaining ones exhibit a quite unpredictable attenuation and frequency selectivity levels. Even more, the same link on a certain car can be measured on different days and the estimated channel responses will present a modest correlation [204]. This result claims for the use of flexible modulation schemes, like multicarrier modulation, with good channel estimation techniques at the receiver side. This situation also appears in PLC systems for indoor applications and, thus, some techniques from them can be successfully applied for in-cars systems [208].

In Table 2.17, a summary of statistics of some behavioral parameters of the measured channel responses is shown for the three evaluated engine states in the second car. In particular, the coherence bandwidth B_C (measured at the 90% of the frequency response correlation) and the logarithmic mean gain $\overline{|H|}$ have been selected. The mean values among all the measured links are given and the standard deviation as well, to emphasize the high variability of the measurements. Here, it is remarkable again the impact that engine state has on the channel behavior: the higher the rpm the worse the channel characteristics (smaller coherence bandwidth, and so higher dispersion, and higher signal attenuation). Similar behaviors hold for the electric cars, where the states of the motors strongly affect the injected noise [205]. It is worth mentioning that reported results from measurements in other cars are essentially aligned with the ones presented here (see for instance [207]).

Regarding the impact of the load variations, in [204] many channel responses are registered when the car is in motion through an urban environment and while electrical devices are being switched on and off. As it could be guessed, the response changes are more important in the indirect links (where the correlation among responses from the same link is around 0.7) than in the direct ones (where the correlation is close to 1). The effect of the connection or disconnection of loads can be comparable to a channel topology change (or to a transmitter/receiver location change) what demands from modems some capability to perform 'on the fly' channel estimation and modulation adaptation.

2.9.2.3 Input Impedance of the Electrical Circuit

Another interesting parameter of the measured channels is the input impedance, especially to design the analog front-end of the transmitters and receivers. Results reveal that in-car input impedance at the access points of the power line is far for being constant over frequency, which complicates (even makes impossible) impedance matching techniques at the transmission or reception side. As an example, in Figure 2.117 the input impedance measured at three different access nodes of the second car is shown (when the engine was turned off). The frequency

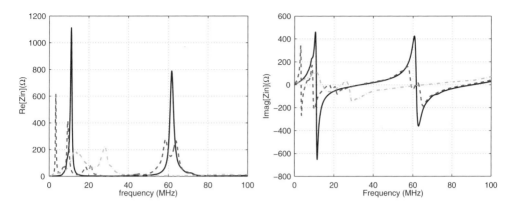

Figure 2.117 Input impedance measured on three different points of the power grid of the second car (adapted from Vallejo-Mora *et al.* 2010 [206]).

selectivity of the impedance is really high and the curve shapes resemble that of several RLC resonant circuits in a parallel setup: there are various resonances of diverse bandwidth. It has been tested that the engine state does not influence notably the result of this parameter. Additional measurements can be found in [207] and in [205] (for a compact electric car), with similar conclusions, thus the usefulness of making efforts in impedance matching techniques is questionable.

2.9.2.4 Noise and Interferences

In the transportation domain, electromagnetic compatibility (EMC) standards have been established by car manufacturers for conducted noise produced by electric or electronic equipment, and also by international bodies to specify upper limits of the electromagnetic field radiated by a vehicle at a given distance [210]. All these standards and the test procedure are defined in the frequency domain. However, to study the robustness of a digital communication, noise characteristics must also be known in the time domain.

From the experimental analysis discussed and assessed in literature [211]–[214] the in-car scenario is affected by a mixture of impulsive and periodic noise, with a background Power Spectral Density (PSD) profile which depends on the vehicle under test, on the equipment and vehicle state (i.e. engine on/off), and on the considered measurement point (near to or far away the main noise source). Two main sources of impulsive and periodic noise are the DC-DC converter (used in both conventional and electric cars), due to its inherent switching activity, and the Controller Area Network (CAN), due to its operability.

An example of an experimental noise PSD, measured at the cigar lighter plug in Figure 2.114 with a spectrum analyzer in the 1–50 MHz band, is shown in Figure 2.118. The high values of the PSD in the low frequency range are due to the presence of impulsive noise. This noise introduced by the DC-DC converter occurs itself on the output voltage as spikes at the switching frequency of 2 MHz, with the maximum PSD being −70 dBm/Hz. One can also note the presence of CAN signals around 12 MHz appearing on the DC wires, with a PSD of −80 dBm/Hz. Beyond 30 MHz, the PSD of the background noise can be estimated with a floor equal to −130 dBm/Hz.

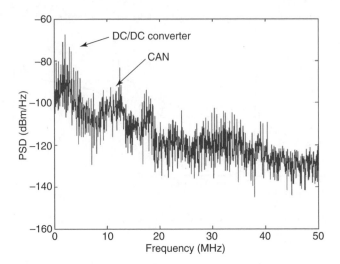

Figure 2.118   Noise PSD at the cigar lighter plug.

In order to highlight the periodic behavior of the noise, the time-domain must be considered. Figure 2.119 shows a record of the time variation of the voltage at the cigar lighter plug during a cruising phase of the car. The periodic impulsive noise of highest amplitude (50 mV) corresponds to the noise generated by the DC-DC converter, while small amplitude pulses due to CAN interference also appear. As a comparison, if we assume that the PSD of the transmitted PLC signal is −50 dBm/Hz and that the path loss is 30 dB, the equivalent rms voltage of the received signal will be 120 mV, i.e. much greater than the peak amplitude of the noise.

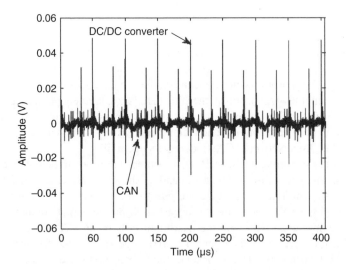

Figure 2.119   Impulsive noise at the cigar lighter plug.

The noise effects are even more pronounced in the electric car scenario, where, beyond the DC-DC converter component, a great deal of noise is injected into the network by the drives that feed the motors, depending on the acceleration status [205].

2.9.3   Power Line Communications On-board Ships

The successful application of PLC technology for in-home automation and networking has led to suggestions of applying PLC in other systems, including ships. The increase of on-board ship electronic equipment, together with the complexity of communication systems, leads to a constantly increasing volume and weight of dedicated cables. In addition, travelers require services as infotainment and communications, which also goes to the direction of increasing cable harness. Regular cruise ships may have thousands of cabins connected by hundreds of kilometers of data cabling. The possibility of creating an integrated system which uses the power grid as a communication channel is obviously appealing, and would lead to a reduction of cables cost, weight and volume, which is fundamental, especially for aerospace and maritime applications. Although, the wireless LAN technology also does not need the wiring costs, it is not suitable in ships because they are commonly constructed by iron plates that mainly prevent the propagation of the radio waves. Several on-board applications could benefit of the use of PLC technology: on-board surveillance, telephone communications, internet connections, control and automation systems, etc.

Utilizing existing power grid for PLC is, in principle, a straightforward procedure. However, the architecture of ship power distribution network is different from that of residential power grid. In fact, the power system of a ship (regardless its nature: cruise, cargo, military, etc.) is a standalone system, which includes power generators, transmission lines, transformers, power electronic converters, bus transfers, circuit breakers, and switches. The system is also characterized by a large variety of loads, as pumps, mechanical loads, emergency systems, kitchens, lighting system, standard and luxury cabin electrical equipment, etc. These characteristics, often with very few voltage levels and with a grounding system different from the most commonly used in residential or industrial environments, make it a unique environment for the PLC technology. In general, studies of the PLC channel in ships indicate the feasibility of the use of PLC technology in this special environment, though different channel performance results are observed, depending on the ship type and dimension. A major issue for ship application is the determination and minimization of the unintentional-radio emissions, as the PLC system must coexist with the emergency, navigation and several other systems. Furthermore, in ship EMC regulations are in general very strict, specially in the wheelhouse.

2.9.3.1   In-ship PLC Literature Review and Power Grid Peculiarities

Despite a recognized growing interest for the application of PLC on-board ships, there is not an extensive literature on the topic. The piloting work [215] reports broadband channel attenuation measurements, and noise floor measurements, using inductive signal coupling, up to 100 MHz on a typical cruise ship. Homeplug 1.0 modems are tested and radio emissions are also measured and reported not in breach of EMC regulations. The link quality of the point to point communication between cabins is found to be unsatisfactory and the use of repeaters is prospected. Signal degradation is imputed to the presence of many switchboards that serially connect the cabins. Another seminal work [216] describes the results of a vast measurement campaign of the transfer function and the noise level up to 100 MHz, over three Asian cargo ships. Most Asian ships use shielded cables, differently from European and American ships, and this make the use of common mode transmission methods feasible. Results in [216] show a large variation of the channel attenuation values, requiring receivers with 100 dB dynamic range at least. Also a large signal attenuation is observed, from 20 to 40 dB, when passing a single switchboard, which has in general 20 or more attached loads. Other works form the same authors further study the use of a proposed common mode transmission method [217], revealing that it performs better using unshielded cables, though EMC issues arise and must be mitigated. In [218] an adaptive method to control the modem power is proposed so that emissions respect EMC regulations, while in [219] it is shown that to cover long links, over 87 m, in a cargo ship, with switchboards in the path with more than 20 loads, a broadband PLC system can usefully use only a reduced frequency band, from 2 MHz to 13 MHz. Channel measurements on-board a luxury yacht in [220] show a similar usable band up to 12.5 MHz, as at higher frequencies the attenuation becomes too high. Channel capacities between 10 and 74 Mbps are observed in U.S. Navy ships [221], using an analytical model. Very good channel theoretical capacity, between 200 and 600 Mbps, is observed on a cruise ship [222] in the band 2–28 MHz with an increase of 85% in the band 2–50 MHz. It is also found that the links between the main panel and the distribution boards have higher capacity than the links between the distribution boards and the room service panels, and this is justified by the fact that in the second case the topology has more branches, causing higher frequency selectivity and fading. An interesting remark in [222] is the prospect of using a MIMO scheme, which is applicable to three phase distribution systems, and yields an approximately double capacity.

Most of the cited works give details of measurement tests, while little work is found on modeling the particular in-ship distribution network topology. In general, the same methods developed for modeling residential PLC can be applied, like bottom-up analytical models based on transmission line equations. However, in-ship distribution networks have an interesting and persistent characteristic, which is observed by many of the cited works. The topology has a typical and preferable star-style configuration, so that electrical nodes are present solely at switchboard panels, where tens of cables are attached. Moreover the number of in series panels tends to be reduced, so that they are mostly connected directly to the main switchboard, where the generators and the heavy loads are attached. The high number of parallel circuits attached to the panel makes the impedance seen by the panel relatively low, and in general it is little affected by switching on or off an attached load. This fact is first observed in real measurements [220], and further analyzed in [104], [223]–[225], where it is shown that when such 'big nodes' are present, the variation of the connected loads produce small variations of the transfer function, making the channel virtually invariant. In the following sections this peculiar property of the ship power grid will be described in more detail.

2.9.3.2   Grid Topology and Measurements On-board a Mega-yacht

In this section the results of an on-board measurement campaign will be shown, as a representative example. The campaign was conducted on the power grid of a six deck mega-yacht, 50 meters long, which was under construction (available by courtesy of the owner) [220]. A portion of the power distribution grid is shown in Figure 2.120. It has two generators of 155 kW and 200 kW, operating at 400 V, 50 Hz, which, in operation, are directly connected to the main panel (divided into two sub-panels, Left and Right, always connected together).

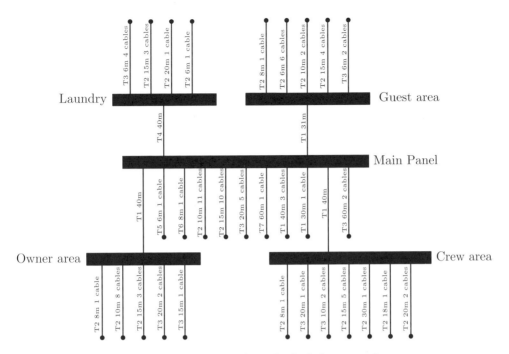

Figure 2.120 Grid topology of a six deck mega-yacht.

Seven types of unshielded cables (T1...T7) are used in the power grid, to connect the loads to the panels or between the main panel and different functional switchboards. All the cables are commercial cables of type LKM-HF with three phases plus neutral and their characteristics have been measured and simulated in [224]. Measurements and simulation results reported in the following sections are based on this representative set of cables for naval applications. For simplicity, four functional areas, located in different decks, are shown in Figure 2.120, each one served by a dedicated switchboard. Cables of the same type and length, connected to the same board, are represented in groups in Figure 2.120. The whole set of cables was disconnected at their far end (open circuit). The measurements have been performed by the use of notebook PCs connected to National Instruments acquisition cards (NI PXI-5422 AWG, and NI PXI-5124 digitizer) whose characteristics are 200 MS/sec and the input and output resistance are set to 50 Ω. A multi-tone excitation signal is used, and the transfer function is obtained by means of online spectrum analysis. The tests show that the attenuation becomes too high above 12.5 MHz, mainly below −40 dB, regardless of the position of the transceivers. In particular, when the transmitter is placed at the switchboard of the crew area and the receiver at the switchboard of the guest area, the path has a length of 80 meters and it goes through the service panel where a large number of branches are connected.

The long term variations of the channel produced by load connections and disconnections are one of the main issues of PLC in general. In order to show this, we focus on the dependency of the channel frequency response from the number and types of cables connected to the main panel.

Figure 2.121 shows the phase-to-neutral frequency responses of two measured channels. Channel 1 is obtained by disconnecting the owner area switchboard from the main panel,

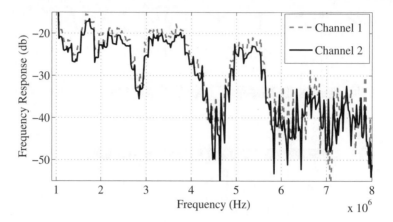

Figure 2.121   Amplitude of the frequency response of two channels.

while Channel 2 is obtained by disconnecting the laundry switchboard (reconnecting owner's area again). It can be seen that the signal attenuation is high through such a channel (a multi-branched node is located in the propagation path), though the transmitted signal can be detected at the receiver, and a similar result is shown in [216]. On the other hand, the topology in the two configurations is significantly different, but the frequency responses are quite similar. This can be justified by the possibly small variation of the low impedance at high frequencies of the multi-branched node, that mitigates the channel variation. In the following we denote such a node as 'big node'.

2.9.3.3   Sensitivity of the Transfer Function to the Node Admittance

The channel response invariance property which has been observed empirically in the previous section, can be further investigated using a theoretical approach. To better evidence the dependency of the transfer function on the big node impedance, a closed form expression of the transfer function can be obtained, referring to the general scheme in Figure 2.122. The distribution network is represented by a three port network, where the transmitter and the receiver, with internal admittance $Y_m = 0.02 \, \Omega^{-1}$, are connected to ports 1 and 2 respectively.

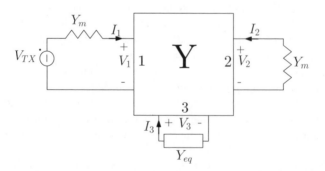

Figure 2.122   Three port equivalent.

At port 3 the equivalent admittance Y_{eq} is connected, representing the parallel admittance of all the circuits connected to the big node that is in the path between the modems. We are looking for an expression of the voltage transfer function $H = V_2/V_{TX}$, to highlight its dependence on the equivalent node admittance Y_{eq}.

The three port network is described in the frequency domain by the matrix $\mathbf{Y} \in \mathbb{C}^{3\times3}$, which imposes the relation between port currents and voltages:

$$
\begin{pmatrix} I_1 \\ I_2 \\ I_3 \end{pmatrix} = \begin{pmatrix} y_{11} & y_{12} & y_{13} \\ y_{21} & y_{22} & y_{23} \\ y_{31} & y_{32} & y_{33} \end{pmatrix} \begin{pmatrix} V_1 \\ V_2 \\ V_3 \end{pmatrix}, \tag{2.58}
$$

where $\mathbf{Y} = \{y_{ij}\}$. It is straightforward to show that under the considered topological conditions we always have $y_{12} = y_{21} \equiv 0$. In fact, a short circuit on port 3 causes the channel from port 1 to port 2 to vanish. Moreover $\mathbf{Y}$ is certainly symmetric, i.e. $y_{13} = y_{31}$ and $y_{32} = y_{23}$. The external circuits connected to the ports impose the following relation between port currents and voltages:

$$
\begin{pmatrix} I_1 \\ I_2 \\ I_3 \end{pmatrix} = \begin{pmatrix} Y_m V_{TX} \\ 0 \\ 0 \end{pmatrix} - \begin{pmatrix} Y_m & 0 & 0 \\ 0 & Y_m & 0 \\ 0 & 0 & Y_{eq} \end{pmatrix} \begin{pmatrix} V_1 \\ V_2 \\ V_3 \end{pmatrix}. \tag{2.59}
$$

From (2.58) and (2.59) we get:

$$
\begin{pmatrix} V_1 \\ V_2 \\ V_3 \end{pmatrix} = \mathrm{inv} \left\{ \begin{pmatrix} y_{11} + Y_m & 0 & y_{13} \\ 0 & y_{22} + Y_m & y_{23} \\ y_{13} & y_{23} & y_{33} + Y_m \end{pmatrix} \right\} \begin{pmatrix} Y_m V_{TX} \\ 0 \\ 0 \end{pmatrix}. \tag{2.60}
$$

The inverse matrix in (2.60) is easily calculated as a symbolic expression, so that for the transfer function the following holds:

$$
H = \frac{M}{K Y_{eq} + L}, \tag{2.61}
$$

where the terms M, K and L do not depend on Y_{eq}, and they are given by:

$$
M = y_{13} y_{23} Y_m,
$$
$$
K = y_{11}(y_{22} + Y_m) + y_{22} Y_m + Y_m^2,
$$
$$
L = y_{11}(-y_{23}^2 + y_{22} y_{33} + y_{33} Y_m) + y_{22}(-y_{13}^2 + y_{33} Y_m) + Y_m(-y_{13}^2 - y_{23}^2 + y_{33} Y_m).
$$

We obtain that the transfer function H in (2.61) has an inverse proportional dependence on the equivalent node admittance Y_{eq}, that for the big node is expected to be high. If we make the assumption that $|KY_{eq}| \gg |L|$, then

$$H \approx \frac{M}{KY_{eq}}, \tag{2.62}$$

and we can derive the sensitivity:

$$\frac{\partial H}{\partial Y_{eq}} = -\frac{M}{KY_{eq}^2}. \tag{2.63}$$

A small variation of the equivalent admittance ΔY_{eq} will produce a variation of the transfer function as $\Delta H = \frac{\partial H}{\partial Y_{eq}} \Delta Y_{eq}$. From (2.62) and (2.63) we obtain the following

$$\frac{\Delta H}{H} \approx -\frac{\Delta Y_{eq}}{Y_{eq}}, \tag{2.64}$$

which states that the relative variation of the transfer function is approximated by the negative relative variation of the node admittance. Considering that the big node admittance Y_{eq}, which is given by the parallel of tens of terminated cables, is in general high, then a single load connection or disconnection from the node will not produce a significant relative variation of the equivalent admittance Y_{eq}. As a consequence, the variation of the transfer function will be small as well.

The analysis carried out only makes assumptions on the topology and on the key condition $\left|\frac{KY_{eq}}{L}\right| \gg 1$, that is necessary for the results (2.61), (2.62) and (2.63) to hold.

In [224] it is shown that, using a set of typical commercial cables for naval applications and considering the representative complex topology of Figure 2.120, the parameter $\left|\frac{KY_{eq}}{L}\right|$ is in general large (approximately 30 times) and almost frequency independent. This result can be then considered of general validity, as the considered cables and topology are well representative of a large category of in-ship power distribution grids.

2.9.3.4   Variation of the Node Admittance and Identification of the Big Nodes

Since the relative variation of the channel transfer function has the same magnitude of the relative variation of the node admittance, the focus has been moved to the characterization of the node admittance variation. If we consider a simple case of N open terminated cables of the same type and length, connected in parallel to the switchboard, the admittance variation given by adding one more cable of the same type and length is given by:

$$\frac{\Delta Y_{eq}}{Y_{eq}} = \frac{Y_{\text{cable}}}{\sum\limits_{N} Y_{\text{cable}}} = \frac{1}{N}$$

In [224] it is shown that the mean value of $\frac{\Delta Y_{eq}}{Y_{eq}}$ tends to $\frac{1}{N}$, even when adding one cable of a random type and length, between seven possible types (T1...T7) of commercial cables, as described above. For a typical big node, with 30 connected cables of different types and average length of 20 m, the admittance variation due to the addiction of one random cable, in the frequency range 2–10 MHz, is about 3% in the average, and the 3σ deviation is about 7%, where σ is the standard deviation. The real impedance variation that occurs mainly for the load connections and disconnections on the far end of a cable connected to the switchboard are in general smaller than the variation due to the addiction of one cable directly to the switchboard, as here considered. Hence the conclusion that the big node admittance is almost time invariant with respect to load changes is further supported. Using further Montecarlo simulations [225] better identifies the case when the node admittance can be considered practically invariant to the load variations, that is when the number of connected loads is lager than 13, with length randomly in the range 6 m–20 m. Then, in [225] it is found that the number of branches has the major impact in determining whether a node is a big node and this result is based on the use of a set of commercial cables for ship power distribution.

It can be concluded that PLC in a typical naval power network (where commonly tens of parallel cables are connected at every switchboard) can assume a substantially time invariant frequency response, though with high attenuations. Then, if some frequency intervals are found where the channel attenuation is not drastic, it is expected that those sub-channels will remain stable, regardless of load connections and disconnections.

2.9.4 Final Remarks

In this section, different and less conventional PLC application scenarios have been considered, namely, the LVDC distribution grid, the in-car and the in-ship scenarios. The main evidence is that for all these environments the channel exhibits frequency selectivity and high attenuation. Indeed, the specific characteristics depend on the underlying network topology structure. Relations with the in-home case can also be found. A comparison in terms of average channel gain, delay spread and coherence bandwidth between the in-home scenario and the in-vehicle environment, i.e. in-car and in-ship, has been done in [187].

Another important aspect that must be taken into account concerns the noise properties. Indeed, in the considered scenarios high noise levels are generated and coupled to the PLC networks. This is particularly true in the in-car scenario where the engine and DC-DC converters are the main source of noise. In the electric car context, even higher noise levels can be experimented due to the high intensity currents that the drives provide to the motors [205]. The results lead to conclude that the channel and noise pose some challenges to the design of reliable PLC techniques. In this respect, two state-of-the-art transmission technologies that can be considered for application in this context are multicarrier modulation (see Section 5.3) and impulsive ultra wide band modulation (see Section 5.5), as discussed in [226].

References

1. J. D. Parsons, *Mobile Radio Propagation Channel*, 2nd ed. John Wiley & Sons Ltd, Chichester, 2000.
2. S. Galli, A. Scaglione, and K. Dostert, Broadband is power: Internet access through the power line network, *IEEE Commun. Mag.*, 41(5), 82–83, May 2003.
3. H. A. Latchman and L. W. Yonge, Power line local area networking, *IEEE Commun. Mag.*, 41(4), 32–33, Apr. 2003.

4. E. Biglieri, S. Galli, Y.-W. Lee, H. V. Poor, and A. J. H. Vinck, Power line communications: Guest editorial, *IEEE J. Sel. Areas Commun.*, 24(7), 1261–1266, Jul. 2006, Special Issue on Power Line Communications.

5. E. Biglieri, Coding and modulation for a horrible channel, *IEEE Commun. Mag.*, 41(5), 92–98, May 2003.

6. L. Lampe, A. M. Tonello, and D. Shaver, Power line communications for automation networks and smart grid, *IEEE Commun. Mag.*, 49(12), 26–27, Dec. 2011.

7. J. Anatory, M. V. Ribeiro, A. M. Tonello, and A. Zeddam, Power-line communications: Smart grid, transmission, and propagation, *J. Electric. Comp. Eng.*, 2013, 1–2, 2013.

8. I. H. Cavdar, Performance analysis of FSK power line communications systems over the time-varying channels: Measurements and modelling, *IEEE Trans. Power Delivery*, 19(1), 111–117, Jan. 2004.

9. F. J. Cañete, J. A. Cortés, L. Díez, and J. T. Entrambasaguas, Analysis of the cyclic short-term variation of indoor power line channels, *IEEE J. Sel. Areas Commun.*, 24(7), 1327–1338, Jul. 2006.

10. S. Barmada, A. Musolino, and M. Raugi, Innovative model for time-varying power line communication channel response evaluation, *IEEE J. Sel. Areas Commun.*, 24(7), 1317–1326, Jul. 2006.

11. F. J. Cañete, L. Díez, J. A. Cortés, and J. T. Entrambasaguas, Broadband modelling of indoor power-line channels, *IEEE Trans. Consumer Electron.*, 48(1), 175–183, Feb. 2002.

12. F. J. Cañete, L. Díez, and J. T. Entrambasaguas, A time variant model for indoor power-line channels, in *Proc. Int. Symp. Power Line Commun. Applic.*, Malmö, Sweden, Apr. 4–6, 2001, 85–90.

13. M. Antoniali and A. M. Tonello, Measurement and characterization of load impedances in home power line grids, *IEEE Trans. Instrum. Meas.*, 63(3), 548–556, Mar. 2014.

14. T.-E. Sung, A. Scaglione, and S. Galli, Time-varying power line block transmission models over doubly selective channels, in *Proc. IEEE Int. Symp. Power Line Commun. Applic.*, Jeju Island, Korea, Apr. 2–4, 2008, 193–198.

15. H. Philipps, Modelling of power line communication channels, in *Proc. Int. Symp. Power Line Commun. Applic.*, Lancaster, UK, Mar. 30–Apr. 1, 1999, 14–21.

16. M. Zimmermann and K. Dostert, A multipath signal propagation model for the power line channel in high frequency range, in *Proc. Int. Symp. Power Line Commun. Applic.*, Lancaster, UK, Mar. 30–Apr. 1, 1999, 45–51.

17. ——, A multipath model for the power line channel, *IEEE Trans. Commun.*, 50(4), 553–559, Apr. 2002.

18. K. Dostert, Propagation channel characterization and modeling: Outdoor power supply grids as communication channels, in *Proc. IEEE Int. Symp. Power Line Commun. Applic.*, Vancouver, Canada, Apr. 6–8, 2005, Keynote Talk.

19. A. M. Tonello, F. Versolatto, and A. Pittolo, In-home power line communication channel: Statistical characterization, *IEEE Trans. Commun.*, 62(6), 2096–2106, Jun. 2014.

20. J. S. Barnes, A physical multi-path model for power distribution network propagation, in *Proc. Int. Symp. Power Line Commun. Applic.*, Tokyo, Japan, Mar. 24–26, 1998, 76–89.

21. T. C. Banwell and S. Galli, A new approach to the modeling of the transfer function of the power line channel, in *Proc. Int. Symp. Power Line Commun. Applic.*, Malmö, Sweden, Apr. 4–6, 2001, 319–324.

22. T. Sartenaer and P. Delogne, Powerline cables modelling for broadband communications, in *Proc. Int. Symp. Power Line Commun. Applic.*, Malmö, Sweden, Apr. 4–6, 2001, 331–337.

23. T. Calliacoudas and F. Issa, Multiconductor transmission lines and cables solver, an efficient simulation tool for PLC networks development, in *Proc. Int. Symp. Power Line Commun. Applic.*, Athens, Greece, Mar. 27–29, 2002.

24. T. Esmailian, F. R. Kschischang, and P. G. Gulak, An in-building power line channel simulator, in *Proc. Int. Symp. Power Line Commun. Applic.*, Athens, Greece, Mar. 27–29, 2002.

25. ——, In-building power lines as high-speed communication channels: Channel characterization and a test-channel ensemble, *International Journal of Communications Systems*, 16(5), 381–400, Jun. 2003, Special Issue: Powerline Communications and Applications.

26. T. C. Banwell and S. Galli, A novel approach to accurate modeling of the indoor power line channel – Part I: Circuit analysis and companion model, *IEEE Trans. Power Delivery*, 20(2), 655–663, Apr. 2005.

27. S. Galli and T. Banwell, A novel approach to the modeling of the indoor power line channel – Part II: transfer function and its properties, *IEEE Trans. Power Delivery*, 20(3), 1869–1878, Jul. 2005.

28. S. Galli and T. C. Banwell, A deterministic frequency-domain model for the indoor power line transfer function, *IEEE J. Sel. Areas Commun.*, 24(7), 1304–1316, Jul. 2006.

29. R. Hashmat, P. Pagani, A. Zeddam, and T. Chonavel, MIMO communications for inhome PLC networks: Measurements and results up to 100 MHz, in *Proc. IEEE Int. Symp. Power Line Commun. Applic.*, Rio de Janeiro, Brazil, Mar. 28–31, 2010, 120–124.

30. A. M. Tonello, F. Versolatto, B. Béjar, and S. Zazo, A fitting algorithm for random modeling the PLC channel, *IEEE Trans. Power Delivery*, 27(3), 1477–1484, Jul. 2012.

31. A. M. Tonello and F. Versolatto, Bottom-up statistical PLC channel modeling – Part II: Inferring the statistics, *IEEE Trans. Power Delivery*, 25(4), 2356–2363, Oct. 2010.

32. ———, Bottom-up statistical PLC channel modeling – Part I: Random topology model and efficient transfer function computation, *IEEE Trans. Power Delivery*, 26(2), 891–898, Apr. 2011.

33. T. C. Banwell and S. Galli, On the symmetry of the power line channel, in *Proc. Int. Symp. Power Line Commun. Applic.*, Malmö, Sweden, Apr. 4–6, 2001, 325–330.

34. S. D'Alessandro and A. M. Tonello, On rate improvements and power saving with opportunistic relaying in home power line networks, *EURASIP J. Advances Signal Process.*, Sep. 2012, 1–17.

35. A. Pittolo and A. M. Tonello, Physical layer security in PLC networks: An emerging scenario, other than wireless, *IET Commun.*, 8(8), 1239–1247, 2014.

36. M. Schwartz, The origins of carrier multiplexing: Major George Owen Squier and AT&T, *IEEE Commun. Mag.*, 46(5), 20–24, May 2008.

37. IEEE P1901 draft standard for broadband over power line networks: Medium access control and physical layer specifications, IEEE Standards Association, Draft Standard, Informative Annex, Chapter 3: Theoretical/Mathematical Channel Models for BPL Systems.

38. H. Li, D. Liu, J. Li, and P. Stoica, Channel order and RMS delay spread estimation for AC power line communications, *Digital Signal Processing*, 13(2), 284–300, Apr. 2003.

39. Y.-H. Kim, H.-H. Song, J.-H. Lee, and S.-C. Kim, Wideband channel measurements and modeling for in-house power line communication, in *Proc. Int. Symp. Power Line Commun. Applic.*, Athens, Greece, Mar. 27–29, 2002.

40. K. H. Afkhamie, H. Latchman, L. Yonge, T. Davidson, and R. E. Newman, Joint optimization of transmit pulse shaping, guard interval length, and receiver side narrow-band interference mitigation in the HomePlugAV OFDM system, in *Proceedings IEEE Workshop on Signal Processing and Advanced Wireless Communications*, New York City, USA, Jun. 5–8, 2005, 996–1000.

41. Powerline channel data, Contribution to ITU-T SG15Q4 Working Group, Geneva, Switzerland, Standard Document B07-05-15 (NIPP-NAI-2007-107R1), Jun. 2007.

42. B. O'Mahony, Field testing of high-speed power line communications in North American homes, in *Proc. IEEE Int. Symp. Power Line Commun. Applic.*, Orlando, USA, Mar. 26–29, 2006, 155–159.

43. S. Galli, A simplified model for the indoor power line channel, in *Proc. IEEE Int. Symp. Power Line Commun. Applic.*, Dresden, Germany, Mar. 29–Apr. 1, 2009, 13–19.

44. A. Schwager, L. Stadelmeier, and M. Zumkeller, Potential of broadband power line home networking, in *Proc. IEEE Consum. Commun. Netw. Conf.*, Las Vegas, USA, Jan. 3–6, 2005, 359–363.

45. F. M. Tesche, B. A. Renz, R. M. Hayes, and R. G. Olsen, Development and use of a multiconductor line model for PLC assessments, in *Proc. Int. Zurich Symp. Electromagn. Compat.*, Zurich, Switzerland, Feb. 18–20, 2003, 99–104.

46. J.-J. Lee, S.-J. Choi, H.-M. Oh, W.-T. Lee, K.-H. Kim, and D.-Y. Lee, Measurements of the communications environment in medium voltage power distribution lines for wide-band power line communications, in *Proc. Int. Symp. Power Line Commun. Applic.*, Zaragosa, Spain, Mar. 31–Apr. 2, 2004, 69–74.

47. P. Amirshahi and M. Kavehrad, High-frequency characteristics of overhead multiconductor power line for broadband communications, *IEEE J. Sel. Areas Commun.*, 24(7), 1292–1303, Jul. 2006.

48. F. Versolatto, A. M. Tonello, C. Tornelli, and D. Della Giustina, Statistical analysis of broadband underground medium voltage channels for PLC applications, in *IEEE Int. Conf. Smart Grid Commun.*, Venice, Italy, Nov. 3–6, 2014, 493–498.

49. M. D'Amore and M. S. Sarto, A new formulation of lossy ground return parameters for transient analysis of multiconductor dissipative lines, *IEEE Trans. Power Delivery*, 12(1), 303–314, Jan. 1997.

50. J. R. Carson, Wave propagation in overhead wires with ground return, *Bell Syst. Tech. J.*, 5(4), 539–554, Oct. 1926.

51. H. Kikuchi, Wave propagation along an infinite wire above ground at high frequencies, *Electrotech. J.*, 2, 73–78, Dec. 1956.

52. T. Sartenaer and P. Delogne, Deterministic modeling of the (shielded) outdoor power line channel based on the multiconductor transmission line equations, *IEEE J. Sel. Areas Commun.*, 24(7), 1277–1291, Jul. 2006.

53. Signalling on low voltage electrical installations in the frequency range 3 kHz to 148.5 kHz, European Committee for Electrotechnical Standardization (CENELEC), Brussels, Belgium, Standard EN 50065-1, 1991.

54. H. Meng, S. Chen, Y. L. Guan, C. L. Law, P. L. So, E. Gunawan, and T. T. Lie, Modeling of transfer characteristics for the broadband power line communication channel, *IEEE Trans. Power Delivery*, 19(3), 1057–1064, Jul. 2004.

55. H. He, S. Cheng, Y. Zhang, and J. Nguimbis, Analysis of reflection of signal transmitted in low-voltage powerline with complex wavelet, *IEEE Trans. Power Delivery*, 19(1), 86–91, Jan. 2004.

56. A. M. Tonello, Wide band impulse modulation and receiver algorithms for multiuser power line communications, *EURASIP J. Adv. Signal Process.*, 2007, article ID 96747.

57. O. G. Hooijen, On the relation between network-topology and power line signal attenuation, in *Proc. Int. Symp. Power Line Commun. Applic.*, Tokyo, Japan, Mar. 24–26, 1998, 45–56.

58. D. Anastasiadou and T. Antonakopoulos, Multipath characterization of indoor power-line networks, *IEEE Trans. Power Delivery*, 20(1), 90–99, Jan. 2005.

59. C. R. Paul, *Analysis of Multiconductor Transmission Lines*. John Wiley & Sons, Chichester, 1994.

60. ——, Decoupling the multiconductor transmission line equations, *IEEE Trans. Microw. Theory Tech.*, 44(8), 1429–1440, Aug. 1996.

61. ——, A SPICE model for multiconductor transmission lines excited by an incident electromagnetic field, *IEEE Trans. Electromagn. Compat.*, 36(4), 342–354, Nov. 1994.

62. S. Galli, T. Banwell, and D. Waring, Power line based LAN on board the NASA space shuttle, in *Proc. IEEE Veh. Technol. Conf.*, 2, Milan, Italy, May 17–19, 2004, 970–974.

63. T. Huck, J. Schirmer, T. Hogenmuller, and K. Dostert, Tutorial about the implementation of a vehicular high speed communication system, in *Proc. IEEE Int. Symp. Power Line Commun. Applic.*, Vancouver, Canada, Apr. 6–8, 2005, 162–166.

64. A. M. Tonello and F. Versolatto, New results on top-down and bottom-up statistical PLC channel modeling, in *Proc. Workshop Power Line Commun.*, Udine, Italy, Oct. 1–2, 2009.

65. M. Tlich, A. Zeddam, F. Moulin, and F. Gauthier, Indoor power-line communications channel characterization up to 100 MHz – Part I: One-parameter deterministic model, *IEEE Trans. Power Delivery*, 23(3), 1392–1401, Jul. 2008.

66. B. Glance and L. J. Greenstein, Frequency-selective fading effects in digital mobile radio with diversity combining, *IEEE Trans. Commun.*, 31(9), 1085–1094, Sep. 1983.

67. K. Dostert and S. Galli, Keynote II: Modelling of electrical power supply systems as communication channels, in *Proc. IEEE Int. Symp. Power Line Commun. Applic.*, Vancouver, Canada, Apr. 6–8, 2005, 137.

68. M. Götz, M. Rapp, and K. Dostert, Power line channel characteristics and their effect on communication system design, *IEEE Commun. Mag.*, 42(4), 78–86, Apr. 2004.

69. M. Arzberger, K. Dostert, T. Waldeck, and M. Zimmermann, Fundamental properties of the low voltage power distribution grid, in *Proc. Int. Symp. Power Line Commun. Applic.*, Essen, Germany, Apr. 2–4, 1997, 45–50.

70. O. G. Hooijen, *Aspects of Residential Power Line Communications*. Aachen, Germany: Shaker Verlag, 1998.

71. M. Arzberger, *Datenkommunikation auf elektrischen Verteilnetzen für erweiterte Energiedienstleistungen*. Logos Verlag Berlin, 1998.

72. T. Waldeck, Einzel- und Mehrträgerverfahren für die störresistente Kommunikation auf Energieverteilnetzen (in German), Ph.D. dissertation, Logos Verlag, Berlin, Germany, 2000.

73. M. Babic, M. Hagenau, K. Dostert, and J. Bausch, Theoretical postulation of PLC channel models, the OPERA IST Integrated Project, Technical Report, 2005.

74. S. Galli, A novel approach to the statistical modeling of wireline channels, *IEEE Trans. Commun.*, 59(5), 1332–1345, Mar. 2011.

75. V. Degardin, M. Lienard, and P. Degauque, Transmission on indoor power lines: from a stochastic channel model to the optimization and performance evaluation of multicarrier systems, *Int. J. Commun. Syst.*, 16(5), 363–379, Jun. 2003, Special Issue: Powerline Communications and Applications.

76. I. C. Papaleonidopoulos, C. N. Capsalis, C. G. Karagiannopoulos, and N. J. Theodorou, Statistical analysis and simulation of indoor single-phase low voltage power-line communication channels on the basis of multipath propagation, *IEEE Trans. Consumer Electron.*, 49(1), 89–99, Feb. 2003.

77. F. J. Cañete, J. A. Cortés, L. Díez, and J. T. Entrambasaguas, A channel model proposal for indoor power line communications, *IEEE Commun. Mag.*, 49(12), 166–174, Dec. 2011.

78. H. Philipps, Performance measurements of power line channels at high frequencies, in *Proc. Int. Symp. Power Line Commun. Applic.*, Tokyo, Japan, Mar. 24–26, 1998, 229–237.

79. D. Liu, E. Flint, B. Gaucher, and Y. Kwark, Wide band AC power line characterization, *IEEE Trans. Consumer Electron.*, 45(4), 1087–1097, Nov. 1999.

80. P. A. Rizzi, *Microwave Engineering: Passive Circuits*. Prentice Hall, 1988.

81. F. J. Cañete, Caracterización y modelado de redes eléctricas interiores como medio de transmisión de banda ancha (in Spanish), Ph.D. dissertation, Universidad de Málaga, Málaga, Spain, 2004.

82. J.-J. Werner, The HDSL environment, *IEEE J. Sel. Areas Commun.*, 9(6), 785–800, Aug. 1991.

83. J. A. Cortés, F. J. Cañete, L. Díez, and J. T. Entrambasaguas, Characterization of the cyclic short-time variation of indoor power-line channels response, in *Proc. IEEE Int. Symp. Power Line Commun. Applic.*, Vancouver, Canada, Apr. 6–8, 2005, 326–330.

84. F. J. Cañete, L. Díez, J. A. Cortés, J. J. Sánchez-Martínez, and L. M. Torres, Time-varying channel emulator for indoor power line communications, in *Proc. IEEE Global Telecom. Conf.*, New Orleans, USA, Nov. 30–Dec. 4, 2008, 1–5.

85. S. Sancha, F. J. Cañete, L. Díez, and J. T. Entrambasaguas, A channel simulator for indoor power-line communications, in *Proc. IEEE Int. Symp. Power Line Commun. Applic.*, Pisa, Italy, Mar. 26–28, 2007, 104–109.

86. F. J. Cañete, Keynote II. Power line channels: frequency and time selective – Part 1: Response of indoor PLC channels, in *Proc. IEEE Int. Symp. Power Line Commun. Applic.*, Pisa, Italy, Mar. 26–28, 2007, 13–14.

87. F. J. Cañete, J. A. Cortés, and L. Díez, Indoor PLC channel generator. Downloadable Matlab program, Communication Engineering Department, University of Málaga. [Online]. Available: www.plc.uma.es

88. F. Versolatto and A. M. Tonello, A MIMO PLC random channel generator and capacity analysis, in *Proc. IEEE Int. Symp. Power Line Commun. Applic.*, Udine, Italy, Apr. 3–6, 2011, 66–71.

89. ——, An MTL theory approach for the simulation of MIMO power-line communication channels, *IEEE Trans. Power Delivery*, 26(3), 1710–1717, Jul. 2011.

90. A. M. Tonello, Indoor PLC channel generator. Downloadable Matlab program. [Online]. Available: http://www.andreatonello.com

91. S. Galli, A simple two-tap statistical model for the power line channel, in *Proc. IEEE Int. Symp. Power Line Commun. Applic.*, Rio de Janeiro, Brazil, Mar. 28–31, 2010, 242–248.

92. J. A. Cortés, F. J. Cañete, L. Díez, and J. L. G. Moreno, On the statistical properties of indoor power line channels: Measurements and models, in *Proc. IEEE Int. Symp. Power Line Commun. Applic.*, Udine, Italy, Apr. 3–6, 2011, 271–276.

93. J. A. Cortés, F. J. Cañete, L. Díez, and L. M. Torres, On PLC channel models: an OFDM-based comparison, in *Proc. IEEE Int. Symp. Power Line Commun. Applic.*, Johannesburg, South Africa, Mar. 24–27, 2013, 333–338.

94. IEEE guide for the functional specification of medium voltage (1 kV–35 kV) electronic series devices for compensation of voltage fluctuations, IEEE WGI1 Power Electronics Equipment Working Group, IEEE Standard 1585, 2002.

95. IEEE guide for the functional specification of medium voltage (1 kV–35 kV) electronic shunt devices for dynamic voltage compensation, IEEE WGI1 Power Electronics Equipment Working Group, IEEE Standard 1623, 2004.

96. Electrical Installation Guide. (2014) Connection to the utility network—electrical installation guide. [Online]. Available: http://www.electrical-installation.org/enw/index.php?title=Connection_to_the_utility_network&oldid=17294

97. S. Robson, A. Haddad, and H. Griffiths, Simulation of power line communication using ATP-EMTP and MATLAB, in *Proc. IEEE PES Innovative Smart Grid Technol. Conf. Europe*, Gothenburg, Sweden, Oct. 11–13, 2010, 1–8.

98. N. Nasiriani, R. Ramachandran, K. Rahimi, Y. P. Fallah, P. Famouri, S. Bossart, and K. Dodrill, An embedded communication network simulator for power systems simulations in PSCAD, in *Proc. IEEE Power Energy Soc. General Meeting*, Vancouver, Canada, Jul. 21–25, 2013, 1–5.

99. M. Wei and Z. Chen, Communication systems and study method for active distribution power systems, in *Proc. Nordic Distribution and Asset Manage. Conf.*, Aalborg, Denmark, Sep. 6–7, 2010, 1–11.

100. A. Cataliotti, D. Di Cara, R. Fiorelli, and G. Tine, Power-line communication in medium-voltage system: simulation model and onfield experimental tests, *IEEE Trans. Power Delivery*, 27(1), 62–69, Jan. 2012.

101. J. Anatory, N. Theethayi, R. Thottappillil, M. Kissaka, and N. Mvungi, The effects of load impedance, line length, and branches in typical low-voltage channels of the BPLC systems of developing countries: Transmission-line analyses, *IEEE Trans. Power Delivery*, 24(2), 621–629, Apr. 2009.

102. F. Gianaroli, F. Pancaldi, and G. M. Vitetta, The impact of load characterization on the average properties of statistical models for powerline channels, *IEEE Trans. Smart Grid*, 4(2), 677–685, Jun. 2013.

103. X. Yang, Z. Tao, B. Zhang, F. Ye, J. Duan, and M. Shi, Research of impedance characteristics for medium-voltage power networks, *IEEE Trans. Power Delivery*, 22(2), 870–878, Apr. 2007.

104. M. Raugi, T. Zheng, M. Tucci, and S. Barmada, On the time invariance of PLC channels in complex power networks, in *Proc. IEEE Int. Symp. Power Line Commun. Applic.*, Rio de Janeiro, Brazil, Mar. 28–31, 2010, 56–61.

105. J. R. Wait, Theory of wave propagation along a thin wire parallel to an interface, *Radio Sci.*, 7(6), 675–679, Jun. 1972.

106. M. D'Amore and M. S. Sarto, Simulation models of a dissipative transmission line above lossy ground for a wide-frequency range – Part I: Single conductor configuration, *IEEE Trans. Electromagn. Compat.*, 7(6), 127–138, May 1996.

107. A. G. Lazaropoulos and P. G. Cottis, Broadband transmission via underground medium-voltage power lines—part I: Transmission characteristics, *IEEE Trans. Power Delivery*, 25(4), 2414–2424, Oct. 2010.

108. A. Cataliotti, V. Cosentino, D. Di Cara, and G. Tine, Measurement issues for the characterization of medium voltage grids communications, *IEEE Trans. Instrum. Meas.*, 62(8), 2185–2196, Aug. 2013.

109. L. T. Berger, A. Schwager, P. Pagani, and D. M. Schneider, MIMO power line communications, *IEEE Commun. Surveys Tutorials*, 17(1), 106–124, First Quarter 2015.

110. A. Schwager, W. Bäschlin, H. Hirsch, P. Pagani, N. Weling, J. L. G. Moreno, and H. Milleret, European MIMO PLT field measurement: Overview of the ETSI STF410 campaign & EMI analysis, in *Proc. IEEE Int. Symp. Power Line Commun. Applic.*, Beijing, China, Mar. 27–30, 2012, 298–303.

111. M. Nassar, K. Gulati, Y. Mortazavi, and B. L. Evans, Statistical modeling of asynchronous impulsive noise in powerline communication networks, in *Proc. IEEE Global Telecom. Conf.*, Houston, USA, Dec. 5–9, 2011, 1–6.

112. M. Zimmermann and K. Dostert, Analysis and modeling of impulsive noise in broad-band powerline communications, *IEEE Trans. Electromagn. Compat.*, 44(1), 249–258, Feb. 2002.

113. R. Pighi, Evoluzione dei sistemi PLC nelle reti elettriche ad alta tensione (in Italian), in *Proc. AEIT National Conf.*, Milan, Italy, Jun. 27–29, 2011.

114. L. L. Grigsby, *Electric power generation, transmission, and distribution*, 3rd ed. CRC Press, 2012.

115. R. Pighi, M. Franceschini, G. Ferrari, and R. Raheli, Fundamental performance limits of communications systems impaired by impulse noise, *IEEE Trans. Commun.*, 57(1), 171–182, Jan. 2009.

116. D. Middleton, Statistical-physical models of electro-magnetic interference, *IEEE Trans. Electromagn. Compat.*, 19(3), 106–127, Aug. 1977.

117. ——, Canonical non-Gaussian noise models: Their implications for measurement and for prediction of receiver performance, *IEEE Trans. Electromagn. Compat.*, 21(3), 209–220, Aug. 1979.

118. M. Ghosh, Analysis of the effect of impulse noise on multicarrier and single carrier QAM systems, *IEEE Trans. Commun.*, 44(2), 145–147, Feb. 1996.

119. R. Pighi and R. Raheli, Linear predictive detection for power line communications impaired by colored noise, in *Proc. IEEE Int. Symp. Power Line Commun. Applic.*, Orlando, USA, Mar. 27–29, 2006, 337–342.

120. R. Pighi, An information rate analysis of power line communications impaired by colored noise, in *Proc. IEEE Int. Symp. Power Line Commun. Applic.*, Udine, Italy, Apr. 3–6, 2011, 434–439.

121. P. S. Maruvada, *Corona Performance on High-Voltage Transmission Lines*. Research Studies Press Ltd., 2000.

122. N. Suljanović, A. Mujčić, M. Zajc, and J. F. Tasič, Corona noise characteristics in high voltage PLC channel, in *Proc. IEEE Int. Conf. on Ind. Tech.*, 2, Maribor, Slovenia, Dec. 10–12, 2003, 1036–1039.

123. ——, Computation of high-frequency and time characteristics of corona noise on HV power line, *IEEE Trans. Power Delivery*, 20(1), 71–79, Jan. 2005.

124. P. Burrascano, S. Cristina, and M. D'Amore, Performance evaluation of digital signal transmission channels on coronating power lines, in *Proc. IEEE Int. Symp. Circuits Syst.*, Espoo, Finland, Jun. 7–9, 1988, pp. 365–368.

125. ——, Digital generator of corona noise on power line carrier channels, *IEEE Trans. Power Delivery*, 3(3), 850–856, Jul. 1988.

126. A. Mujčić, N. Suljanović, M. Zajc, and J. F. Tasič, Power line noise model appropriate for investigation of channel coding methods, in *IEEE EUROCON*, Ljubljana, Slovenia, Sep. 22–24, 2003, 299–303.

127. S. Cristina and M. D'Amore, Analytical method for calculating corona noise on HVAC power line carrier communications channels, *IEEE Trans. Power App. Syst.*, 104(5), 1017–1024, May 1985.

128. P. Burrascano, S. Cristina, and M. D'Amore, Digital generator of corona noise on power line carrier channels, in *IEEE Trans. Power Delivery*, 3(3), 850–856, Jul. 1988.

129. S. Haykin, *Adaptive Filter Theory*, 4th ed. Prentice-Hall International Editions, New York: 2001.

130. J. P. Burg, Maximum entropy spectral analysis, in *Proc. Meeting Soc. Explor. Geophysicists*, Oklahoma City, USA, 1967, 34–41.

131. R. Pighi and R. Raheli, Linear predictive detection for power line communications impaired by colored noise, *EURASIP J. Advances Signal Process.*, 2007, pp. 1–12, Jun. 2007.

132. L. Yonge, J. Abad, K. Afkhamie, L. Guerrieri, S. Katar, H. Lioe, P. Pagani, R. Riva, D. M. Schneider, and A. Schwager, An overview of the HomePlug AV2 technology, *J. Electric. Comp. Eng.*, 2013, 1–20, 2013.

133. G. J. Foschini and M. J. Gans, On limits of wireless communications in a fading environment when using multiple antennas, *Wireless Personal Commun.*, 6(3), 311–335, Mar. 1998.

134. Wireless LAN medium access control (MAC) and physical layer (PHY) specifications amendment 5: Enhancements for higher throughput, IEEE WG802.11 Wireless LAN Working Group, IEEE Standard 802.11n-2009, 2009.

135. LTE advanced, European Telecommunications Standards Institute, 3GPP, 2012.

136. Powerline telecommunications (PLT); MIMO PLT; Part 1: Measurement methods of MIMO PLT, European Telecommunications Standards Institute, Tech. Rep. ETSI TR 101 562-1, 2012.

137. NFPA 70: National electrical code, National Fire Protection Association, Standard, 2009.

138. IEC60364-1 low-voltage electrical installations – Part 1: Fundamental principles, assessment of general characteristics, definitions, International Electrotechnical Commission, Standard, 2005.

139. C. L. Giovaneli, B. Honary, and P. G. Farrell, Space-frequency coded OFDM system for multi-wire power line communications, in *Proc. IEEE Int. Symp. Power Line Commun. Applic.*, Vancouver, Canada, Apr. 6–8, 2005, 191–195.

140. A. Schwager, D. Schneider, W. Bäschlin, A. Dilly, and J. Speidel, MIMO PLC: Theory, measurements and system setup, in *Proc. IEEE Int. Symp. Power Line Commun. Applic.*, Udine, Italy, Apr. 3–6, 2011, 48–53.

141. R. Hashmat, P. Pagani, A. Zeddam, and T. Chonavel, A channel model for multiple input multiple output in-home power line networks, in *Proc. IEEE Int. Symp. Power Line Commun. Applic.*, Udine, Italy, Apr. 3–6, 2011, 35–41.

142. D. Veronesi, R. Riva, P. Bisaglia, F. Osnato, K. Afkhamie, A. Nayagam, D. Rende, and L. Yonge, Characterization of in-home MIMO power line channels, in *Proc. IEEE Int. Symp. Power Line Commun. Applic.*, Udine, Italy, Apr. 3–6, 2011, 42–47.

143. D. Schneider, A. Schwager, W. Bäschlin, and P. Pagani, European MIMO PLC field measurements: Channel analysis, in *Proc. IEEE Int. Symp. Power Line Commun. Applic.*, Beijing, China, Mar. 27–30, 2012, 304–309.

144. Powerline telecommunications (PLT); MIMO PLT; Part 3: Setup and statistical results of MIMO PLT channel and noise measurements, European Telecommunications Standards Institute, Tech. Rep. ETSI TR 101 562-3, 2012.

145. A. Tomasoni, R. Riva, and S. Bellini, Spatial correlation analysis and model for in-home MIMO power line channels, in *Proc. IEEE Int. Symp. Power Line Commun. Applic.*, Beijing, China, Mar. 27–30, 2012, 286–291.

146. D. Rende, A. Nayagam, K. Afkhamie, L. Yonge, R. Riva, D. Veronesi, F. Osnato, and P. Bisaglia, Noise correlation and its effect on capacity of in-home MIMO power line channels, in *Proc. IEEE Int. Symp. Power Line Commun. Applic.*, Udine, Italy, Apr. 3–6, 2011, 60–65.

147. A. Paulraj, R. Nabar, and D. Gore, *Introduction to Space-Time Wireless Communications*. Cambridge University Press, 2003.

148. R. P. Clayton, *Analysis of Multiconductor Transmission Lines*. John Wiley & Sons, Chichester, 1994.

149. J. Anatory, N. Theethayi, and R. Thottappillil, Power-line communication channel model for interconnected networks – Part II: Multiconductor system, *IEEE Trans. Power Delivery*, 24(1), 124–128, Jan. 2009.

150. ——, Power-line communication channel model for interconnected networks – Part I: Two-conductor system, *IEEE Trans. Power Delivery*, 24(1), 118–123, Jan. 2009.

151. A. M. Tonello and T. Zheng, Bottom-up transfer function generator for broadband PLC statistical channel modeling, in *Proc. IEEE Int. Symp. Power Line Commun. Applic.*, Dresden, Germany, Mar. 29–Apr. 1, 2009, 7–12.

152. J. C. Clements, P. R. Clayton, and A. T. Adams, Computation of the capacitance matrix for systems of dielectric-coated cylindrical conductors, *IEEE Trans. Electromagn. Compat.*, 17(4), 238–248, Nov. 1975.

153. F. Versolatto and A. M. Tonello, PLC channel characterization up to 300 MHz: Frequency response and line impedance, in *Proc. IEEE Global Telecom. Conf.*, Anaheim, USA, Dec. 3–7, 2012, 3525–3530.

154. Powerline telecommunications (PLT); MIMO PLT; Part 2: Measurement methods and statistical results of MIMO PLT EMI, European Telecommunications Standards Institute, Tech. Rep. ETSI TR 101 562-2, 2012.

155. R. Hashmat, P. Pagani, T. Chonavel, and A. Zeddam, Analysis and modeling of background noise for inhome MIMO PLC channels, in *Proc. IEEE Int. Symp. Power Line Commun. Applic.*, Beijing, China, Mar. 27–30, 2012, 316–321.

156. P. Pagani, R. Hashmat, A. Schwager, D. Schneider, and W. Bäschlin, European MIMO PLC field measurements: Noise analysis, in *Proc. IEEE Int. Symp. Power Line Commun. Applic.*, Beijing, China, Mar. 27–30, 2012, 310–315.

157. M. Nassar, J. Lin, Y. Mortazavi, A. Dabak, I. H. Kim, and B. L. Evans, Local utility power line communications in the 3–500 kHz band: Channel impairments, noise, and standards, *IEEE Signal Process. Mag.*, 29(5), 116–127, Sep. 2012.

158. F. J. Cañete, J. A. Cortés, L. Díez, J. T. Entrambasaguas, and J. L. Carmona, Fundamentals of the cyclic short-time variation of indoor power-line channels, in *Proc. IEEE Int. Symp. Power Line Commun. Applic.*, Vancouver, Canada, Apr. 6–8, 2005, 157–161.

159. M. Katayama, T. Yamazato, and H. Okada, A mathematical model of noise in narrowband power line communication systems, *IEEE J. Sel. Areas Commun.*, 24(7), 1267–1276, Jul. 2006.

160. G. Marubayashi, Noise measurements of the residential power line, in *Proc. Int. Symp. Power Line Commun. Applic.*, Essen, Germany, Apr. 2–4, 1997, 104–108.

161. A. Kawaguchi, H. Okada, T. Yamazato, and M. Katayama, Correlations of noise waveforms at different outlets of a power line network (in Japanese), *IEICE Trans. Fund. Electr. Commun. Comput. Sci.*, J90-A(11), 851–860, Nov. 2007.

162. A. V. Oppenheim and R. W. Schafer, *Discrete-Time Signal Processing*, 3rd ed. Prentice Hall, 2009.

163. M. Nassar, A. Dabak, I. H. Kim, T. Pande, and B. L. Evans, Cyclostationary noise modeling in narrowband powerline communication for smart grid applications, in *Proc. IEEE Int. Conf. Acoustics, Speech and Sig. Proc.*, Kyoto, Japan, Mar. 25–30, 2012, 3089–3092.

164. W. A. Gardner and L. Franks, Characterization of cyclostationary random signal processes, *IEEE Trans. Inf. Theory*, 21(1), 4–14, Jan. 1975.

165. W. Gardner, *Cyclostationarity in Communications and Signal Processing*. IEEE, New York, 1994.

166. K. F. Nieman, J. Lin, M. Nassar, K. Waheed, and B. L. Evans, Cyclic spectral analysis of power line noise in the 3–200 kHz band, in *Proc. IEEE Int. Symp. Power Line Commun. Applic.*, Johannesburg, South Africa, Mar. 24–27, 2013, 315–320.

167. D. Middleton, Canonical and quasi-canonical probability models of Class A interference, *IEEE Trans. Electromagn. Compat.*, 25(2), 76–106, May 1983.

168. S. Miyamoto, M. Katayama, and N. Morinaga, Performance analysis of QAM systems under Class A impulsive noise environment, *IEEE Trans. Electromagn. Compat.*, 37(2), 260–267, May 1995.

169. N. Gonzalez-Prelcic, C. Mosquera, N. Degara, and A. Currais, A channel model for the Galician low voltage mains network, in *Proc. Int. Symp. Power Line Commun. Applic.*, Malmö, Sweden, Apr. 4–6, 2001, 365–370.

170. J. Lin, M. Nassar, and B. L. Evans, Impulsive noise mitigation in powerline communications using sparse Bayesian learning, *IEEE J. Sel. Areas Commun.*, 31(7), 1172–1183, Jul. 2013.

171. M. H. L. Chan and R. W. Donaldson, Amplitude, width, and interarrival distributions for noise impulses on intrabuilding power line communication networks, *IEEE Trans. Electromagn. Compat.*, 31(3), 320–323, Aug. 1989.

172. L. T. Tang, P. L. So, E. Gunawan, S. Chen, T. T. Lie, and Y. L. Guan, Characterization of power distribution lines for high-speed data transmission, in *Proc. Int. Conf. on Power System Technology*, 1, Perth, Australia, Dec. 4–7, 2000, 445–450.

173. V. Degardin, M. Lienard, A. Zeddam, F. Gauthier, and P. Degauque, Classification and characterization of impulsive noise on indoor powerline used for data communications, *IEEE Trans. Consumer Electron.*, 48(4), 913–918, Nov. 2002.

174. J. A. Cortés, L. Díez, J. J. Cañete, and J. J. Sánchez-Martínez, Analysis of the indoor broadband power-line noise scenario, *IEEE Trans. Electromagn. Compat.*, 52(4), 849–858, Nov. 2010.

175. D. Umehara, H. Yamaguchi, and Y. Morihiro, Turbo decoding in impulsive noise environment, in *Proc. IEEE Global Telecom. Conf.*, 1, Dallas, USA, Nov. 29–Dec. 3, 2004, 194–198.

176. L. Di Bert, P. Caldera, D. Schwingshackl, and A. M. Tonello, On noise modeling for power line communications, in *Proc. IEEE Int. Symp. Power Line Commun. Applic.*, Udine, Italy, Apr. 3–6, 2011, 283–288.

177. H. Meng, Y. L. Guan, and S. Chen, Modeling and analysis of noise effects on broadband power-line communications, *IEEE Trans. Power Delivery*, 20(2), 630–637, Apr. 2005.

178. A. Voglgsang, T. Langguth, G. Körner, H. Steckenbiller, and R. Knorr, Measurement, characterization and simulation of noise on powerline channels, in *Proc. Int. Symp. Power Line Commun. Applic.*, Limerick, Ireland, Apr. 5–7, 2000, 139–146.

179. D. W. Rieken, Periodic noise in very low frequency power-line communications, in *Proc. IEEE Int. Symp. Power Line Commun. Applic.*, Udine, Italy, Apr. 3–6, 2011, 295–300.

180. A. Dabak, B. Varadrajan, I. H. Kim, M. Nassar, and G. Gregg, Appendix for noise channel modeling for IEEE P1901.2, Standards Contribution IEEE P1901.2, Jun. 2011.

181. A. D. Spaulding and D. Middleton, Optimum reception in an impulsive interference environment – Part-I: Coherent detection, *IEEE Trans. Commun.*, 25(9), 910–923, Sep. 1977.

182. S. Miyamoto, M. Katayama, and N. Morinaga, Receiver design using the dependence between quadrature components of impulsive radio noise (in Japanese), *IEICE Trans. Commun.*, J77-BII(2), 63–73, Feb. 1994.

183. Y. Hirayama, H. Okada, T. Yamazato, and M. Katayama, An adaptive receiver for power-line communications with the estimation of instantaneous noise power, *IEICE Trans. Fund. Electr. Commun. Comput. Sci.*, E88-A(3), 755–760, Mar. 2005.

184. A. Kawaguchi, H. Okada, T. Yamazato, and M. Katayama, Correlations of noise waveforms at different outlets in a power-line network, in *Proc. IEEE Int. Symp. Power Line Commun. Applic.*, Orlando, USA, Mar. 26–29, 2006, 92–97.

185. IEEE 1901.2-2013 for low-frequency (less than 500 kHz) narrowband power line communications for smart grid applications, IEEE Standards Association, Active Standard IEEE 1901.2-2013, 2013. [Online]. Available: http://standards.ieee.org/findstds/standard/1901.2-2013.html

186. A. M. Tonello, S. D'Alessandro, and L. Lampe, Cyclic prefix design and allocation in bit-loaded OFDM over power line communication channels, *IEEE Trans. Commun.*, 58(11), 3265–3276, Nov. 2010.

187. A. M. Tonello, A. Pittolo, and M. Girotto, Power line communications: Understanding the channel for physical layer evolution based on filter bank modulation, *IEICE Trans. Commun.*, E97-B(8), 1494–1503, Aug. 2014.

188. F. Aalamifar, A. Schlögl, D. Harris, and L. Lampe, Modelling power line communication using network simulator-3, in *Proc. IEEE Global Telecom. Conf.*, Atlanta, USA, Dec. 9–13, 2013, 2969–2974, source code available at http://www.ece.ubc.ca/~lampe/PLC.

189. G. Marrocco, D. Statovci, and S. Trautmann, A PLC broadband channel simulator for indoor communications, in *Proc. IEEE Int. Symp. Power Line Commun. Applic.*, Johannesburg, South Africa, Mar. 24–27, 2013, 321–326, source code available at http://plc.ftw.at.

190. F. Gruber and L. Lampe, On PLC channel emulation via transmission line theory, in *Proc. IEEE Int. Symp. Power Line Commun. Applic.*, Austin, TX, USA, Mar. 29–31, 2015, source code available at http://www.ece.ubc.ca/lampe/MIMOPLC.

191. T. Kaipia, P. Salonen, J. Lassila, and J. Partanen, Possibilities of the low-voltage DC distribution systems, in *Proc. Nordic Distribution and Asset Manage. Conf.*, Stockholm, Sweden, Aug. 21–22, 2006, 1–10.

192. ——, Application of low voltage DC-distribution system – a techno-economical study, in *Int. Conf. on Electricity Distribution*, Vienna, Austria, May 21–24, 2007, 1–4.

193. Low voltage directive LVD 73/23/EEC, European Commission, Brussels, Belgium, European Commission Directive, 1973.

194. P. Nuutinen, P. Salonen, P. Peltoniemi, T. Kaipia, P. Silventoinen, and J. Partanen, Implementing a laboratory development platform for an LVDC distribution system, in *Proc. IEEE Int. Conf. Smart Grid Commun.*, Brussels, Belgium, Oct. 17–20, 2011, 84–89.

195. P. Nuutinen, P. Peltoniemi, and P. Silventoinen, Short-circuit protection in a converter-fed low-voltage distribution network, *IEEE Trans. Power Delivery*, 28(4), 1587–1597, Apr. 2013.

196. T. Kaipia, P. Nuutinen, A. Pinomaa, A. Lana, J. Partanen, J. Lohjala, and M. Matikainen, Field test environment for LVDC distribution – Implementation experiences, in *CIRED Workshop 2012*, Lisbon, Portugal, May 29–30, 2012, 1–4.

197. A. Pinomaa, J. Ahola, A. Kosonen, and P. Nuutinen, Noise analysis of a power-line communication channel in an LVDC smart grid concept, in *Proc. IEEE Int. Symp. Power Line Commun. Applic.*, Johannesburg, South Africa, Mar. 24–27, 2013, 41–46.

198. A. Pinomaa, J. Ahola, and A. Kosonen, PLC concept for LVDC distribution systems, *IEEE Commun. Mag.*, 49(12), 55–63, Dec. 2011.

199. P. Salonen, T. Kaipia, P. Nuutinen, P. Peltoniemi, and J. Partanen, An LVDC distribution system concept, in *Nordic Workshop Power Ind. Electron.*, Espoo, Finland, Jun. 9-11, 2008, 1–7.

200. A. Pinomaa, J. Ahola, and A. Kosonen, Power-line communication-based network architecture for low voltage direct current distribution system, in *Proc. IEEE Int. Symp. Power Line Commun. Applic.*, Udine, Italy, Apr. 3–6, 2011, 358–363.

201. A. Haidine, B. Adebisi, A. Treytl, H. Pille, B. Honary, and A. Portnoy, High-speed narrowband PLC in smart grid landscape state-of-the-art, in *Proc. IEEE Int. Symp. Power Line Commun. Applic.*, Udine, Italy, Apr. 3–6, 2011, 468–473.

202. M. K. Lee, R. E. Newman, H. A. Latchman, S. Katar, and L. Yonge, HomePlug 1.0 powerline communication LANs—protocol description and performance results, *Int. J. Commun. Syst.*, 16(5), 447–473, May 2003.

203. J. Ahola, Applicability of of power-line communication to data transfer of on-line condition monitoring of electrical drives, Ph.D. dissertation, Lappeenranta University of Technology, Lappeenranta, Finland, 2003.

204. M. O. Carrion, M. Lienard, and P. Degauque, Communication over vehicular DC lines: Propagation channel characteristics, in *Proc. IEEE Int. Symp. Power Line Commun. Applic.*, Orlando, USA, Mar. 27–29, 2006, 2–5.

205. M. Antoniali, M. De Piante, and A. M. Tonello, PLC noise and channel characterization in a compact electrical car, in *Proc. IEEE Int. Symp. Power Line Commun. Applic.*, Johannesburg, South Africa, Mar. 24–27, 2013, 29–34.

206. A. B. Vallejo-Mora, J. J. Sánchez-Martínez, F. J. Cañete, J. Cortés, and L. Díez, Characterization and evaluation of in-vehicle power line channels, in *Proc. IEEE Global Telecom. Conf.*, Miami, USA, Dec. 6–10, 2010, 1–5.

207. M. Mohammadi, L. Lampe, M. Lok, S. Mirabbasi, M. Mirvakili, R. Rosales, and P. van Veen, Measurement study and transmission for in-vehicle power line communication, in *Proc. IEEE Int. Symp. Power Line Commun. Applic.*, Dresden, Germany, Mar. 29–Apr. 1, 2009, 73–78.

208. V. Degardin, M. Lienard, P. Degauque, and P. Laly, Performances of the HomePlug PHY layer in the context of in-vehicle powerline communications, in *Proc. IEEE Int. Symp. Power Line Commun. Applic.*, Pisa, Italy, Mar. 26–28, 2007, 93–97.

209. M. Lienard, M. O. Carrion, V. Degardin, and P. Degauque, Modeling and analysis of in-vehicle power line communication channels, *IEEE Trans. Veh. Technol.*, 57(2), 670–679, Mar. 2008.

210. CISPR-25: 2002 - Limits and methods of measurement of radio disturbance characteristics for the protection of receivers used on board vehicles, International Electrotechnical Comission (IEC) and International Special Committee on Radio Interference (CISPR) 25, Draft standard, 2002.

211. J. A. Cortés, M. Cerdá, L. Díez, and F. J. Cañete, Analysis of the periodic noise on in-vehicle broadband power line channels, in *Proc. IEEE Int. Symp. Power Line Commun. Applic.*, Beijing, China, Mar. 27–30, 2012, 334–339.

212. A. Schiffer, Statistical channel and noise modeling of vehicular DC-lines for data communication, in *Proc. IEEE Veh. Technol. Conf.*, Tokyo, Japan, May 15–18, 2000, 158–162.

213. V. Degardin, M. Lienard, P. Degauque, E. Simon, and P. Laly, Impulsive noise characterization of in-vehicle power line, *IEEE Trans. Electromagn. Compat.*, 50(4), 861–868, Nov. 2008.

214. Y. Yabuuchi, D. Umehara, M. Morikura, T. Hisada, S. Ishiko, and S. Horihata, Measurement and analysis of impulsive noise on in-vehicle power lines, in *Proc. IEEE Int. Symp. Power Line Commun. Applic.*, Rio de Janeiro, Brazil, Mar. 28–31, 2010, 325–330.

215. E. Liu, Y. Gao, G. Samdani, O. Mukhtar, and T. Korhonen, Powerline communication over special systems, in *Proc. IEEE Int. Symp. Power Line Commun. Applic.*, Vancouver, Canada, Apr. 6–8, 2005, 167–171.

216. S. Tsuzuki, M. Yoshida, Y. Yamada, H. Kawasaki, K. Murai, K. Matsuyama, and M. Suzuki, Characteristics of power-line channels in cargo ships, in *Proc. IEEE Int. Symp. Power Line Commun. Applic.*, Pisa, Italy, Mar. 26–28, 2007, 324–329.

217. S. Tsuzuki, M. Yoshida, Y. Yamada, K. Murai, H. Kawasaki, K. Matsuyama, T. Shinpo, Y. Saito, and S. Takaoka, Channel characteristic comparison of armored shipboard cable and unarmored one, in *Proc. IEEE Int. Symp. Power Line Commun. Applic.*, Jeju Island, South Korea, Apr. 2–4, 2008, 7–12.

218. S. Tsuzuki, S. Tatsuno, M. Takechi, T. Okabe, H. Kawasaki, T. Shinpo, Y. Yamada, and S. Takaoka, An adaptive power control method to electromagnetic environment for PLC in cargo ships, in *Proc. IEEE Int. Symp. Power Line Commun. Applic.*, Dresden, Germany, Mar. 29–Apr. 1, 2009, 131–136.

219. J. Nishioka, S. Tsuzuki, M. Yoshida, H. Kawasaki, T. Shinpo, and Y. Yamada, Characteristics of 440V power-line channels in container ships, in *Proc. IEEE Int. Symp. Power Line Commun. Applic.*, Dresden, Germany, Mar. 29–Apr. 1, 2009, 217–222.

220. S. Barmada, L. Bellanti, M. Raugi, and M. Tucci, Analysis of power-line communication channels in ships, *IEEE Trans. Veh. Technol.*, 59(7), 3161–3170, Sep. 2010.

221. A. Akinnikawe and K. L. Butler-Purry, Investigation of broadband over power line channel capacity of shipboard power system cables for ship communication networks, in *Proc. IEEE Power & Energy Soc. General Meeting*, Calgary, Canada, Jul. 26–30, 2009, 1–9.

222. M. Antoniali, A. M. Tonello, M. Lenardon, and A. Qualizza, Measurements and analysis of PLC channels in a cruise ship, in *Proc. IEEE Int. Symp. Power Line Commun. Applic.*, Udine, Italy, Apr. 3–6, 2011, pp. 102–107.

223. T. Zheng, M. Raugi, and M. Tucci, Analysis of transmission properties of naval power line channels, in *Proc. IEEE Int. Symp. Ind. Electron.*, Bari, Italy, Jul. 4–7, 2010, 2955–2960.

224. S. Barmada, M. Raugi, M. Tucci, and T. Zheng, Analysis of time-varying properties of power line communication channels in ships, in *Proc. IEEE Int. Symp. Power Line Commun. Applic.*, Udine, Italy, Apr. 3–6, 2011, 72–77.

225. T. Zheng, M. Raugi, and M. Tucci, Time-invariant characteristics of naval power-line channels, *IEEE Trans. Power Delivery*, 27(2), 858–865, Apr. 2012.

226. M. Antoniali, M. Girotto, and A. M. Tonello, In-car power line communications: Advanced transmission techniques, *Int. J. Automotive Technol.*, 14(4), 625–632, Aug. 2013.

3

Electromagnetic Compatibility

H. Hirsch, M. Koch, N. Weling, and A. Zeddam

3.1 Introduction

Electromagnetic compatibility (EMC) is an important and non-trivial consideration for the development and operation of power line communication (PLC) systems. First, ingress of impulsive and narrowband disturbance into the power distribution network is a challenge for transmission of communication signals over power lines. Second, the imperfect symmetry of power lines and PLC installations leads to conversion of the desired symmetric (differential mode) signals into asymmetric (common mode) signals, which may cause interference to nearby operated radio receivers.

The disturbance models for a PLC device being a victim and a source of interferences are shown in Figures 3.1 and 3.2, respectively. In its role as a victim of interference, a PLC device is subjected to several man-made and natural electromagnetic phenomena. Devices like switched power supplies, photovoltaic converters, frequency converters, or switches connected to the power distribution grid mainly produce impulsive noise. Signals from radio transmitters appear as an asymmetric signal on the line, which is converted in part into a symmetric signal, which then may interfere with the desired PLC signal. From reciprocity arguments it follows that a symmetric PLC signal can be converted into an asymmetric signal and then be radiated from the power line, if a certain line length is exceeded (see Figure 3.2). This may lead to interference at nearby operated radio receivers. In addition to this effect, some radio receivers with a rod antenna and weak common mode decoupling at the mains connector may be disturbed directly by the asymmetric voltage.

Following these disturbance models, this chapter discusses the EMC challenges for PLC installations in some detail. First, Section 3.2 describes the parameters for characterizing the symmetry of a network, the relation between the forward power and the resulting field strength, and electric and magnetic field theory pertinent for radiation effects through PLC. Section 3.3 then explains ways how to measure electromagnetic emission from PLC installations, while Section 3.4 elaborates on electromagnetic susceptibility of PLC devices. Following these

Power Line Communications: Principles, Standards and Applications from Multimedia to Smart Grid, Second Edition. Edited by Lutz Lampe, Andrea M. Tonello, and Theo G. Swart.
© 2016 John Wiley & Sons, Ltd. Published 2016 by John Wiley & Sons, Ltd.

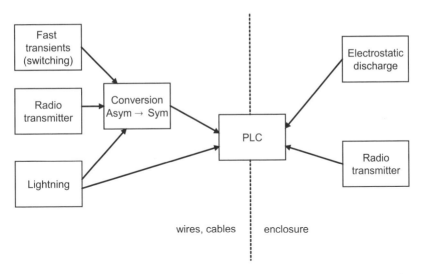

Figure 3.1 Illustration of the EMC problem with the PLC device being victim to interference.

elementary considerations, Section 3.5 discusses acceptable emission levels and how to approach PLC EMC regulation from a technical point. It also explains the relatively new approach of cognitive notching. Section 3.6 elaborates on the current EMC standardization and regulatory situation, with a focus on regulation in Europe. Finally, in Section 3.7 the coupling between power line and other wireline communications systems is studied.

3.2   Parameters for EMC Considerations

3.2.1   EMC Relevant Transmission Line Parameters

In low voltage (LV) installations, multi-wired cables with sector shaped or round wires are used. Both types of cables typically have a weak twist to make sure that mechanical forces during

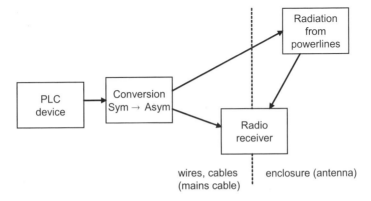

Figure 3.2 Illustration of the EMC problem with a radio receiver being victim to interference.

installation will not damage the structure of the cable. Signal propagation in the frequency range of less than 100 MHz is comparable to that for other cables used for communication when symmetric transmission is used. In this case the electromagnetic wave is guided by the conductors within the well controlled insulation. For asymmetric signals (between line and ground), the guided wave travels in the surrounding of the cable. The propagation properties are therefore strongly influenced by objects in the environment.

Medium voltage (MV) cables typically have a shield to control the electric field inside the cable. Modern cables are built as radial field cables, i.e. with a coaxial structure. The main insulator is made of polyethylene, which is also used in coaxial cables for high frequency signal transmission (measurement cables, antenna cables, etc.). The transmission characteristics of MV cables are therefore very similar to those for other coaxial cables.

Overhead lines in LV and MV distribution networks behave very similar to cables used in LV power installations in buildings. The main differences are the insulation, which is air instead of plastic material, and the distance between the conductors.

If such power cables would be used only for point-to-point communication between two PLC modems, EMC of PLC (considering interference produced by PLC) would be unchallenged. However, a PLC network includes numerous interconnection points, switches, etc., which cause unsymmetries[1] and thus differential-to-common mode conversions. With regard to EMC a common mode signal may produce interference to radio receivers operated in the vicinity of or attached to the power line. This interference is several orders of magnitudes higher than that caused by symmetrical signals.

Although modern MV cables are built as coaxial cables, a common mode signal may exist on the cable, since the shield and its connection are designed to minimize power losses. Techniques like cross-bonding, where the shields of the three separate phase cables are cyclically interchanged along the cable tray, are very common.

For the characterization of EMC of networks with its attached equipment the symmetry properties of the network must be known. For telecommunication lines it is an established method to characterize the longitudinal conversion loss (LCL) of the line and to simulate this value by a measurement network (impedance stabilization network (ISN)). To measure the LCL, a common mode signal is injected into a network port and the differential mode signal U_{LCL} is measured at the same port (see Figure 3.3, top left). The LCL is the relation of the injected and the measured voltage and provides a measure for the unsymmetry of the network. However, a one-port method may not be appropriate to characterize PLC installations. If there is an attenuation on the line and a local point of unsymmetry somewhere inside the network, the measured LCL value would be higher than the true LCL, if the measurement was made directly at this unsymmetry point.

Therefore, the transverse conversion transfer loss (TCTL) becomes an important quantity. It defines the relation of the launched symmetric PLC signal U_0 to the asymmetric signal U_{TCTL} occurring at another port of the network (see Figure 3.3, bottom). In the case of PLC, this accounts for the situation that, for example, a PLC modem injects a signal at one socket and a radio receiver is plugged into another socket of the same power line network. The radio receiver may then experience a common mode voltage traveling along its power cable. Since especially lower-priced radio receivers with a built-in rod antenna use the power cable as part

---

[1] For the exact definitions of asymmetric and unsymmetric, please refer to [1].

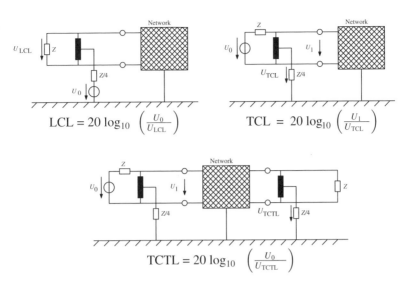

$$\text{LCL} = 20 \log_{10} \left(\frac{U_0}{U_{\text{LCL}}} \right) \qquad \text{TCL} = 20 \log_{10} \left(\frac{U_1}{U_{\text{TCL}}} \right)$$

$$\text{TCTL} = 20 \log_{10} \left(\frac{U_0}{U_{\text{TCTL}}} \right)$$

Figure 3.3   Some important quantities to characterize the symmetry of networks.

of the receiving antenna (at least as foot-point), this common mode voltage may produce radio interferences. A statistical evaluation of a large number of TCTL measurements performed in private households in several European countries is shown in Figure 3.4. It can be seen that the TCTL is only for 20% of all cases 45 dB or below. In other words, in 80% of all cases the converted common mode signal somewhere in the network is more than 45 dB below

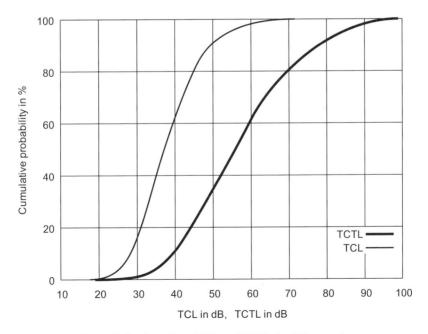

Figure 3.4   Statistics of TCL and TCTL for LV networks.

the injected signal level. If the common mode signal is measured at the same port as used for injection (one-port-method), then the transverse conversion loss (TCL) is obtained (see Figure 3.3, top right). Figure 3.4 also shows the statistic for the TCL obtained from the same measurement locations used for the TCTL. We note that the one-port-quantity TCL provides a worst-case estimate for the unsymmetry of networks, which in terms of PLC corresponds to the scenario that a PLC modem and radio receiver are connected to the same power outlet.

3.2.2 Coupling Factor

When a communication signal is coupled into the mains, the relation between the forward power of this signal and the resulting electric field strength at a certain distance to the mains installation lines is called the 'coupling factor'.

There exist a number of different definitions for the coupling factor in literature. The following useful definition for the coupling factor has been introduced by the PLCforum [2]:

$$k(f) = \frac{E(f)}{\sqrt{P(f)}},$$

(3.1)

where $E(f)$ (in µV/m) and $P(f)$ (in mW) are the electric field strength and the forward power at frequency f, respectively. Relation 3.1 reads in decibel scale:

$$k_{dB}(f) = E_{dB}(f) - P_{dB}(f),$$

(3.2)

where

$$E_{dB}(f) = 20 \log_{10} \left(\frac{E(f)}{1 \mu V/m} \right)$$

(3.3)

and

$$P_{dB}(f) = 10 \log_{10} \left(\frac{P(f)}{1 \, mW} \right) = PSD_{dB(mW/Hz)} + 10 \log_{10}(9000)$$

(3.4)

as the forward power is usually measured with a bandwidth of 9 kHz and $PSD_{dBmW/Hz}$ is the signal power spectral density (PSD) in dB(mW/Hz).

The coupling factor at a real installation depends on various parameters such as the cabling, the topology of the power grid, etc. Therefore, the PLCforum performed measurement campaigns in Europe to create a statistical database for the coupling factor. Although individual measurements revealed a strong frequency dependency of the coupling factor, the overall statistical evaluation indicated that the coupling factor is essentially flat over frequency. Outside of buildings, the 80th percentile for the coupling factor at a distance of 10 m was found as

$$k_{dB} = 50 \, dB(\mu V/m/mW).$$

(3.5)

Assuming, for example, a PSD mask of -45 dB(mW/Hz), the radiated field strength in a 10 m distance of the power line can be estimated from (3.4) and (3.5} as

$$E_{dB} = P_{dB} + k_{dB}$$
$$= -45\,dB(mW/Hz) + 39.54\,dB(Hz) + 50\,dB(\mu V/m/mW)$$
$$\approx 45\,dB(\mu V/m). \tag{3.6}$$

3.2.3 Electric and Magnetic Field

Considering the Hertz dipole of length $\Delta\ell$ traversed by an alternating current $i(t) = \Re\left\{Ie^{j2\pi ft}\right\}$, the electric and magnetic field components are given by (e.g. [3, p. 483])

$$E_\vartheta = \frac{A\sin\vartheta}{j\omega\epsilon_0 r^3}\left(1 + j\frac{\omega r}{c} - \left(\frac{\omega r}{c}\right)^2\right)e^{-j\frac{\omega r}{c}}, \tag{3.7}$$

$$E_r = \frac{2A\cos\vartheta}{j\omega\epsilon_0 r^3}\left(1 + j\frac{\omega r}{c}\right)e^{-j\frac{\omega r}{c}}, \tag{3.8}$$

$$H_\varphi = \frac{A\sin\vartheta}{r^2}\left(1 + j\frac{\omega r}{c}\right)e^{-j\frac{\omega r}{c}}, \tag{3.9}$$

$$H_r = H_\vartheta = E_\varphi = 0, \tag{3.10}$$

where c is the speed of light (in vacuum), $\omega = 2\pi f$, and $A = \frac{I\Delta\ell}{4\pi}$. With the far-field condition $r \gg \lambda/(2\pi)$, we have

$$\frac{\omega r}{c} \gg 1, \tag{3.11}$$

and thus E_ϑ is the dominating electric field component, which can be simplified to

$$E_\vartheta = \frac{j\omega\mu_0 A\sin\vartheta}{r}e^{-j\frac{\omega r}{c}}. \tag{3.12}$$

Also, the magnetic field strength can be simplified to

$$H_\varphi = \frac{j\omega\mu_0 A\sin\vartheta}{r}\sqrt{\frac{\epsilon_0}{\mu_0}}e^{-j\frac{\omega r}{c}}. \tag{3.13}$$

Dividing 3.12 by 3.13 leads to the field resistance

$$\frac{|E|}{|H|} = Z_0 = \sqrt{\frac{\mu_0}{\epsilon_0}} \approx 377\,\Omega. \tag{3.14}$$

Since in PLC the frequency range is typically between a few kHz and 87 MHz and the measurement distances are typically between 1 m and 30 m, condition (3.11) is not fulfilled. Thus, it is not possible to measure the magnetic field and to convert the measured values simply

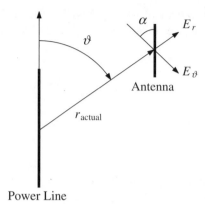

Figure 3.5   Measuring the electric field strength with a biconical antenna.

by (3.14) into electric field strength, but the electric field needs to be measured directly in order to enable a comparison between measurement result and limits for field strength, which is typically defined in terms of the electric field, e.g. [4], [5].

In order to measure the electric field strength, a biconical antenna can be used. Depending on its orientation the biconical antenna measures the superposition of E_ϑ and E_r as illustrated in Figure 3.5, where α denotes the angle between the antenna and E_ϑ. The field strength at a distance r_{actual} is given by

$$E(r_{\text{actual}}) = E_\vartheta(r_{\text{actual}}) \cos \alpha + E_r(r_{\text{actual}}) \sin \alpha, \tag{3.15}$$

where $E_\vartheta(r)$ and $E_r(r)$ follow from 3.7 and 3.8, respectively. Accordingly, the measured field strength can be extrapolated to the normative measurement distance by

$$E_{\text{norm}} = E_{\text{measured}} \frac{E_\vartheta(r_{\text{norm}}) \cos \alpha + E_r(r_{\text{norm}}) \sin \alpha}{E_\vartheta(r_{\text{actual}}) \cos \alpha + E_r(r_{\text{actual}}) \sin \alpha}, \tag{3.16}$$

where, for example, [4], [5] define $r_{\text{norm}} = 3$ m.

In the special case of $\alpha = 90°$, only E_r is measured. Then, (3.16) simplifies to

$$E_{\text{norm}} = E_{\text{measured}} \frac{r_{\text{actual}}^3 \sqrt{1 + \frac{r_{\text{norm}}^2 \omega^2}{c^2}}}{r_{\text{norm}}^3 \sqrt{1 + \frac{r_{\text{actual}}^2 \omega^2}{c^2}}} \tag{3.17}$$

The extrapolation factor for $r_{\text{norm}} = 3$ m is shown in Figure 3.6 for frequencies $f = 2$ MHz and 30 MHz. Figure 3.7 shows the extrapolation factor for the other extreme case of $\alpha = 0°$ (only E_ϑ is measured).

The above computation of the extrapolation factor has been based on a Hertz dipole with length $\Delta\ell \ll r$. In a real PLC radiation measurement scenario, the electric field strength is measured along a power wire with a much longer length. Hence, the above results are only

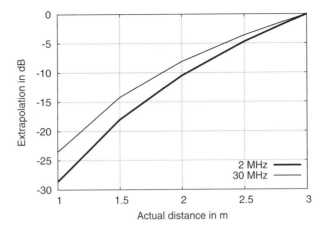

Figure 3.6   Extrapolation factor for $r_{norm} = 3$ m and $\alpha = 90°$.

coarse estimates, and, for example, method-of-moments field computations are required to obtain the extrapolation factor, cf. e.g. [6].

3.3   Electromagnetic Emission

The main aim for PLC EMC regulation is to define methods for compliance testing and assessing interference cases. The most obvious approach would be to directly measure the radiation coming from a power line. Unfortunately, this approach has disadvantages in practice, as it is difficult to reproduce results of near-field measurements. Furthermore, the measured radiation does not only depend on parameters (forward power) of the PLC device, but also on many other parameters such as the cable characteristics or obstacles (e.g. trees, etc.) around the measurement object, which influence especially the electric field component. Therefore, the

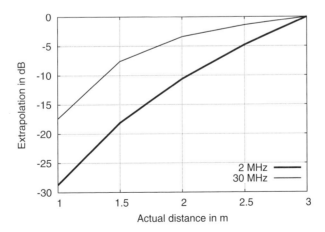

Figure 3.7   Extrapolation factor for $r_{norm} = 3$ m and $\alpha = 0°$.

(a)   (b)   (c)

Figure 3.8 Electrical field strength measurement. Equipment used include (a) biconal antenna, (b) biconal antenna measurement setup and (c) radio dummy.

PLC community uses the concept of coupling factors (see Section 3.2.2), which allows the derivation of the expected radiation from the output power of a PLC modem. However, the International Special Committee On Radio Interference (CISPR)[2] of the International Electrotechnical Commission (IEC) prefers the measurement of conducted emissions and the extrapolation of the possible radio interference (including both the conducted and the radiated path, see Figure 3.2). The new European PLC-Emission-Standard EN 50561-1 combines both approaches. In special sensitive frequency bands and for asymmetric signals injected by a PLC-modem the CISPR approach is followed. In the other bands used for the PLC signal transmission a limit for the symmetric signal level (in principle it is the modem's output power) is defined. In the following, we discuss the measurement of radiation and the measurement of conducted emissions.

3.3.1   Radiated Emissions

Broadband PLC systems typically operate in the frequency range from 2 MHz to 30 MHz, with possible extension to frequencies of up to 87.5 MHz currently under consideration. As discussed in Section 3.2.3, the simplified conversion between magnetic and electric field strength is only allowed for far field measurements, which might not be the case in a practical measurement setup. Therefore the question is whether to measure the electric or the magnetic field strength.

The electrical field strength can be measured by a small biconal antenna (typically it is an active antenna) shown in Figure 3.8(a) with a setup illustrated in Figure 3.8(b). Although it delivers an accurate and reliable value for the electrical field strength, it is challenged if the values represents the real disturbance situation. Therefore, a radio dummy (see Figure 3.8(c)) can be used to practically simulate the operation of a shortwave receiver with an attached whip antenna. Its principle corresponds to onboard-measurements for vehicles as defined in CISPR 25 [7]. Inside the dummy housing is an amplifier acting as input impedance converter. The

---

[2] Comité International Spécial des Perturbations Radioélectriques

Figure 3.9   Loop antenna.

radio dummy allows the evaluation of effects from, e.g. the connection between radio dummy housing and mains. The results from field strength measurements with biconal antennas and using a radio dummy depend very strongly on the angle between the measuring antenna and the power line. This effect could be stronger for the radio dummy depending on the length of its antenna.

The magnetic field strength is measured with a loop antenna (sometimes also referred to as 'frame') as shown in Figure 3.9. The advantage over electric field measurements is that the results are less dependent on the measurement environment formed by trees, plants, etc., which have a stronger influence on the electric field than on the magnetic for the frequencies, where PLC operates. The disadvantage is that for most evaluations, the measurement results need to be converted into electrical field strength, which is not possible without assumption of far field conditions.

The measurement results of a single scan are typically strongly frequency dependent, with differences of easily 40 dB in the electric (magnetic) field strength measured at different frequencies. But when considering the average of measurements obtained from many locations, the field strength is found to be flat over frequencies, i.e. field strength does not decrease or increase towards higher frequencies.

The measurement results are usually compared to limits provided by regulations and standards. These limits are defined for a normative measurement distance, which may or may not be applied in practice. Therefore, often an extrapolation factor is used to normalize measurement results to the normative distance. As described in Section 3.2, this extrapolation factor is a function of frequency. However, a flat extrapolation factor is quite commonly assumed in measurement evaluations by regulators. For example, [8] defines the extrapolation factor as 40 dB/decade, while [4] applies a factor of 20 dB/decade. Based on analytical considerations, simulations, and measurements reported in e.g. [6] and within CISPR discussions, a value of 33 dB/decade appears to be a good compromise for the frequency range above 2 MHz. Similar research has been done for the frequency range 150–500 kHz in [9].

3.3.2 Conducted Emissions

Experience shows that radiation measurements in the frequency range below 30 MHz lack reproducibility. Therefore, indirect methods based on the measurement of conducted signals have been developed and are now established in almost all EMC product standards. If the radio frequency behavior of the cabling structure, which normally is attached to the connection port of the equipment under test (EUT), is known, adequate limits for the voltage or current injected from the EUT into the line can be defined. If an EUT fulfills these limits, a sufficient protection of radio services is assumed.

Conducted emission measurement is performed by operating the EUT at a defined network. In its simplest form, such a network consists of a low-pass filter for the power signal and a high-pass filter to guide the relevant signal parts to the measurement receiver. The implementation of such network for power lines is called artificial mains network (AMN) and shown in Figure 3.10(a). It measures the unsymmetric disturbance voltage, which can be interpreted as a superposition of an asymmetric and a symmetric voltage as illustrated in Figure 3.11. Since radiation is dominated by the asymmetric voltage, the measurement of unsymmetric voltage represents a worst-case evaluation.

Since a symmetric signal component is desired in PLC systems, the AMN method would considerably overestimate the interference potential. Furthermore, PLC modems typically need a second modem for normal operation, which would be prevented by the built-in low-pass filter. Therefore, a special type of T-shaped ISN (T-ISN) shown in Figure 3.10(b) is used to measure both, the unwanted asymmetric signal and the wanted symmetric signal. The Z-transformer separates the asymmetric part of the injected signal. The impedances $Z_{\text{Termination}}$ and Z_{asy} provide a well defined symmetric and asymmetric impedance seen from the EUT's side. Through an attenuator a connection to a second PLC modem can be established. The additional attenuator with the transformers allow the measurement of the symmetric part. The impedance Z_{unsym} may optionally be used to realize a well-defined unsymmetry accounting for mode conversion (symmetric to asymmetric) experienced in power line installations. The TCL values presented in Section 3.2.1 are a good starting point for this unsymmetry.

The median asymmetric impedance of an LV power distribution network is on the order of 150 Ω, which is respected by the corresponding immunity standard (IEC 61000-4-6). However, due to historical reasons CISPR uses an asymmetric impedance of 25 Ω for two wire, 12.5 Ω for four wire and 10 Ω for five wire mains lines.

In addition to the measurements of asymmetric and symmetric signal contents the network can be used to validate the proper functioning of automatic power management by using an adjustable attenuator between the modem under test (EUT) and its communication partner (second modem).

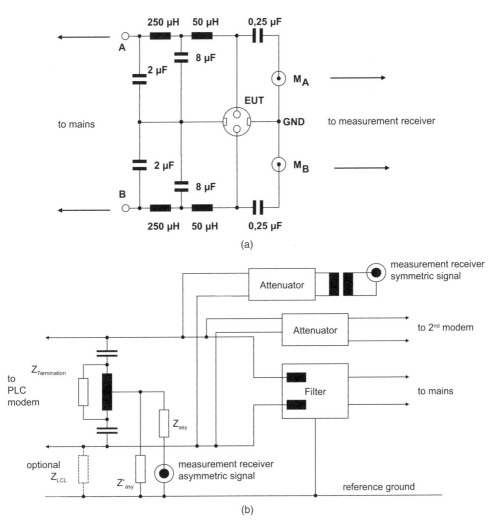

Figure 3.10 Measurement of conducted emission. (a) AMN for the measurement of unsymmetric voltages. (b) T-ISN for conducted disturbance measurements of PLC systems (Z'_{asy} and Z_{asy} for adjustment of the asymmetric impedance, $Z_{Termination}$ for the adjustment of the symmetric impedance and Z_{LCL} for the adjustment of a certain TCL or LCL value).

The use of T-ISNs only enables measurement of disturbance voltages. Disturbance currents cannot be measured by simply using a current probe (clamp-on current transformer) at the connection line between EUT and T-ISN. A variant of the network with the unsymmetry impedance (Z_{unsym}) placed in series in one of the power lines avoids this problem. However, it is argued that this type of T-ISN will not reflect correctly the mode conversion, since the measurement signal only depends on the common mode through the EUT, which can be significantly reduced by use of a proper common mode choke.

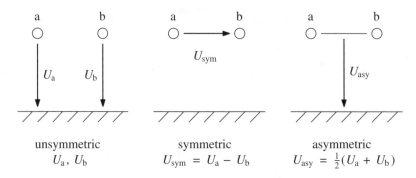

Figure 3.11   Unsymmetric, symmetric and asymmetric voltages.

3.4 Electromagnetic Susceptibility

As already mentioned in the introduction of this chapter (see Figure 3.1), there are a number of electromagnetic phenomena that potentially disturb the operation of PLC systems. In this section, we discuss the most important ones in some detail.

- Electrostatic discharge: A person can accumulate electric charges, if humidity is not too high and clothes separate charges by friction, and cause electric sparks when touching a grounded object. The discharge introduces a highly transient current into the object, where rise times of below 1 ns and pulse amplitudes of several amperes can be reached. This current pulse can disturb or damage sensitive electronic components. As any other electronic systems, PLC devices need to be made sufficiently immune to electrostatic discharge (e.g. by adequate isolation of the housing).
- Fast transients (burst): Fast transients are produced by switching activities in a power installation. Mechanical switches with arcing effects can produce steep voltage impulses, which due to reignition are grouped in impulse clusters (burst). The duration of the pulses, the slope, the pulse amplitude, and the repetition frequency are random variables. Typical parameters are a pulse duration of 50 ns, a rise time of 5 ns, an amplitude of up to 2 kV pulse amplitudes, and a repetition frequency of 5 kHz [10].
- Surge: Lightning often causes transient over-voltages in power grids due to lightning strikes into overhead lines or nearby objects, which produce a potential rise in the ground. Lightning strikes typically occur at some distance from PLC devices, at which they are experienced as over-voltages of some kV. The voltage pulse due to lightning has a lower rise time and larger duration than the burst pulse. The surge can occur as asymmetric or symmetric voltage. Manufacturers of electronic products typically add a varistor or surge arrester in the power supply circuit to absorb a major part of the pulse energy. For PLC devices the large capacitance of varistors will attenuate the PLC signal. Therefore special designs are required for the protection circuitry.
- Radio frequency electromagnetic fields: Radio frequency electromagnetic fields produced by stationary and mobile radio transmitters can induce a common mode signal to the power line. This undesired signal can be separated from the differential mode PLC signal through application of a sufficient common mode rejection in the input stages (e.g. common mode

choke) of the PLC receiver. However, if there is a local point of unsymmetry in the installation, a part of the disturbance signal is converted into a differential mode signal. Fortunately, signals from strong radio transmitters are narrowband signals. Therefore, narrowband filtering avoids undue interference. For example, in a multicarrier system a number of sub-carriers can be switched off if a narrowband interference is detected. Of course, this filtering in the digital domain does not prevent a loss in receiver sensitivity due to the interference power experienced in the analog front-end.

- There are two standardization projects with TC77 of IEC to draft standards for immunity tests with broadband signals. One project deals with the frequency range below 150 kHz and the other project covers 150 kHz to 80 MHz. These standards should reflect disturbances produced on power lines by PLC and other broadband signal sources (e.g. frequency converters, power electronics for LED lamps). The two projects will end up with two new basic standards. Product committees may decide to use or not use these standards in the product standards under their responsibility.

3.5   EMC Coordination

PLC EMC regulation is a controversial issue. Like in everyday life, when multiple parties share the same resources and need to coexist, also in PLC EMC regulation different interest groups make firm statements contradicting each other. This section sheds some light on the issue of PLC emission limits, starting with a discussion of the compatibility level and followed by calculating forward power limits applying indirect emission measurement techniques. The use of cognitive radio techniques for PLC to meet EMC requirements is also discussed.

3.5.1   Compatibility Level

EMC coordination refers to the process of defining adequate emission and immunity requirements for equipment working in a certain environment to guarantee their electromagnetic compatibility [11]. As a starting point for this coordination, the electromagnetic compatibility level (CL) needs to be defined, which is the specified electromagnetic disturbance level used as a reference in the setting of emission and immunity limits [12]. As illustrated in Figure 3.12, two CLs are needed for radio services. The immunity limit for PLC devices needs to be higher than CL1, with a certain safety margin, while the PLC emission limits must not exceed CL2.

Narrowband PLC systems operating at frequencies below 150 kHz are mostly unregulated with regards to protecting radio services. For example, the European standard for narrowband PLC [13] is based on EMC coordination without taking into account radio services. In part as a result of this, narrowband PLC devices are allowed to produce significantly higher signal amplitudes than broadband PLC systems, and EMC problems with other equipment as susceptible devices could occur.

In the frequency range above 150 kHz the emission and immunity limits for non-radio equipment have been defined some 50 years ago with only few adjustments since then. While higher sensitivity of radio receivers would allow the reduction of both CL1 and CL2, proponents of PLC technology suggest an increase of CL2. A quick solution of this dilemma is not in sight and emission limits for new PLC products are likely to be oriented on already defined

field strength or voltage

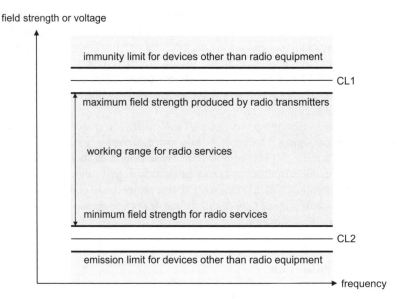

immunity limit for devices other than radio equipment

CL1

maximum field strength produced by radio transmitters

working range for radio services

minimum field strength for radio services

CL2

emission limit for devices other than radio equipment

frequency

Figure 3.12   EMC coordination including radio services.

emission limits. Product committees of standardization bodies are asked to demonstrate that new limits provide at least the same level of protection of radio services as the established emission limits.

3.5.2   Definition of Limits

For broadband PLC systems immunity limits can be taken from CISPR 24 [14] and the immunity test methods referred to in CISPR 24 are applicable to PLC. The definition of emission limits for broadband systems is postponed for the moment within CISPR committees.

For conducted emissions CISPR 22 defines limits for the mains port and the telecommunication port. A direct comparison of these limits is not possible, since the network used for measurement at the mains port (AMN) differs in structure and in asymmetric impedance from the network used for measurements at the telecommunication port (T-ISN). If the allowed asymmetric current is derived from the voltage limit and the asymmetric impedance of the network, it can be seen from the values in Table 3.1 that there is not much difference between these two limits. It seems reasonable to apply the current limits also for broadband PLC systems.

As for radiated emissions, it is reasonable to apply the asymmetric current limits defined in CISPR 22 for radiation from cables (e.g. telephone lines) also for broadband PLC systems.

The above approach only takes into account the launched common mode signal, e.g. produced by clock frequencies or frequencies of a switched power supply inside the PLC modem coupled to the mains port. As discussed in Section 3.2.1, part of the injected symmetric signal is converted into an asymmetric signal at local points of unsymmetry within the power

Table 3.1 CISPR 22-Limits for ITE class B products (average detector, see [15])

Frequency MHz	Mains Port AMN (with an asymmetric impedance of 25 Ω seen from the EUT port)		Telecommunication Port T-ISN with $Z_{asy} = 150$ Ω	
	Voltage Limit dB(μV)	Current Limit dB(μA)	Voltage Limit dB(μV)	Current Limit dB(μA)
0.15 to 0.5	56 to 46	28 to 18	74 to 64	30 to 20
0.5 to 5	46	18	64	20
5 to 30	50	22	64	20

installation. The limit for the desired symmetric signal can be derived from the statistical data for TCTL shown in Figure 3.4. That is, a maximum PSD in dB(mW/Hz) can be calculated as

$$PSD_{limit} = Voltage_{limit} + TCTL - 10\log_{10}(BW) - k \tag{3.18}$$

where the voltage limit for mains $Voltage_{limit} = 50$ dB(μV) between 5 MHz and 30 MHz, the measurement bandwidth BW = 9 kHz, and $k = 100$ dB(mW/μV) is the conversation factor between voltage and power considering a 100 Ω load. Assuming, for example, that TCTL = 45 dB (see Section 3.2.1), then a limit of -54.5 dB(mW/Hz) results. Considering a measurement bandwidth of 9 kHz and again 100 Ω load this value corresponds to a voltage of 94.5 dB(μV). In the new European standard EN 50561-1 a value of 95 dB(μV) is defined for measurement with the average detector.

A higher degree of protection for radio services, and thus higher emission limits, are possible using advanced signal processing features in PLC modems, such as power management or notching of frequency bands, which are part of EN 50561-1.

3.5.3 Cognitive Radio Techniques

Radio frequency bands are a scarce resource. As discussed before, although PLC is using the power line grid for data transmission, there is a small amount of energy converted into radiation due to the non-perfect symmetry of the power line systems. This radiated signal may interfere with other radio services. The goal of cognitive radio techniques is to avoid interferences, while at the same time the usage of a frequency band can be allocated dynamically and thus more efficiently.

The first broadband PLC modem generations were just using the frequency band from 2 to 30 MHz. To prevent interferences with amateur radio bands it was sufficient to permanently exclude these bands from the PLC PSD. With the recently released new European standard EN 50561-1 (see Section 3.5.2 and 3.6.3.1) the list of permanently notched frequencies has been extended to include the aeronautical frequency bands. This can lead to a significant performance degradation for PLC, since a number of sub-carriers cannot be used for data transmission anymore.

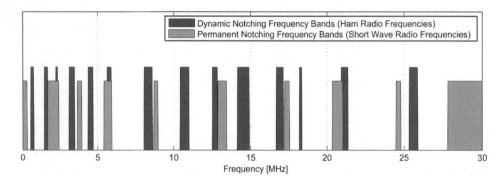

Figure 3.13   Illustration of permanent and dynamic notching frequency bands according to EN 50561-1.

Currently, new PLC notching implementations are being specified to improve spectral efficiency of notching, such as e.g. HomePlug AV2.0. While today's PLC modems require removing additional three carriers at both sides of each notch, the new implementations may do with less. Due to the notable number of permanent notches, this change has a significant impact on the PLC performance. More specifically, the rate loss caused by the extended permanent notched frequency list is in the range of 14% to 18% when using today's notching methods. This can be reduced to about 8% to 12% in the case of efficient notching methods.

The European standard EN 50561-1 introduces two new measurement methods and procedures for PLC ports. The first is one called 'Cognitive frequency exclusion' and the second is referred to as 'Dynamic power control'. Both of these methods are based on cognitive radio techniques, which is a very useful approach in all cases where radio resources are limited and may be shared by different technologies. The basic idea behind this concept is to check whether a certain frequency band is already in use by a service. If the frequency band is not in use, it can be used to transmit data. If it is already in use, then transmissions in that frequency band should be avoided or the transmit level should be reduced so that no harmful interference can occur.

The cognitive frequency exclusion method specified in EN 50561-1 is intended to reduce the potential interference risk with short wave broadcasting services. Therefore the standard contains a list of frequency bands which need to be permanently or dynamically excluded. Figure 3.13 depicts the existing amateur radio and the new aeronautical notches as well as the new broadcasting bands in the cognitive exclusion list.

To implement cognitive frequency exclusion methods it is required that the PLC modem is capable to detect if a broadcasting service is present or not. The signal of broadcasting radio stations creates ingress into the power line grid. This ingress can be detected in different ways. A first one is to simply measure the noise floor by using the PLC devices' A/D-converters. The disadvantage of this method is that the PLC transmissions need to be stopped during this measurement to avoid measuring the PLC signal itself instead of the ingress caused by the broadcasting stations. A second, more sophisticated detection method is to implement an algorithm which is capable to detect specific patterns in the signal to noise ratio (SNR). If a radio station is causing ingress into the power line grid, the SNR will show a sharp notch at this frequency. Figures 3.14 and 3.15 show examples for the detection of several short wave radio stations. The big advantage of this method is that the detection algorithm can run without interruption and the PLC traffic does not need to be stopped during this measurement.

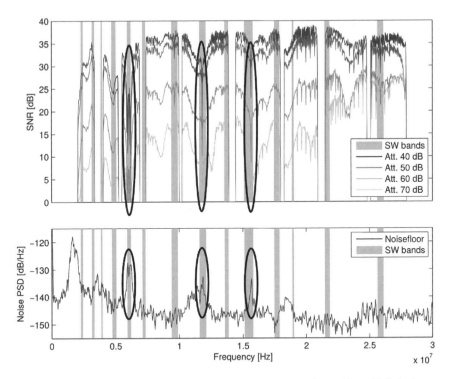

Figure 3.14   Sample SNR and noise PSD measurements performed by a PLC device.

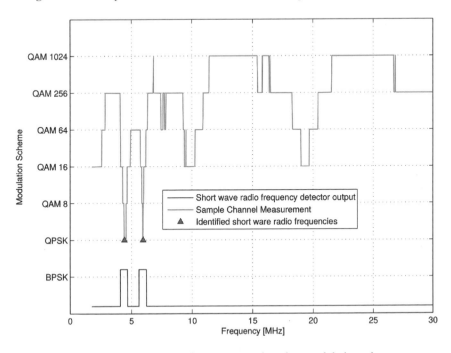

Figure 3.15   Sample channel measurement based on modulation schemes.

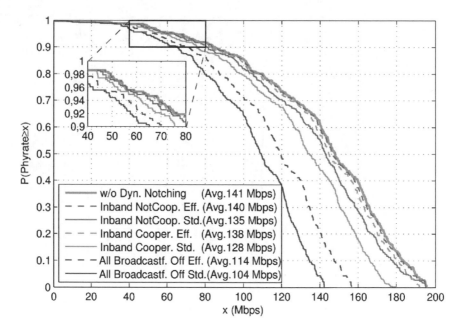

Figure 3.16 Performance with different dynamic notching scenarios and methods.

In several countries field tests have been performed to analyze the potential performance impact caused by the cognitive frequency exclusion method required by EN 50561-1. In Figure 3.16 the performance losses in terms of physical layer data rate for different implementations of notching are compared. The top trace, labeled 'w/o Dyn. Notching', is the performance result of the field test if no dynamic notching is applied. This reflects the performance result of today's PLC devices. All other traces show the results of different kinds of implementation. The traces labeled 'Eff.' are the results if efficient notching is used. In this implementation only the carrier itself or one adjacent carrier of the detected broadcasting frequency will be removed. The traces labeled 'Std.' are using today's standard way of notching where it is required to remove three carriers at both sides of the notch which is not that efficient. Further tests have been evaluated where a so-called cooperative method was used. These traces are labeled 'Cooper.' in contrast to 'NotCoop.'. These cooperative methods will force all modems to create notches even if only one modem has detected a broadcasting station. This method requires a new protocol to exchange this information through the PLC network. The two last traces labeled 'All Broadcastf. Off' are the result of the worst-case scenario in which all broadcasting frequencies were detected and all dynamic notches would be active. In this case the performance loss can go up to more than 25%. More details can be found in [16].

The intention of the 'dynamic power control' method is to reduce the transmit power in all cases where the signal level is higher than required. Today's PLC devices transmit their signals with a constant transmit power independent of the distance between the modems. In all the cases where the distance between the modems is short, or more technically speaking, the attenuation between the devices is low, the transmit power can be reduced without negative impact on the PLC performance. Table 3.2 shows how the maximum transmit PSD level should

Table 3.2   Maximum transmit PSD: CISPR 22 CIS/I/301/CD

DM-Insertion loss EUT-AE (dB)	>40	40	30	20
Maximum Transmit PSD (dBm/Hz)	−55	−55	−63	−73

be adjusted depending on the insertion loss between the transmitting (EUT) and the receiving modem (AE). Reducing the transmit power will result in lower radiation and thus free up radio resources. Different implementations and test results can be found in [17].

Power control can also be done on a subband basis using a tool to configure the PSD mask of a PLC device. This tool can be used to avoid data transmission in frequency bands which may be used by other local services already. The example configuration in Figure 3.17 shows that PLC will only use carriers in the frequency range up to 20 MHz. Furthermore it is also possible to adjust the transmit level individually instead of completely switching of carriers. This can be seen in the frequency range from 10 to 12 MHz.

3.6   EMC Standardization and Regulation in Europe

In this section, we describe the current status of EMC standardization and regulation applicable to PLC technology in the European Union (EU). Some of the principles of the pertinent EMC Directive [18] and the PLC Recommendation [19] have been adopted in part or entirely by other countries such as European non-EU-member states and countries in Asia and Africa.

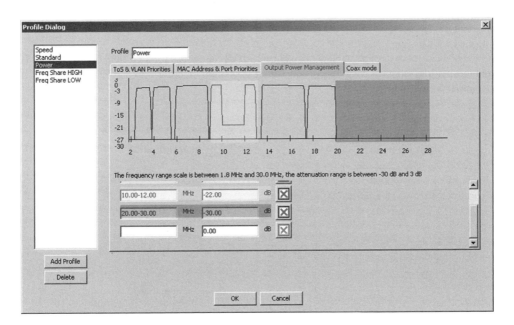

Figure 3.17   Graphical user interface for output power management.

3.6.1   Distinction Between Standardization and Regulation in the EU

In the EU, the regulatory framework is set by Directives from the European Commission (EC), while specific limits are set by official ('de jure') standardization platforms. The EU recognizes three standardization platforms officially: ETSI, CENELEC and CEN. When signing the Treaty of Rome (EU constitution), the EU members agreed '... to abolish existing trade barriers created through legislation and standardization' [20]. The approach to achieve this goal is seen in separation between legislation, which is politically driven, and standardization, which should be market-driven. The EU legislation with its legal instruments such as Directives that are setting the legal framework is referring to European Harmonized Standards for technical details. These European Harmonized Standards are technical specifications that have been adopted by one of the three official platforms and afterwards have been published by the European Commission in the Official Journal of the European Union. Therefore, there is a close link between standardization and technical regulation in Europe, whereas the technical regulation should be clearly separated from EU trade policy.

Standards that are specified by European Telecommunications Standards Institute (ETSI), Comité Européen de Normalisation Electrotechnique (CENELEC) and Comité Européen de Normalisation (CEN) can be categorized to be primarily intended for market growth (e.g. inter-working standards so that equipment of different manufacturers can inter-work) or for regulatory usage (e.g. electrical safety or EMC standards). Figure 3.18 gives an overview on the approach taken in the EU.

For market access of PLC solutions, the most important EU legal instruments are:

- Directive 2004/108/EG (EMC directive).
- Commission Recommendation of 6 April 2005 on broadband electronic communications through power lines.

As the EMC directive refers to European Harmonized Standards for technical details, the EMC standardization aspect is very important for PLC. This is handled in CENELEC and will be discussed subsequently.

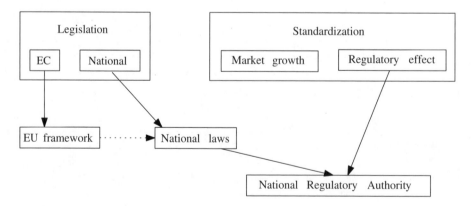

Figure 3.18   Link between standardization and regulation in the EU.

3.6.2 EMC Regulation for PLC

The EC follows the principle that regulation for telecommunication devices should be applicable regardless of the underlying technology. Therefore, PLC systems are only allowed to radiate at the same level as other telecommunication technologies such as, e.g. digital subscriber lines, and similar measurement procedures are applied, taking into account that mains and telecommunication port are identical for PLC devices.

Generally, PLC equipment is telecommunication equipment and regulated in the EU according to the EMC Directive 2004/108/EC[3] [18]. Another important directive within this context is the Radio and Telecommunications Terminal Equipment (R&TTE) Directive 1999/5/EC, which becomes relevant for access PLC systems and PLC products that bridge to wireless systems such as WiFi or ZigBee. The EMC Directive describes the legal framework and the principles to be applied. It distinguishes between installations and consumer products. It does not provide technical aspects such as measurement setups and/or limits. The EMC Directive regulates both market access (please refer to Section 3.6.2.1) as well as interference complaint handling (please refer to Section 3.6.2.2). The R&TTE Directive refers to the EMC Directive for EMC aspects. Therefore, the EMC rules of the EMC Directive apply to all types of PLC.

3.6.2.1 Market Access

Market access is regulated according to the mentioned EMC or R&TTE Directive, with the EMC relevant part covered in the EMC Directive. The objectives of the EMC Directive are:

- free movement of apparatus, installations and networks throughout the EU,
- and maintaining of an acceptable EMC environment
  - to ensure that electromagnetic disturbances produced by electrical and electronic apparatus do not affect the functioning of other apparatus as well as radio and telecommunication networks, related equipment and electricity distribution networks,
  - and to ensure that the apparatus has an adequate level of intrinsic immunity to enable it to operate as intended in the presence of electromagnetic disturbances.

Thus, the protection requirements consist of emission and immunity requirements.

Market access is granted, if products are compliant with the EMC regulation. To this end, product specific compliant tests are required. Directive 2004/108/EC is a so called 'New Approach Directive' that requires manufacturers' self-declaration. A manufacturer is allowed to assess the EMC of its products either by compliance with a European Harmonized Standard, which is EN 50065 [13] for narrowband PLC, or by doing an EMC assessment based on own procedures and methods. The latter option often includes the certification of these procedures and methods by an independent Notified Body (as defined by the EMC Directive). Usually these procedures and methods are based on discussions in the relevant standardization platforms such as CISPR. Independent of which option has been chosen, the manufacturer must provide a declaration of conformity to assert the compliance of its products with the

---

[3] EMC Directive 2004/108/EC has been officially published in December 2004 and replaced the former EMC Directive 89/336/EEC.

protection requirements according to the EMC Directive. Product compliance with all applicable directives (e.g. safety, etc.) is shown by the 'Conformité Européenne' (CE) marking.

Access PLC networks and equipment are installations in the sense of the EMC Directive 2004/108/EC. Differences in their applications such as usage as part of the public telecommunication infrastructure or in Smart Grids as part of a corporate utility network have an impact on the general regulatory framework but not with regards to the EMC aspect. The regulations for market access for installed access PLC devices such as head-ends or repeaters are different from those for access PLC end-user modems (CPEs). According to the EMC Directive, installed access PLC devices do not need to be marked with the CE mark. But since the CE mark implies conformance to all applicable directives for a specific product and therefore includes compliance with EMC requirements, most PLC operators (e.g. power utilities) require the CE mark in their procurement rules and hence only accept CE marked products for installation in their power grid. Access PLC CPEs need CE marking if sold to consumers. Since CPEs are network terminals they could fall under the regulation of the R&TTE Directive 1999/5/EC. But since the R&TTE Directive is referring to the EMC Directive for EMC aspects, this is irrelevant as far as EMC is concerned.

The EC has released a Commission Recommendation on broadband electronic communications through power lines in April 2005 [19]. The key element concerning market access in this recommendation is that ex-ante conformity assessment is not required. That is, the basic assumption is that PLC products are legal and do not need specific ex-ante regulation. Only in the interference complaint case, ex-post regulation applies.

3.6.2.2   Regulation in the Case of Interference Complaints

Each broadband system (DSL, CaTV, Ethernet, etc.) causes radiation which can potentially interfere with radio services. In 2001, the EC gave mandate 313 [21] to ETSI/CENELEC/CEN to define a harmonized standard for radiation limits which should apply for all telecommunications technologies including PLC. Therefore, the radiation limits shall be technology-neutral for all broadband telecommunications technologies. National limits on network radiation as e.g. the German NB30 [5] should be avoided. Mandate 313 is carried out in the ETSI/CENELEC Joint Working Group. While the standard is pending, the mandate has not been given back to the EC. Therefore, individual national measures (as e.g. the German NB30 regulation) are not legal.

Another document that is frequently used as reference in the context of EMC regulation is the Recommendation (05)04 by the Electronic Communication Committee (ECC) [22]. This recommendation is without any normative regulatory impact and only for information purposes.

When approaching an interference complaint, the first point to be evaluated is whether interference is harmful or not. The EC Recommendation on PLC [19] states in Article 4:

> Where it is found that a power line communications system is causing harmful interference that can not be resolved by the parties concerned, the competent authorities of the Member State should request evidence of compliance of the system and, where appropriate, initiate an assessment.

Accordingly, it is not considered harmful interference, if, for example, a radio receiver has been placed on purpose directly besides a street-cabinet with PLC equipment. It is also not appropriate to request the PLC operator for interference mitigation when the radio user does not use state of the art equipment. If interference is deemed harmful, the mitigation measures should be proportionate. The EC Recommendation on PLC states in Article 5:

> If the assessment leads to an identification of non-compliance of the power line communications system, the competent authorities should impose proportionate, non-discriminatory and transparent enforcement measures to ensure compliance.

For example, it is proportionate to request mitigation only at the very location and frequency at which interference occurred.

Misunderstandings by EU member states when implementing the EC recommendation is minimized by the control function of the EC. The EC Recommendation on PLC states in Article 7:

> Member States should report to the Communications Committee on a regular basis on the deployment and operations of power line communications systems in their territory. Such reports should include any relevant data about disturbance levels (including measurement data, related injected signal levels and other data useful for the drafting of a harmonised European standard), interference problems and any enforcement measures related to power line communications systems.

In conclusion, ETSI and CENELEC have not yet completed the standard requested by the EC mandate 313. Hence, the situation in the case of an interference complaint is open within the limits given by the EMC Directive. Although the EC Recommendation attempts for a coordinated approach within the EU, interference complaints might be dealt with differently by different national regulatory authorities.

3.6.3   EMC Standardization for PLC

3.6.3.1   CENELEC

CENELEC is organized in platforms related to a technology and on horizontal issues such as EMC. For PLC, SC205A is the CENELEC technology related platform and TC210 deals with EMC issues. TC210 is known for converting CISPR standards into European standards such as CISPR 22 into EN 55022.

CENELEC's normative specifications are categorized as ES and EN. ES is not to be confused with ETSI ES, as the approval procedure is different and involves the national mirror committees whereas in ETSI company members approve. An EN might become regulatory important if listed by the EC.

The EMC standardization in CENELEC distinguishes PLC solutions by the frequency range in which PLC data are transmitted.

3.6.3.1.1   PLC in the Frequency Range from 3 to 148.5 kHz

SC205A's scope is 'Mains Communicating Systems'. Its success story is the narrowband PLC standard EN 50065, published first in 1991. EN 50065 is a European Harmonized Standard and covers the frequency range from 3 to 148.5 kHz. Before it was published, different EU

Table 3.3   Overview of the main regulations in EN 50065

Band	Frequency [kHz]	Max. Transmission Level (Single Phase) [dBµV]	Access Protocol	Usage
A	3–9	134	—	Utility
	9–95	Logarithmic decrease over frequency from 134 till 120	—	Utility
B	95–125	2 classes of devices: 122 and 134	—	Private
C	125–140	2 classes of devices: 122 and 134	CSMA carrier at 132 kHz	Private
D	140–148.5	2 classes of devices: 122 and 134	—	Private

countries (e.g. Germany) allowed free usage of this spectrum with the maximum of 5 mW. Since the EC listed EN 50065, only compliant PLC systems are allowed.

EN 50065 distinguishes the four frequency bands A, B, C and D. The band A (sometimes referred to as CENELEC A band) is reserved for utility usage only. Bands B, C and D are for private usage. As band B and D do not require any access mechanism, there is the risk of interference by multiple users on the line. Band C requires a CSMA protocol for coexistence between different users. Table 3.3 [13] provides an overview.

3.6.3.1.2   *PLC in the Frequency Range above 1.6065 MHz*

In spring 1999, SC205A established a working group (WG10) to specify broadband PLC systems in the frequency range between 1.6 and 30 MHz. The work item is called 'Power line communication apparatus and systems used in low-voltage installations in the frequency range 1.6 MHz to 30 MHz'. Results will be published as EN 50412-x-y. So far only EN 50412-2-1:2004 'Immunity requirements for power line communication apparatus and systems used in low-voltage installations in the frequency range 1.6 MHz to 30 MHz – Part 1: Residential, commercial and industrial environment' has been successfully approved. Compliance to EN 50412-2-1 is checked for the Declaration of Conformity.

In 2010, the EC sent a letter to CENELEC TC210 to mandate CENELEC to draft a harmonized product standard for PLC apparatus of all types. In reaction, CENELEC established a working group (WG11) called 'PLT apparatus standard working group'. WG11 decided to draft several specifications for different frequency ranges and for different user scenarios. Priority was given to PLC apparatus for in-home use in the frequency range 1.6065–30 MHz. This work was successfully approved by CENELEC in November 2012 as EN 50561-1.

EN 50561-1 requires the following key points:

- Out of band emissions and without PLC data transmission shall be tested according to EN 55022, class B, mains port.
- Introduction of dynamic power control, i.e. reduction of the maximum allowed PSD level, if the channel attenuation between two PLC apparatus is low.
- Introduction of static frequency exclusion for specific radio frequencies.

- Introduction of dynamic frequency exclusion for protection of radio broadcast services in specified frequencies. Please refer to Section 3.5.3 where the cognitive frequency exclusion as one option to fulfill the requirement is described.
- The asymmetric disturbances of the PLC port with data transmission shall be measured using an ISN with a high symmetry (LCL shall be at least 55 dB).

Work on access PLC apparatus in the frequency range 1.6065–30 MHz started and should lead to EN 50561-2. As present in-home PLC products already also use frequencies above 30 MHz, EN 50561-3 is planned to deal with this frequency range. Both, EN 50561-2 and -3, have not been approved yet, by July 2013.

3.6.3.1.3   PLC in the Frequency Range 150–500 kHz

This frequency range is in particular interesting for Smart Grid applications. The work on an EMC PLC apparatus standard for this frequency range is in the scope of CENELEC TC210 WG11 (see Section 3.6.3.1). But work has not started yet, by June 2014. However, internationally a first proposal has been made by IEEE 1901.2 which has been approved in December 2013. IEEE 1901.2 proposes to follow the principles as specified in EN 50561-1, but adapted to the lower frequency range and taking into consideration that industrial environment shall also be covered. The later led to a definition of class A and B limits, following the principle of EN 55022.

3.6.3.2   ETSI-CENELEC Joint Working Group

The ETSI-CENELEC JWG has been established in June 2000. Participants are from several ETSI and CENELEC platforms, such as ETSI TC EMC, CENELEC TC210 for EMC aspects, ETSI TC PLT and CENELEC SC205A for PLC issues. In August 2001, the European Commission empowered ETSI and CENELEC by mandate 313 to specify 'EMC harmonized standards for telecommunication networks' (please refer to [21]). Mandate 313 states that the standard should be technically neutral. The mandate was shifted into the ETSI-CENELEC JWG.

Discussions in the ETSI-CENELEC JWG were contentious. On one side, participants required that the EMC standard for the network should be in line with the EMC product standard (in the case of DSL or Cable: EN 55022). Otherwise a CE certified product might not be allowed to be operated in a network. On the other side, participants required a high protection level for radio services. PLC was always in a particular focus of discussion as there was not any existing EMC product standard for broadband PLC. Referring to the technological neutrality as required in the mandate 313, the same radiated limits were proposed for all network types including the power grid. Figure 3.19 [23] gives an overview on the limit proposals. As one can see, the proposals differ by about 60 dB! As each party tried to provide substantial reasoning for its proposal, a conclusion or compromise was not possible.

Since 2008, a different approach has been tried. Now, the ETSI-CENELEC JWG is drafting three specifications, where DSL, Cable and PLC issues are handled separately. The idea behind this is that technical neutrality can also be achieved by using different, technology dependent methods. PLC should be covered by the planned EN 50529-3 'Conducted transmission

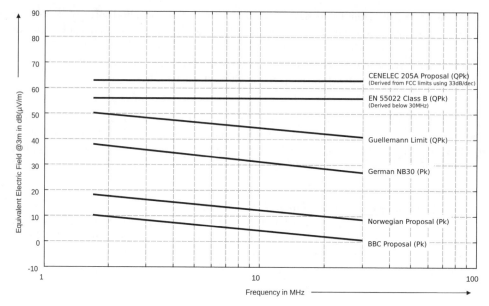

- For comparison purposes limits scaled to 3m using 20dB/dec except CENELEC 205A proposal (33dB/dec)
- H-Field limits converted to E-field using the far field conversion of 51.5dB

Figure 3.19 EMC limits proposed for the network in the ETSI-CENELEC JWG. Order of lowest to highest limits: BBC Proposal, Norwegian Proposal, German NB30, Gueleman Limit, EN 55022 Class B, CENELEC 205A Proposal.

networks – Part 3: Power line communication (mains network-based)' which is currently under revision of a former draft from November 2008.

3.6.3.3   International EMC Product Standardization

This section intends to give an overview about the history of international EMC product standardization which basically takes place within the IEC CISPR platform [24]. EMC product standardization might be important (depending on the national regulation) for product compliance testing and so for the market access of products. It must not be confused with EMC network standardization which is relevant for the operation.

CISPR is a special technical committee of IEC. It is organized in seven subcommittees. For PLC issues, the most relevant subcommittee is CISPR/I. It deals with EMC of information technology equipment (ITE), multimedia equipment and receivers. In particular CISPR/I is responsible for the standard CISPR 22. CISPR 22 was converted by CENELEC into EN 55022.

Opponents of PLC technology have stated for many years that CISPR 22 should be applied as it is for PLC products. They refer to the mains ports tests so that a PLC product should be evaluated in the same way as a power supply of any telecommunication equipment, despite that the characteristic of the signals is crucially different and the measurement set-up cannot be applied without adaptations (please refer to Section 3.3.1). As the working procedures of IEC require public voting of the IEC members (that are states) when launching new work items, this biased opinion on how to interpret CISPR 22 has been publicly contradicted since 1999, when the CISPR Project 'CISPR 22 amendment for PLC products' started. This project has lasted till 2005. The main working documents were:

- 03.2000: CISPR/G/179/CD
    - Requiring to use a T-ISN with an LCL dropping from 30 dB down to 6 dB over the frequency range of 2 MHz till 30 MHz.
- 05.2001: CISPR/G/218/CDV
    - Requiring to use a V-network (AMN) with the LCL of 6 dB; the existing CISPR 22 mains port limits were required.
        - This would again result in testing the PLC products just like power supplies of any telecommunication equipment.
        - 07.2001: CISPR/I withdraws 218/CDV because of the reasoning as described above.
- 07.2002: CISPR/I/44/CD
    - Requiring to use a T-ISN with the LCL of 36 dB and the common mode impedance of 150 Ω; the existing CISPR 22 telecommunication ports limits were required.
- 11.2003: CISPR/I/89/CD
    - Revision of 44/CD with the main difference to use a T-ISN with the LCL = 30 dB.
    - Since 2003 till present, millions of PLC products have been tested for EMC compliance based on this CD. As the world-wide number of radio interference complaints is negligible, it appears that this approach is sound. However, PLC opponents were always fighting against it.

Another project entitled 'Amendment of the CISPR 22 with appropriate limits and methods for PLT devices (CISPR/I/145/NP)' ran from 2005 till 2010 without coming to a conclusion. The main working documents were:

- 02.2008: CISPR/I/257/CD
    - Requires an ISN with LCL = 24 dB and Z_{asy} = 25 Ω, referring to the existing CISPR 22 mains port limits.
- 02.2008: CISPR/I/258/DC
    - Showing mitigation measurements such as notching and power management.
- 08.2009: CISPR/I/301/CD
    - This CD defines two types of PLC modems.
        - Type 1 should not exceed a PSD mask of −55 dBm/Hz and should provide power management capabilities depending on the attenuation between two nodes as well as notching capabilities.
        - Type 2 modems are tested with an ISN with LCL = 16 dB, referring to a new set of limits.

In conclusion, for more than ten years of CISPR standardization on PLC EMC aspects showed a confusing discussion between

- the methodology (T-ISN versus AMN),
- the limits (mains port limits versus telecommunication port limits versus new proposals),

- the symmetry characterization of the power grid (LCL range between 6 and 36 dB, or even denying the LCL as applicable),
- the network impedance (range between 25 and 150 Ω).

Therefore, the whole discussion appears more politically inspired where the relatively small PLC industry lobby has difficulties to get enough support, although the experience of PLC products in operation during more than ten years without major radio interference complaints should be convincing. After the second project failed, IEC CISPR did not launch a new project on the PLC issue. This gave room for CENELEC to start work on the same issue, because double-work should be avoided between the two organizations.

3.7 Coupling Between Power Line and other Wireline Communications Systems

This section is dedicated to the aspects of EMC dealing with the coupling between a PLC system and other wireline communications systems inside the home environment. In particular, it focuses on the characterization of the coupling between PLC and VDSL2 systems. While PLC uses the existing electricity power lines, VDSL2 uses the existing telecommunication lines. As the lines of the two networks (the electricity and the telecommunication) are relatively close to each other, it is very important to consider the mutual impact of PLC and VDSL2. VDSL2 technology is described in the ITU-T Recommendation G993.3 (International Telecommunication Union – Telecommunication standardization sector) [25], and telephone companies are concerned about the influence of the PLC transmission on the delivery of services over VDSL2, where the two technologies overlap in frequency range (band plans used by both technologies are 1.8 MHz–30 MHz for the PLC and 138 kHz–30 MHz for VDSL2).

In order to ensure the coexistence of both technologies in home networks, it is important to characterize the interferences produced on the telephone lines by PLC and those produced on the electric lines by VDSL2. When the coupling between the two networks is known, it becomes possible to study and implement mitigation techniques allowing a transmission with good performances and acceptable QoS for both PLC and VDSL2 systems.

In the following, Section 3.7.1 discusses the ways coupling can happen and their relative importance. Section 3.7.2 then describes the influence of a PLC system on a VDSL2 transmission by paying particular attention to the impact on the Quality of Service (QoS) for the disturbed system. It is also explained how the different parameters may influence the coupling between the two technologies. Section 3.7.3 focuses on the impact of a VDSL2 transmission on the PLC performances. Finally, Section 3.7.4 elaborates on mitigation techniques allowing a transmission with good performances and acceptable QoS for both PLC and VDSL2 systems.

3.7.1 Characterization of the Coupling Between Power Line and Telecommunications Line Inside the Home Environment

Electromagnetic interference occurs when a signal generated by one system is transferred through electromagnetic coupling to another system. Interference generally appears in the

Table 3.4   Different types of coupling

Conductive	Occurs through a common path. Conductive coupling occurs on different circuits with a common reference conductor system.
Capacitive	Occurs through the electric field. Capacitive coupling occurs between neighboring circuits such as between power circuits and signal lines.
Inductive	Occurs through the magnetic field. Inductive coupling is caused in parallel running lines such as cables and cable ducts.
Wave or Radiation influence	Occurs through the electromagnetic field.

form of electrical voltage, current, electrical and magnetic fields. The different coupling mechanisms are summarized in Table 3.4 [26].

Interfering signals are generally transferred along a conductive wire (guided interference) or through space (radiated interference). These two types of interferences can be found together and be coupled to inputs, outputs, and the power supply and communications lines. The relative importance of each coupling is driven by the ratio between wavelengths of the interfering signal and the characteristic measurements of the system components under study. If the wavelength of the interfering signal is greater than characteristic measurements of the system conductive, capacitive and inductive coupling may occur independently of each other. Otherwise, the influence mechanisms are governed by:

- Wave influence, which is a combination of capacitive and inductive coupling of parallel lines. In this case, a propagation wave constitutes the source of interference.
- Radiated coupling through space, which leads to the transfer of energy across the electro-magnetic field to the receivers.

The principal types of interference are common mode and differential mode interference. Usually, interference occurs as a common mode signal which then produces an interfering differential mode signal due to asymmetries of the circuit (see Section 3.2.1). This differential-mode signal is then superposed to the (differential-mode) useful signal. Any unbalance in the electrical circuits will create such common mode–differential mode conversion, which will have an impact on the performances of the transmission and acceptable QoS for both PLC and VDSL2 systems.

The amount of coupled interferences and their impact on the useful signals depend on different parameters like length of and distance between electrical and communications lines, electrical network topology, impedances of the disturbed lines, etc. The reduction of the electromagnetic effects is generally undertaken by adopting EMC measures which are based on the following two principles:

- Measures taken at sources of interference in order to limit their transmission;
- Measures taken in order to limit the spreading of interferences.

3.7.2   Influence of the PLC Transmission on the Delivery of Services over VDSL2

The influence of a PLC system on a VDSL2 transmission is studied considering the following parameters, which may influence the coupling between PLC and VDSL2 systems:

- power line load,
- signal frequency,
- cable lengths,
- unbalance of the telephone cable (bad connections in sockets, impedance discontinuities, presence of RC circuit, etc.), which affects the electromagnetic immunity of the cable,
- separation between power cable and telephone cable (affects the electromagnetic coupling),
- coupling length (length the two cables are in close proximity of each other),
- category of telephone cable (twisted pairs and rate of twists), which affects the electromagnetic immunity of the cable,
- and crosstalk noise in the telephone cable (induced noise by adjacent pairs/cables), which is also affected by the category of the telephone cable.

There are only few contributions reported in the literature that address the coexistence of both technologies in home networks, also since the VDSL2 technology is not yet extensively deployed [27], [28], [29], [30], [31]. However, the recent works [30], [31], [32] have started to address the possible degradation of QoS and appropriate interference mitigation techniques for both PLC and VDSL2 systems. In the following, the main results of the extensive studies on the coexistence between PLC and VDSL2 systems in [30], [31], [32] are presented and discussed. In order to highlight the coupling effect, two approaches have been followed: laboratory measurements, studying the PLC influence on VDSL2 considering different coexistence scenarios, and field test measurements in several houses, where coupling noises were measured on telephone outlets when PLC signals were transmitted through electric plugs.

3.7.2.1   Laboratory Tests

In the laboratory tests, a realistic 'Fiber To The Cabinet (FTTCab)' configuration was considered. Figure 3.20 shows the experimental setup, which includes a 12 m PLC link in contact with a VDSL2 link leading to a strong electromagnetic coupling.

The telephone line consists of a 28-pair shielded cable connected to a 4-pairs cable (category 3). This latter part of the telephone cable is close to the power line over a length of 12 m. The power line cable is connected to the electrical network via a power strip and has an HPAV compliant PLC modem at each end. The flow of data between the two modems is ensured by a pair of laptops with Ethernet flow generator software. In the FTTCab architecture, an MDU (Multi Dwelling Unit) supposed to be installed in a street cabinet is emulated and connected to the VDSL2 CPE (Customer Premise Equipment) by means of a telephone cable. The VDSL2 CPE is connected to the same power strip as the PLC modems in order to take into account the conducted interferences via its power supply. The VDSL2 CPE also provides the IPTV (Internet Protocol Television) service to the Home Gateway (Orange LiveBox in

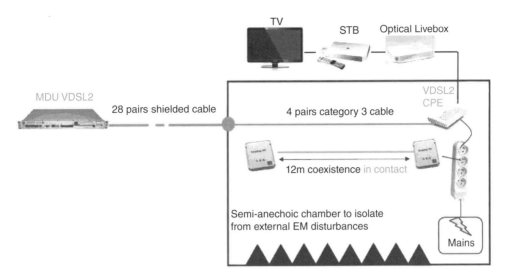

Figure 3.20   Experimental test-bed setup.

this case) connected to a Set Top Box (STB). In order to isolate the test bed from external disturbances, the coexistence area is placed into a semi-anechoic chamber. The measurements were conducted using balanced connections to inject and measure signals into the power cable and from the telephone line. The independent electrical network of this chamber protects the systems from potential noise generated by external devices. Also, a shielded cable connects the MDU to the semi-anechoic chamber entrance. In this way, the measured mutual impacts only come from the test bed and not from any other external disturbances.

3.7.2.1.1   *Influence of the Power Line Load*

The power load influence was not considered in [30] as the study was carried out for a given load. However, the effect of loads has been analyzed in [29] and references therein. A very interesting result reported there was that the loads connected to the power line have a clear effect, especially in the regime where the conducted field dominates. When the radiated field dominates, the overall interference level is independent of load, but the position of the peaks of the PLC signal is governed by the value of the load. As the characteristic impedance of the line is also frequency dependent, the line may show reduced transmission quality at some frequencies and improved quality at other tones. The same effect may occur for the properties of the interference channel.

3.7.2.1.2   *Influence of the Unbalance of the Telephone Cable*

In real installations, the phone line unbalance is generally due to a complex topology of the home networking, bad connections in sockets, impedance discontinuities, and presence of RC circuit (e.g. a 3-pin device typical in France). As seen in Section 3.7.1, such unbalance creates a common-to-differential mode conversion. Here, the unbalance of the telephone line is created by adding a third wire in the phone plug.

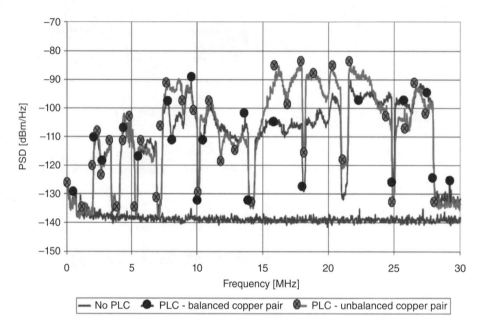

Figure 3.21   PSD induced by PLC on the adjacent copper pair – CPE side.

The curves of Figure 3.21 show the PSD of the stationary noise measured on the phone cable at customer side when:

- no PLC signal is carried by the electrical line (i.e. noise floor),
- a PLC signal is transmitted on the electrical line in contact with a balanced telephone cable,
- a PLC signal is transmitted on the electrical line close to an unbalanced telephone cable.

It can be observed that the stationary noise level increases when the PLC signal is transmitted and can increase by 50 dB in the case of a balanced telephone phone line and by 55 dB for the unbalanced one. The induced noise measured at customer side has an impact on the VDSL2 downstream channels (flow from MDU to VDSL2 modem), whereas noise measured at MDU side has an impact on upstream channels (flow from VDSL2 modem to MDU). Additional tests have shown that the level of the induced noise by PLC on phone lines at the customer side is the same regardless of the length of the telephone cable.

3.7.2.1.3   Influence of the Telephone Cable Length
Since the cable length are on the order of wavelength for MHz PLC signals, the dominant source of interference is electromagnetic radiation. In order to have a qualitative characterization of the role of telephone cable length on the QoS degradation due to PLC-signal coupling, the VDSL2 link carries an IPTV service and an PLC flow is launched. Table 3.5 show the impact of PLC on VDSL2 as a function of phone line length. In this table, '++' means that PLC transmission has no impact on IPTV (no pixels, no freezing), and '− −' means that an interruption of IPTV service occurs due to a synchronization loss of the VDSL2 link. It is interesting to confirm the considerations made in the previous sections on the role of the unbalance of the line with

Table 3.5   PLC impact on VDSL2 service (IPTV)

VDSL2 line's length (m)	50	125	200	275	350
Balanced copper pair	++	++	++	− −	− −
Unbalanced copper pair	− −	− −	− −	− −	− −

regard the level of interferences and their impact on VDSL2 transmission. For example, it can be seen that unbalanced lines are much more sensitive to the disturbances generated by the PLC link. For a VDSL2 access with a target noise margin of 10 dB, even short lines are impacted by the neighboring PLC transmission (loss of synchronization of VDSL2 line and interruption of IPTV service). For a balanced line with a length smaller or equal to 200 m, the PLC transmission has no influence on TV service. When the line length is greater than 200 m, the attenuation of the VDSL2 signal become important and consequently the induced signal will have more impact on IPTV.

Table 3.6 shows an example of the impact on VDSL2 performances for a telephone line length of 200 m. This impact is expressed in terms of decrease of noise margin (in dB) and decrease of bit rate (in %). For the balanced line, the values presented just above the decrease of the noise margin are related to the gap between VDSL2 performances at initialization (without PLC) and with PLC (the VDSL2 link is still turned on when the PLC flow is launched). For the unbalanced line, the values presented above the decrease of the noise margin and the bit rate are related to the variation of VDSL2 performances before and after the re-synchronization generated by the PLC system. It is important to mention that even in the case where the PLC system has no impact on IPTV service, it may still affect the signal carried on the VDSL2 line. For example, it can be seen that in presence of a balanced line the upstream noise margin decreases by 15.5 dB, which will make the VDSL2 line more sensitive to its electromagnetic environment (typically to impulse noises).

3.7.2.1.4   *Influence of the Coupling Length*
The coupling length is the distance over which the telephone cable and power line are close to each other or in the same duct. Depending on the customer premises configuration, this coupling length is a variable parameter. Measurements were conducted using a 275 m long balanced telephone cable with different coupling lengths, ranging from 30 cm to 12 m. The interference values are shown in Figure 3.22, and Tables 3.7 and 3.8 refer respectively to the

Table 3.6   PLC impact on VDSL2 Performances – 200 m VDSL2 line

	Downstream		Upstream	
	Noise Margin	Bit Rate	Noise Margin	Bit Rate
Balanced VDSL2 line	−0.9 dB 9.8 ⇒ 8.9 dB		−15.5 dB 16.9 ⇒ 1.4 dB	
Unbalanced VDSL2 line	+0.2 dB 9.8 ⇒ 10 dB	−28% 104 ⇒ 74.8 Mbit/s	−4.7 dB 14.7 ⇒ 10 dB	−2% 58 ⇒ 56.8 Mbit/s

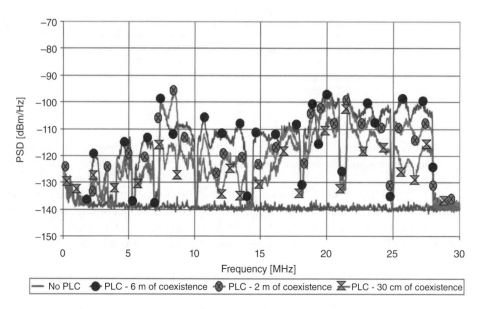

Figure 3.22 Influence of the coupling length on the interference level.

Table 3.7 PLC impact on VDSL2 service (IPTV) – influence of the coupling length

VDSL2 line's length (m)	200	275
Balanced telephone cable – Coupling length: 12 m	++	– –
Balanced telephone cable – Coupling length: 6 m	++	– –
Balanced telephone cable – Coupling length: 2 m	++	– –
Balanced telephone cable – Coupling length: 30 cm	++	++

Table 3.8 PLC impact on VDSL2 performances – influence of the coupling length

Coupling Length	Downstream		Upstream	
	Noise Margin	Bit Rate	Noise Margin	Bit Rate
12 m	+0.3 dB	−26.8%	−1.4 dB	−0.5%
	10 ⇒ 10.3	104 ⇒ 76 Mbits/s	11.7 ⇒ 10.3	38.9 ⇒ 38.7 Mbits/s
6 m	+0.2 dB	−16.6%	−2.1 dB	−2%
	10 ⇒ 10.2	104 ⇒ 86 Mbits/s	12.5 ⇒ 10.4	38.4 ⇒ 37 Mbits/s
3 m	−0.2 dB	−5.5%	−2.1 dB	−0.1%
	10.2 ⇒ 10	104 ⇒ 98 Mbits/s	12.4 ⇒ 10.3	37.7 ⇒ 37.6 Mbits/s
30 cm	+0.4 dB	—	−7.7 dB	−2%
	10 ⇒ 10.4	104 Mbits/s	12 ⇒ 4.5	38 ⇒ 37 Mbits/s

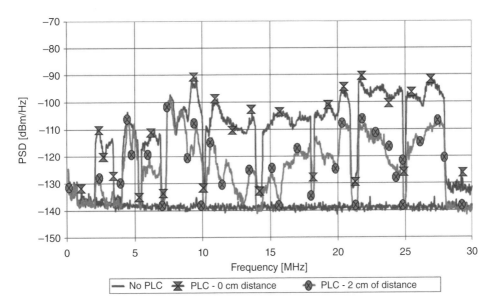

Figure 3.23   Influence of the separation distance on the interference level.

associated impacts on IPTV service and on VDSL2 performances as a function of the coupling length.

As expected, the level of interference induced by PLC on the telephone cable decreases for shorter coupling length. However, even at a small coupling length of around 2 m, interference represents a strong challenge for the impact on VDSL2 service. When the coupling length is equal to 30 cm, the QoS of VDSL2 is improved (no major impact of PLC on VDSL2 performance).

3.7.2.1.5   *Influence of the Separation Distance Between the two Cables*

The distance between the power cable and the telephone cable is also an important feature of an in-door network to predict the mutual impacts of both technologies. This influence is measured using a coupling length of 12 m and a 200 m long telephone cable. The results are shown in Figure 3.23, corresponding to the case where the interference is measured on the balanced telephone cable without PLC and with PLC transmission on the power cable located at a distance of 0 and 2 cm, respectively.

The level of interference decreases when the distance between wires increases. As seen in Section 3.7.1, such behavior is influenced by the guided interference contribution (capacitive and inductive coupling) and the radiated field contribution. The contribution of the capacitive and inductive coupling vanishes after a certain distance and the radiated field becomes the dominant contribution. The impact of the separation distance on the VDSL2 service is shown in Table 3.9 for two lengths of the telephone cable. It can be observed that the VDSL2 service is not impacted by PLC as soon as the cables are 2 cm away from each other. Indeed, adding some distance between cables may constitute a measure to take in order to reduce the impact of PLC on VDSL2 transmission and to improve the QoS.

Table 3.9 PLC impact on VDSL2 service (IPTV) – influence of the separation distance

VDSL2 line's length (m)	200	275
Balanced telephone cable – Separation distance: 0 cm	++	– –
Balanced telephone cable – Separation distance: 2 cm	++	++
Balanced telephone cable – Separation distance: 10 cm	++	++

3.7.2.1.6 *Influence of the Category of the Telephone Cable*

Telephone lines are twisted to decrease the interference that may affect the signal carried on these lines. The category of the cable refers to the rate of twists and some other characteristics of the cable. Such category has a strong impact on the electromagnetic immunity of the cable since it governs the level of the interfering signal due to the common mode–differential mode conversion which is related to the symmetry of the circuit. In this part, two categories of telephone cables are considered in the customer premises: a category 3 and a category 5 cable. The influence of the category of the cable on the coupling between PLC and VDSL2 links is shown in Figure 3.24. As expected the coupling is more pronounced for a category 3 cable, which has a weak balance in comparison to that of a category 5 cable. This behavior is more evident for frequencies higher than 10 MHz.

Table 3.10 shows PLC impact on the IPTV service for different VDSL2 line lengths. With a balanced line and a category 5 cable, PLC has no more impact on IPTV service, whatever line length. With an unbalanced line, the category 5 cable does not improve the quality of IPTV service. Indeed, as observed in Figure 3.21, when unbalance of the line is created by adding a third wire in the phone plug, the coupling is quite the same for category 3 and category 5 cables.

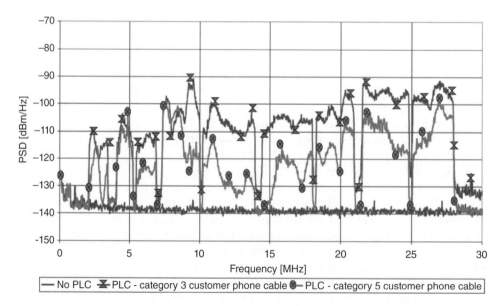

Figure 3.24 Influence of the category of the telephone cable.

Table 3.10 PLC impact on VDSL2 service (IPTV) – influence of the category of the cable

VDSL2 line's length (m)	50	125	200	275	350
Cat. 3 balanced copper pair	++	++	++	− −	− −
Cat. 5 balanced copper pair	++	++	++	++	++
Cat. 3 unbalanced copper pair	− −	− −	− −	− −	− −
Cat. 5 unbalanced copper pair	− −	− −	− −	− −	− −

3.7.2.1.7 *Influence of Crosstalk Noise in the Telephone Cable*

The category of the cable will also influence the interference induced through inductive and capacitive couplings by the signals carried on other pairs of the telephone cable. In other words, it is one of the parameter which governs the level of the noise induced on the pair carrying the VDSL2 signal by the other signals carried on the adjacent pairs. In order to evaluate the PLC coupling effect on VDSL2 transmission in the presence of crosstalk, the following two VDSL2 neighboring scenarios were considered:

- Scenario 1: Two VDSL2 links located on quads different from the third quad carrying the reference signal VDSL2.
- Scenario 2: One VDSL2 link located on the same quad carrying the reference signal VDSL2 and one VDSL2 link located on another quad.

In both scenarios, the VDSL2 link is synchronized by taking into account the crosstalk noise, after that the PLC signal is transmitted. Table 3.11 shows the IPTV QoS for different VDSL2 line lengths and for the considered scenarios. It is interesting to note that there is no significant influence of crosstalk noise on the PLC impact on VDSL2. (This is clearly shown in the first line of the table corresponding to VDSL line length of 50 m.) For greater lengths the impact on QoS is due to the attenuation of the VDSL2 signal as seen previously in Table 3.5. For the 350 m line length, there is no loss of synchronization in presence of PLC transmission with scenario 2. Such behavior can be explained by the fact that the noise induced by PLC transmission is lower than the crosstalk noise induced in the downstream frequency bands. Therefore the PLC transmission has no impact on VDSL2 QoS.

Figure 3.25 shows different noise measurements for the VDSL2 link and a 600 m telephone line length. It can be noted that in some downstream frequency bands the noise induced by crosstalk (Scenario i and ii curves) can be greater than the noise induced by PLC transmission.

Table 3.11 PLC impact on VDSL2 service (IPTV) – influence of the crosstalk

VDSL2 line length	No crosstalk	Scenario 1	Scenario 2
50 m	++	++	++
350 m	− −	− −	++
600 m	− −	− −	− −

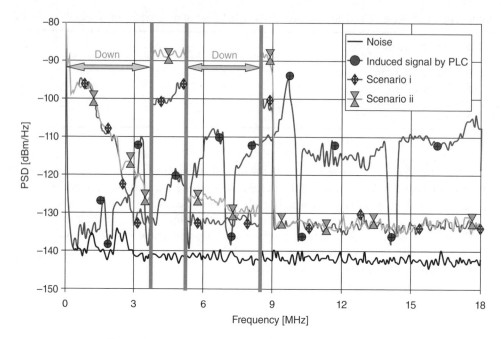

Figure 3.25   Influence of the crosstalk in the telephone cable.

3.7.2.2   Field Test Measurements

To complement the laboratory tests, this section presents the results of coupling noise measurements carried out at 31 different customer premises. The coupling noises were measured at telephone outlets in the presence of a PLC signals transmitted through electric plugs. This measurement campaign allowed a statistical analysis of the PLC interferences and assessment of their impact on VDSL2 data rates [31].

Figure 3.26 shows the experimental setup, where an Arbitrary Waveform Generator is used as a PLC transmitter. A PLC signal (HPAV-compatible) is continuously injected into an electric outlet using a passive coupler. The interferences coupled into the customer's telephone copper pairs are directly measured in the telephone outlets with a digital oscilloscope. The measurements were conducted using balanced connections to measure the induced signals on the telephone line. To minimize the influence of measurement equipment on the noise measurements, filtering extensions are used to isolate the Arbitrary Waveform Generator and the oscilloscope from the power line network.

3.7.2.2.1   Statistical Analysis of the Interference

For each telephone outlet in the 31 different customer premises, first the stationary noise without PLC transmission was measured, followed by measurements with PLC transmission. A total of 478 noise measurements were obtained.

The empirical cumulative probability density function of the 478 measured noises is shown in Figure 3.27. This is obtained by considering the mean value over frequency of each measured PSD. It can be seen that for 90% of cases, the mean values of the noise without PLC

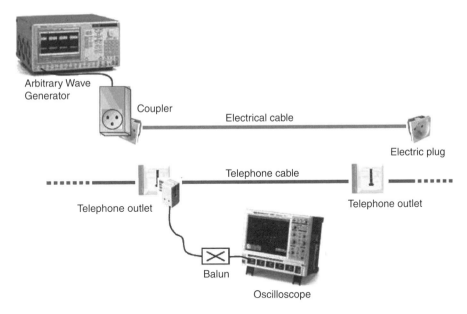

Figure 3.26   Measurement setup.

transmission are lower than −143 dBm/Hz, in comparison to −124 dBm/Hz when the PLC transmission is present.

3.7.2.2.2   PSD of a Typical Noise

The statistical analysis has also shown that there is a great difference ($\sim$ 50 dB) between the mean and maximum values of the noise PSD. This difference can be explained by the fact that the coupling varies greatly with frequency. In other words, this means that even in the case where the mean value of noise is low, the maximum value can have an impact on the VDSL2 performances. As an example, Figure 3.28 shows the PSD of a typical noise without and with PLC transmission. Obviously, the PLC coupling is stronger when the telephone and electrical outlets are close to each other, but it is important to note that the levels of induced noises are comparable to those obtained in the laboratory tests (see Section 3.7.2.1).

3.7.3   Influence of a VDSL2 Transmission on PLC

In this section, the impact of a VDSL2 transmission on PLC is studied. Figure 3.29 shows the noise PSD for the electrical line with and without VDSL2 signal transmitted on a balanced phone line of different length (50 m, 200 m and 350 m). The coupling length is equal to 12 m. It can be noticed that the level of the induced noise decreases when the length of the VDSL2 line increases. In particular, as the VDSL2 signal decreases in the downstream channel when the length of the link is increasing (due to the line attenuation), the level of the induced signal on the PLC link decreases.

Table 3.12 shows the impact of VDSL2 transmission on the PLC throughput. The reference throughput attainable without VDSL2 transmission is 40 Mbit/s. It can be seen that even for

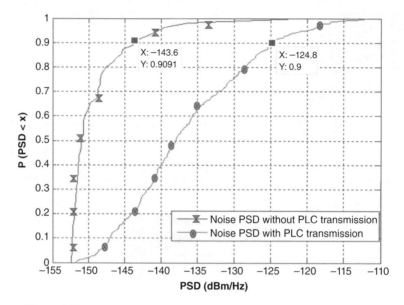

Figure 3.27   Cumulative distributions of the mean values of the noise PSD.

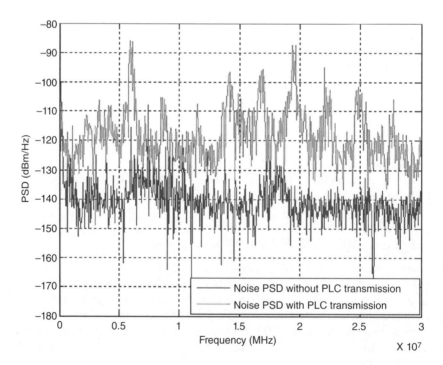

Figure 3.28   Measured noise PSD without and with PLC transmission.

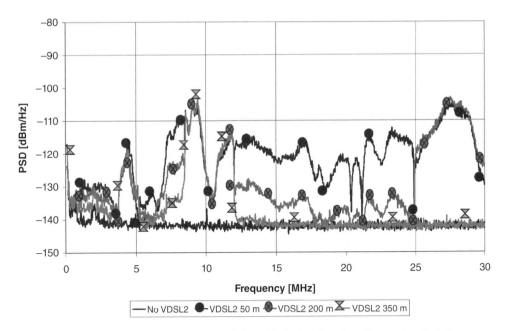

Figure 3.29   PSD of the interferences induced by VDSL2 on the adjacent electrical line.

a balanced line, VDSL2 transmission has an impact on the performances of the adjacent PLC link. This effect is more pronounced when the VDSL2 line is shorter than 200 m. Additional measurements have shown that the decrease of PLC throughput is proportional to the coupling length and in the case of an unbalanced VDSL2 line, the impact remains important for all coexistence lengths.

3.7.4   Summary and Ways to Mitigate Impacts

The main conclusions from the laboratory and field test measurements can be summarized as follows.

- When the VDSL2 link is built with a balanced copper pair:
  - PLC transmission has an impact on the VDSL2 IPTV service when the copper pair is longer than 200 m.

Table 3.12   PLC impact on VDSL2 service (IPTV) – influence of the crosstalk

PLC throughput decrease in Mbit/s (Initial PLC throughput = 40 Mbit/s)					
VDSL2 line length (m)	50	125	200	275	350
Balanced VDSL2 line	−22	−18	−13	−12	−3
Unbalanced VDSL2 line	−26	−26	−17	−14	−5

- VDSL2 transmission has an impact on PLC throughput when the copper pair is shorter than 200 m.
- With a well balanced copper cable (category 5) or with a separation distance of 2 cm or more between the two cables, the impact of the coupling between the two systems is negligible regardless of the copper line length;
- The coupling length or the presence of crosstalk has no significant influence on PLC and VDSL2 coexistence.
- When the VDSL2 link is built with an unbalanced copper pair, the mutual impact of PLC and VDSL2 is very important regardless of the different parameters considered for the experimental configuration.

In EMC, the mitigation measures to implement in order to solve the emission and immunity problems traditionally resort to techniques of filtering or shielding [26]. However, in the field of broadband services over wired telecommunications, the concept of 'cognitive EMC' [33] can be applied following the example of the cognitive radio [34]. The idea consists of lessening the EMC by means of a system capable of perceiving its environment, of interpreting it, of making decisions suited to react according to the constraints related to the electromagnetic environment. Section 3.5.3 gives examples where cognitive radio techniques are applied to avoid interferences while at the same time the usage of a frequency band can be allocated dynamically and thus more efficiently.

Some examples of solutions to mitigate interference between PLC and VDSL2 can be found in [35], [36] and references therein. Among the proposed solutions, splitter filters can be used to reduce the crosstalk of PLC signal to telephone cables produced in the home wiring. The concept of the splitter can be improved by adding an adaptive filter to suppress the PLC signal coupled on the telephone wires. Other promising solutions to mitigate the effect of the coupling are based on the spectral management in the downstream bands.

3.8   Final Remarks

The EMC behavior of PLC technology has been researched for more than 25 years. As the implementation of PLC products followed those research results, today's PLC products fit very well into the current EMC environment. However, when surfing for 'PLC' on the Internet, one will find many hits that report about EMC issues. This demonstrates the difficulty of acceptance for a relatively new technology and that issues where different stakeholders have opposite perceptions need time to be resolved. In this context, it is interesting to recall that the EMC standard for street cars has only been ratified almost 100 years after the first street car line has commenced operation.

New cognitive radio techniques have shown to be a good solution to efficiently use the available frequency resources while at the same time protecting already established services such as broadcasting radio stations. To further improve the spectral efficiency a state of the art dynamic power management is supported by the power line modems to decrease the transmit power in case of low attenuated power line links. With regard to EMC issues for the coexistence of PLC and VDSL2 technologies, it has been shown that it is important to characterize the coupling between the electric and telecommunication lines in order to implement mitigation techniques allowing a transmission with good performances and acceptable QoS for both PLC and VDSL2 systems.

References

1. CISPR 16-2-1: Specification for radio disturbance and immunity measuring apparatus and methods – Part 2-1: Methods of measurement of disturbances and immunity – conducted disturbance measurements, International Electrotechnical Comission (IEC) and International Special Committee on Radio Interference (CISPR) 16, Standard, 2008.

2. PLCforum, Measurement results of radiated emissions from PLC by means of the coupling factor, conference presentation 2000.

3. K. Küpfmüller, *Einführung in die theoretische Elektrotechnik*. Springer Verlag, 1990 (in German).

4. Verordnung zum Schutz von öffentlichen Telekommunikationsnetzen und Sende- und Empfangsfunkanlagen, die in definierten Frequenzbereichen zu Sicherheitszwecken betrieben werden (Sicherheitsfunk-Schutzverordnung-SchuTSEV, Bundesnetzagentur für Elektrizität, Gas, Telekommunikation, Post und Eisenbahnen, Bundesgesetzblatt Nr. 26, May 2009 (legislation, in German).

5. Frequenznutzungsplan, zitierte Nutzungsbestimmungen, Bundesnetzagentur für Elektrizität, Gas, Telekommunikation, Post und Eisenbahnen, NB30, May 2006 (in German).

6. M. Koch, H. Hirsch, and M. Heina, Derivation of the extrapolation factor for powerline communications radiation measurements, in *Proc. IEEE Int. Symp. Power Line Commun. Applic.*, Pisa, Italy, Mar. 26–28, 2007, 336–341.

7. Vehicles, boats and internal combustion engines – Radio disturbance characteristics – Limits and methods of measurement for the protection of on-board receivers, International Electrotechnical Comission (IEC) and International Special Committee on Radio Interference (CISPR) 25, Draft standard, 3rd edition, 2008.

8. Carrier current systems, including broadband over power line systems, amendment of Part 15 regarding new requirements and measurement guidelines for access broadband over power line systems, U.S. Federal Communication Commission, FCC 04-29, 2004.

9. M. Wächter, M. Koch, C. Schwing, and H. Hirsch, The extrapolation factor for PLC radiation measurements in the 150–500 kHz frequency range, in *Proc. IEEE Int. Symp. Power Line Commun. Applic.*, Johannesburg, South Africa, Mar. 24–27, 2013, 220–224.

10. IEC 61000-4-4: Electromagnetic compatibility (EMC) – Part 4-4: Testing and measurement techniques – electrical fast transient/burst immunity test, International Electrotechnical Comission (IEC), Geneva, Standard, 2004.

11. IEC 61000-1-1: Electromagnetic compatibility (EMC) – Part 1-1: Application and interpretation of fundamental definitions and terms, International Electrotechnical Comision (IEC), Geneva Standar, 2004.

12. IEC 60050-161: International electrotechnical vocabulary, chapter 161: Electromagnetic compatibility, International Electrotechnical Comission (IEC), Geneva, Standard, 1990.

13. Signalling on low voltage electrical installations in the frequency range 3 kHz to 148.5 kHz, Part 1: General requirements, frequency bands and electromagnetic disturbances, European Committee for Electrotechnical Standardization (CENELEC), Standard EN 50065-1, 2001.

14. Information technology equipment – immunity characteristics – limits and methods of measurement, International Electrotechnical Comission (IEC) and International Special Committee on Radio Interference (CISPR) 24, Standard, 1997.

15. Information technology equipment – radio disturbances characteristics – limits and methods of measurement, International Electrotechnical Comission (IEC) and International Special Committee on Radio Interference (CISPR), Geneva 22, Standard, 2008.

16. N. Weling, SNR-based detection of broadcast radio stations on powerlines as mitigation method toward a cognitive PLC solution, in *Proc. IEEE Int. Symp. Power Line Commun. Applic.*, Beijing, China, Mar. 27–30, 2012, 52–59.

17. ——, Field analysis of 40.000 PLC channels to evaluate the potentials for adaptive transmit power management, in *Proc. IEEE Int. Symp. Power Line Commun. Applic.*, Rio de Janeiro, Brazil, Mar. 28–31, 2010, 201–206.

18. EMC Directive 2004/108/EC, European Commission, Directive, 2004.

19. Commission recommendation on broadband electronic communications through powerlines, European Commission, Recommendation, Apr. 2005.

20. Treaty of Rome, European Union, Article 100, 1957.

21. Standardisation mandate, European Commission, Mandate 313, Aug. 2001. [Online]. Available: http://cq-cq. eu/M313.pdf

22. Criteria for the assessment of radio interferences caused by radiated disturbances from wire-line telecommunication networks, Electronic Communication Committee (ECC) within the European Conference of Postal and Telecommunications Administrations (CEPT), ECC Recommendation (05)04, Jun. 2005.

23. Draft questionnaire to the national standardisation organisations on setting the limits for permissible radiated disturbance emissions from telecommunication networks, CENELEC/ETSI JWG on EMC of wire-line telecommunications networks, Draft, 2003.

24. Electromagnetic compatibility zone webpage. [Online]. Available: http://www.iec.ch/zone/emc/

25. Measurement results of radiated emissions from PLC by means of the coupling factor, ITU-T, Recommendation G.993.2, International Telecommunications Union, Nov. 2000.

26. P. Degauque and A. Zeddam, *Compatibilité Electromagnétique 1 – des concepts de base aux applications (in French)*. Hermes Science Publications, 2007, (in French).

27. K. Kerpez, Broadband powerline (BPL) interference into VDSL2 on drop wires, in *DSLForum 2007*, Beijing, China, May 2007.

28. A. Bergaglio, U. Eula, M. Giunta, and A. Gnazzo, Powerline effects over VDSL2 performances, in *Proc. IEEE Int. Symp. Power Line Commun. Applic.*, Jeju Island, Korea, Apr. 2–4, 2008, 209–212.

29. M. Bshara, L. van Biesen, and J. Maes, Potential effects of power line communication on xDSL inside the home environment, in *Int. Seminar Electr. Metrology*, João Pessoa, Brazil, Jun. 17–19, 2009, 7–11.

30. F. Moulin, P. Péron, and A. Zeddam, PLC and VDSL2 coexistence, in *Proc. IEEE Int. Symp. Power Line Commun. Applic.*, Rio de Janeiro, Brazil, Mar. 28–31, 2010, 207–212.

31. B. Praho, M. Tlich, F. Moulin, A. Zeddam, and F. Nouvel, PLC coupling effect on VDSL2, in *Proc. IEEE Int. Symp. Power Line Commun. Applic.*, Udine, Italy, Apr. 3–6, 2011, 317–322.

32. B. Praho, R. Razafferson, M. Tlich, A. Zeddam, and F. Nouvel, Study of the coexistence of VDSL2 and PLC by analyzing the coupling between power line and telecommunications cable in home network, in *URSI General Assembly and Scientific Symp.*, Istanbul, Turkey, Aug. 13–20, 2011, 1–4.

33. A. Zeddam, Environnement électromagnétique & télécommunications: vers une CEM cognitive, in *Colloque International et Exposition sur la Compatibilité Electromagnétique*, Paris, France, May 20–23, 2008 (in French).

34. J. Mitola and G. Q. Maguire, Jr., Cognitive radio: Making software radios more personal, *IEEE Personal Commun.*, 6(4), 13–18, Aug. 1999.

35. B. Praho, Application de la compatibilité électromagnétique cognitive dans un contexte courant porteur en ligne, Ph.D. dissertation, Institut National des Sciences Appliquées de Rennes, France, Jan. 2012 (in French).

36. Powerline telecommunications (PLT); study on signal processing improving the coexistence of VDSL2 and PLT, European Telecommunications Standards Institute, Tech. Rep. ETSI TR 102 930 V1.1.1, Sep. 2009.

4

Coupling

C. J. Kikkert

4.1 Introduction

In this chapter, we consider coupling of power line communication (PLC) signals to and from low voltage (LV), medium voltage (MV) and high voltage (HV) power lines. Typically HV covers voltages from 66 kV and above, MV covers from 7.2 kV to 33 kV and LV covers 110 V to 400 V. A PLC coupler allows us to connect a PLC modem to HV/MV/LV power lines and inject a PLC signal to be transmitted onto the power line, with a low coupling loss. The same coupler also allows us to recover a PLC signal to be received from the power line with a sufficient signal-to-noise ratio (SNR) to permit the error free decoding of the PLC signal. The power lines used for PLC can be above ground power lines on towers or poles and they can be underground cables. Figure 4.1 shows a basic PLC system for automated meter reading (AMR), including the couplers. The coupler has to protect both humans and the PLC modems from the mains voltages, which vary from 110 V for LV power lines to 1000 kV for HV power lines.

Power lines are a very noisy communications medium and to obtain the best SNR at the receiving modem, the insertion loss of the transmitter couplers should be as low as possible. That requires the impedance of the power line port of the coupler to be matched to the power line, and the impedance of the injection ports to be matched to the cable connected to the coupler and the PLC modem hardware. Often this impedance is 75 Ω or 50 Ω, so that a low loss, low cost coaxial cable can be used in-between the coupler and the modem. For receiving PLC signals, the insertion loss of the coupler is less critical since normally the man made noise is much larger than the thermal noise, so that the received SNR stays the same, even with a reasonable amount of coupler loss. Since each coupler is normally used for both transmitting and receiving, a low coupling loss is important.

A PLC system for LV lines is different from a system on MV or HV lines. LV lines cover short distances and have many branches. There is one modem operated by the electricity distributor which connects to up to a hundred customer modems for smart metering applications as shown

---

Power Line Communications: Principles, Standards and Applications from Multimedia to Smart Grid, Second Edition.
Edited by Lutz Lampe, Andrea M. Tonello, and Theo G. Swart.
© 2016 John Wiley & Sons, Ltd. Published 2016 by John Wiley & Sons, Ltd.

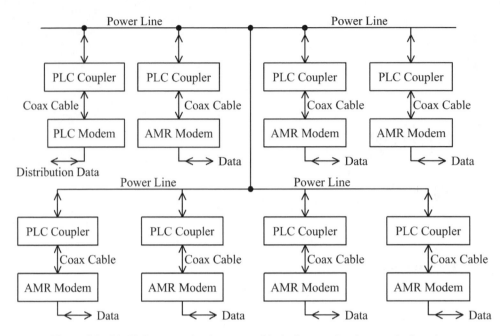

Figure 4.1   LV PLC communication system block diagram showing coupler locations.

in Figure 4.1. HV power lines are long and normally have no branches between switch yards. There is one modem at each end of the line and a different PLC system may cover the next HV power line at the other side of the switch yard.

In Figure 4.1, the PLC modem at the top left is the modem controlled by the electricity distributor and that sends data to all of the customer PLC modems, which have been labeled AMR modems, to distinguish them from the modem operated by the electricity distributor. For LV PLC systems, the AMR modem and PLC coupler are normally integrated, however since this is a chapter on PLC couplers, the modem is shown as a separate unit connected via a coaxial cable.

Depending on the country, the LV mains frequency voltage is 110 V to 230 V at 50 Hz or 60 Hz (phase to neutral) for LV lines. The common mode or zero sequence characteristic impedance for LV underground power cables is typically 10 Ω to 50 Ω, depending on the line current carrying capability and around 225 Ω for overhead lines on cross-arms [1]. For LV lines there normally are many branches, many PLC couplers and several hundred domestic and industrial appliances connected to the power line. As a result the impedance seen on the power line varies significantly with time and frequency. Cavdar [2] showed that in the 10 kHz to 150 kHz frequency range, the measured line impedance varies between 1 Ω to 20 Ω and tends to increase with frequency. Kim [3] measured the impedance of power lines in the 1 MHz to 30 MHz frequency range and found the impedance to vary widely from < 10 Ω to > 300 Ω, with an average impedance of about 50 Ω. Because of this wide variation in power line impedance and the time and frequency impedance variation of the loads connected to it, it is very difficult to obtain a low loss coupling and thus maximum power delivery of the PLC signals onto the power line without providing some impedance adjustment settings as part of the coupler design.

For the system shown in Figure 4.1 the impedance of the transmitting modem is ideally matched to the characteristic impedance of the transmission line, to provide a low coupling loss onto the line and put as much PLC signal onto the LV power line. Consider an LV power line with no branches containing a PLC network, with one modem transmitting at one end of that line and say 20 modems receiving the PLC signal at various locations along the power line. If the power line at PLC frequency has a 50 Ω characteristic impedance, then a transmitting modem with a 50 Ω will be well matched to put its power onto the power line. If a receiving modem presents the same 50 Ω impedance across the power line, the first modem absorbs half the PLC power on the line, the next modem absorbs half of the remaining power, leaving a quarter of the PLC power, the next modem absorbs half of that power, leaving one eight of the power and so on. The 8th modem will thus only be able to absorb 1/256 of the transmitted power and the power available at the 20th modem on the line will be -60.2 dB of the transmitted power. That is unlikely to be sufficient signal for error free operation of the modem. If the receiving modem causes a 500 Ω impedance across the power line, the first modem only takes 1/11 of the transmitted power, so that $10/11 = 0.909$ of the transmitted power is available at the second modem. The signal power at the 20th modem is then 0.148 of the transmitted power or -8.27 dB of the transmitted power. This should be a big enough signal to permit error free reception. Each modem absorbs a PLC signal -10.4 dB of the signal power at the modem, so that the last modem absorbs a signal power of -18.7 dB below the transmitted signal power. Restricting the receiving power at each modem is normally not a problem, since the signal and man-made noise are many times larger than the thermal noise, so that SNR is not reduced with this -18.7 dB loss. For LV PLC systems, it is thus normal practice to have a low output impedance for a transmitting modem and a higher input impedance for a receiving modem.

For LV operation, a coupler must thus have a good frequency response for both a low modem source impedance and a high modem load impedance. In addition the coupler must be designed such that it has a low coupling loss for transmitting the PLC signal onto the transmission line and the load presented by the coupler during reception should not cause much attenuation to the PLC signal on the LV line. For inductive coupling, the LV modems and coupling networks must thus be designed such that, the impedances that each receiving modem connects in series with the transmission line is much smaller than the impedance required for a low coupling loss. Similarly, for capacitive coupling the impedance placed by each receiving modem across the power line should be much larger than the impedance required for a low coupling loss.

For MV and HV lines there are few receiving modems for every transmission modem and the modem's transmitting and receiving impedances can be selected such that both result in a low coupling loss from the coupling network.

In the example in Figure 4.2, the PLC signals are of the order of 10 V on the power line, while the mains frequency signals for a 275 kV power line are 160 kV, line to ground, corresponding to a -84 dB wanted PLC signal to unwanted mains signal ratio. The coupler needs to protect humans and the PLC equipment by filtering out the mains frequency signals and its harmonics, such that they are much smaller than the PLC signals. This then allows the modem to detect the PLC signals without interference and demodulate those signals without errors. To ensure that all the Data 1 PLC signal, that is injected onto the line by the middle PLC coupler in the figure, travels to the right PLC coupler and that all the PLC energy on the line is then directed

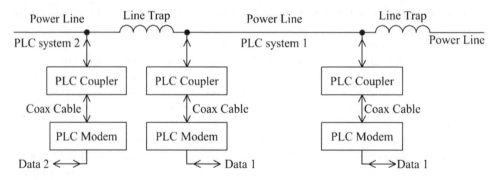

Figure 4.2   HV PLC communication system block diagram showing coupler and line trap locations.

into the right PLC coupler, without continuing along the power line, large inductors or line traps are inserted into the line as shown. These line traps present a high impedance and also prevent PLC signals from PLC system 1 to couple into PLC system 2.

One of the major challenges of PLC system design is the variation of the communication channel parameters, such as the channel frequency response, the characteristic impedance, the transmission losses and the level of man-made noise on the power line at PLC frequencies. These channel parameters determine the level of PLC signal required to obtain reliable transmissions. Usually power line channels have high and variable noise levels. This noise is often synchronized to the mains frequency. For LV networks, this noise is generated by electronic appliances, power inverters, power tools and other devices. For MV and HV networks the dominant noise sources are

(a) corona noise, which is an impulse-like discharge occurring every half mains cycle that varies with dust and moisture levels;
(b) impulsive noise produced by switchgear operation;
(c) arching and lighting, and
(d) converter noise produced by HVDC converter stations.

Much of this man-made impulsive noise is synchronized with the mains frequency and produces a line spectrum with mains frequency spacing at PLC frequencies. The power lines are also long wire antennae and long and medium wave AM radio transmissions will couple onto the lines if phase to ground coupling is used. There are significant frequency components from all these noise sources that occur in the PLC signal transmission band. The noise in PLC communication channels is thus time varying man-made noise rather than thermal noise. These high and variable noise and interference levels can cause a severe reduction in SNR and degradation of overall network reliability if the system is not properly designed.

Because LV lines can not be terminated in the correct impedance at PLC frequencies, reflections occur from the ends of the power lines and the many loads along those power lines, causing deep frequency notches at some frequencies. HV power lines are typically overhead lines, mounted on towers and the common mode characteristic impedances are well behaved and typical values are 400 Ω for phase-to-earth coupling and 600 Ω for phase to phase coupling on single conductor lines and 300 Ω for phase-to-earth coupling and 500 Ω for phase to phase coupling on bundle conductor lines [4]. Because these lines are typically several hundred kilometers long, they also make good antennae, so that PLC transmission on frequencies used by nearby AM radio stations should be avoided. Such interference between PLC and AM radio broadcasts is much less of a problem on LV lines.

4.2   Coupling Networks

This section describes the fundamental requirements and principles of coupling networks and how increasing mains frequency rejection can be achieved by implementing higher order coupling networks. Conventional Capacitive and Inductive couplers are described as well as a novel Inductive Shunt Coupler. For this section some knowledge of passive filters and their frequency responses is assumed. For those readers, who want to refresh their knowledge, Zverev [5], Huelsman [6] and others [7] provide this information.

4.2.1   Requirements

Basically a coupling network is a highpass filter that rejects the 50 Hz/60 Hz AC mains and passes the modulated power line signal. PLC coupling networks must firstly provide a low insertion loss for the PLC signal and secondly provide sufficient attenuation to the mains frequency voltages, so that these firstly do not present lethal voltages to humans and equipment or secondly degrade the ability of the modem to demodulate the PLC signal. For narrowband PLC (N-PLC), covering 9 kHz to about 500 kHz, the bandwidths required are: 35 kHz to 91 kHz for the CENELEC A band and 98 kHz to 122 kHz for the CENELEC B band in Europe, 155 kHz to 403 kHz for the ARIB band in Japan and 155 kHz to 487 kHz for the FCC band for the US and some other countries. In addition signals can be transmitted in the Broadband over power lines (BPL) or broadband PLC (B-PLC) frequency range of 1.7 MHz to 100 MHz. Throughout this chapter, this is denoted as B-PLC. There can be significant radiation from B-PLC signals on overhead power lines [1], and B-PLC systems on overhead lines may not satisfy the relevant IEC, FCC or EEC regulations relating to interfere to existing radio services. The radiation from broadband PLC on underground power lines is negligible. Due to the skin effect, the PLC attenuation on a power line increases with the square root of the frequency [1] (see also Chapter 2), thus limiting the coverage area of BPL and B-PLC.

The designs and frequency plots for coupling networks presented in this section cover the N-PLC (9–500 kHz) frequency range, with most networks and frequency plots covering a useable operation of 3 decade bandwidth. However many commercial couplers [8] cover about one decade of operation. Many B-PLC systems like HomePlug [9], which uses a 2–28 MHz bandwidth, require just over one decade bandwidth. For B-PLC, the PLC frequency is 2 decades higher than N-PLC and the mains frequency filtering requirements are eased. Since N-PLC has the wider percentage bandwidth and more severe mains rejection requirement, that frequency range is used for the coupler designs presented in this chapter. For B-PLC designs, the coupling networks presented in this chapter can easily be changed in frequency and mains rejection, to cover B-PLC operation.

The coupler needs to provide electrical isolation between the power line and the output of the coupler. Normally this is done by including an isolation transformer as part of the coupler.

This transformer can also provide an impedance transformation between the power line and the coupler output and thus reduce the insertion loss of the coupler. For a transformer with an $N_1 : N_2$ turns ratio, the impedance seen at terminal 2 is

$$Z_2 = \left(\frac{N_2}{N_1}\right)^2 Z_1. \tag{4.1}$$

For a low insertion loss of the coupler, the turns ratio of the isolation transformer is chosen to match the ratio of the line impedance and the required modem impedances, so that standard doubly terminated filter designs can be used for the coupler. If the transformer is placed in the middle of the high pass or bandpass filter then an impedance scaling needs to be applied to the components on one side of the transformer to match the impedance transformation of the transformer.

The standard EN 61000-3-2 [10] specifies the current harmonics for domestic appliances connected to the mains. For lighting, the 39th current harmonic must be less than 3%. For other appliances absolute and power related current limits apply. Assuming that 1 A is drawn by an appliance, then a maximum harmonic content of less than 7.1% is permitted at the 39th harmonic. Even harmonics are generally much less than the odd harmonics and voltage harmonics are generally less than the current harmonics. Induction coupling injects PLC currents onto the line, so that the couplers must be able to accommodate these mains harmonic signals. Figure 4.3 shows the attenuation obtained by different order high pass filters, used as PLC couplers. For convenience both the input and output impedances are assumed to be the same and Butterworth high pass filters are used [5]. The cut off frequency of the filters has been set to 30 kHz to cover the CENELEC-A, CENELEC-B, ARIB and FCC bands. The resulting attenuation provided by these different order couplers at the mains frequency and at the 40th harmonic have been tabulated in Table 4.1.

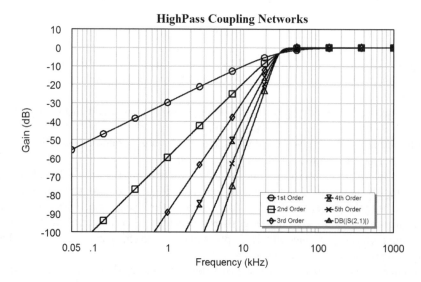

Figure 4.3 Frequency response of different order couplers with 30 kHz cut-off frequency

Table 4.1 Mains frequency attenuation (dB) versus coupler filter order for a 30 kHz cut-off frequency

Filter order	50 Hz	60 Hz	2.5 kHz	3 kHz
1	55.6	54.0	21.6	20.0
2	111	108	43.2	40
3	167	162	64.8	60
4	222	215	86.3	80
5	278	270	108	100
6	333	323	130	120

Table 4.2 shows the PLC to mains voltage ratio in dB obtained for typical LV, MV and HV voltages and modem sensitivities at the input to PLC couplers for N-PLC. Note that the PLC/Mains ratio in dB is negative since the mains voltage is much larger than the PLC signals. The coupler, together with any filters in the modem, must attenuate the mains voltage to get a PLC signal sufficiently larger than the mains voltage before demodulation by the modem. From Table 4.2, a 230 V mains voltage needs to be attenuated by 77.2 dB, to provide, for example, a 10 dB PLC/Mains signal ratio. Comparing Table 4.1 and Table 4.2, it can be seen that a second order filter is required to obtain that attenuation. To save costs, it is possible to use a first order filter, by

- increasing the cut-off frequency and using higher frequency N-PLC signals,
- relying on the power line inductive coupler or the RF transformer that provides electrical isolation (see e.g. Figure 4.4) to provide the extra attenuation required at the mains frequency, or
- one can rely on additional high pass filtering or AC coupling incorporated in the PLC modem to reduce the mains frequency voltages to the required level.

Alternately one can rely on the inductive impedance of an LV power line that is much shorter than a quarter wavelength and has low load impedances connected to it, to resonate with

Table 4.2 Typical SNR for various mains voltages

System	Vac	PLC Vmin	PLC/Mains dB
LV	110	0.1	−60.8
LV	230	0.1	−67.2
MV	3300	1	−70.4
MV	33000	1	−90.4
HV	66000	1	−96.4
HV	275	1	−108.8
HV	765	1	−117.8

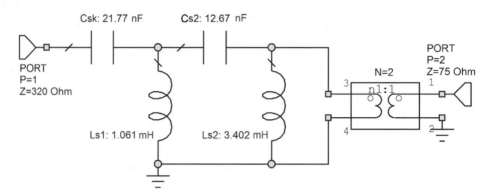

Figure 4.4   Fourth order coupling network, for HV lines.

the PLC coupler capacitance and provide a bandpass response to provide the required mains frequency rejection.

A greater than 10 dB return loss, indicating a good match between the coupler and the modem is required, to ensure that:

- there is a low TX/RX cross talk in the TX/RX hybrids that are included in most MV and HV PLC modems,
- there are no attenuation notches in the signal transfer from the coupler to the modem, and
- the coupler has a low return loss so that nearly all the PLC signal from the modem is coupled to the power line.

A 3 dB return loss results in a 3 dB coupler loss. PLC systems can work well with more than 20 dB coupling losses. However a 3 dB return loss will cause the TX/RX hybrid to dump half the transmitter power into the receiver input. This may cause serious problems. Details of transformer hybrids and their operation are shown in [11, Chapter 3].

4.2.2   Capacitive Coupling

Networks which couple voltages to and from the power line are either high pass filters, like the ABB A9BS couplers as shown in [12, Figure 10], or bandpass filters like the ABB A9BP coupler as shown in [12, Figure 11]. Figure 4.4 shows a typical simplified high pass filter similar to the one used by the ABB A9BS [12] capacitive coupler. To save cost the magnetizing inductance of the isolation transformer can be made to be the same as $Ls1$ or $Ls2$ and the transformer can then be used in place of that inductor. Capacitive couplers are convenient to use when the active conductor of the power line is readily accessible, as is the case for overhead power lines. To fit the coupler, the port 1 terminal of the coupler in Figure 4.4 is connected directly to the active line of a power line. For LV and MV lines specialized line crews can fit the couplers live and no power supply interruption is required. For HV lines, the couplers and associated line traps are normally mounted when the line or

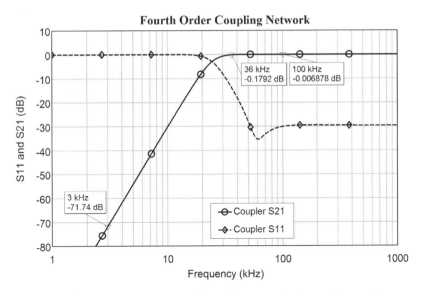

Figure 4.5   Frequency response of coupling network of capacitive coupler.

substation is built. For a line impedance of 300 Ω, the attenuation of the coupler in Figure 4.4 is less than 0.2 dB for all frequencies between 36 kHz and 1 MHz, so that a low coupling loss is obtained. The coupling network of Figure 4.4 is optimized for a 320 Ω transmission line and a 75 Ω impedance between the coupling network and the modem. A 2:1 turns ratio transformer is used, which matches the 75 Ω coaxial cable to a 300 Ω impedance for the power line. Any large variation from this 300 Ω impedance requires a change in transformer turns ratio or it will result in an increase in the insertion loss of the coupler. There is however a large range of impedances over which a return loss of more than 15 dB can be obtained. For the coupler with a 2:1 turns ratio of the transformer, transmission lines of 210 Ω to 420 Ω can be matched with a better than 15 dB return loss over a 36 to 500 kHz frequency range.

The frequency response of the circuit of Figure 4.4 is obtained using NI-AWR's Microwave Office [13] circuit simulation software and is shown in Figure 4.5. To reduce the mains voltage to less than 1 V for a 275 kV power line a 104 dB attenuation is required at 50/60 Hz. The mains signals are attenuated by more than 200 dB, effectively removing the mains frequency signal from the PLC signal output, even if the coupler is connected to a 275 kV power line. The 50th harmonic of the mains frequency occurs at 2.5 kHz for a 50 Hz system and 3 kHz for a 60 Hz system. The attenuation of the coupler is more than 71 dB for signals at 3 kHz. It is very likely that the 50th harmonic of the mains voltage is more than 20 dB below the fundamental, so that all harmonics will be much less than the PLC signal.

These couplers use an ideal transformer and have been optimized for frequency response, without considering the voltages across each element, or the reactive power drawn, both of which need to be considered for HV couplers. Practical RF transformers are described in Section 4.2.4 and the voltages across the elements are considered in Section 4.4.

Figure 4.6   Commercial PLC Inductive coupler [8].

4.2.3   Inductive Coupling

When the active conductor of a power line has an insulation material covering, it is more difficult to safely connect a capacitive coupler to the power line. In that case an inductive coupler is easier to fit to the power line. The inductive coupler consists of a split ferrite toroid, with a coil wound on it. That coil is connected to a BNC connector as shown in Figure 4.6. There are several commercial manufacturers of inductive couplers [8], [14], [15]. The toroid is clamped over the power line. In many cases, it is not necessary to remove power from the line. Inductive couplers have an advantage over capacitive couplers in this regard. The inductive coupler is basically a single turn transformer that is placed in series with the power line, as shown in Figure 4.7, and is thus called a coupling transformer in this chapter. The high pass filter network consisting of *L1*, *L2* and *C1* is attached across the secondary winding of the coupling transformer as shown in Figure 4.7. The inductance *L1* can be the magnetizing inductance of the coupling transformer, or it can be an external inductor. Using the correct value of magnetizing inductance has a cost advantage. At the mains frequency inductor *L1* causes a low impedance between ports P1 and P3, so that the voltage drop in the power line across the coupling transformer is negligible.

Figure 4.7 shows that at the desired PLC coupling frequency, the high pass filter reflects the terminating impedance, of typically 75 Ω, at the PLC port 2 across the coupling transformer. This impedance is normally larger than the impedance of the mains power lines at the desired PLC coupling frequency and most of the PLC signal at port P2 of Figure 4.7 is coupled onto the line. The coupling loss and the reflection coefficient looking into port P2 depends on the turns-ratio of the coupling transformer and the line impedances. Since these couplers are often

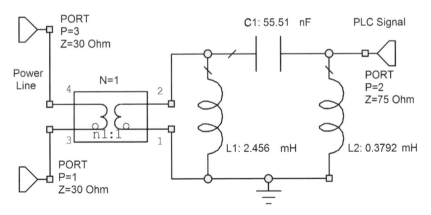

Figure 4.7 Inductive coupler schematic.

used on underground power lines, an impedance of 30 Ω is used for the network in Figure 4.7. For a 75 Ω PLC impedance at port 2, and a 30 Ω impedance at ports 1 and 3, a 1:1 turns ratio results in a 3 dB coupling loss from the power line to the PLC port.

4.2.4 Real RF Transformers

The circuits of Figures 4.4, and 4.7 use ideal RF transformers. Figure 4.8 shows the circuit model of a real RF transformer, such as those used for the isolation transformers in PLC couplers. A real transformer using ferrite cores typically has a ratio between the upper and lower −3 dB points of 300:1. For the inductive coupler circuit of Figure 4.7, a real transformer is included as part of the circuit and the component values are optimized to give a better

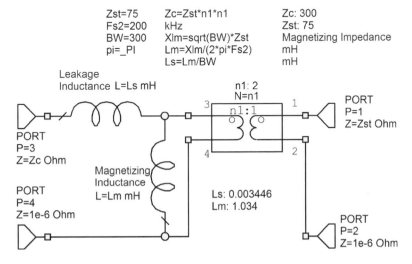

Figure 4.8 Real RF transformer model.

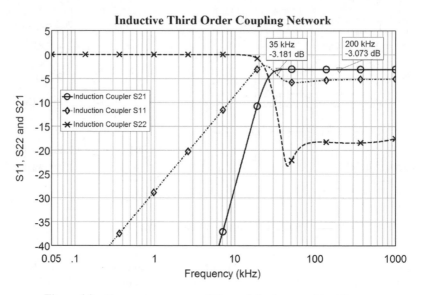

Figure 4.9 Frequency response of Figure 4.7 using a real transformer.

than 15 dB return loss and a flat frequency response from 35 kHz to 1 MHz. The resulting frequency response is shown in Figure 4.9. At 50 Hz the attenuation from port P1 to port P2 is 164 dB so that the mains frequency voltage is attenuated very effectively and will not cause any problems in the PLC modem. At 2.5 kHz, the 50th harmonic of the 50 Hz mains frequency, the attenuation is 62 dB, so that all harmonics of the mains frequency will also be reduced to acceptable levels.

For an ideal inductive coupling transformer, mains frequency currents cause $1/N$ times that current to flow through inductor $L1$ of Figure 4.7, where N is the turns ratio of the coupling transformer. To prevent the mains frequency current causing the coupling transformer to saturate, the transformer is made using a gapped toroid, with the two halves of the toroidal core being separated with a thin spacer. Such gapped transformers typically have a one or two decade bandwidth. Increasing the gap thickness increases the current carrying capability of the core at mains frequency before saturation occurs and reduces the bandwidth of the transformer. By designing the transformer to have a center frequency corresponding to the center of the desired PLC band, a significant attenuation will occur between the primary and secondary windings at mains frequencies. In commercial designs, a detailed mains frequency analysis needs to be carried out to ensure that neither the coupling transformer or $L1$, if it is a separate inductor, saturate. The voltage drop at 50/60 Hz across the primary winding, due to the mains frequency currents flowing in the power line must also be evaluated, to ensure that this voltage drop across all couplers on a power line is small enough so that the mains supply voltage to all consumers to the power line remains inside the allowable voltage limits.

To show the effect of real RF transformers on capacitive couplers, a basic coupler consisting of a single capacitor together with this transformer model is used, as shown in Figure 4.10. This capacitive coupler is placed across a power line, represented by the 20 Ω transmission line. The transformer is used to provide the 75 Ω to 20 Ω impedance transformation to match the

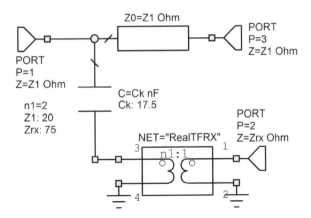

Figure 4.10 Second order capacitive coupler with an RF transformer.

modem to the power line. For the ideal transformer the capacitance Ck is tuned to give a 3 dB corner frequency of the terminal voltage gain at 30 kHz, resulting in a $Ck = 1$ μF capacitor. For the real transformer, both the capacitor value and the transformer center frequency and hence the magnetizing inductance Lm are tuned to give a flat passband response and a 30 kHz cut-off frequency. This resulted in $Lm = 0.138$ mH and $Ck = 187$ nF. The corresponding transformer center frequency is 400 kHz. The resulting frequency response of this coupling network for both an ideal and a real transformer model is shown in Figure 4.11. The only difference between the two networks is the transformer. The magnetizing inductance of the

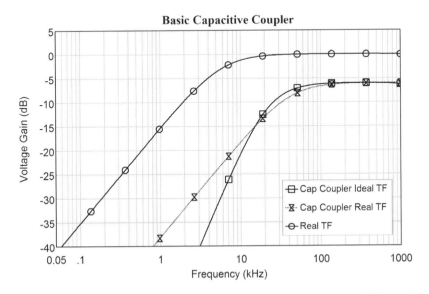

Figure 4.11 Frequency response of the circuit of Figure 4.10 with ideal and real RF transformers.

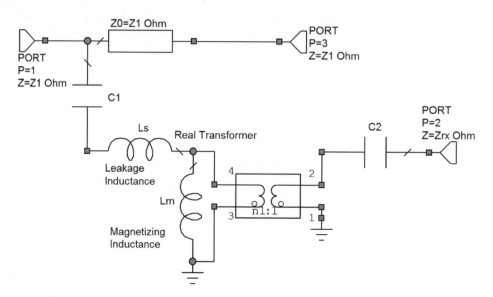

Figure 4.12   Third order capacitive coupler with a real RF transformer providing the filter inductance.

RF transformer results in a flatter frequency response in the 40–100 kHz frequency region and the transformer leakage inductance Ls results in a 6.5 MHz, −3 dB upper cut-off frequency. For the circuit with the ideal transformer, the terminal voltage attenuation at 50 Hz is 50 dB and as shown in Table 4.2 is not sufficient for operation at 230 V. For the circuit with the real transformer, the terminal voltage attenuation is 106 dB and is sufficient attenuation for operation on LV power lines.

In this example, the magnetizing inductance of the isolation transformer is used as part of the high pass filter in the coupling network. For most practical PLC coupling networks covering the 30 kHz to 2 MHz frequency band, the magnetizing inductance of the coupling transformer can be used as the inductance $L1$ in Figure 4.7, required for the high pass filter in the PLC coupler. The bandwidth of the RF isolation transformer is limited due to the transformer having a gap, which reduces the magnetizing inductance but does not change the inherent leakage inductance. The frequency response of the RF isolation transformer is thus a critical part in the design of the PLC coupler. Figure 4.12 shows the circuit a third order capacitive coupler using the magnetizing inductance of the isolation transformer for the inductance of the high pass filter. A significantly higher mains frequency attenuation is obtained at the minor expense of one capacitor $C2$.

4.2.5 Resistive Shunt

PLC coupling by injecting a voltage across a resistive shunt is also used for low current applications. Such shunts have resistances of < 10 mΩ to ensure that the voltage drop across them is negligible. In use, the resistive shunt is placed across ports 1 and 3 in Figure 4.7. Since most of the mains current flows through the shunt resistor, saturation in the isolation transformer can be avoided. However when injecting PLC signals onto the power line, most of the PLC current will simply loop through the transformer and resistive shunt and poor PLC coupling onto the power line is achieved. To increase the coupling efficiency, the impedance of the shunt needs to be larger than the load impedances connected to the power line, so that most of the PLC current will flow onto the power line rather than flow through the shunt only. This is achieved using the Inductive Shunt.

4.2.6   Inductive Shunt

Inductive coupling requires the power line to be the primary winding for a coupling transformer with a ferrite core as shown in Figure 4.7, to be clamped on the power line. Together with the high pass filtering network, which provides mains frequency isolation, this transformer forms the network shown in Figure 4.7. The high level mains frequency currents that normally flow in power lines may cause saturation of the coupling transformer. This saturation causes an increase in the PLC signal loss across the coupler. For many coupling networks where digital data communications are transmitted through the coupling network, some saturation can be tolerated. For other applications, such as ones where the PLC coupler is used to measure impedances of the power network and its loads at PLC frequencies, any saturation causes errors.

A resistive shunt is often used to measure mains frequency currents, however the small resistance of the resistive shunt results in a poor SNR and makes it difficult to inject signals onto the power line. When inserting a shunt in series with the power line, any reduction in mains voltage due to the shunt must be less than about 0.5% to ensure that the effect of inserting the shunt is minimal. Having a higher impedance across the shunt at the desired PLC frequency, results in larger voltages across the shunt and thus a better SNR for coupling PLC signals on or off the power line. Having a higher impedance of the shunt also results in a better coupling efficiency for injecting signals onto the power line. Since PLC frequencies are much higher than the mains frequency, having an inductance in series with the resistive shunt achieves this. Since the power factor of most loads are more than 0.8. The reactive impedance of the inductance in the inductive shunt causes a voltage drop that is close to 90 degrees angle with respect to the supply voltage. If the voltage across the resistive part of the shunt is 0.2%, then a 7.7% inductive voltage drop, together with the resistive voltage drop, will only cause less than a 0.5% supply voltage drop if a unity power factor load is assumed. For a 230 V LV supply with 100 A current, that corresponds to $Rsh = 4.6$ mΩ, and a mains frequency impedance for the inductive part of the shunt of 0.177 Ω, corresponding to an inductance $Lsh = 5.644$ mH. The circuit diagram of an inductive shunt with a 50 Ω load across it is shown in Figure 4.13. The 50 Ω load provides a constant impedance across the inductive shunt at PLC frequencies. The impedance seen across the inductive shunt is shown in the dotted line of Figure 4.14. Figure 4.15 shows the Inductive Shunt used as a PLC coupler. A real transformer model is used for the isolation transformer. The circuit includes a third order high pass filter consisting of $C1$, Lm and $C2$. Ports 1 and 2 are connected in series with the power line and port 3 is the PLC modem port. The coupling capacitor $C1$ in Figure 4.12 can be made up from two series capacitors $C11$ as shown in Figure 4.15 to provide additional electrical isolation and reduce the voltage stress on the isolating transformer. This PLC coupler has a 187 dB

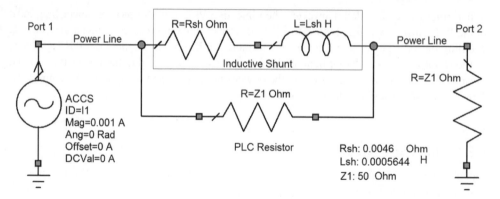

Figure 4.13   Basic inductive shunt.

attenuation at 50 Hz. A 50 Ω impedance is used in this illustration, as this coupler was designed to inject signals from a signal generator onto a power line, but other impedances can be used. A 1:1 turns ideal ratio transformer is included to provide electrical isolation. The number of turns on the isolation transformer windings is chosen to make the magnetizing inductance of this transformer to be the required inductance for the high pass filter. The capacitor *C11* is a high impedance at mains frequency and ensures that the mains currents do not cause any saturation of the isolation transformer.

To illustrate the use of the inductive shunt for PLC couplers, Figure 4.16 shows the frequency response of two PLC couplers mounted adjacent to each other or connected at either end of a

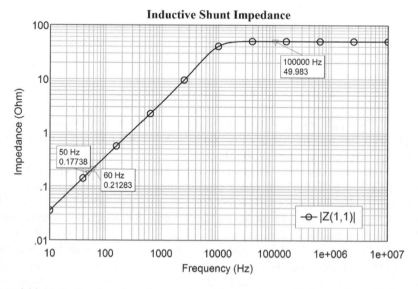

Figure 4.14   Inductive shunt impedance and transfer impedance of inductive shunt PLC coupler.

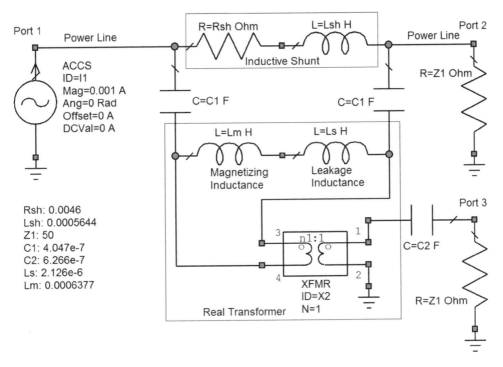

Figure 4.15   Inductive shunt PLC coupler with a real transformer.

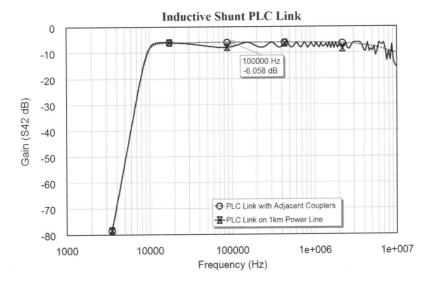

Figure 4.16   Transfer function of inductive shunt PLC links.

1 km long power line. A 50 Ω line impedance is assumed. That is larger than the impedance of underground power lines but smaller than the impedance of overhead lines. The standing waves on the 1 km long power line caused by the insertion of the inductive shunt into the transmission lines are evident. Figure 4.16 shows that the inductive shunt works well as a coupler from 20 kHz to the limit imposed by the attenuation of the power line or reflections due to impedance mismatches on it. By changing the impedance that the inductive shunt inserts in series with the power line at the desired PLC frequency, different coupling ratios can be obtained.

The inductive shunt makes a very accurate low loss PLC coupler that overcomes the saturation problems associated with inductive couplers. However like a resistive shunt, the inductive shunt needs to be inserted in series with the power line.

4.2.7   Modem TX and RX Impedances

For LV power lines there are many modems connected to the power line. For valid signal operation, one modem will be transmitting and all the others will be listening to the transmissions in order to determine if that transmission is addressed to that modem. For maximum power transfer, the impedance of the transmitting modem must be matched through the coupler to the power line. For maximum signal to be received by all the receiving modems, the impedance of all the modems in parallel must be matched to the power line through the couplers. The PLC modem must thus include different coupling conditions and different modem impedances for transmitting and receiving. For an LV N-PLC system (9 kHz to 500 kHz), the transmitter (TX) hardware has a low output impedance and the input impedance of the receiver (RX) hardware is a high impedance. Having different transmitter and receiving impedances and coupling transfer impedance requirements makes it more difficult to obtain a good frequency response from the PLC coupler. But a good system power transfer can be obtained as shown in the following example.

In this example an ideal design is assumed, where all the modem transmitted power is coupled onto the power line and all the PLC signal on the power line is coupled equally to each of the many receiver modems. A third order capacitive coupled modem connected to a 50 Ω source is used, as shown in Figure 4.17. The transmitter is assumed to have a 1 Ω,

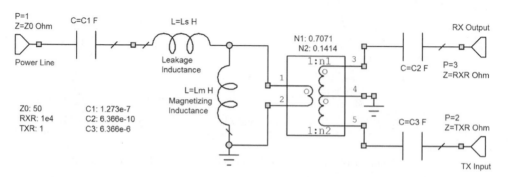

Figure 4.17   Circuit for coupler with different transmitter and receiver impedances.

(*TXR*), output impedance when transmitting and be an open circuit when not transmitting. This corresponds to typical modem hardware. The receiver hardware input impedance is 10 kΩ (*RXR*). The inductance of for the third order high pass filter is provided by the magnetizing inductance of the isolation transformer, by selecting the correct number of turns for the primary winding of the transformer.

The output impedance of the coupler to the power line when driven from a modem transmitter should thus be 50 Ω, *Z0*, for maximum power transfer. Firstly, the turns ratio of the secondary transmitter winding *N2* of the isolation transformer and the capacitor *C3* are selected to provide a low loss Butterworth high pass filter frequency response for the transmitter (Port 2 with an impedance *TXR*) to the power line (Port 1 with an impedance *Z0*) path, when the modem is transmitting. Each receiving modem should only couple one percent of the power from the power line, allowing 100 receiving modems to be used on the line. That corresponds to a 20 dB insertion loss. Then the turns ratio of the secondary receiver winding *N1* of the isolation transformer and the capacitor *C2* are selected to provide a Butterworth high pass filter frequency response, with the required 20 dB insertion loss, for the power line (Port 1 with an impedance *Z0*) to receiver (Port 3 with an impedance *RXR*) path, when the modem is receiving and the transmitter is an open circuit. Figure 4.18 shows the corresponding frequency response. Since the impedance of the receiver hardware reflected at the power line is 5 kΩ, it is normally not necessary to switch the receiver out of circuit when the transmitter operates. If isolation between the transmitter and receiver terminals is required, then a special transformer hybrid as described in [11, Chapter 3] can be used. Different impedance values can easily be accommodated.

In this example 100 receiving modems were assumed, with each modem using 1 percent of the PLC signal power. Since the noise on the power line is much higher than thermal noise, designing a PLC system for 100 receiving modems will result in each modem still having the same SNR in the PLC channel as a 10 receiving modem system, where each modem uses

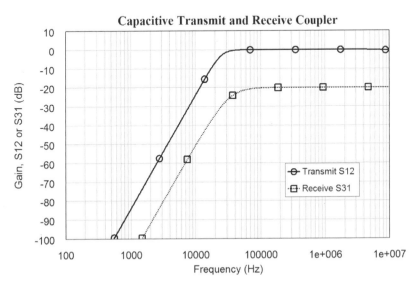

Figure 4.18   Frequency response of coupler in Figure 4.17.

10 percent of the PLC signal power. These transmit and receive impedance ratios assume an ideal transmission medium. Since the impedance of a real power line at PLC frequencies varies depending on the length of the power line and its branches and the impedances connected to it, in practice the frequency response of the coupler is not as well behaved as shown in Figure 4.18. Some power line branches will have lower level of PLC signals on them, requiring more PLC signal to be coupled from the power line for acceptable data bit rates. By providing a multiple-tapped secondary winding for the receiving winding, $N1$ and the corresponding value of $C2$, can be varied in the field to obtain the desired PLC receive power, thus giving the best performance. This example shows that a good PLC frequency response, a high mains frequency rejection and the correct transmitting and receiving coupling losses can be obtained by correct design.

4.2.8   Transformer Bypass Coupling

To minimize the cost of PLC distribution in an urban network, it is desirable to inject the PLC signals onto the MV network and pass the signals from the MV to the LV network and vice versa. Reference [16] showed it is possible to pass N-PLC signals through a transformer, but this causes a typical 40 dB loss. Reference [17] designs a coupler that connects to both the MV and LV lines of a transformer and allows a PLC signal to bypass a transformer with low loss, without affecting the system at mains frequency as shown in Figure 4.19. This coupler must present high impedances at its terminals at the mains frequency and thus prevent the coupler

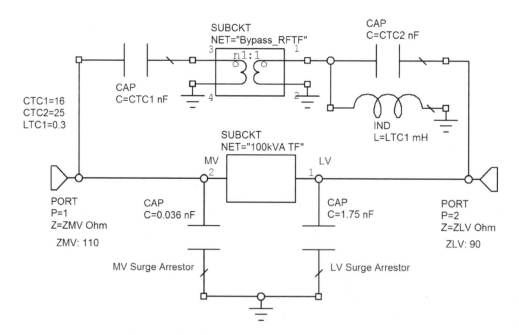

Figure 4.19   Circuit for a transformer bypass coupler.

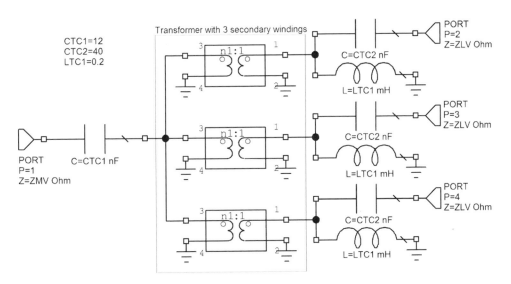

Figure 4.20   Circuit for a 3 phase transformer bypass coupler.

loading the mains. For a capacitively coupled bypass coupler, third or fifth order filters can be used, with capacitors being located at the ends of the coupler as shown in Figure 4.19. The transformer used in this figure is a model [17] of a 100 kVA SWER line isolation transformer with a 33 kV MV voltage and a 240 V LV voltage. The line impedances used in this figure are $ZMV = 260\ \Omega$ for the MV line impedance to ground and $ZLV = 110\ \Omega$ for the LV line impedance to ground. This corresponds to the overhead line impedances in a SWER network. To minimize costs, a 12 nF MV capacitor $CTC1$ is used and that is the smallest value that provides a good performance.

For a three-phase network, one MV line can couple onto 3 LV lines as shown in Figure 4.20. For underground cables, it is more convenient to use inductive coupling. The network of Figure 4.21 shows a circuit diagram for an inductive coupled transformer bypass coupler. Low

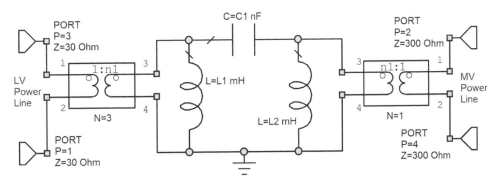

Figure 4.21   Circuit for an inductive coupler transformer bypass coupler.

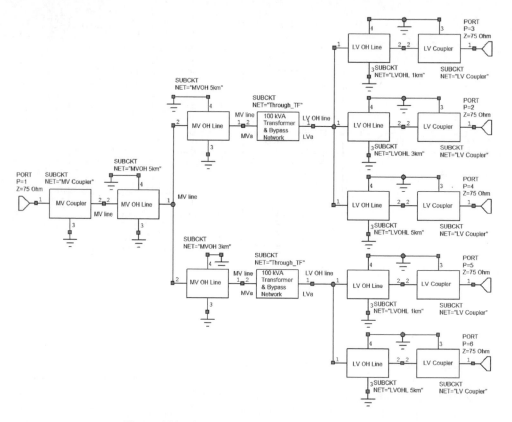

Figure 4.22   Circuit for a transformer bypass coupler.

impedance current paths should be able to exist for the MV and LV lines, where they connect into the transformer. Surge arrestors typically provide a 36 pF capacitance to ground at MV terminals and a 1.75 nF capacitance to ground at LV terminals of the transformer. If needed additional capacitances can be placed between the transformer's line terminals or between the line terminals and ground to provide a sufficiently low impedance PLC path. For the MV input, a transformer is capacitive at PLC frequencies [18], [19] and provides a low impedance path between the transformer's MV terminals. For the LV input of a transformer, the terminal impedance at PLC frequencies can be quite high and a low impedance path must be provided by placing a capacitor between the active lead and ground.

Example 4.1:   The reduction of the MV-LV path loss can be shown by considering the PLC network of Figure 4.22 [20], which contains both MV and LV transmission lines and 2 MV-LV transformers. The network of Figure 4.22 is used to demonstrate the benefits of the bypass coupling networks. The network uses accurate models of transmission lines and models of single phase SWER transformers [17] were used for convenience. Port 1 is connected to an

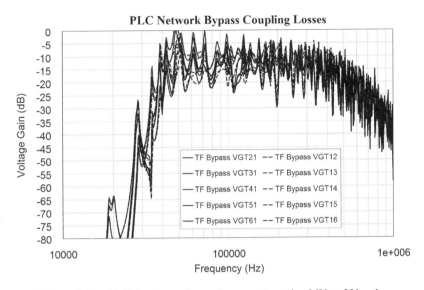

Figure 4.23   PLC signal transformer bypass attenuation MV ⇔ LV paths.

MV coupler, Ports 2 to 6 are connected to LV couplers and the TF bypass network is the circuit shown in Figure 4.19. The network has a branch in the MV network, each of which connect to a transmission line and distribution transformer, which then connects to an LV network with several branches. The resulting signal path attenuation with and without using bypass couplers at the distribution transformers is shown in Figures 4.23 and 4.24, respectively.

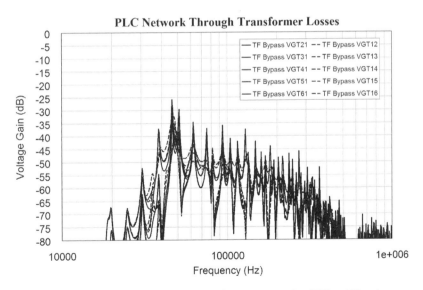

Figure 4.24   PLC signal through transformer attenuation MV ⇔ LV paths.

Comparing Figures 4.23 and 4.24 shows the effect of the bypass coupler on the signal loss between the MV port and all the LV ports and between all the LV ports and the MV port. A PLC signal with an ideal 5 way power split has a 15.6 dB loss for each of the 5 output ports. Figure 4.23 shows that for all the MV to LV and LV to MV connections the average attenuation is close to this ideal 15.6 dB attenuation over the 35.9 kHz to 90.6 kHz frequency band, resulting in a good signal quality when bypass couplers are used. Figure 4.24 shows that for the same MV to LV network without the transformer bypass couplers, the attenuation is an average of about 55 dB over the 35.9 kHz to 90.6 kHz frequency band. This is close to 40 dB more attenuation than the same network with transformer bypass couplers, and results in a poor signal quality. For mains frequency the voltage gain from MV to LV or LV to MV is determined by the transformer turns ratio and is very different in both directions, due to conventional transformer action. But for PLC frequencies, the loss in both directions is similar, as shown in Figures 4.23 and 4.24, where the solid lines are for MV to LV transmissions and the dotted lines are for LV to MV transmissions.

4.2.9   Reactive Power and Voltage and Current Ratings

The coupling capacitor $C1$ in Figure 4.17 draws reactive power as given by

$$Q = \omega CV^2 = 2\pi f CV^2 . \tag{4.2}$$

For Figure 4.17, the 127 nF capacitor $C1$ at 230 V and 50 Hz absorbs a reactive power of 2.1 VAR, which is negligible compared to all the other loads connected to the power line. For MV and HV power lines the reactive power can be significant and must be evaluated as part of the design. For example a 13 nF capacitor used for $C1$ of Figure 4.17 on a 275 kV line (159 kV line to ground) results in a 103 kVA reactive power, which is significant. Depending on the values of the components used in the coupler and the terminating impedances, the capacitor Csk of Figure 4.4 can have a voltage that is up to 50% greater than the power line voltage across it. Those voltages may exceed the capacitor's ratings and the coupler design may need to be changed to reduce the voltage across the components. This is discussed in more detail in the section on HV couplers. Many power lines carry currents of several hundreds of Amperes. If inductive coupling is used, the inductive couplers, like the one shown in Figure 4.6 must be capable of carrying those currents without saturating.

4.2.10   Uncertainties

So far in this chapter we have assumed a known line impedance. However in practice the LV line impedance at both N-PLC and B-PLC frequencies changes over time, as consumers switch devices on and off. The mains frequency impedance connected to the power line is inversely related to the power load applied to the network. The night-time power demand is much smaller than the early evening load. The impedances seen at N-PLC and B-PLC frequencies, depend on the impedances at those frequencies of the loads connected to the power line and that will thus also vary significantly during the day. Since many of these loads also generate high frequency switching noise, the noise level will also vary widely during the day.

Since many loads contain switch mode power supplies or rectifiers, large current pulses occur at the peak of the mains voltage and the impedance seen at PLC frequencies may be much smaller near the peak voltage than near the zero crossing of the voltage.

The network topology can also cause large impedance variations with frequency. The combination of power line length and connected load impedances on a branch power line, may cause a short circuit to be reflected on the main line, at a frequency inside the PLC band, thus making communications at that frequency impossible. The impedance variations result in increased coupling losses and the changes in noise level may effect the bit error rates of the channel. Despite these uncertainties, practical couplers can tolerate insertion losses of up to 80 dB and as a result practical couplers can perform very well, even under these adverse conditions.

4.2.11   Summary

This section described the principles, requirements and different network configurations that can be used for PLC couplers. The designs and frequency plots for coupling networks presented in this section cover the 30–500 kHz N-PLC frequency range, with most networks and frequency plots covering a useable operation of 3 decade bandwidth. By a simple frequency scaling process, these networks can be adapted to operate in the 1-100 MHz B-PLC frequency band. Circuits for passing PLC signals around distribution transformers have been described. Such circuits allow for many LV PLC modems to communicate with modems operated by electricity suppliers on the MV network. The following sections describe how these principles are applied to specific market areas, such as LV Smart Grid and Power line Local Area Network (LAN) applications and MV and HV protection and switching circuits. In the following sections implementations by certain manufacturers are described. These are just examples where some manufacturers have made their hardware details available either on the web, or directly to the author.

4.3   LV Coupling

4.3.1   Introduction

In Europe, a Smart Grid system [21] is predominantly being implemented using N-PLC. This will allow remote meter readings to be done using Advanced Meter Infrastructure (AMI) and will allow users or electricity providers to improve supply efficiency and reduce costs by smart scheduling of power demands, like delaying the use of dishwashers, swimming pool filters or electric car charging to periods of low electricity cost. Much of such PLC communication will be done on the LV network. In Europe and many other countries the mains power supply is generally 230 V and a hundred customers can be supplied from one distribution transformer. The one N-PLC or B-PLC network can thus cover many households. In the USA the LV is 110 V and because of the higher current, each distribution transformer normally services a few houses only, making LV PLC less easy to implement than in Europe.

B-PLC couplers are readily available to provide a LAN for Internet access in different rooms using the mains power cabling in a house [9]. Most of these units conform to the HomePlug Power Alliance specifications [9] or the HD-PLC Alliance [22]. Both these systems conform to the IEEE 1901 [23] standards. In this section all these devices will be called Power line devices, as that is what many manufacturers call them. These Power line applications operate

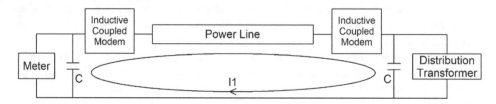

Figure 4.25   Two inductive coupled modems used in a PLC system.

in the 2–28 MHz frequency band. It is likely that fridges, washing machines, air-conditioners, dishwashers and other household appliances will include such Power line devices in the near future, to allow them to become part of a smart home automation control system. Power line devices will thus be the dominant PLC modems in the near future. Since these devices are consumer devices, the PLC couplers in them must be low cost, reliable and safe.

4.3.2   N-PLC Couplers

For Smart Grid and AMI applications, G3-PLC [24] and PRIME PLC [25] both operate in part of the CENELEC-A frequency band, which is set aside for exclusive use by energy suppliers. The frequency band from 35.9 kHz to 90.6 kHz is used by G3-PLC and PRIME uses 41.992 kHz to 88.867 kHz in Europe. G3-PLC can also use the FCC 155 kHz to 487 kHz frequency band.

The full 35.9 kHz to 487 kHz bandwidth used by these systems is just in excess of one decade, so that gapped inductive couplers can be used. The gapped inductive couplers can simply be clamped over the power line, without requiring any additional connections. A current transformer injects a PLC current onto the power line. To ensure that most of the energy coupled onto the line by the PLC transmitter is absorbed by the PLC receiver, this current should flow relatively unimpeded and the impedance of the loads connected to the power line at PLC frequencies need to be small. This can be achieved by having capacitors across the power line at strategic places, as shown in Figure 4.25. Inductive couplers are very suitable for clamping onto underground cables near distribution transformers. The surge arrestors placed on the LV terminals of transformers provide the required capacitance shown in Figure 4.25.

Capacitive couplers are very simple to connect to the power line, they apply a voltage to the power line. If a capacitive coupler is placed next to an electricity meter for an AMI application, then the capacitive coupler will cause a current, $I1$ to flow into the meter, as shown in Figure 4.26. As shown in Figure 4.26, a current $I2$ will flow along the power line. For good

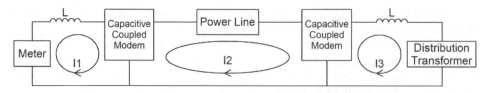

Figure 4.26   Two capacitive coupled modems used in a PLC system.

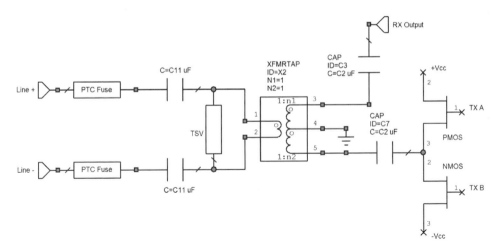

Figure 4.27 Capacitive coupled PLC coupler with protection devices and different input and output impedances.

coupling efficiency, *I2* should be much larger than *I1* or *I3*. As shown in Figure 4.26, inductors can be placed between the capacitive coupler and the meter or between the capacitive coupler and the distribution transformer to ensure that both *I1* and *I3* are small.

Figure 4.27 shows a typical PLC coupler used for AMI. This circuit is very similar to Figure 4.17, but includes a typical output stage and hardware protection devices. The element *PTC Fuse* is a Positive Temperature Coefficient (PTC) fuse. Such a fuse opens when the device gets too hot due to excessive current, but as the device cools down it resets itself. These fuses provide protection but do not need to be replaced each time a fault occurs. The element *TSV* is a transient voltage protection device, like a gas discharge device or two back-to-back Zener diodes of the appropriate rating, such that they only conduct if the normal voltage is exceeded. The transient voltage protection together with the overcurrent protection provided by the PTC fuse ensures that the PLC coupler survives temporary overload conditions. Sometimes a modem connection is made at a power point. Since in Europe and the USA, the power outlets are not strongly polarized, having two PTC fuses and capacitors *C11* provides extra isolation and reduces the possible voltage between the primary and secondary windings of the isolation transformer if the active and neutral line are swapped. In Figure 4.27, different secondary windings and different capacitors are used to provide the different impedance matching for transmission and reception as described in Figures 4.17 and 4.18. This will result in efficient coupling for both the transmitter and receiver. For applications where it is certain which line is the active line and which line is the neutral, some costs can be saved by using one capacitor and one PTC fuse. In practice [26] more complex bandpass filtering networks are used instead of capacitors *C2* and *C3* in Figure 4.27.

4.3.3 B-PLC Couplers

Most B-PLC couplers are used for power line applications and the typical operating frequency is 2–28 MHz. These devices include a B-PLC modem and LAN hardware to either allow the

Figure 4.28 Inductors used in power line coupler [27].

device to be a WiFi hub, or to allow hardwired LAN connections to be made to computers, TV's or other devices. Since these LAN couplers are normally plugged into two or more different power points, the block diagram of Figure 4.26 applies. If the power line coupler does not include a power socket, then current $I1$ and $I2$ in Figure 4.26 is zero and the inductors shown are not required. If the power line modem includes an AC pass through socket, so that a mains appliance can be plugged in as well, then inductors must be included between the AC pass through socket and the PLC coupler, to ensure that $I1$ and $I3$ are minimized. Figure 4.28, shows these inductors, the PTC fuse, capacitors and transformer used for a typical PLC coupler [27]. Data rates of 600 Mbps can easily be achieved on normal house wiring.

The PLC couplers for power line applications normally are capacitive couplers similar to the one shown in Figures 4.27 and 4.17, but designed for the 2–28 MHz frequency range. Different transmit and receive impedances are incorporated by having multiple secondary windings on the isolation transformer connected to different filter networks as described in Section 4.2.7.

4.3.3.1 Impedance Matching

For accurate coupler design, the PLC frequency impedance of the LV lines on which the couplers are placed should be known. The PLC frequency power line impedance varies with

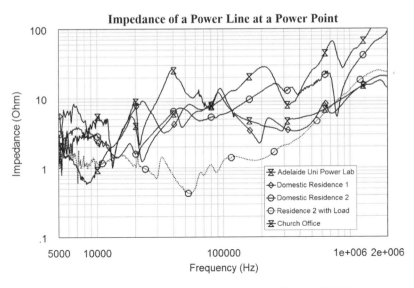

Figure 4.29 Measured impedances of power lines as N-PLC.

time as different loads are connected to the grid, and also depends on the length of the power line and its branches. Not many power line impedance measurements are published. One of the difficulties is that these impedances must be accurate and carried out while the network is powered up. Cavdar [2] and Kim [3] show a capacitor being used to isolate the mains frequency, but do not show how this capacitor affects their measurement or the accuracy of their measurements. Cavdar reported impedance measurements in Europe between 1 Ω and 20 Ω for N-PLC frequencies. Kim reported measurements in the 1–30 MHz frequency range and showed impedance values from about 5 Ω and 220 Ω. The PRIME Alliance TWG [25] has been doing impedance measurements over various places in Spanish electricity grid, Lu [28], shows the LV impedances measured at several Spanish substations, impedances range from 0.1 Ω to 1.5 Ω, however details of the measurement system are not provided. Figure 4.26 shows that without isolation inductors L, the impedance of the power line in parallel with the distribution transformer will be seen by the right capacitive coupler modem. By comparing these results with those by Cavdar, it appears that the impedance measured is that of the transformers. Using isolation inductors L allows On-line impedance measurements to be performed on only the power line or the only the distribution transformer. Figure 4.29 shows some power line impedance measurements made by Kikkert from 5 kHz to 2 MHz, using an Inductive Shunt Impedance Analyzer [29]. The solid line measurements were made at power points without any equipment being plugged into other power-points nearby. The dotted curve shows the measured impedance when typical office equipment is connected to the same power-point. Figure 4.29 shows that the impedance of the office equipment dominates in the N-PLC frequency band. The measured impedance is much lower than the power line characteristic impedance because there are many devices connected in parallel across the power line at each house, factory or office. These measurements from Cavdar, Lu and Kikkert and the block diagrams in Figures 4.25 and 4.26, suggest that because of the very low transformer and equipment impedances, inductive couplers will couple much more PLC signal power onto the power line than capacitive couplers, unless isolation inductors L are used.

The B-PLC impedances are generally higher than the N-PLC impedances and they vary more rapidly over the frequency band, since at B-PLC frequencies many power lines are close to a quarter wavelength and reflect an open circuit at the end of that line as a short circuit at the other end. Since the impedances of appliances for 110 V LV systems are one quarter those for 220 V systems, the impedances at PLC frequencies will also be smaller for 110 V systems than for 220 V systems. Because the connected loads, and thus the line impedance, are time varying, it is difficult to have a precise impedance match for the PLC coupler. However modern PLC frequency communication systems have sufficient performance headroom that communication with low errors is achieved, even with significant impedance mismatches. Switched line impedance selection can also be incorporated in PLC coupler designs to overcome these impedance variations.

4.3.4   Phase-to-phase Coupling

In some countries or some other applications such as street lighting, there is no neutral or ground wire available for power line connections. In this case, phase-to-phase coupling is required to connect the PLC modem to or from the power line. The circuit for phase to phase coupling is the same as is shown in Figure 4.27 but the Line+ is connected to one active phase and the Line- terminal is connected to the other active phase. In this situation, two capacitors $C11$ as shown in Figure 4.27 are used to reduce the voltage stress between the primary and the secondaries of the isolation transformer.

4.3.5   Single Phase Coupling

Most houses in 220 V countries and many in 110 V countries have a single phase power supply. As a result any PLC distribution over the LV network must be distributed such that PLC signals are carried on all three supply phases. Having three identical PLC couplers with the same modem connection, as shown in Figure 4.30, can supply all three phases from one data source. To accommodate phase to phase coupling for any of the three phases, it is possible to generate the three PLC signals with a 120 degree phase difference between each power line, just like the three-phase mains supply. Alternately, if the PLC distribution comes from one phase of the MV supply, a three phase bypass coupler as shown in Figure 4.20 can be used.

4.4   HV Coupling

PLC has been used on HV lines for more than 100 years [30], to provide voice and data communication for power line operators located at both ends of these long power lines. The initial PLC system used analogue transmissions using amplitude modulation (AM), later systems used single sideband modulation (SSB), modern digital systems use quadrature amplitude modulation (QAM), or digital multicarrier modulation, such as OFDM. A PLC system on an HV transmission line typically has a block diagram as shown in Figure 4.2. These lines can be up to 1000 km long and still accurately support PLC data transmissions. The HV lines

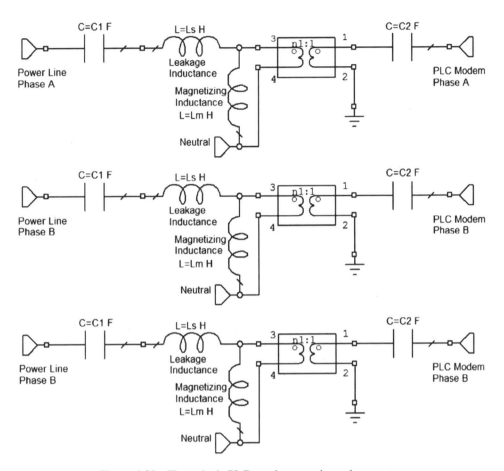

Figure 4.30 Three single PLC couplers on a three-phase system.

normally do not have branches between the switch-yards at each end of the line. As a result the characteristic impedance of the line is well defined and the impedance seen by the modems is primarily that of the HV line. The characteristic impedance of each phase of an HV line is typically 400 Ω for single conductors and 300 Ω for bundle-conductors. The end of the transmission line is terminated in a switching bus-bar which then may be connected to an HV power transformer. The HV impedance of such transformers is capacitive at PLC frequencies [19], so that a line trap must be used to ensure that the PLC signals are coupled into the modem. Figure 4.31 shows HV PLC line-traps and modems in a switch-yard being constructed. These line-traps are very large and expensive, and it is an advantage if a PLC system can be designed so that no line traps are required.

Amperion [31] is just one of the companies that make capacitively coupled MV and HV PLC couplers for broadband PLC covering the 1.7 MHz to 10 MHz frequency range for transmission line protection communication. The broadband PLC system uses a very wide bandwidth, so that channel allocations can be made without channels needing to be re-used. They claim their system does not need line traps. The 1–10 MHz frequency band includes AM

Figure 4.31   HV switch-yard with line-traps and couplers, courtesy ABB.

broadcast transmissions and the broadband PLC system could create significant radiation and interference if used on overhead HV power lines. With capacitive couplers and no line traps, there will also be a significant mismatch at the end of the transmission lines due to the bus-bars in the switch-yard and the capacitive impedance of HV transformers at PLC frequencies, causing reflections or multi-path signals at the end of the transmission lines, thus degrading the PLC data error rate. With inductive coupling, these low impedances assist in creating a current loop for the PLC communication. Inductive coupling is more expensive to implement on very high voltage transmission lines than capacitive coupling. If the CENELEC-A band is used, then the bandwidth is much smaller and channels need to be re-used, thus requiring line traps in the PLC system, regardless of end of line reflections.

Tables 4.1 and 4.2 show that for HV lines, a mains frequency attenuation about 10 dB greater than 117.8 dB is required from the PLC coupler. This can be obtained by either a third or a fourth order coupler network. ABB [12] for their A9BS and A9BT high pass coupling units use a fourth order coupling network, similar to Figure 4.4. The A9BP and A9BR bandpass coupling units [12] contain a combination of a second order low-pass followed by a second order bandpass network as shown in Figure 4.32. The bandpass response will reject high

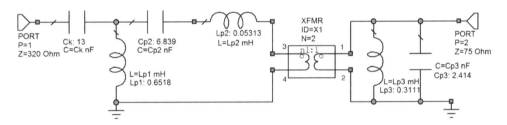

Figure 4.32 HV bandpass coupler circuit.

frequency signals as well as low frequency signals. The upper cut-off frequency must be chosen to be less than 1 MHz for the A9BP and A9BR units. For both these coupling units, the capacitor connected to the HV line, *Ck* in Figures 4.4 and 4.32, is mounted external to the coupling unit and this capacitor also includes an HV insulator as part of the capacitor mount. Figure 4.33 shows the construction details of this capacitor on the left and it shows the ABB A9BS/A9BP (MCD80 type) coupling module [12] on the right. The inductor *Ls1* in Figure 4.4 or *Lp1* in Figure 4.32 can be adjusted by selecting the tapping points in the MCD80 module.

Using two of these couplers and a hybrid transformer to generate differential signals, differential PLC signals can be placed on two lines of a three phase system. The ABB A9BT and A9BR couplers include such differential hybrid transformers. In addition to ABB, there are other manufacturers of such HV couplers [32] and line traps, such as Alstom, Arteche, Siemens, HilKar, Trench and others [33]–[37]. In the following description the couplers from ABB are used with their permission, purely to describe the principles of operation.

The circuit diagram of the high-pass A9BS and A9BT coupling units and the bandpass A9BP and A9BR units are shown in Figures 4.4 and 4.32 respectively. However, ABB use

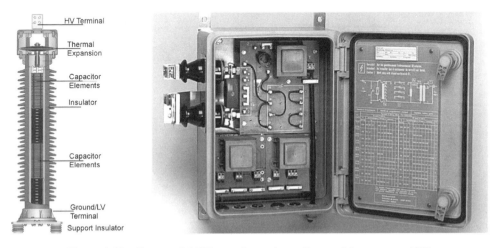

Figure 4.33 Commercial HV capacitor and coupling module, courtesy ABB.

different component values to what is shown in these figures. The HV capacitor Ck is the most expensive component in this circuit. Reducing the value of this capacitance, reduces the cost of this capacitor and also reduces the reactive power flowing through it. Depending on the component values used in the PLC couplers, the voltage across Ck at the mains frequency can be larger than the power line voltage. Higher voltage capacitors are more expensive, so it is desirable to ensure that the voltage across capacitor Ck is at worst only slightly larger than the line to neutral voltage. The coupler design must thus be optimized to fulfill all the following requirements:

(a) The smallest possible value of the high voltage capacitor Ck should be used. This implies that the highest possible high-pass corner frequency should be used for the modem.
(b) The mains frequency voltage across the capacitor Ck should be less than 5% greater than the line to neutral voltage, even when component values of the coupler circuit or the impedances connected to the coupler change a reasonable amount.
(c) The coupler should have a high attenuation at the mains frequency to prevent damage to humans or equipment.
(d) The coupler should have a low insertion loss at PLC frequencies.

These aims can be achieved as shown in the following example. Selecting a lower corner frequency of 52 kHz and a 13 nF capacitor for Ck, and then optimizing the bandpass coupler circuit of Figure 4.32, results in the component values shown in Figure 4.32. The frequency response of this coupler and the corresponding optimized low pass coupler, also using a value of $Ck = 13$ nF is shown in Figure 4.34. The voltage across Ck is not larger than the coupler input voltage at any frequency and there are no resonances which may cause voltage ratings to be exceeded. The return loss is better than 20 dB from 60 kHz to 400 kHz for the

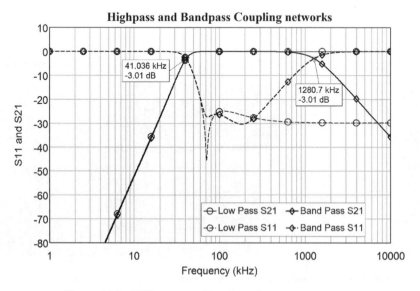

Figure 4.34   HV lowpass and bandpass frequency responses.

bandpass coupler and 60 kHz to more than 10 MHz for the high-pass coupler. The attenuation at 60 Hz is 230 dB, so that people and equipment are well protected from the mains frequency voltage.

4.5   MV Coupling

MV lines are used to distribute power in local areas and are normally connected to an HV transformer at a substation and many LV distribution transformers along the MV line. Since on-line monitoring of LV distribution transformers has not traditionally been done, PLC on MV lines does not have a long operational history. This situation is changing. The Smart Grid roll-out in Europe using PLC provides an opportunity for MV PLC to firstly monitor MV power transformers and secondly using MV to LV transformer bypass coupling, as described in Section 4.2.8 or transmitting through the MV-LV transformer [16], to provide easy access to LV power lines for PLC applications such as AMR.

There are two different markets for PLC on MV power lines.

Broadband PLC: B-PLC covers the frequency band above 1.7 MHz and up to 100 MHz, though most systems use frequencies below 30 MHz. Reference [38] describes the use of MV couplers for B-PLC applications. For MV underground power lines, inductive coupling using couplers like the one shown in Figure 4.6 [8] can be used. Amperion [31] is proposing to use B-PLC for transmission line protection communication for both HV and MV lines. Since they claim that no line traps are needed, their system can be cost effective for an MV PLC system. MV B-PLC couplers are readily available, but the coupling capacitances and inductances used in these commercial B-PLC couplers are too small for them to be used for N-PLC applications. For example the 24 kV capacitive couplers for B-PLC produced by Power Plus Communications [39] have a 1.2 nF capacitor for the high voltage capacitor like Csk in Figure 4.4 or Ck in Figure 4.34, compared with the 13 nF value used by ABB or 21.77 nF used in the optimized coupling network of Figure 4.4. MV B-PLC couplers have a very high insertion loss for 35–500 kHz N-PLC frequencies and are thus unsuitable for N-PLC applications.

Narrowband PLC: G3 PLC and PRIME PLC are 35–500 kHz PLC systems currently being rolled out to millions of homes in France and Spain for smart grid applications. ADD GRUP and Maxwell technologies [40], [41] make suitable MV capacitors and coupling networks. The ADD GRUP coupling capacitor is rated at 8.67 kV, for a 15 kV 3-phase system, and has an 8 nF capacitance. The Maxwell Condis MV capacitors have capacitances from 2–500 nF and voltages from 10 to 52 kV. These capacitors allow suitable MV coupling networks to be constructed.

For inductive coupling, circuits like the third order coupler shown in Figure 4.7, provide sufficient mains frequency rejection for MV voltages. Adding a capacitor in series with the 75 Ω output port will make this coupler a fourth order filter thus providing an extra 50 dB of mains frequency attenuation at low cost. For overhead MV lines a coupler configuration similar to Figure 4.4 is used. Since the voltages at MV are lower than those at HV, the cost of the MV capacitors is significantly less than the HV capacitors.

MV PLC communication networks can be an economical means for providing an MV infrastructure, for monitoring and distributing PLC signals to LV lines, either using direct transmission through the MV-LV transformer or by using MV-LV transformer bypass couplers. MV PLC allows the real time performance monitoring of HV to MV and MV to LV power distribution transformers. Monitoring transformer temperatures allows the maximum safe power distribution to MV and LV lines to be determined and together with Smart Grid power control, will prevent transformer overload failure as domestic loads increase with electric vehicle charging.

4.6 Summary

The correct design of a PLC coupler is critical in obtaining a reliable PLC network. This chapter has shown that PLC couplers can be designed to:

(a) Provide a low insertion loss for coupling the transmitted signal onto the power line, even when the line impedances change with the daily variation of electrical load.
(b) Provide the desired load impedance at PLC frequency to the power line to obtain the correct amount of coupling from the power line to enable all receiving modems on the power line to demodulate the PLC signal with a low bit error rate, even under the high noise conditions normally experienced on power lines.
(c) Provide sufficient isolation of the mains frequency voltages and currents, to protect both humans and the modems connected to the coupler.
(d) Provide a flat frequency response and low modem port return loss.
(e) Provide all these functions with a very high reliability and at the lowest possible cost.

The material presented in this chapter has shown that these criteria can be met for HV, MV and LV power lines and that each of these different power lines have different priorities on the above requirements.

References

1. C. J. Kikkert, Calculating radiation from power lines for power line communications, in *Matlab for Engineers – Applications in Control, Electrical Engineering, IT and Robotics*, K. Perutka, Ed. InTech, 2011, ch. 9, 221–246.
2. I. H. Cavdar and E. Karadeniz, Measurements of impedance and attenuation at CENELEC bands for power line communications systems, *Sensors*, 8(12), 8027–8036, Dec. 2008.
3. Y.-S. Kim and J.-C. Kim, Characteristic impedances in low-voltage distribution systems for power line communication, *J. Electr. Eng. Technol.*, 2(1), 29–34, Jan. 2007.
4. Planning of (single-sideband) power line carrier systems, *International Electrotechnical Commission (IEC)*, Geneva, Switzerland, Technical Report CEI/IEC 663: 1980, 1980.
5. A. I. Zverev, *Handbook of Filter Synthesis*. John Wiley & Sons, Chichester 1967.
6. L. P. Huelsman, *Active and Passive Analog Filter Design: An Introduction*. McGraw-Hill International Editions, 1993.
7. Butterworth filter, 2013. [Online]. Available: http://en.wikipedia.org/wiki/Butterworth_filter
8. Premo, products, PLC, PLC Accessories, MICU 300A-S/LF, Premo Smart Grid, 2013. [Online]. Available: http://www.grupopremo.com/in/product/254/features/plc/plcaccessories/micu300amediumvoltageinductivecoupling units300a.html

9. Resources and white papers, HomePlug Alliance, 2014. [Online]. Available: http://www.homeplug.org/tech-resources/resources/

10. *Harmonic Current Emissions: Guidelines to the Standard EN 61000-3-2ID*, European Power Supply Manufacturers Association, Nov. 2010. [Online]. Available: http://www.epsma.org/PFC Guide_November 2010.pdf

11. C. J. Kikkert, *RF Electronics: Design and Simulation*. James Cook University, Townsville, Queensland, Australia, 2013.

12. *MCD80 — Power Line Carrier Coupling Devices*, ABB Switzerland Ltd., Mar. 2011, brochure. [Online]. Available: http://new.abb.com/network-management/communication-networks/power-line-carriers/mcd80

13. NI-AWR design environment, Microwave Office, National Instruments-AWR. [Online]. Available: http://www.awrcorp.com/products/microwave-office

14. PLC/BPL couplers for MV, Arteche, 2014. [Online]. Available: http://www.arteche.com/en/products-and-solutions/category/plc-bpl-couplers-for-mv

15. Inductive coupler for PLC, Mattron, 2014. [Online]. Available: http://www.mattrone.com/eng2/product/productic.html

16. K. Razazian, M. Umari, A. Kamalizad, V. Loginov, and M. Navid, G3-PLC specification for powerline communication: Overview, system simulation and field trial results, in *Proc. IEEE Int. Symp. Power Line Commun. Applic.*, Rio de Janeiro, Brazil, Mar. 28–31, 2010, 313–318.

17. C. J. Kikkert, MV to LV transformer PLC bypass coupling networks for a low cost smart grid rollouts, in *Proc. IEEE PES Innovative Smart Grid Technol. Asia*, Perth, Australia, Nov. 13–16, 2011, 1–6.

18. ——, Power transformer modelling and MV PLC coupling networks, in *Proc. IEEE PES Innovative Smart Grid Technol. Asia*, Perth, Australia, Nov. 13–16, 2011, 1–6.

19. ——, A PLC frequency model of 3 phase power distribution transformers, in *Proc. IEEE Int. Conf. Smart Grid Commun.*, Tainan, Taiwan, Nov. 5–8, 2012, 205–210.

20. ——, Effect of couplers and line branches on PLC communication channel response, in *Proc. IEEE Int. Conf. Smart Grid Commun.*, Brussels, Belgium, Oct. 17–20, 2011, 309–314.

21. Roadmap 2010-18 and detailed implementation plan 2010-12, The European Electricity Grid Initiative, May 2010. [Online]. Available: http://www.smartgrids.eu/documents/EEGI/EEGI_Implementation_plan_May 2010.pdf

22. What's HD-PLC? HD-PLC Alliance, 2014. [Online]. Available: http://www.hd-plc.org/modules/about/hdplc.html

23. IEEE standard for broadband over power line networks: Medium access control and physical layer specifications, IEEE Standards Association, IEEE Standard 1901-2010, Sep. 2010. [Online]. Available: http://grouper.ieee.org/groups/1901/

24. Narrowband orthogonal frequency division multiplexing power line communication transceivers for G3-PLC networks, ITU-T, Recommendation G.9903, May 2013. [Online]. Available: http://www.itu.int/rec/T-REC-G.9903

25. Narrowband orthogonal frequency division multiplexing power line communication transceivers for PRIME networks, ITU-T, Recommendation G.9904, Oct. 2012. [Online]. Available: http://www.itu.int/rec/T-REC-G.9904-201210-I/en

26. PRIME-based PLC solutions, ATPL230A, Atmel, 2014. [Online]. Available: http://www.atmel.com/devices/ATPL230A.aspx

27. 600Mbps powerline kit with Gigabit Ethernet – NP507, NetCommWireless. [Online]. Available: http://www.netcommwireless.com/product/powerline/np507

28. X. Lu, I. H. Kim, and R. Vedantham, Implementing PRIME for robust and reliable power line communication (PLC), Texas Instruments, White paper, Jul. 2013. [Online]. Available: http://www.ti.com/general/docs/lit/getliterature.tsp?baseLiteratureNumber=SLYY038

29. S. Zhu, C. J. Kikkert, and N. Ertugrul, A wide bandwidth, online impedance measurement method for power systems, based on PLC techniques, in *IEEE Int. Symp. Circuits Syst.*, Melbourne, Australia, Jun. 1–5, 2014, 1167–1170.

30. M. Schwartz, Carrier-wave telephony over power lines: Early history, *IEEE Commun. Mag.*, 47(1), 14–18, Jan. 2009.

31. BPLC for HV and MV systems, Amperion, 2013. [Online]. Available: http://www.amperion.com/solutions.php

32. Trench MV and HV AC capacitors, Trench Group, 2013. [Online]. Available: http://www.trenchgroup.com/en/Products-Solutions/Instrument-Transformers/Capacitors/AC-Capacitors

33. Line traps air core, dry type up to 800 kV, Alstom. [Online]. Available: http://www.alstom.com/grid/products-and-services/high-voltage-power-products/Instrument-Transformers/Line-traps-air-core-dry-type-up-to-800-kV/

34. Line traps, Arteche, 2014. [Online]. Available: http://www.arteche.com/en/products-and-solutions/category/line-traps

35. Coil products, Siemens. [Online]. Available: http://www.energy.siemens.com/hq/en/power-transmission/high-voltage-products/coil-products.htm

36. PLC line traps, Hilkar. [Online]. Available: http://www.hilkar.com/plclinetraps.html

37. Trench HV PLC line traps, Trench Group, 2013. [Online]. Available: http://www.trenchgroup.com/en/Products-Solutions/Coil-Products/Line-Traps/node_670

38. N. Sadan, M. Majka, and B. Renz, Advanced P&C applications using broadband power line carrier (B-PLC), in *DistribuTECH Conf. and Exhibition*, San Antonio, USA, Jan. 24–26, 2012.

39. Broadband powerline communications, Power Plus Communications, 2014. [Online]. Available: http://www.ppc-ag.de/5-1-BPL-Products.html

40. ADD Grup MV capacitive and inductive coupler units, ADD GRUP, 2013. [Online]. Available: http://addgrup.com/products/index/parent/8

41. Maxwell, condis medium voltage capacitors and voltage dividers, Maxwell, 2013. [Online]. Available: http://www.maxwell.com/products/high_voltage/docs/mv_capacitor_and_voltage_divider_ds.pdf

5

Digital Transmission Techniques

K. Dostert, M. Girotto, L. Lampe, R. Raheli, D. Rieken, T. G. Swart,
A. M. Tonello, A. J. H. Vinck, and S. Weiss

5.1 Introduction

Digital transmission is at the core of any communication system, including PLC systems. When a new communication medium or channel is investigated, there are several contenders amongst the known modulation and coding techniques which can be considered. Unusual channels such as the PLC channel may require adaptation of the known techniques or inspire research into developing new modulation and coding techniques with better performance. There may be more than one suitable technique, and in fact under different channel conditions or for different applications, a unique contender may outperform the others. For the unpredictable and widely varying PLC channel, robustness under different conditions is of prime importance.

In this chapter, we first discuss (Section 5.2) single carrier modulation techniques which have been proposed and used in first generation narrowband PLC systems. In this section, we emphasize recent insight and research focusing first on frequency/phase shift keying combined with permutation coding. Then, spread spectrum techniques and their properties are described in the context of PLC. Section 5.3 covers multicarrier modulation which is at the heart of latest narrowband and broadband PLC systems. Several known and more recent schemes are here described including OFDM, FMT, Pulse-shaped OFDM, Wavelet OFDM, OQAM-OFDM, and Cyclic block FMT. In Section 5.4 voltage and current modulation is discussed as a simple and effective means to transmit data at low speed and over long distances. The usage of carrierless ultra wide band impulsive modulation is instead discussed in Section 5.5. If multiple wires are available, multiple input multiple output (MIMO) transmission schemes can be used to exploit spatial diversity. This is the subject of Section 5.7. Noise mitigation techniques are also very important as discussed in Section 5.6. Finally, it is notable that there is a multitude of candidate error control coding techniques which can be adapted for PLC as described in Section 5.8. We present error control techniques currently found in various PLC

Power Line Communications: Principles, Standards and Applications from Multimedia to Smart Grid, Second Edition.
Edited by Lutz Lampe, Andrea M. Tonello, and Theo G. Swart.
© 2016 John Wiley & Sons, Ltd. Published 2016 by John Wiley & Sons, Ltd.

protocols and standards, as well as other promising techniques that are still only found in the literature.

5.2 Single Carrier Modulation

5.2.1 Frequency Shift Keying

Low complexity modulation schemes, like On-Off Keying (OOK), PSK and FSK are candidates for implementation of power line modems. Under ideal AWGN circumstances, On-Off keying has a 3 dB performance loss as compared to BPSK. Furthermore, the receiver needs a detector with an adaptive threshold to compensate for the unknown and varying channel attenuation. For BPSK, we transmit information using the phase of the signal. Practical measurements showed phase changes in the order of 10°. Hence, phase tracking is necessary to obtain a low detection error rate. For AWGN channels, there is a 3 dB difference in performance between BPSK and BFSK. In an access environment, where distances are of the order of 100–500 meters and attenuation in the order of 10–100 dB per kilometer, this difference does not play a role of importance. In this scenario, robustness of the transmission scheme is more relevant. We will concentrate on FSK as the basic modulation scheme for power line communications, since it is a well studied and robust modulation technique. FSK has the advantage of a constant envelope signal, demodulation in a coherent as well as a non-coherent way and a non-adaptive zero threshold, leading to low complexity transceivers.

Modulation schemes which use constant envelope signal waveforms, such as BPSK with rectangular shaping pulse, BFSK and M-ary FSK (M-FSK), are in agreement with the CENELEC norms, EN 50065.1, part 6.3.2. For narrowband power line communications, we considered the problems caused by: frequency dependent background noise, impulsive noise; coupling and network losses leading to high attenuation. In addition to these disturbances, narrowband interference from, for instance, television sets or radio service stations cannot be neglected. Without specific error control methods or redundancy, there is no reliable communication possible. A modulation/coding scheme that incorporates frequency- and time diversity can be expected to be robust against narrow- and broadband interference. The 'spread-FSK' (S-FSK) introduced by Schaub [1], is an example where the modem can still be used even in the presence of a narrowband disturbance. Both frequencies of the FSK are chosen relatively far apart such that a narrowband disturbance only destroys one of the two transmission frequencies and demodulation can still take place using the undisturbed channel on an OOK basis. We need an adaptive threshold to be used in the OOK and also a mechanism to detect the presence of a disturbance. We show that we can use the FSK system in such a way, that we do not need the channel state information in order to detect the transmitted message.

The combination of M-FSK modulation and coding can provide for a constant envelope modulation signal, frequency spreading to avoid bad parts of the frequency spectrum, and time spreading to facilitate correction of frequency disturbances and impulse noise simultaneously. M-FSK has been used in early modem design by Dostert [2].

In an M-FSK modulation scheme, symbols are modulated as one of the sinusoidal waves described by

$$s_i(t) = \sqrt{\frac{2E_s}{T_s}} \cos(2\pi f_i t), \quad 0 \le t \le T_s, \tag{5.1}$$

where $i = 1, 2, \ldots, M$, E_s is the signal energy per modulated symbol, T_s is the symbol interval and

$$f_i = f_0 + \frac{i-1}{T_s}, \quad 1 \le i \le M.$$

The signals are orthogonal and for non-coherent reception the frequencies are spaced by the transmission rate $1/T_s$ Hz. To avoid abrupt switching from one frequency to another, the information bearing signal may modulate a single carrier whose frequency is changed continuously. The resulting frequency-modulated signal is phase continuous and is called continuous-phase FSK. Details regarding this type of modulation and demodulation can be found in [3]. We restrict ourselves to the ideal M-FSK modulation as given in (5.1) and do not further consider spectral properties of the modulation scheme. According to [3] the theoretical measure of bandwidth efficiency, in bits/s/Hz, of M-FSK modulation is given by

$$\rho = \frac{\log_2 M}{M}.$$

For large M, M-FSK is spectrally inefficient. The *symbol* error probability for transmission over an AWGN channel with single sided noise power spectral density N_0, at high values of E_s/N_0, can be approximated as

$$P_s \approx \frac{1}{2} e^{-\frac{E_s}{2N_0}},$$

where $E_s = E_b \log_2 M$ and E_b is the energy per information bit. For AWGN channels the probability of bit error can be made arbitrarily small by increasing M, provided that the signal-to-noise ratio (SNR) per bit E_b/N_0 is greater than the Shannon limit of -1.6 dB [4]. The cost for increasing M is the bandwidth required to transmit the signals.

Since we want to use M-FSK modulation, we consider codes with an M-ary output. We use the integers $1, 2, \ldots, M$ to represent the M frequencies, i.e. the integer i represents f_i. A message is encoded as a code word of length M with the integers $1, 2, \ldots, M$ as symbols. The symbols of a code word are transmitted in time as the corresponding frequencies. An interesting problem is the design of codes and the effect of coding on the transmission efficiency. Let $|C|$ denote the cardinality of the code.

Definition 5.2.1 *A permutation code C consists of $|C|$ code words of length M, where every code word contains M different symbols.*

As an example, for $M = 4$ and cardinality of the code $|C| = 4$, the code words are $(1, 2, 3, 4)$, $(2, 1, 4, 3)$, $(3, 4, 1, 2)$ and $(4, 3, 2, 1)$. The code has four differences between any two code words. As an example, message 3 is transmitted in time as the series of frequencies (f_3, f_4, f_1, f_2). Note that the code C has four words with the property that two words always differ in four positions.

The code as given in Table 5.1 has 12 words, each with $M = 4$ different numbers and a minimum difference between any two words or *minimum Hamming distance* $d_{\min}$ of three. For $M = 3$ we have the code books as given in Table 5.2. An interesting problem is the design

Table 5.1 Twelve code words for $M = 4$, $d_{\min} = 3$

1, 2, 3, 4	1, 3, 4, 2	1, 4, 2, 3
2, 1, 4, 3	2, 4, 3, 1	2, 3, 1, 4
3, 1, 2, 4	3, 4, 1, 2	3, 2, 4, 1
4, 2, 1, 3	4, 3, 2, 1	4, 1, 3, 2

Table 5.2 Two code books for $M = 3$

$d_{\min} = 2$		$d_{\min} = 3$
1, 2, 3	1, 3, 2	1, 2, 3
2, 3, 1	2, 1, 3	2, 3, 1
3, 1, 2	3, 2, 1	3, 1, 2

of codes and the effect of coding on the transmission efficiency. In Table 5.3 we give the code construction results for $M < 6$.

If we have an information transmission rate of b bits per second, we obtain a signal duration time

$$T_s = \frac{1}{b}\frac{\log_2 |C|}{M}.$$

The bandwidth required is thus approximately

$$B = M\frac{bM}{\log_2 |C|},$$

and the bandwidth efficiency of this coded M-FSK scheme is defined by

$$\rho = \frac{b}{B} = \frac{\log_2 |C|}{M^2}. \tag{5.2}$$

To maximize the efficiency, we have to find the largest $|C|$ for a given M and $d_{\min}$. It is easy to see that for a code with code words of length M each having M different numbers $d_{\min}$ is

Table 5.3 Code book sizes for $M = 2, 3, 4, 5$

	$d_{\min}$			
M	2	3	4	5
2	2			
3	6	3		
4	24	12	4	
5	120	60	20	5

always ≥ 2. The cardinality $|C|$ of a code with $d_{\min} = 2$ is $M!$. Hence, the bandwidth efficiency can be defined as

$$\rho = \frac{\log_2 M!}{M^2} \approx \frac{\log_2 M}{M},$$

for large M. This is the same efficiency as uncoded M-FSK.

The next theorem gives an upper bound on the number of code words in a permutation code.

Theorem 5.2.2 *For a permutation code of length M with M different code symbols in every code word and minimum Hamming distance $d_{\min}$, the cardinality is upper bounded by*

$$|C| \leq \frac{M!}{(d_{\min} - 1)!}. \tag{5.3}$$

For $d_{\min} = 2$, we always have equality in (5.3) for any M. For $d_{\min} = M - 1$ and M prime, the code has $M(M - 1)$ code words. It can be shown mathematically, that for $M = 6$ and $d_{\min} = 5$ the upper bound (5.3) cannot be met with equality.

Blake [5] uses the concept of sharply k-transitive groups to define permutation codes with distance $M - k + 1$. The structure of all sharply k-transitive groups is known for $k = 2$. In [5] it is also shown that the group of even permutations on M symbols is a permutation code with $|C| = M!/2$ code words and $d_{\min} = 3$. To find good codes in general appears to be quite difficult. The codes described in [5] are 'simple' examples. If we assume that codes exist that meet (5.3) with equality, then from (5.2) it follows that the bandwidth efficiency defined for a code with minimum Hamming distance $d_{\min}$ and increasing length M approaches

$$\rho \approx \frac{M - d_{\min} + 1}{M} \frac{\log_2 M}{M}.$$

Next we discuss the modification of the demodulator and the use of the modified demodulator output in the decoding of permutation codes.

The sub-optimum non-coherent demodulator computes M envelopes using $2M$ correlators, two per signal waveform preceded by an automatic gain control unit (AGC). It outputs as estimate for the transmitted frequency the one that corresponds to the largest envelope. See Figure 5.1 for a general envelope detector.

An optimum decision rule can be derived using the knowledge of the SNR per sub-channel for a particular frequency. In practical schemes the SNR per sub-channel can be obtained with the help of a well defined preamble or by using the output of the correlators, see also [1].

The output r_k can be normalized with respect to the noise variance σ_k^2. The probability density function for the normalized output $y_k := r_k/\sigma_k$ at time j is given by [6], [3]

$$\Delta_{k,j} := p(y_k | \text{frequency } k \text{ transmitted})$$

$$= y_k \exp\left\{-\frac{y_k^2 + 2E_k/\sigma_k^2}{2}\right\} I_0\left(y_k\sqrt{2E_k/\sigma_k^2}\right),$$

$$\nabla_{k,j} := p(y_k | \text{frequency } k \text{ not transmitted})$$

$$= y_k \exp\left\{-\frac{y_k^2}{2}\right\},$$

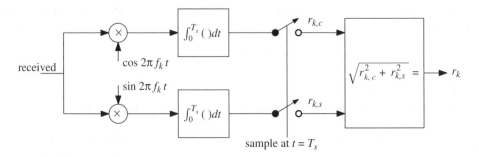

Figure 5.1 Envelope detector for frequency f_k.

where E_k and σ_k^2 are the received symbol energy and the noise variance for the particular channel k, respectively and I_0 is the modified Bessel function of order zero.

As was shown in [1], this type of FSK demodulation is not optimum in case of a simplified model of a frequency selective channel as is the power line channel, where also signal distortions are neglected. In addition, the presence of narrowband noise may cause large envelopes and thus errors to occur at the demodulator output. Impulse noise has a broadband character and thus could lead to a multiple of large envelopes. To be able to handle these types of noise processes, we propose to modify the demodulator in such a way that the detected envelopes can be used in the decoding process of a special class of error control codes, called permutation codes.

For the permutation code, we know that a particular frequency must be present only once per code word. This fact can be used to estimate the received symbol energy as well as the noise variance per sub channel. We assume that we have these values available at the receiver.

We define four types of detector/decoder combinations. For that purpose we use the $M \times M$ matrix Y.

(a) The *classical* detector with column wise hard decisions.

In the matrix Y the element $(i,j) = 1$ if y_i is the largest envelope detector output at time j; otherwise $(i,j) = 0$.

The value $(i,j) = 1$ corresponds to the assumption that frequency f_i has been transmitted. Hence, the permutation decoder compares its code words with the corresponding frequencies and outputs the code word at minimum distance.

(b) The *modified classical* detector with column wise soft decisions.

In the matrix Y the element $(i,j) = 1$ if $\Delta_{i,j}/\nabla_{i,j}$ is the largest density at time j; otherwise $(i,j) = 0$.

Again, the permutation decoder compares its code words with the corresponding frequencies and outputs the code word at minimum distance.

(c) The column and row wise *hard-decision threshold detector*.

In this case we use a threshold T_i for every envelope detector. The position of the threshold values T_i can be optimized depending on E_i and σ_i^2. A practical value could be $0.6\sqrt{E_i}$.

The elements for the column j are given by $(i,j) = 1$ if $y_i > T_i$; otherwise $(i,j) = 0$.

The permutation decoder compares its code words with the corresponding frequencies and outputs the code word at minimum distance.

(d) The column and row wise *soft-decisions threshold detector*.

We now put in the elements of Y the values $(i, j) = \Delta_{i,j}/\nabla_{i,j}$ for all i and j. The permutation decoder computes for a particular code word k with a frequency f_i at time j, the value

$$F_k = \prod_{i,j=1}^{2} \frac{\Delta_{i,j}}{\nabla_{i,j}}; \quad k = 1, 2.$$

It can be shown that the maximum *a posteriori* (MAP) probability demodulator outputs the value of k that maximizes F_k. In fact, we calculate the normalized probability that we receive a certain matrix Y given a code word k. For optimum performance, we need to know E_i and σ_i.

There are several channel disturbances that degrade the performance:

- Non-coherent demodulation with largest envelope detection is not optimum in case of a simplified model of a frequency selective channel as is the power line channel.
- Narrowband noise may cause large envelopes and thus errors to occur at the demodulator output.
- Impulse noise has a broad band character and thus could lead to a multiple of large envelopes.
- Interference may cause envelopes to disappear.
- In addition, the detectors using 'soft values' require exact channel state knowledge and thus are not useful if we want to detect without channel state knowledge.

Detector (c) contains more information about the received signal and will be investigated further under non-AWGN circumstances. We adopt the following decoding rule.

Decoding rule: Output the message corresponding to the code word that has the maximum number of agreements with the symbols (frequencies) at the demodulator output.

The introduced thresholds in the modified demodulator in combination with the permutation code allow the correction of incorrect demodulator outputs caused by narrowband noise, impulse noise, signal fading or background noise.

- Narrowband noise at frequency f_i may cause the values of $[y_{i,j}] = 1$ for $j = 1, 2, \ldots, M$.
- Impulse noise at time interval j may give as a result that $[y_{i,j}] = 1$ for $i = 1, 2, \ldots, M$.
- Background noise degrades performance by introducing unwanted (called insertions) demodulator outputs or by causing the absence (called deletions) of a transmitted frequency in the demodulator output.
- The absence of a frequency in the demodulator output always reduces the number of agreements between a transmitted code word and the received code word by one. The same is true for the other code word having the same symbol at the same position. If the symbols are different, the number of agreements does not change.
- The appearance of every unwanted output symbol may increase the number of agreements between a wrong code word and the received code word by one. It does not decrease the number of agreements between a transmitted code word and the received code word.

The effect of the different kinds of noise on the multi-valued detector output can be seen from Figure 5.2. We assume that $M = 4$ and transmit the code word $(1, 2, 3, 4)$ as frequencies (f_1, f_2, f_3, f_4).

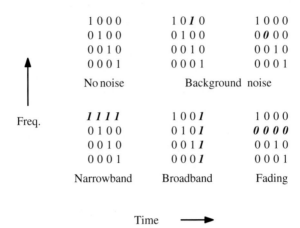

Figure 5.2   Several types of disturbances (in italic) in the channel.

For the example with $M = 4$ and $d_{min} = 4$, a permanent disturbance (narrowband noise) present at the sub-channel for frequency f_4 and transmission of code word $\{3, 4, 1, 2\}$ could lead to a demodulator output $\{(3, 4), (4), (1, 4), (2, 4)\}$. The decoder compares the demodulator output with all possible transmitted code words. It outputs the code word for which the maximum number of agreements with the symbols at the demodulator output occurs. For the example, all symbols corresponding to code word 3 are present and thus correct decoding follows. Since code words are different in at least d_{min} positions, $d_{min} - 1$ errors of this type still allow correct decoding. The example code has $d_{min} = 4$ and hence, we can tolerate the presence of three permanent disturbances present in the demodulator output. We give some general remarks on the influence of the different types of disturbance.

For the S-FSK, where we use only two frequencies, the permutation code has two code words, $(1, 2)$ and $(2, 1)$, respectively. Using the same principle as indicated in Figure 5.2, one can see that any single error event can be detected and corrected. Application of the permutation coding, makes S-FSK a very robust system to be used in practice.

Impulse noise has a broad frequency band character and impulses have a duration of typically less than 100 μsec. Measurements in networks indicate that the inter-arrival times are independent and 0.1 to 1 second apart. For a signaling scheme using a signaling rate of 10 kHz, we have a symbol duration of 100 μsec, which is in the range of the impulse duration. So, impulse noise, affecting at least two adjacent symbols cannot be excluded. Due to the broadband character, impulse noise may cause the demodulator to output the presence of all frequencies. This type of noise can be seen as erasures. Hence, two affected adjacent transmissions may reduce the minimum distance of the code by two. At a signaling rate of 10 kHz, a code with $d_{min} = 3$ is thus capable of correcting two permanent disturbances or to give a correct output in the presence of an impulse. For higher signaling rates, more symbols may be affected and thus a code with a larger minimum distance is needed. Suppose that we transmit the example code word $\{3, 4, 1, 2\}$. If an impulse noise causes all envelopes to be present at three symbol transmissions, then we may have as a demodulator output $\{(1, 2, 3, 4), (1, 2, 3, 4), (1, 2, 3, 4), (2)\}$. Comparing this output with the possible transmitted code words gives a difference (distance) of zero to the correct code word and one to all other code words. Thus, even if three of these multi-valued outputs occur, we are still able to find back the correct code word since there is always a remaining symbol that gives a difference of one to the incorrect code words.

Background noise degrades performance by introducing unwanted (called insertions) demodulator outputs or by causing the absence (called deletions) of a transmitted frequency in the demodulator output. Note that for this type of 'threshold' demodulation, the decoding is still correct for $d_{min} - 1$ errors of the insertion/deletion type.

Signal fading, or the absence of a frequency in the demodulator output always reduces the number of agreements between a transmitted code word and the received code word by one. The same is true for the other code words having the same symbol at the same position. If the symbols are different the number of agreements does not change.

The appearance, or *insertion*, of every unwanted output symbol may increase the number of agreements between a wrong code word and the received code word by one. It does not decrease the number of agreements between a transmitted code word and the received code word.

The disappearance, or *deletion*, of an output symbol may decrease the number of agreements between a transmitted code word and the received code word. It does not decrease the number of agreements between a wrong code word and the received code word by one.

In conclusion, we can say that the introduced thresholds in the modified demodulator in combination with the permutation code allow the correction of $d_{min} - 1$ incorrect demodulator outputs caused by narrowband noise, impulse noise, signal fading or background noise.

For the background noise, with assumed constant noise power density we can give an estimate of the detection error probability in the un-coded situation as

$$P_e \leq e^{\log_2 M \left[\ln 2 - \frac{E_b}{2N_0} \right]},$$

where E_b is the energy per transmitted information bit, see [7]. If we use a permutation code with minimum distance d_{min}, the *coded* error probability can be estimated as

$$P_e \leq e^{\log_2 |C| \left[\ln 2 - \frac{d_{min} E_b}{2MN_0} \right]},$$

where we used the fact that for a permutation code of length M, $ME_s = \log_2 |C| E_b$. The cardinality for permutation codes is upper bounded as

$$|C| \leq \frac{M!}{(d_{min} - 1)!}.$$

Assuming equality, for $d_{min} = M - 1$, we have for large values of M

$$\lim_{M \to \infty} P_e \leq e^{2 \log_2 M \left[\ln 2 - \frac{E_b}{2N_0} \right]},$$

which shows that coding can give a great improvement in the error exponent. However, the power line channel is not a nice AWGN channel and the calculations are probably not very realistic. The final judgment of the system performance depends on the knowledge of the channel, the parameters of the disturbances and the modulation/coding scheme used.

5.2.2   Spread Spectrum Modulation

Spread spectrum techniques (SST) were originally developed for military communication, in order to achieve resistance against intentional or inadvertent interferers by high spectral redundancy. A further benefit of SST is that – due to very low power spectral density – transmitted signals become almost 'invisible' for outsiders so that eavesdropping can be successfully excluded if the spreading code can be kept confidential.

With respect to PLC, SST systems are able to provide robustness against selective fading and narrowband interference. At the same time they mitigate problems of electromagnetic compatibility (EMC), and allow multiple access without coordination, due to their so-called 'graceful' degradation property. This way, media access can be elegantly accomplished by code division multiple access (CDMA) schemes.

In the past, SST systems were characterized through high costs, so that their usage remained rather restricted. Due to recent progress in the field of microelectronic systems, SST have, however, become feasible for almost any application, including PLC. In this section, the types of SST, which are appropriate for PLC, are briefly introduced and analyzed with respect to their possible performance. Various ways of adaptation of SST to the properties of typical PLC links are investigated, including bandwidth requirements as well as aspects of synchronization. Eventually, experience with practical applications is evaluated, pointing out guidelines for improvements and possible future developments.

As the name already indicates, the core element of SST is plenty of bandwidth. In fact, there must be much more spectral resources available than would be needed for the transmission of a usually modulated signal. Thus SST can only do well with tremendous spectral redundancy. Let us assume that a conventionally modulated signal occupies the bandwidth B_m around a carrier with the frequency f_0. Then, in a next step band-spreading is performed so that the resulting bandwidth B_{SP} extends over a range which is in practice between approximately 10 and several thousand times B_m. The ratio $P_G = B_{SP}/B_m$ is denoted as processing gain. As we will see in the following sections, the essential benefit of applying SST is to achieve a trade-off between bandwidth occupation and signal-to-noise ratio (SNR). In more detail, the improvement of SNR is approximately equal to the processing gain P_G. As in practice P_G figures below 10 do not make much sense, it is clear that SST – although they are true broadband techniques – are not necessarily high-speed techniques in the sense of great data throughput and spectral efficiency. On the contrary, robustness both against hostile channel properties and various kinds of interference is always in the foreground when thinking about SST applications, and data throughput is of minor importance. Following this rule, the role of SST for PLC is in fact difficult to determine today. For, on one hand, with properly selected SST schemes, extremely bad channels can still be used for very slow but reliable data transmission. On the other hand, if high spectral efficiency is the goal, then other broadband technologies such as OFDM can offer better results due to their high spectral efficiency, although they cannot guarantee the same degree of robustness, especially against interference.

5.2.2.1   Types of SS Technologies: Direct Sequencing Spread Spectrum (DSSS)

Direct Sequencing Spread Spectrum (DSSS) can be considered as the 'father' of all band-spreading technologies. It has been one of the first technologies in this field, mainly due to the fact that it is so easy to handle. This is particularly true for the spreading modulation and demodulation procedures.

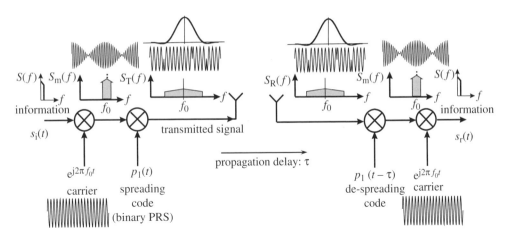

Figure 5.3   Basic overview of a complete DSSS system.

As illustrated in Figure 5.3, a single carrier with the frequency f_0 is modulated with the information signal $s_i(t)$ in a conventional way, so that the narrowband spectrum $S_m(f)$ with approximately twice the bandwidth of $s_i(t)$ results. Now a second 'high-speed' modulation follows, using a binary pseudo-random sequence (BPRS) with 'chip' intervals sufficiently small, specified by the spreading code $p_1(t)$. The specific feature of DSSS is that this binary spreading code causes 180° phase hops at each of its chip edges. This process is illustrated in Figure 5.4.

The envelope and shape of the resulting 'spread' spectrum $S_T(f)$ as depicted in Figure 5.3 is determined by certain properties of the spreading code which are investigated in the following sections. For the moment it is sufficient to keep in mind that the bandwidth is approximately given by twice the clock frequency of the BPRS.

The resulting broadband signal is now transmitted over the communication channel. At the receiver the same sequence $p_1(t)$, synchronized with the received signal, must be available in order to execute de-spreading, of course delayed by the signal propagation time τ between transmitter and receiver, i.e. $p_1(t - \tau)$ is necessary. In a first mixer (being identical with the corresponding device at the transmitter) the rapid phase hops are now removed and the spectrum $S_R(f)$ is restored (see Figure 5.3). A conventional demodulator follows for recovery of the information.

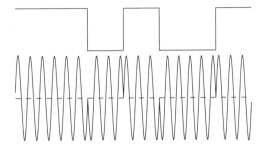

Figure 5.4   180° phase hops initiated by a binary pseudo-random sequence.

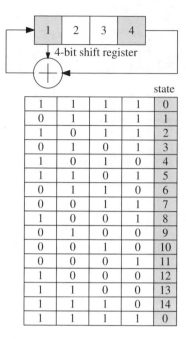

Figure 5.5   Example for generating a 15-bit BPRS with shift register feedback.

The simplicity of spreading and de-spreading modulator technology can certainly be regarded as a great benefit of DSSS which made this technology a primer in the field of spread spectrum applications. In fact, passive double balanced mixers, being small and inexpensive and widely available, can excellently do the job. Moreover, also the generation of BPRS is easily accomplished by simple shift registers with appropriate feedback over XOR gates, as depicted in Figure 5.5. Here, by selecting two or more taps of the shift register (one of which must always be the last stage), and feeding back the modulo-2 result into the first stage, so-called m-sequences are generated. The 'm' stands for maximum length, meaning that for an n-stage shift register a sequence of the length $L = 2^n - 1$ results. Thus, each of the possible 2^n states of the register contents occurs once, until the sequence repeats, except the state of all zeros, which would of course permanently lock the register in this state.

With respect to the properties of the spreading code in the form of BPRS, it was already mentioned that the resulting bandwidth is approximately given by twice the clock frequency f_c of the shift register used for code generation. Let us now investigate the more sophisticated properties of m-sequences, i.e. the detailed shape of the spectrum and their periodic autocorrelation function. The periodic autocorrelation function (ACF) of a binary sequence $p(i)$, as it is found in the second last column of the table in Figure 5.5, can be calculated by

$$\mathrm{ACF}(\tau) = \sum_{i=0}^{2^n-2} p(i)p(i + \tau),$$

(5.4)

where both i and τ represent discrete time steps given by the shift register's clock period.

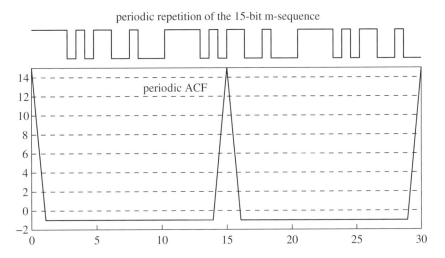

Figure 5.6   Illustration of the specific periodic ACF-property of m-sequences.

As only pure digital signals are involved, we get the proper result by adding '1' for each coincident pair of bits and subtracting '1' in case of different bits. Thus, (5.4) turns into

$$\text{ACF}(\tau) = \sum_{i=0}^{2^n-2} \overline{p(i) \oplus p(i+\tau)} - p(i) \oplus p(i+\tau).$$

In Figure 5.6 the result is illustrated for the 15-bit m-sequence from Figure 5.5.

Obviously, m-sequences exhibit ideal autocorrelation properties, i.e. no side-lobes in excess of '1' will occur, whenever they are repeated periodically without gaps in-between. However, as soon as the requirement of periodicity is not fulfilled, the ACF will show side-lobes which can take on considerable amplitudes.

The only class of binary codes which maintain their ideal ACF properties also in the non-periodic case are the well-known Barker codes. Unfortunately no codes with a length greater than 13 could be found until today, so that they are not of interest for the spreading applications considered here. In contrast, a major benefit of m-sequences is their length. Using for example a shift register with 89 stages and performing the feedback by modulo-2 addition of the stages 89, 6, 5 and 3 would in fact result in a code length of

$$618\,970\,019\,642\,690\,137\,449\,562\,112 \text{ bits,}$$

see e.g. [8]. Especially for military spread spectrum applications it was of utmost importance to hide the spreading code as much as possible, so that extremely low repetition rates were paramount. Moreover, as we will see below, the code length determines the corresponding waveform energy, i.e. the peak of the ACF, as depicted in Figure 5.6, directly represents the code length in bits.

In order to analyze the spectral properties in more detail, we consider the Fourier transform of the ACF, which – according to the Wiener-Khintchine theorem – describes the energy or power spectral density (PSD). As Figure 5.6 indicates, the envelope of the ACF of an m-sequence can be described by a triangular function, with a peak equal to the length of the sequence and reaching almost zero at $\pm 1/f_c$, with f_c being the shift register's clock frequency. It may be convenient to normalize the ACF so that its peak is 1. Likewise, the horizontal axis can be scaled so that $1/f_c$ becomes 1. For sufficiently large n, this normalized ACF can be approximated by the ideal triangular function

$$\Lambda(t) = \begin{cases} 1 - |t|, & |t| \leq 1, \\ 0, & \text{otherwise.} \end{cases}$$

The Fourier transform pair

$$\Lambda(t) \circ\!\!-\!\!\bullet \ \text{sinc}^2(\pi f)$$

can be used to obtain a good approximation of the power spectral density of DSSS signals.

As shown in Figure 5.7, the power spectral density of a DSSS transmission signal is not evenly distributed over frequency but sinc^2-shaped. This means that the maximal power is concentrated in the neighborhood of the carrier, while zeros appear at multiples of the BPRS

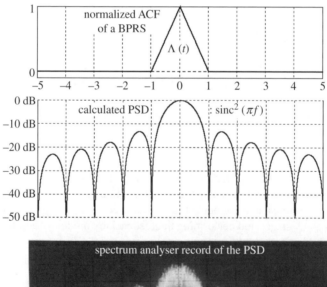

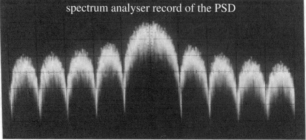

Figure 5.7   ACF of a BPRS and PSD of a bi-phase modulated carrier.

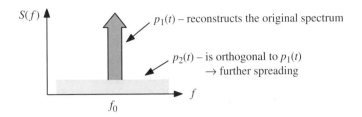

Figure 5.8   De-spreading results with orthogonal codes.

generator's clock frequency f_c. The power of the side-lobes is slowly decaying (the ones next to the main lobe are attenuated by only 13.5 dB), so that significant portions of 'out-of-band' interference may be produced. In practice, this calls for filtering, so that only the main lobe spectrum is transmitted. The uneven distribution is also a drawback, making e.g. the spectral range around the carrier more vulnerable with respect to interference than the 'outer' parts.

The calculated result of the PSD in Figure 5.7 obviously fits very well to the envelope of the spectrum analyzer record (bottom of figure). With a closer look, however, the photo indicates a 'fine structure' of the spectrum, i.e. although the spectrum may appear contiguous at first glance, it exhibits spectral lines at a distance of the spreading code's repetition rate. Also this fact points out the necessity of rather long spreading sequences in order for the real spectrum to be well approximated by the $sinc^2$ shape. In practical applications, the common length of m-sequences as spreading codes is in the range of 1023 to 8191, i.e. 10 to 13-bit shift registers are used.

Another important aspect which calls for a certain code length is introduced by the multiple access feature, which is explained by means of Figure 5.8. Within a multi-user environment, DSSS offers the possibility to access the same spectral resources without coordination. This is accomplished by the implementation of orthogonal spreading codes $p_1(t)$, $p_2(t)$, ..., etc.

As the spectral sketches in Figure 5.8 indicate, a second participant, to whom the orthogonal spreading code $p_2(t)$ has been assigned, cannot perform the spectral compression of $S_R(f)$, as it is spread by the code $p_1(t)$. Thus for $p_2(t)$ the received spectrum $S_R(f)$ remains almost unchanged. In contrast, for the correct code $p_1(t)$ the desired spectral compression occurs, so that the desired narrowband signal, which contains the information, is obtained and can be properly filtered before demodulation is performed.

A similar situation is given for narrowband interferers. In this case, the interferer's spectrum is subjected to the spreading process, so that only a small portion, corresponding to the message bandwidth $S_m(f)$, can impair the desired signal. Obviously, the degree of suppression is given by the processing gain P_G.

For the mentioned multi-user scenario, of course more than two participants may randomly access a channel in parallel without coordination. A prerequisite for this feature is that each of them uses a spreading code, which is orthogonal to all others. Such a configuration is described by the term 'code division multiple access' – CDMA. Thus, each active participant occupies the entire frequency range, however, only slightly disturbing the others. In fact, even if the spreading codes are perfectly orthogonal, the de-spreading process – as explained with Figure 5.8 – always leaves a small portion of the other user's spectral power within the recovered spectrum $S_m(f)$. Therefore, the more participants become active, the higher is the level of mutual disturbance. The situation is comparable with a room in which numerous people are talking to each other at the same time but in different languages. This will work well only up to a certain limit. Thus, in practice an adaptation of the permissible number of active participants to the channel capacity is needed. This particular feature of a properly designed CDMA system is also denoted as 'graceful degradation', which means that the signal quality is slowly deteriorating, but still remains acceptable for all users.

The crucial figure in this context is again the mentioned processing gain P_G. As a general rule, the number of participants must always remain smaller than P_G, otherwise any robustness against other kinds of interference would be totally lost.

Besides the mentioned drawback of the uneven envelope of the PSD, the DSSS scheme also requires some effort for synchronization. The required timing precision is approximately 1/10 of the spreading code's clock period, i.e. the necessary precision increases with P_G. In practical DSSS systems the effort for synchronization may comprise approximately 2/3 of the complete receiver.

Usually, acquisition and tracking are designed as separate functions, whereby matched filters for the spreading codes are used for fast acquisition and delay locked loop technology performs tracking [8]. In the past, surface acoustic wave tapped delay lines have been successfully implemented as matched filters, containing complete m-sequences or at least certain selected portions of the respective spreading sequence – see e.g. [9], [10]. Today, these analog and inflexible[1] devices have been widely replaced by digital matched filters e.g. in the form of correlator banks, which can easily be adapted to any spreading code.

The use of DSSS for applications on power line channels was considered in the 1980s at rather low frequencies in the 100 kHz range and at low data rates up to some kbits/s – see e.g. [11], [12]. Benefits were mainly expected from the robustness of this technology against selective fading and narrowband interference. The overall success was not very impressive in comparison with usual narrowband modulation schemes, such as FSK or BPSK. In spite of the early pitfalls, attempts were made to use DSSS also for fast data transmission at higher frequencies. However, not surprising, also in this field no success could be reported.

The following list of reasons will explain why the transportation and adaptation of highly successful DSSS systems from military applications to power lines did not work, and cannot be brought to proper operation, even with tremendous effort.

- DSSS needs significant portions of contiguous and rather 'flat' bandwidth, i.e. no large gaps are allowed. On the other hand, narrow notches would not be critical. However, note that:
  - Due to the inherent spectral redundancy only a small portion (e.g. less than 1/10) of the available bandwidth can be used for data transportation – this leads to low and unsatisfactory throughput figures.
  - In contrast to wireless channels, power line channels always exhibit low-pass behavior, so that significant parts of 'expected' bandwidth are simply not usable because attenuation is too high – in particular, due to strong frequency selective fading effects, no extended 'flat' portions are present.
- DSSS is phase-sensitive, i.e. if the channel does not provide a linear phase response over the whole bandwidth, severe degradation must be expected. Note that especially in multi-path environments strong phase fluctuation occur around notches.

---

[1] as the spreading code is directly mapped into the structure

- DSSS is rather sensitive to multi-path propagation. Specifically, in multi-path channels rake reception is required which on one hand it offers multi-path diversity gains, on the other hand it imposes more stringent requirements in terms of synchronization and channel estimation since this requires to lock to the right channel paths.
- With respect to interference DSSS can indeed overcome narrowband impairments, which is, however, only one of the possible noise classes in power lines.

From the above considerations, it is clear that no advantages can be expected for background noise suppression and – even more important – for impulsive noise combat. The latter follows from the fact that impulsive noise usually exhibits broadband character, so that only a modest reduction of the PSD is entailed by the receiver de-spreading operation.

5.2.2.2 Types of SS Technologies: Frequency Hopping (FH)

This section discusses Frequency Hopping (FH) in detail as it forms the basis for a large number of system concepts for current and future power line communications. The reasons for this attention towards FH will be explained gradually together with some discussions of applications.

FH is a classical spread spectrum method, which has been widely used during the past and still today in various military applications. FH makes no use of a carrier signal with fixed frequency, but instead uses a large number of waveforms at various frequencies – in military applications sometimes in excess of 100 000.

The FH signal changes frequency in quick succession, with the hop rate h_r. The shorter the dwell time $T_h = 1/h_r$ at one frequency, the less determinate is the FH signal, i.e. it appears more noise-like. The dwell time T_h is also called frequency validity interval or chip duration, depending on the context. An FH waveform $s_{FH}(t)$ with amplitude A and instantaneous frequency f_m is described by

$$s_{FH}(t) = A \, \text{rect} \left(\frac{t}{T_h} \right) \sin(2\pi f_m t). \tag{5.5}$$

The waveform according to (5.5) occupies not only the spectral line f_m, but a continuous spectrum

$$S_{FH}(f) = A \, T_h \text{sinc} \left[\pi T_h (f - f_m) \right],$$

because the frequency f_m is occupied only for a short time T_h. $S_{FH}(f)$ is symmetric with respect to f_m. Assuming matched filter reception, a transmission bandwidth B_g can be occupied by a maximum of

$$N_{FH} = \lceil B_g T_h \rceil - 1 \tag{5.6}$$

FH waveforms, where $\lceil x \rceil$ is the integer portion of x. This means that the smallest admissible frequency distance equals the hop rate h_r. We seek a set of orthogonal waveforms, because a receiver equipped with matched filtering is perfectly capable of receiving any desired waveform from this set, i.e. it can supply the maximum of the autocorrelation function for the desired waveform, while all other waveforms of the set are perfectly suppressed. An orthogonal FH waveform set, which is grouped around a frequency f_0, located in the center of the transmission band with bandwidth B_g, for example with odd N_{FH} according to (5.6), is described by

$$s_{FHi}(t) = A \, \text{rect} \left(\frac{t}{T_h} \right) \sin \left(2\pi \left(f_0 + \left[i - \frac{N_{FH} + 1}{2} \right] h_r \right) t \right), \tag{5.7}$$

with $i \in \{1, \ldots, N_{FH}\}$. For example, the waveform $s_{FH1}(t)$ has its spectrum symmetrically shaped with respect to the lowest frequency, $f_0 - [(N_{FH} - 1)/2]h_r = f_0 - (B_g - h_r)/2$, located at the lower band end. The waveform with the frequency $f_0 + [(N_{FH} - 1)/2]h_r = f_0 + (B_g - h_r)/2$ is located at the upper band end, accordingly with $i = N_{FH}$.

FH can be thought of as an expansion of FSK modulation to more, or even much more than two frequencies. The following example explains the relationship. In this example, FH uses not only two frequencies, but five to transmit a data bit. We can imagine that the information (one data bit) is present in five separate discrete spectral positions. The benefits are obvious: interference or the complete deletion in one or two of these positions cannot corrupt the data transmission. The data bit can be easily reconstructed in the receiver with a simple majority decision (three out of five).

Figure 5.9 represents an example with an 'H' and an 'L' data bit (in the upper part) and the respectively allocated frequency sequences (in the lower part). In this example, the 'H' bit is represented by a (symbolic) sequence of ascending frequencies, f_1, f_2, f_3, f_4, and f_5, which are sent consecutively in fixed time slots (chips), corresponding to one fifth of the bit duration. Likewise, the 'L' bit is represented by the sequence of frequencies f_2, f_3, f_4, f_5, and f_1.

The frequency changes abruptly at the chip limits, without a transient and without a phase hop. To better understand, we have selected the frequency values for this example so that the number of oscillations fitting into a time slot increases by one from one time slot to the next.

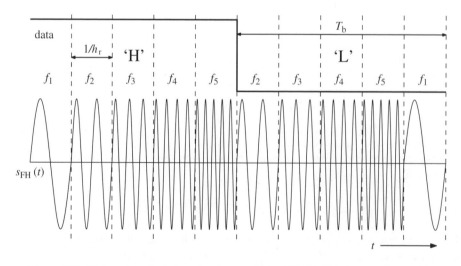

Figure 5.9   Data and transmission signal example for FH with five frequencies per bit.

This means that the first time slot contains one oscillation and the fifth contains five. Identical frequencies were used for the 'L' bit, but in another time sequence, i.e. f_2, f_3, f_4, f_5, and f_1 in the example above. Note that this is not the only option, because when thinking of the five frequencies as binary variables, we can compose $2^5 = 32$ different combinations, of which only two are used here. This redundancy is in the end responsible for the robustness of this band-spreading modulation. When selecting the two combinations for the 'L' and 'H' bits, however, we have to observe that the frequencies in the respective time slots of the data bits are always different to achieve optimum resistance to interference. Regarding this side condition we still obtain more than two combinations, so that we could transmit more than one bit with the five frequencies, without compromising the resistance to interference.

The use of more than five carriers, or further spreading, can achieve an even higher resistance to interference. In general, FH offers the possibility of a very broad spectral spreading, where the spectrum does not necessarily have to be coherent, and where no pseudo-noise sequence with high clock rate is required, in contrast to DSSS. This simplifies the synchronization problem enormously. Due to this fact, mains networks allow even synchronization with the mains voltage as a globally available reference. Some success of FH and related technologies for power line communications is ascribed mainly to this possibility.

FH can be thought of as a method with fair frequency economy despite the high redundancy, as shown in the following example, provided that global synchronization is possible. This means that also the transmitters are operating in synchronism with the same reference (e.g. in form of the mains voltage) as the receivers. Figure 5.10 shows the total spectrum under full utilization of a certain bandwidth B_g. If a hop rate of $h_r = B_g/100$ is selected in (5.7), then

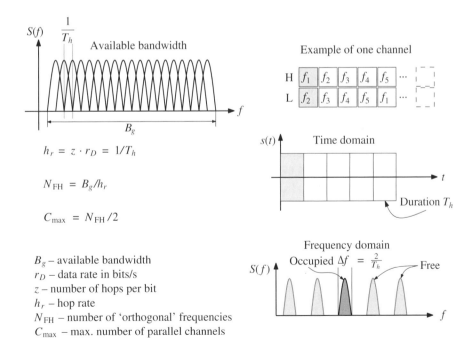

Figure 5.10   Example demonstrating the spectral efficiency of FH.

we obtain 99 orthogonal FH waveforms with frequency distance h_r which, together, occupy the total bandwidth $B_g = 100h_r$. Figure 5.10 shows the good spectral efficiency of FH. We see an almost flat envelope of the spectrum $S(f)$ which means that the available bandwidth can be evenly and well utilized by FH. In addition, FH lets us occupy non-contiguous portions of the spectrum, for example to avoid interference in a targeted way, or to exclude ranges with strong attenuation, or to skip certain frequency ranges in view of regulation and frequency allocation rules. This observation indicates the rather close relation of FH with OFDM, the well-known preferred method for power line communications.

It is important to fully understand that, in order to fill a bandwidth B_g by use of FH, a hop rate of $h_r = B_g/2$ is by no means required, as one may assume in connection with DSSS; in fact, $h_r = B_g/100$ or less is sufficient.

Finally, we will look at the difference between 'fast' and 'slow' frequency hopping (FH). Fast hopping distributes the information carried by a data bit over the available frequency band so that z FH waveforms in one orthogonal block are sent consecutively during the data bit duration T_b. 'H' and 'L' bits are generally distinguished by the sequence of these z frequencies. In fast hopping, the hop rate h_r is always an integer multiple z of the data rate $r_D = 1/T_b$.

In contrast, slow hopping transmits several data bits in one 'frequency slot' with duration T_h. To achieve resistance against interference, the same data bits are repeated at one or several other frequencies. When using slow FH over critical channels like power lines, there is a risk that phase fluctuations in the received signal can occur within one chip duration T_h – which is now fairly long in comparison with fast hopping. This leads to a degradation of the useful signal in the receiver, and possibly to complete disappearance of the desired autocorrelation maximum in the worst case [2], [13]. This is the reason why fast FH is the preferred method in mains networks, in particular for the low-voltage level, despite its higher requirements in terms of synchronization accuracy.

When robust data transmission over power lines is the issue, then the generally lower requirements to the synchronization precision are solely not sufficient to argue in favor of FH versus other methods, such as DSSS. The transmission properties and the interference load of a channel supply additional arguments in favor of FH.

If we look at a power line channel and its typical transmission properties, and if we assume that white Gaussian noise with a zero mean value is the only interference that can occur, then we can achieve an equally good transmission quality with both DSSS and FH, provided that matched filter receivers are used and sufficient synchronization precision is guaranteed. If, in contrast, non-white noise and/or non-Gaussian interferers are present, then FH proves to be superior. In such cases, which are most common in power lines, there is a time-variant and frequency-dependent signal-to-noise ratio at the receiver input. FH allows to beneficially utilize the statistical differences of the signal quality – expressed by the signal-to-noise ratio – at different frequencies. The following example demonstrates how this is possible.

A DSSS receiver converts each interferer approximately into white noise by the spectral compression of the desired signal. This means that DSSS does normally not allow the complete elimination of a received interferer's effect. The power of a strong sinusoidal interferer with frequency f_{int}, for example, is distributed symmetrically with respect to f_{int} over a bell-shaped range after despreading at the DSSS receiver. As already pointed out in the previous section, some amount of the interference always falls into the range of the spectrally compressed useful signal and contributes to the degradation of the signal-to-noise ratio at the receiver output. When transmitting digital information, the interferer increases the probability of bit errors.

The closer f_{int} is to the center frequency of the desired signal, the higher is obviously the interference effect. In contrast, FH allows to distribute a message bit over v FH waveforms at various frequencies within the transmission band. No bit error will occur, as long as at least $\lceil v/2 \rceil + 1$ undisturbed waveforms reach the receiver. As the impact of a narrowband interferer remains limited to a narrow frequency range in the FH receiver, it can never disturb more than one out of v waveforms. Consequently, an FH system can be absolutely resistant to interference from up to $\lceil v/2 \rceil$ narrowband noise sources at different frequencies.

In contrast, when using DSSS, a single but very strong sinusoidal interferer could cause the message transmission to fail. The superiority of FH becomes obvious when looking at the typical interference scenario in mains networks at the medium and low-voltage levels. However, it does not make sense for an FH transmitter to densely occupy the available transmission frequency band, because this would mean that numerous frequencies would be subject to the same negative impact, for example in positions with frequency-selective attenuation. On the other hand, it would be an advantage to distribute the information of a message bit over distant frequencies to avoid the concentrated impact of selective attenuation and/or selective interference power maxima. DSSS would not offer these or similar advantages.

A 'historical' example showing a surprising multiple access capacity with the usage of FH at power lines – as indicated by Figure 5.10 – has been treated in detail in [14] and [15]. There it is shown that in fact 142 FH channels could operate concurrently in the A band according to the EN 50065 standard [16], each of them providing the data rate $r_D = 60$ bits/s.

However, even if a certain robustness can be guaranteed, from today's point of view the provided data rate of that example appears far too low. Moreover, in modern applications there is no mandatory need for multiple access, neither in the implementation of the energy-related value-added services, nor in building automation. For media access the master-slave principle generally proves to be superior whenever data rates in excess of some 10 kbits/s are envisaged to exploit the quite narrow spectral resources given by EN 50065. In fact, the lack of bandwidth, together with synchronization precision requirements has ruled out the usage of FH on power lines, also in the low frequency range. For high-speed communication at data rates in the Mbits/s range, FH had never been seriously under consideration, although some benefits in comparison with DSSS can be identified:

- FH can deal with non-contiguous spectral portions – gaps of arbitrary size are not a problem; note that:
  - With FH – similarly as with OFDM – certain parts of the spectrum can be excluded from usage, e.g. due to high attenuation, heavy noise or regulation issues.
- With FH the transmitted signals always exhibit a fixed and constant envelope (with DSSS fluctuations occur due to modulation and filtering at the transmitter output, as spectral limitation to the main lobe is necessary).
- Using slow hopping can considerably mitigate the synchronization effort.

As OFDM is able to maintain most of the essential advantages of FH and at the same time offers high spectral efficiency, a total movement to this modulation technique has taken place during the past ten years in all fields of power line communications. In fact, the only serious disadvantage of OFDM versus FH is the strongly fluctuating transmission signal amplitude. Therefore, in OFDM systems sophisticated methods of crest factor control are needed, as well as more effort for the design of transmitter power stages and coupling equipment.

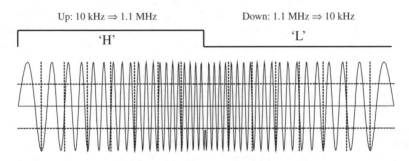

Figure 5.11   Up and down chirps as orthogonal waveforms for binary data transmission.

5.2.2.3   Types of SS Technologies: Chirp

A chirp usually consists of a sinusoidal signal with constant amplitude and frequency that linearly increases over time. Depending on the sign of the so-called chirp-parameter μ, we have an up-chirp (with +) or a down-chirp (with −). Although the envelope over time must not necessarily be linear, in most technical applications this kind of chirp is preferred for reasons which will become obvious in the following. A usual linear chirp signal is described by

$$s_{CH}(t) = A \operatorname{rect}(t/T) \cos \left[2\pi(f_0 t \pm 1/2\mu t^2) \right], \tag{5.8}$$

where A is the (constant) amplitude, T the overall duration and f_0 the starting frequency. The chirp-parameter μ was already introduced above. Differentiating the argument of the cosine in (5.8) with respect to time delivers the instantaneous frequency f_i of the chirp, i.e.

$$f_i(t) = f_0 \pm \mu t. \tag{5.9}$$

Equation (5.9) clearly indicates the meaning of the term 'linear' frequency modulation.

Chirp signals exhibit excellent resistance against various kinds of interference and also against hostile channel transfer functions. This statement is explained in more detail by a simple example given in Figure 5.11.

Before discussing the data transmission example from Figure 5.11, let us have a look at the spectrum of a chirp. A typical outcome of the Fourier transform of (5.8) is depicted in Figure 5.12. The amplitude frequency response is almost flat with steep edges on both sides.

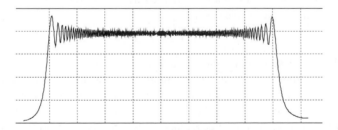

Figure 5.12   Typical envelope of a chirp spectrum.

Extending the bandwidth by raising the parameter μ leads to an improvement of both flatness and steepness of the curve, so that in the end almost rectangular shapes can be achieved.

In summary, from Figure 5.11 and Figure 5.12 we recognize the following salient features of chirps:

- In the time domain, chirps exhibit a constant envelope so that a full exploitation of the admissible transmission power is permanently possible, i.e. the maximal waveform energy is sent to a receiver.
- Chirps occupy the spectrum very evenly, which means that the information to be transmitted is optimally distributed.
- By selecting a chirp duration which significantly exceeds the data bit duration, error-free transmission is possible even under extremely bad SNR conditions. This benefit can be fully exploited e.g. by modern direct digital synthesis (DDS) technology, which can guarantee perfect linearity of a chirp over extended frequency ranges.

A chirp-based data transmission system with peculiar robustness can be built, when – for binary data transmission according to Figure 5.11 – e.g. an up-chirp is used for the 'H'-bit and a down-chirp for the 'L'-bit. In this case the chirp duration equals the data bit duration. This way of course very high data rates are not achieved, but extreme robustness can be guaranteed when matched filter technology is used at the receiver. The robustness follows from the band-spreading effect, i.e. it is determined by the size of the frequency range which is occupied during the transmission of a data bit. More details are shown in the following example depicted in Figure 5.13.

From Figure 5.13, in connection with Figure 5.11, it can be seen that a data rate of 5 kbits/s could be implemented with the chirp duration of 100 µs. The resolution with respect to time is in the range of 200 ns which equals approximately the width of the main lobe of the chirp's ACF. As the cross-correlation function (CCF) remains at very low values, it is quite obvious

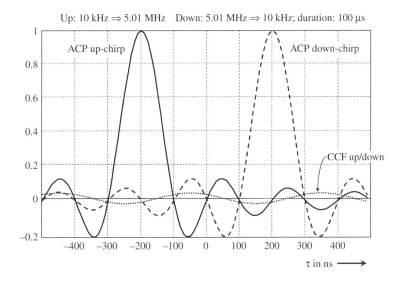

Figure 5.13 Matched filter reception of chirps; up- and down-chirp are orthogonal.

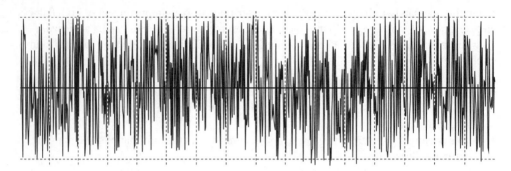

Figure 5.14 Time domain extract of an up-chirp (10 kHz to 5 MHz) at an SNR of −18 dB.

that up- and down-chirp represent quasi orthogonal signals of good quality. Furthermore, due to the fact, that the time resolution of the ACF exceeds the data rate by approximately a factor of ten, the throughput could be enhanced by sending several time-shifted chirps at minimum distances of 200 ns for this example. Such kind of 'chirp position modulation' is used in PLC systems based on commercial chipsets – see [17]–[19].

To conclude this section, let us have a more detailed look at the robustness provided by the chirp system example proposed above. Toward this aim we assume a very bad SNR ≈ −18 dB.

In the time domain it is no longer possible to recognize a chirp in the record of Figure 5.14. Investigating, however, the ACFs of an up- and down-chirp similar to Figure 5.13, we get the result depicted in Figure 5.15. Despite the heavy disturbance, both peaks are clearly visible at the correct positions and with amplitudes significantly above the side-lobes and the cross-correlation level. Therefore, a proper bit decision is easily possible and would still be feasible even at much lower SNR values.

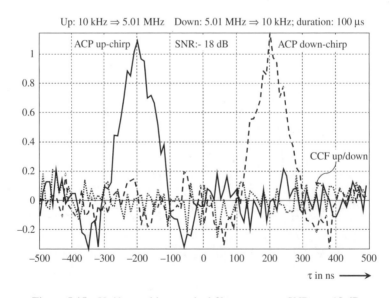

Figure 5.15 Up/down-chirp matched filter outputs at SNR = −18 dB.

In order to enhance the throughput, the bandwidth of the chirp can be extended, so that the ACF peaks become narrower. Thus, a closer staggering is possible. For further improvement of the robustness in a very straightforward manner, the chirp duration can be extended, so that the ACF peak amplitude grows. In summary, by the 'design' of a chirp, a fair adaptation to the desired transmission quality can be easily achieved.

At typical power line channels both in the low (< 500 kHz) and the high (> 1.5 MHz) frequency range, the ideal properties of chirps can unfortunately not be fully exploited for the following reasons:

- Whenever a channel exhibits low-pass character, so that significant parts of 'expected' bandwidth are not usable, a chirp's bandwidth is effectively restricted to portions where the attenuation does not exceed certain limits. Note that any attenuation leads to a degradation of the ACF peak.
- Chirps need a contiguous spectrum with no gaps, so that 'forbidden' regions (e.g. through regulation issues) represent a major obstacle for their application.
- For perfect signal detection at a receiver the channel must exhibit linear phase behavior over the complete chirp bandwidth. Multi-path and the resulting fading effects lead to significant phase distortion, so that the necessary channel coherence which the chirp needs is no longer present. As a result, enhanced side-lobes appear in the ACF and the main peak decreases.
- In multi-path environments the usage of the mentioned 'chirp position modulation' is heavily restricted as echoes may erroneously become mixed with transmitted data.

5.2.2.4 Evaluation of Benefits and Drawbacks SS Technologies for PLC

This section summarizes why or why not band-spreading technologies are appropriate for power line communication applications.

Summary of general benefits:

- SST offer robustness against narrowband selective fading (notches) and narrowband interference. Both can be issues on power line channels.
- Due to the low levels of power spectral density (PSD), electromagnetic compatibility as well as the respect of radiation limits can be easily achieved.
- SST offer multiple access features and exhibit 'graceful' degradation in multi-user environments.
- With the use of DSSS and chirps, excellent autocorrelation properties are available, so that signal detection is possible at very low SNR figures. The width of an autocorrelation peak is determined by the available bandwidth and the height is given by the waveform duration. Thus, in general, very flexible adaptation to various requirements is feasible. Mainly for these reasons such waveforms are ideal candidates for detection and ranging tasks in wireless environments, for example in radio aircraft detection and ranging (RADAR) or global positioning systems (GPS).
- In addition to the above mentioned benefits, especially frequency hopping (FH) allows a highly redundant distribution of information to arbitrary locations within the transmission bandwidth. Thus spectral gaps – either given by bad channel properties or fixed by regulations – are not an obstacle for FH applications.

Summary of general drawbacks:

- Practically all links within a power line network exhibit low-pass character, so that significant parts of the theoretical bandwidth (which could e.g. be provided by the blank lines themselves) are not usable.
- Due to the general lack of bandwidth on power line channels, the usage of band-spreading technologies appears contra-productive. Moreover, as the dominant interference exhibits also broadband character, the gain achievable with the spreading modulation is vanishing.
- For DSSS and FH systems, enhanced effort for synchronization is necessary, i.e. the timing precision exceeds by far the precision which would be needed for bit detection. This will for example rule out the possibility to use the zero-crossings of the line voltage for global synchronization.
- In view of today's requirements toward acceptably high data rates (both in the low and the high frequency range of power line networks) it appears contra-productive to equip communication systems with tremendous spectral redundancy. As outlined above, processing gains below ten are of almost no practical use, which means that approximately 90% of the available bandwidth would have to be 'sacrificed' for a spreading procedure. The necessary spectral resources are simply not available in most cases.

5.2.2.5   Practical Applications of SS Technologies in PLC Systems

In the following, let us evaluate performance, possibilities and limitations of SS technologies. As contiguous bandwidth of equal quality is generally not available in all kinds of power supply networks, the usage of band-spreading modulation technologies is usually ruled out. Even if a moderate low-pass character would still allow a sufficiently large bandwidth, a nonlinear phase response turns out to be prohibitive for successful SST applications, at least for DSSS and chirp.

For the future, it can be expected that solely technologies which both exhibit robustness against the typical channel properties and offer high spectral efficiency will be able to survive. As mentioned above, multicarrier methods such as OFDM are ideal candidates, even for the range of low frequencies, where multi-path propagation and the corresponding echoes are not an issue. Especially here, major benefits of OFDM with respect to robust and sufficiently precise synchronization can be recognized. As in general with multicarrier systems we have symbols with a duration which exceeds the data bit duration by far – this mitigates the synchronization effort.

As an example, for DSSS the required synchronization precision is high, because the maximal error should not exceed 10% of spreading clock period. Usually matched filters are needed for acquisition and delay locked loops are implemented for tracking, so that the synchronization hardware normally is the largest and most expensive functional block of a DSSS receiver.

Within FH systems the hop-rate exceeds the bit rate, so that enhanced precision for proper synchronization is required. Using e.g. the zero-crossings of the line voltage would be applicable for data rates of several hundred bits/s only, even if not more than five frequencies per bit are used for spreading – see [2], [13] or [20].

Chirp waveforms are interesting candidates as preambles for OFDM system synchronization. They allow proper synchronization under heavy noise, as demonstrated by Figure 5.15.

At a multicarrier receiver separate synchronization hardware for chirp processing is needed. The robustness for system synchronization should be approximately 10 dB better than for data detection. Therefore the implementation of a dedicated synchronization hardware usually pays off.

In summary, for SS-based PLC in the low frequency range from 9 kHz to approximately 500 kHz it is expected that chirp and modified frequency hopping systems as proposed in [20] and [21] can have a role. As they offer low data rates and exhibit low spectral efficiency, there is not much future for them in competition with multicarrier approaches. Another technique, referred to as impulsive modulation (see Section 5.5), might also be considered.

In general SST may cover certain niche-applications – no further extensions seem feasible. This statement is also true for SS-based PLC in the high frequency range from approximately 1.5 MHz to 80 MHz. Here, from today's point of view, solely chirp systems with restricted data rates as proposed e.g. in [17] may have a chance to survive in some restricted niches. In comparison with the other SS techniques, impulse modulation may exhibit benefits, especially with respect to moderate data rate applications. As, however, practical experience in the field is small until now, further investigations need to be carried out.

5.3   Multicarrier Modulations

Multicarrier (MC) systems deploy a transmission technique where a high rate information signal is transmitted through a broadband channel by simultaneous modulation of a set of parallel signals at low rate. The technique is referred to as MC modulation. The parallel signals are obtained by the serial-to-parallel (S/P) conversion of the input information signal. The idea dates back 50 years ago [22] and it was originated by the goal of simplifying the equalization task in highly frequency selective channels that introduce severe inter-symbol interference (ISI). This is made possible because the broadband channel is divided into a number of narrowband channels that exhibit a nearly flat frequency response when a sufficiently high number of sub-channels is used. If the sub-channels do not experience cross talks (inter-carrier interference, ICI), the equalizer simplifies into a single tap filter. Furthermore, MC modulation is a practical approach to achieve channel capacity since the available transmission power can be optimally allocated among the sub-channels using the water filling algorithm followed by bit-loading when finite size constellations are used [23].

In this section we describe a general MC architecture using a filter bank (FB) approach in the time domain. The unified analysis allows discussing several solutions and highlight their commonalities and distinctive points, namely, orthogonal frequency division multiplexing (OFDM) [24], pulse-shaped OFDM [25], filtered multitone modulation (FMT) [26], Offset Quadrature Amplitude Modulation OFDM (OQAM-OFDM) [27], discrete wavelet multitone (DWMT) modulation [28], and discrete cosine transform OFDM (DCT-OFDM) [29]. We will also briefly describe other MC schemes that have been proposed in the literature in Section 5.3.4, namely multicarrier CDMA, concatenated OFDM-FMT, and cyclic block FMT which uses a cyclic filter bank instead of a linear filter bank.

All these MC solutions have practical relevance for PLC and have been in part adopted by existing broadband commercial systems. The main ones are currently those developed by the High Definition PLC Alliance [30], the HomePlug Powerline Alliance [31], and the Universal Powerline Association [32]. All three technologies use MC modulation with specific and

Table 5.4 Specifications of existing broadband commercial systems

	HD-PLC	HPPA (AV/AV2)	UPA
Modulation	Wavelet OFDM (DWMT)	Pulse-shaped / windowed OFDM	Pulse-shaped / windowed OFDM
Channel coding	Reed Solomon, convolutional, low density parity check codes	Parallel-concatenated turbo code, convolutional code	Reed Solomon concatenated with Trellis coded modulation
Constellation	Up to 16 PAM	Up to 1024/4096 QAM	Up to 1024 DPSK
Max number of carriers	512 up to 2048	1536/3455	1536
MIMO	no	no/yes	no
Sampling frequency $(2B)^*$	62.5 MHz	75/200 MHz	>60 MHz
Effective band	4–28 MHz, 2–28 MHz optional	2–28/1.8–86.13 MHz	0–30 MHz, 0–20 MHz optional
Max PHY Rate	190 Mbps	200/2024 Mbps	200 Mbps

$^*B$ bandwidth

proprietary solutions. In Table 5.4 we summarize the main specification. In particular, CEPCA uses a DWMT scheme, while both HPPA and UPA deploy a pulse-shaped OFDM scheme. These systems have been the baseline for the development of the broadband standards IEEE P1901 and ITU-T G.9960 known as G.hn (home network). For more details see Table 5.5 and Chapter 7. Since two physical layers have been included in P1901, their coexistence is granted through a mechanism called Inter-System Protocol (ISP) [33]. Not only broadband PLC uses MC modulation, but also recently standardized narrowband PLC systems operating in the spectrum 3–490 kHz deploy OFDM, namely IEEE P1901.2 and ITU-T G.9902 (known as G.hnem). For more details see Table 5.5 and Chapter 8.

5.3.1 Multicarrier Modulation as a Filter Bank

In MC modulation a high rate data signal is split into M parallel data signals $b^{(k)}(\ell N)$, $k = 0, \dots, M - 1$, where N denotes the sub-channel symbol period (Figure 5.16) assuming a

Table 5.5 Specifications of existing narrowband and broadband standards.

	G.hnem	G.hn	P1901.2	P1901
Category	NB-PLC	BB-PLC	NB-PLC	BB-PLC
Standard body	ITU	ITU	IEEE	IEEE
Modulation	OFDM	OFDM	OFDM	OFDM / W-OFDM
Coding	RS, convolutional	LDPC	RS, convolutional	RS, convolutional, LDPC
Constellation	Up to 16 QAM	Up to 4096 QAM	Up to 16 QAM	Up to 4096 QAM
Tones	Up to 256	4096	Up to 256	Up to 3072
Effective band	3–148.5 kHz (EU), 9–490 kHz (US)	1.8–80 MHz	3–148.5 kHz (EU), 9–490 kHz (US)	2–28 MHz
Max PHY Rate	1 Mbps	1 Gbps	0.5 Mbps	540 Mbps

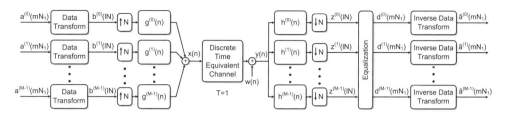

Figure 5.16 Multicarrier modulation interpreted as a filter bank.

unit sampling time T in the system. Each data signal is interpolated by a factor N and filtered with a sub-channel pulse $g^{(k)}(n)$. Therefore, the discrete-time MC signal can be written as the output of a synthesis filter bank (FB) as follows

$$x(n) = \sum_{k=0}^{M-1} \sum_{\ell \in \mathbb{Z}} b^{(k)}(\ell N) g^{(k)}(n - \ell N). \tag{5.10}$$

In the most common MC solutions the sub-channel pulse is obtained by the modulation of a prototype pulse. Modulation can be accomplished either with an exponential function or a cosine function. The former choice leads to the so called exponentially modulated FB. The latter choice leads to the cosine modulated FB. They are also referred to respectively as the Discrete Fourier Transform (DFT) filter bank, and the Discrete Cosine Transform (DCT) filter bank, since the efficient implementation is performed using a DFT or a DCT as discussed in the following. Significant examples of DFT FB modulation are FMT, OFDM, pulse-shaped OFDM, and OQAM-OFDM. DWMT and DCT-OFDM belong to the DCT FB category.

We further note that in Figure 5.16, the data streams $b^{(k)}(\ell N)$ are obtained from a transform of the information data streams $a^{(k)}(\ell N_1)$ whose symbols belong to the constellation sets of the pulse amplitude modulation (PAM), or phase shift keying (PSK), or quadrature amplitude modulation (QAM) and have normalized symbol period N_1. The transform, as it will be clear in the following, is added to allow a unified description of the various FB modulation schemes. For instance, in FMT $N_1 = N$, while in OQAM-OFDM $N_1 = M = 2N$.

The FB modulation scheme is referred to as critically sampled (CS) if $N_1 = M$, while as non critically sampled (NCS) if $N_1 > M$.

The signal $x(n)$ is transmitted over a channel with impulse response $g_{CH}(n)$ and additive noise $w(n)$ to yield the received signal $y(n)$. The received signal is analyzed with a filter bank having sub-channel pulses $h^{(k)}(n)$. The outputs are sampled at rate $1/N$ to obtain

$$z^{(k)}(\ell N) = \sum_{n \in \mathbb{Z}} y(n) h^{(k)}(\ell N - n)$$
$$= b^{(k)}(\ell N) g_{EQ}^{(k)}(0) + \text{ISI}^{(k)}(\ell N) + \text{ICI}^{(k)}(\ell N) + w^{(k)}(\ell N), \tag{5.11}$$

where $g_{EQ}^{(k)}(0) = g^{(k)} * g_{CH} * h^{(k)}(0)$ is the amplitude of the data of interest ($*$ denotes the convolution operator), $w^{(k)}(\ell N)$ is the noise contribution, $\text{ISI}^{(k)}(\ell N)$ and $\text{ICI}^{(k)}(\ell N)$ denote respectively the inter-symbol and inter-carrier interference components. They are in general present as a result of transmitting through a frequency selective channel which can also exhibit time variations. It is possible to design an FB with the perfect reconstruction (PR) property when transmission is through an ideal channel. This is accomplished when the FB satisfies a bi-dimensional Nyquist criterion

$$g^{(k)} * h^{(k')}(\ell N) = V_0 \delta_\ell \delta_{k-k'}, \quad k, k' \in \{0, \dots, M-1\}, \ \ell \in \mathbb{Z}$$

for some $V_0 > 0$, where δ_ℓ is the Kronecker delta function. The FB is orthogonal when the analysis pulse is matched to the synthesis pulse, i.e. $h^{(k)}(n) = g^{(k)*}(-n)$.

We emphasize that although the FB is designed to be with PR, the channel time dispersion and/or its time variations may destroy the orthogonality. Therefore, ISI and ICI may be present and they have to be mitigated with some form of equalization. The FB design aims at reaching a tradeoff between ISI and ICI. While the presence of both ICI and ISI requires a multi channel equalizer, the presence of only ISI allows using sub-channel equalization.

In the following we describe several FB modulation schemes.

5.3.2 DFT Filter Bank Modulation Solutions

In an exponentially modulated filter bank the sub-channel pulses are defined as

$$g^{(k)}(n) = g(n)e^{j2\pi f_k n}, \quad h^{(k)}(n) = h(n)e^{j2\pi f_k n},$$

where $f_k = k/M$ is the normalized frequency of the k-th sub-carrier. With these pulses (5.10) represents the low pass complex MC signal. We first show that any DFT FB can be realized via the use of a DFT and low rate sub-channel filtering with pulses that are obtained by the polyphase decomposition of the prototype pulse. Then, we discuss in detail several examples of significant practical interest.

5.3.2.1 Efficient Realization

The realization of both the synthesis and analysis FB is complex if it is done by the direct implementation of (5.10) and (5.11). This is because they require the realization of M interpolation and decimation modulated filters. It is therefore of great importance to devise a more efficient implementation. Herein, we follow the approach in [34] and show that all the exponentially modulated FB solutions can be realized in a common way. The resulting block diagram is depicted in Figure 5.17.

The synthesis stage is obtained by computing the polyphase decomposition with period M_2 of the signal $x(n)$ assuming $M_2 = \mathrm{lcm}(M, N) = K_2 M = L_2 N$ where $\mathrm{lcm}(.,.)$ denotes the least common multiple. The n-th polyphase component reads

$$x^{(n)}(mM_2) = x(n + mM_2) \quad n = 0, \dots, M_2 - 1, \ m \in \mathbb{Z}$$

$$= \sum_{\ell \in \mathbb{Z}} \sum_{k=0}^{M-1} b^{(k)}(\ell N) e^{j\frac{2\pi}{M}k(n-\ell N)} g(n + mM_2 - \ell N) \tag{5.12}$$

$$= \sum_{\ell \in \mathbb{Z}} B^{(n)}(\ell N) g_P^{(n)}(mL_2 N - \ell N),$$

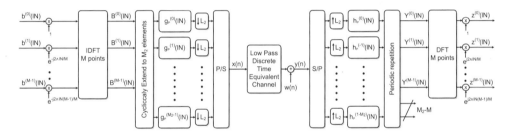

Figure 5.17 Efficient realization of DFT based filter bank modulation.

where the block $\{B^{(n)}(\ell N)\}_{n=0,\ldots,M_2-1}$ is obtained by the M-point inverse discrete Fourier transform (IDFT) of $\{b^{(k)}(\ell N)e^{-j\frac{L\pi}{M}k\ell N}\}_{k=0,\ldots,M-1}$ followed by a cyclic extension with $M_2 - M$ elements (K_2 times periodic repetition). The M_2 streams of coefficients are filtered with the polyphase components of the prototype pulse, i.e. $g_P^{(n)}(\ell N) = g(n + \ell N)$. The filter outputs are sampled by a factor L_2, and finally parallel-to-serial (P/S) converted. It should be noted that in Figure 5.17, the phase rotation $e^{-j\frac{2\pi}{M}k\ell N}$ of the data symbols is introduced because we use a modulated sub-channel pulse in a carrier-less implementation.

Similarly, the analysis filter bank can be implemented as

$$
z^{(k)}(\ell N) = \sum_{n=0}^{M_2-1} \underbrace{\left(\sum_m y^{(n)}(mL_2N)h_P^{(-n)}(\ell N - mL_2N) \right)}_{Y^{(n)}(\ell N)} e^{j\frac{2\pi}{M}(\ell N - n)k}
$$

$$
= e^{j\frac{2\pi}{M}\ell Nk} \sum_{n=0}^{M-1} \left(\sum_{m=0}^{K_2-1} Y^{(n+mM)}(\ell N) \right) e^{-j\frac{2\pi}{M}nk}.
$$

(5.13)

That is, the received signal is S/P converted to obtain M_2 streams of coefficients. The streams are filtered with the interpolation pulses $h_P^{(-n)}(\ell N) = h(\ell N - n)$ that are obtained by a polyphase decomposition of the prototype analysis pulse. Each block of M_2 coefficients $Y^{(n)}(\ell N)$ is made periodic with period M. Then, an M-point DFT is applied to obtain the output samples for the M sub-channels. Again, the phase rotation $e^{j\frac{2\pi}{M}k\ell N}$ is introduced in sub-channel k because we use modulated sub-channel pulses.

We point out that a slightly different realization of the DFT modulated FB has been described in [35]. Another realization can be found in [26] whose main difference is that it uses periodically time-variant sub-channel polyphase pulses.

It should be noted that commercial indoor broadband power line systems are currently deployed in the band 2–28 MHz. The real band pass transmitted signal can be obtained by interpolating the complex signal, and shifting the signal to the desired center carrier. Alternatively, we can directly synthesize the real band pass signal applying a $2M$-point IDFT to a block of input data symbols that are Hermitian symmetrical [36]. This base band implementation of the OFDM system is also known as Discrete Multitone Modulation (DMT).

5.3.2.2 Filtered Multitone Modulation (FMT)

FMT was originally proposed for application in very high speed digital subscriber lines (VDSL) [26]. Then, studied for multi-user wireless communications in [37]. More recently, it has been investigated for power line channels [38].

In FMT the complex data symbols $a^{(k)}(\ell N_1)$, with $N_1 = N$, belong to the QAM signal set and are mapped into the symbols $b^{(k)}(\ell N) = a^{(k)}(\ell N)e^{j\frac{2\pi}{M}k\ell N}$ if we use an FB with carrier modulation, while they remain unchanged, i.e. $b^{(k)}(\ell N) = a^{(k)}(\ell N)$ if we use a carrier-less FB. This is just a formal detail that does not change the substance of the scheme. The sub-channel symbol period is $N \geq M$ and the analysis pulse is matched to the synthesis pulse. It follows that the overall transmission rate is $R = M/NT$ symbols/second.

A distinctive characteristic of FMT is that the prototype pulse is designed to obtain high frequency confinement. Both orthogonal and non-orthogonal solutions are possible. Orthogonal solutions with good sub-channel frequency confinement require the FB to be NCS [39].

The efficient implementation architecture that we have derived in the previous section allows devising the PR conditions in matrix form. Since the DFT is an orthogonal transform, PR is obtained if the signals at the output of the IDFT block $B^{(i)}(\ell N)$ (in Figure 5.17) and the signals at the input of the DFT block $Y^{(i)}(\ell N)$ are identical. Now, if we perform an L_2 order polyphase decomposition, the relation between the input and output signals becomes

$$
Y_{p'}^{(i)}(mM_2) = \sum_{p=0}^{L_2-1} \sum_{n' \in \mathbb{Z}} B_p^{(i)}(n'M_2) \sum_{k=0}^{K_2-1} \sum_{n \in \mathbb{Z}} g_{Mk+i-Np}(nM_2 - n'M_2)
$$
$$
\times\, h_{-Mk-i+Np'}(mM_2 - nM_2),
$$

(5.14)

where

$$
Y_p^{(i)}(mM_2) = Y^{(i)}(mM_2 + p),
$$
$$
B_p^{(i)}(nM_2) = B^{(i)}(nM_2 + p),
$$

are the p-th components of the L_2-order polyphase decomposition of the signals $Y^{(i)}(\ell N)$, $B^{(i)}(\ell N)$, and

$$
g_i(nM_2) = g(nM_2 + i),
$$
$$
h_i(nM_2) = h(nM_2 + i),
$$

(5.15)

are the i-th components of the M_2-order polyphase decomposition of the prototype filters $g(n)$, $h(n)$. In the Zeta domain (5.14) becomes

$$
Y_{p'}^{(i)}(z) = \sum_{p=0}^{L_2-1} B_p^{(i)}(z) \sum_{k=0}^{K_2-1} H_{-Mk-i+Np'}(z)\, G_{Mk+i-Np}(z)
$$

where $G_i(z)$ and $H_i(z)$ are the z-transform of the polyphase components of prototype pulses in (5.15).

The perfect reconstruction condition can then be written as

$$\sum_{k=0}^{K_2-1} H_{-Mk-i+Np'}(z)\, G_{Mk+i-Np}(z) = \delta_{p-p'},$$

and in matrix form as

$$\boldsymbol{H}_{-i}(z)\, \boldsymbol{G}_i(z) = \boldsymbol{I}_{L_2}, \quad i \in \{0, 1, \dots, M-1\}, \tag{5.16}$$

where $\boldsymbol{I}_{L_2}$ is the $L_2 \times L_2$ identity matrix, $\boldsymbol{G}_i(z)$ is the $K_2 \times L_2$ sub-matrix defined as

$$\boldsymbol{G}_i(z) = \begin{bmatrix} G_i(z) & G_{M+i}(z) & \cdots & G_{(K_2-1)M+i}(z) \\ G_{i-N}(z) & G_{M+i-N}(z) & \cdots & G_{(K_2-1)M+i-N}(z) \\ \vdots & \vdots & \ddots & \vdots \\ G_{i-N(L_2-1)}(z) & G_{M+i-N(L_2-1)}(z) & \cdots & G_{(K_2-1)M+i-N(L_2-1)}(z) \end{bmatrix}^{\mathrm{T}}$$

and $\boldsymbol{H}_{-i}(z)$ is the $L_2 \times K_2$ analysis sub-matrix that, to satisfy the orthogonality constraint, is given by

$$\boldsymbol{H}_{-i}(z) = \boldsymbol{G}_i^\dagger(1/z^*), \tag{5.17}$$

where $\dagger$ denotes the conjugate and transpose operator. The orthogonal conditions are expressed by (5.16) and (5.17). We note that for $N = M$ the sub-matrices are squared, therefore the only possible solution is to choose the prototype filter with length M. Sub-matrices are polynomial so their inverse is polynomial if and only if each sub-matrix component is a monomial. Therefore, a plausible solution is to use a rectangular pulse of length M, which yields the OFDM scheme.

If $N > M$, the sub-matrices become rectangular enhancing the degrees of freedom in the choice of the shape and the length of the prototype filter. The design of the prototype pulse that yields an orthogonal FB can be done via the factorization of the matrices $\boldsymbol{G}_i(z)$ using Givens rotations or Householder transforms [40].

Quasi perfect reconstruction can be achieved with an FIR prototype pulse obtained by windowing a root-raised cosine (RRC) pulse. An example is shown in Figure 5.18 for an RRC pulse with roll-off $\alpha = N/M - 1 = 0.2$. Another approach is to design FIR pulses via sampling a root-Nyquist pulse [34] in the frequency domain. In general the pulse is designed such that the in-band to out-of-band energy ratio is maximized. This allows getting negligible ICI even in the presence of a dispersive channel. However, some ISI is present depending on the ratio between the sub-channel band and the channel coherence band. It can be counteracted with simple sub-channel equalization [26].

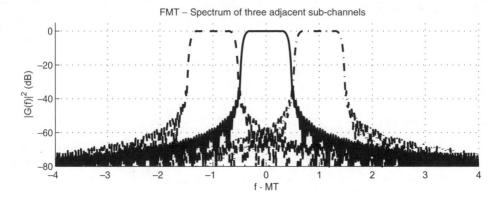

Figure 5.18 Spectrum of FMT modulation.

5.3.2.3 Orthogonal Frequency Division Multiplexing (OFDM)

OFDM [24] is among the most popular MC schemes. It uses the following synthesis pulse

$$g^{(k)}(n) = \text{rect}\left(\frac{n}{N}\right) e^{j\frac{2\pi}{M}nk},$$ (5.18)

while the analysis pulse is

$$h^{(k)}(n) = \text{rect}\left(-\frac{n+\mu}{M}\right) e^{j\frac{2\pi}{M}nk},$$ (5.19)

where the rectangular pulse is defined as rect $(n/L) = 1$ for $n = 0, \ldots, L-1$, and zero otherwise. The analysis pulse, whose duration is M, is not matched to the synthesis pulse, whose duration is $N = M + \mu$. This is done to cope with the channel time dispersion at the expense of a data rate and signal-to-noise ratio penalty. The idea is known as cyclic prefix (CP) insertion. It can be easily understood if we look at the efficient realization as shown in the following. First, we note that the M normalized sub-carrier frequencies are $f_k = k/M$, and the complex data symbols $a^{(k)}(\ell N_1)$ with $N_1 = N$ belong to the QAM signal set and are mapped into the symbols $b^{(k)}(\ell N) = a^{(k)}(\ell N)e^{-j\frac{2\pi}{M}k\mu}$. It follows that the overall transmission rate is $R = M/NT$ symbols/s.

Now, let us further decompose the signal in (5.12) as

$$x^{(n+pN)}(mM_2) = \sum_{\ell \in \mathbb{Z}} B^{(n+pN)}(mM_2 + pN - \ell N)g_p^{(n)}(\ell N),$$

for $n = 0, \ldots, N-1$, $p = 0, \ldots, L_2 - 1$, $m \in \mathbb{Z}$. Then, we can also write for $p \in \mathbb{Z}$, $n = 0, \ldots, N-1$

$$x^{(n)}(pN) = \sum_{\ell \in \mathbb{Z}} B^{(n+pN)}(pN - \ell N)g_P^{(n)}(\ell N).$$ (5.20)

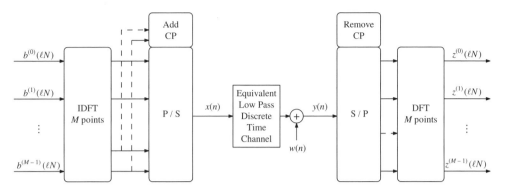

Figure 5.19 OFDM efficient realization.

Since the synthesis pulse is rectangular with duration N the polyphase components in (5.20) have a single coefficient equal to one. It follows that

$$x^{(n)}(pN) = B^{(n+pN)}(pN) = \sum_{k=0}^{M-1} a^{(k)}(pN)e^{j\frac{2\pi}{M}k(n-\mu)},$$

that is, the block $\{x^{(n)}(pN)\}_{n=0,\dots,N-1}$ of N coefficients is simply obtained by computing the M point IDFT of an input data symbol block, and extending it by periodicity to length N. In other words, each IDFT block output is padded with a prefix equal to the last $N - M$ IDFT coefficients, as Figure 5.19 shows.

The analysis FB is realized by discarding the CP in each received block of N coefficients, and then applying an M-point DFT (Figure 5.19). This can be easily proven from the efficient realization described by (5.13). In fact, (5.13) can be rearranged in the following way

$$z^{(k)}(\ell N) = \sum_{n=0}^{N-1} \left(\sum_{p\in\mathbb{Z}} y^{(n)}(\ell N - pN)h_P^{(-n)}(pN)e^{j\frac{2\pi}{M}pNk} \right) e^{-j\frac{2\pi}{M}nk}. \qquad (5.21)$$

Since the analysis pulses are those defined in (5.19) the first μ polyphase components are zero, while the remaining components are equal to one. Therefore,

$$z^{(k)}(\ell N) = \sum_{n=\mu}^{N-1} y^{(n)}(\ell N)e^{-j\frac{2\pi}{M}nk}$$
$$= \sum_{n=0}^{M-1} y(\ell N + \mu + n)e^{-j\frac{2\pi}{M}(n+\mu)k}. \qquad (5.22)$$

Assuming the channel to be a filter of duration shorter than $N - M$, i.e. the CP length, the samples after the CP in each received block are obtained from the circular convolution of the transmitted block and the channel impulse response. Therefore, the outputs of the DFT in (5.22) are

$$z^{(k)}(\ell N) = G_{\text{CH}}^{(k)} a^{(k)}(\ell N)$$

in the absence of noise, i.e. each sub-channel output corresponds to the data symbol weighted by the DFT of the channel impulse response.

The insertion of the CP allows an elegant and simple PR solution. The drawback is that some data rate penalty is introduced. To minimize such a penalty the number of carriers has to be high with respect to the CP length. In turn this increases complexity.

Furthermore, the increase in the number of tones yields an increase in the duration of the OFDM symbol. This has an impact when transmission is over a time-variant channel, since channel time variations within an OFDM symbol may introduce a loss of sub-carrier orthogonality, therefore inter-carrier interference [41]. PLC channels exhibit a periodic time-variant behavior over a main cycle, that fortunately is sufficiently slow. To increase data rate, the CP can be shortened. Clearly this introduces a loss of orthogonality. If this is significant it is required to mitigate it. A first proposed technique is to deploy a time domain equalizer (TEQ) [42], [43] that has the objective of shrinking the channel impulse response before the application of the DFT at the receiver. Another approach is to deploy a multi-channel equalizer at the output of the DFT. The multi-channel equalizer can be either a linear multi-channel equalizer [44] or a maximum *a posteriori* (MAP) multi-channel equalizer [37]. The MAP equalizer can be sub-optimally implemented via iterative (turbo) detection strategies that iteratively cancel the interference components [37].

Despite its attractive simplicity, OFDM suffers from the poor sub-channel spectral containment since the sub-channel frequency response is a sinc function. Therefore, significant overlapping of the sub-channels exists. This translates into high sensitivity to time misalignments, carrier frequency offsets and channel time selectivity which cause sub-channel frequency shifting/spreading. Furthermore, to compact the overall spectrum, as well as to implement notching it is necessary to switch off a large number of tones which diminishes the transmission rate. To partially overcome such limitations, it has been proposed to use pulse shaping as described in the next section.

5.3.2.4 Pulse-Shaped OFDM and Windowed OFDM at the Transmitter

Better spectrum containment can be obtained by substituting the rectangular synthesis pulse in OFDM with a Nyquist window [25]. This yields the Pulse-shaped OFDM solution. It can be viewed as an FMT scheme where however, containment is privileged in the time domain rather than in the frequency domain. The synthesis sub-channel pulse of (5.18) is obtained from a prototype pulse that can be for instance a raised cosine pulse (in the time domain) with an integer roll-off α and duration $N + \alpha$ samples where $N = M + \mu + \alpha$ is the sub-channel symbol period as shown in Figure 5.20.

The analysis pulse is as in OFDM

$$h^{(k)}(n) = \text{rect}\left(-\frac{n + \mu + \alpha}{M}\right) e^{j\frac{2\pi}{M}nk}.$$

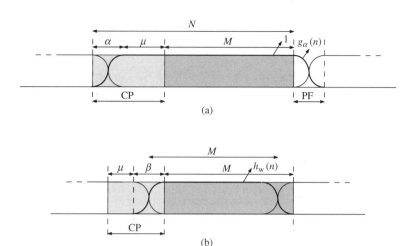

Figure 5.20 (a) OFDM pulse-shaping at the transmitter, and (b) OFDM windowing at the receiver.

It should be however noted that the symbol period is extended by α samples which implies that the CP is extended by α. The symbols $b^{(k)}(\ell N)$ used in Figure 5.16 are set equal to $a^{(k)}(\ell N)e^{-j\frac{2\pi}{M}k(\mu+\alpha)}$.

The realization can be obtained from (5.20). The polyphase components of the pulse are

$$g^{(n)}(\ell N) = \begin{cases} g_\alpha(n), \ell = 0, \ 1 - g_\alpha(n), \ell = 1, \text{ for } n = 0, \ldots, \alpha - 1, \\ 1, \ell = 0, \qquad 0, \ell = 1, \qquad\qquad \text{for } n = \alpha, \ldots, N - 1, \end{cases}$$

where $g_\alpha(n)$ is the raised cosine portion of the window (Figure 5.20). It follows that

$$x^{(n)}(pN) = \begin{cases} B^{(n+pN)}(pN)g^{(n)}(0) + B^{(n+pN)}((p-1)N)g^{(n)}(N), & \text{for } n = 0, \ldots, \alpha - 1, \\ B^{(n+pN)}(pN), & \text{for } n = \alpha, \ldots, N - 1, \end{cases} \quad (5.23)$$

which corresponds (Figure 5.20) to generate a stream of OFDM symbols with a CP of length $\mu + \alpha$, and append a cyclic postfix (PF) of length α. Each extended OFDM symbol is windowed with the raised cosine pulse, and an overlap and add operation with period N is then applied according to (5.23).

If the sub-channel symbol period is set equal to $M + \mu + 2\alpha$, then the overlap and add operation is not required. This is referred to as windowing at the transmitter. Clearly, this diminishes data rate further.

Provided that the channel has a discrete time impulse response shorter than μ, PR is achieved. This is simply proved by observing that the analysis pulse is identical to that used in OFDM, and it collects energy only over the flat portion of the synthesis window.

The advantage relies in the capability of reducing the out-of-band power spectral density (PSD). However, the sub-channel confinement does not significantly change with regards to OFDM without windowing. The spectrum is shown in Figure 5.21. As a result, to obtain

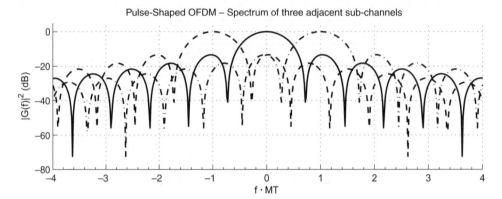

Figure 5.21   Spectrum of pulse-shaped OFDM.

enhanced immunity against narrowband interference, optimal windows have to be designed [45], [46].

5.3.2.5   Windowed OFDM at the Receiver

Windowed OFDM applies a window at the receiver side [25]. The idea is similar to windowing at the transmitter. First, the cyclic prefix is set equal to $\mu + \beta$, such that the symbol period is $N = \mu + \beta + M$. Then, the received signal is passed through an FB whose analysis prototype pulse has impulse response defined by the window $h_w(n)$, e.g. a raised cosine, with duration $\beta + M$, as shown in Figure 5.20. The outputs from (5.21) read

$$z^{(k)}(\ell N) = \sum_{n=\mu}^{N-1} \underbrace{y^{(n)}(\ell N) h_w(n)}_{y_w^{(n)}(\ell N)} e^{-j\frac{2\pi}{M} nk}.$$

This shows that the filtering operation is simply implemented by windowing the received signal to obtain $y_w^{(n)}(\ell N)$. Then, aliasing of the received windowed block is performed in the time domain to obtain a block of M coefficients that are finally transformed by an M-point DFT, i.e.

$$z^{(k)}(\ell N) = \sum_{n=0}^{\beta-1} \left[y_w^{(n+\mu)}(\ell N) + y_w^{(n+M+\mu)}(\ell N) \right] e^{-j\frac{2\pi}{M}(n+\mu)k} + \sum_{n=\beta}^{M-1} y_w^{(n+\mu)}(\ell N) e^{-j\frac{2\pi}{M}(n+\mu)k}$$

$$= \sum_{n=0}^{M-1} y_{w,a}^{(n)}(\ell N) e^{-j\frac{2\pi}{M} nk},$$

where $y_{w,a}^{(n)}(\ell N)$ is the aliased signal. If the channel length is shorter than μ, no inter-block interference is present. Further, no ICI is present because the Nyquist criterion in the time domain is satisfied.

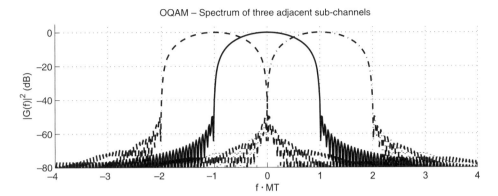

Figure 5.22   Spectrum of OQAM-OFDM.

We emphasize that pulse-shaping at the transmitter and receiver windowing can be independently or jointly applied. Both the HPPA and the UPA systems deploy pulse-shaping and windowing. The effect of windowing is to analyze the received signal with sub-channel pulses that exhibit some better frequency confinement than the rectangular window. To obtain some benefit in the presence of narrowband interference optimal windows have to be designed [47].

5.3.2.6   Offset-QAM OFDM

OQAM-OFDM is another DFT FB modulation architecture originally proposed in [27]. The application in PLC has been proposed in [48]. The key idea is to generate QAM data streams whose changes of the real and imaginary part are staggered in time and between adjacent sub-channels. This is accomplished by the data transform block in Figure 5.16 as we explain in the following. The input QAM signals have symbol period $N_1 = M$. The sub-carrier frequencies are $f_k = k/M$ yielding a CS solution. The data transform block maps the data symbols into a stream $b^{(k)}(\ell N)$ with symbol period that is half of the input one, i.e. $N = M/2$ according to

$$b^{(k)}(\ell N) = \mathrm{j}^{(k+1)\%2} \sum_m \left[a_I^{(k)}(mM)\delta_{\ell-2m} + \mathrm{j}a_Q^{(k)}(mM)\delta_{\ell-2m-1} \right],$$

where $a_I^{(k)}(mM) = \Re\left(a^{(k)}(mM)\right)$, $a_Q^{(k)}(mM) = \Im\left(a^{(k)}(mM)\right)$, and % denotes the modulo operation. It follows that the real and imaginary part have an offset in time equal to half the symbol period. Further, the real/imaginary changes are also staggered between adjacent sub-channels. Similarly to FMT the prototype pulse has a concentrated frequency response, however, the nominal and normalized sub-channel pulse bandwidth is $1/N = 2/M$, i.e. twice that of FMT. Perfect reconstruction FIR OQAM filter banks can be designed [49]. As an example, let us consider the use of a root-raised cosine prototype pulse with normalized Nyquist band $1/N$ (Figure 5.22). Then, the analysis FB uses the same prototype pulse and the outputs are sampled at rate $1/N$ to obtain

$$\Re\left(z^{(k)}(2\ell N)\right) = a_I^{(k)}(\ell M), \Im\left(z^{(k)}((2\ell+1)N)\right) = a_Q^{(k)}(\ell M), k \text{ odd},$$
$$\Im\left(z^{(k)}(2\ell N)\right) = a_I^{(k)}(\ell M), \Re\left(z^{(k)}((2\ell+1)N)\right) = -a_Q^{(k)}(\ell M), k \text{ even}.$$

Therefore, although the sub-channels significantly overlap, staggering the real and imaginary parts across sub-channels allows obtaining PR. However, the high overlapping of adjacent sub-channels may induce significant ICI when transmission is over a non ideal channel. Therefore, equalization both in time and in frequency (across sub-channels) has to be applied. The efficient realization of the scheme can be done according to the structure in Figure 5.17. However, processing has to be done at twice the speed. Therefore, the complexity is twice that of FMT if we assume the same number of tones and prototype pulse length. An interesting aspect of OQAM-OFDM is that the signal phase transitions never cross zero. This translates in a better peak-to-average power ratio, which eases the amplifier design.

5.3.3 DCT Filter Bank Modulation Solutions

A cosine modulated FB can be realized via the use of DCTs and low rate sub-channel filtering. Both CS and NCS solutions have been extensively investigated for application in image coding [40]. They are also referred to as lapped transform and extended lapped transform FB. We herein discuss two critically sampled solutions that have gained attention for digital communications. The former has been proposed in [28] under the name of DWMT originally for DSL applications. The HD-PLC commercial PLC modem uses this modulation scheme. The latter is known as DCT-OFDM.

5.3.3.1 Discrete Wavelet Multitone (DWMT)

This FB modulation scheme has normalized sub-carriers frequencies $f_k = k/2M$. The sub-channel symbol period is $N_1 = N = M$, and the cosine modulated sub-channel pulses are

$$g^{(k)}(n) = g(n) \cos \left[\frac{\pi}{M}(k + 0.5) \left(n - \frac{L_g - 1}{2} \right) - \theta^{(k)} \right],$$

$$h^{(k)}(n) = g(n) \cos \left[\frac{\pi}{M}(k + 0.5) \left(n - \frac{L_g - 1}{2} \right) + \theta^{(k)} \right],$$

where $0 \leq n \leq L_g - 1$, $\theta^{(k)} = (-1)^k \pi/4$, and $L_g = 2KM$ is the pulse length [40]. The PR solution is possible and the orthogonality conditions are similar to those in OQAM-OFDM [49]. As an example, we plot the spectrum in Figure 5.23. It should be noted that this FB architecture uses real constellations such that (5.10) represents the discrete-time real MC signal. Further, the scheme is critically sampled and achieves a symbol rate equal to $R = 2/T$ real symbols/second.

Since the sub-channels have a high degree of overlapping, the orthogonality can be lost after transmission over a dispersive channel. Several solutions to the problem have been proposed in the literature and they include time domain pre-equalization [28], multi-channel equalization as the post-combiner structure in [28], or recent simplified blind solutions in [50] and with training in [51].

5.3.3.2 DCT-OFDM

If we fix $K = 1$, and we choose $g(n) = \text{rect}(n/2M)$, i.e. a rectangular time domain pulse with duration $L_g = 2M$, the scheme simplifies into a simple IDCT-DCT structure [29]. The plot of

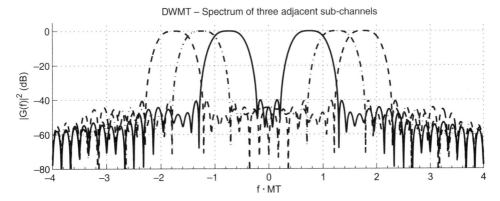

Figure 5.23   Spectrum of DWMT.

the spectrum for this case is shown in Figure 5.24. The orthogonality of the DCT is clearly lost after transmission through a dispersive channel. A possible solution is to zero pad each transmitted block to avoid the inter block interference. Then, the inter-carrier interference can be managed with a minimum-mean-square-error (MMSE) block equalizer, a solution that has been proposed to equalize time variant frequency selective fading channels [52].

5.3.4   Other MC Schemes

There are other MC schemes in the literature that can have a relevance for PLC.

A first scheme is known as multicarrier CDMA [53]. Essentially it is an OFDM scheme where the data symbols are spread over the sub-channels with a spreading code. In wireless it has attracted considerable attention since it has the potentiality of providing frequency diversity exploitation since each data symbol is spread over sub-channels that exhibit independent fading. Further, it can support multiple access if the spreading codes are assigned to different users. However, it requires multi-channel equalization to restore code orthogonality that is destroyed by the frequency selective channel. More general multicarrier CDMA architectures

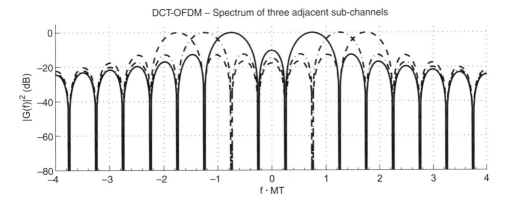

Figure 5.24   Spectrum of DCT-OFDM.

are also referred to as linearly pre-coded OFDM and they have been investigated for PLC in [54].

Another scheme, referred to as concatenated OFDM-FMT, uses an inner (with respect to the channel) FMT modulator to accomplish frequency division multiplexing by the assignment of different FMT sub-channels to different users. OFDM is then used over each FMT sub-channel to cope with the residual sub-channel frequency selectivity. The scheme provides high robustness in the asynchronous multiple access channel [55].

More recently, a different filter bank modulation approach has emerged. The key idea is to replace the linear convolutions in the filter bank with cyclic convolutions [56]. The scheme is referred to as cyclic block filtered multitone modulation (CB-FMT). It is described in the following section.

5.3.4.1   Cyclic Block Filtered Multitone Modulation

Cyclic Block Filtered Multitone Modulation is a multicarrier scheme derived starting from the filter bank philosophy with the objective of merging the strengths of both OFDM and FMT [56], [57]. Similarly to conventional FMT, CB-FMT aims at generating well frequency localized sub-channels. However, differently from FMT, CB-FMT transmits data symbols in blocks. Furthermore, the filter bank does not use linear convolutions but cyclic convolutions. Similarly to OFDM, the block transmission can reduce latency, but the sub-channel frequency confinement is much higher, more similarly to FMT. This translates in higher spectral selectivity, more confined power spectral density and possibly lower Peak-to-Average-Power Ratio (PAPR) than in OFDM given the same target spectral efficiency. CB-FMT allows a better usage of the spectrum than OFDM when the spectral masks specified by the standards have to be fulfilled.

Another important aspect is the implementation complexity. Compared to FMT it is significantly lower than that in FMT with the same number of sub-channels and even when longer pulses are used. In fact, an efficient realization can be devised if both the synthesis and the analysis filter banks in CB-FMT are implemented in the Frequency Domain (FD) via a concatenation of an inner (w.r.t. the channel) Discrete Fourier Transform (DFT) and a bank of outer DFTs. Such an FD architecture enables the use of an FD sub-channel equalizer designed according to the Zero Forcing (ZF) or the Minimum Mean Square Error (MMSE) criterion. In particular, the ZF solution will restore perfect orthogonality if a cyclic prefix (similarly to OFDM) is appended to each block of signal coefficients that are transmitted over a dispersive (frequency selective) channel. This equalization scheme is capable to coherently collect the sub-channel energies so that frequency and time diversity, offered by the fading channel, can be exploited. Consequently, this can provide better performance, i.e. lower symbol error rate and higher achievable rate, than OFDM.

The application of CB-FMT in PLC has been described in [58] focusing on broadband high speed transmission. The application of CB-FMT to realize a flexible spectrum solution that can efficiently adapt the transmission spectrum (as required by cognitive communication applications) has been discussed in [59].

5.3.4.1.1   System model

The block diagram that depicts the CB-FMT scheme is shown in Figure 5.25. Herein, the low data rate data sequences are interpolated by a factor N and, then, filtered with a prototype pulse. Differently from conventional MC modulation, the convolutions in the filter bank are circular.

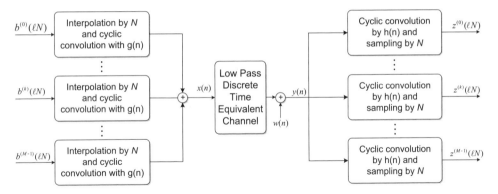

Figure 5.25 CB-FMT transceiver scheme.

The filter outputs are multiplied by a complex exponential to obtain a spectrum translation. Finally, the M modulated signals are summed together yielding the transmitted discrete time signal.

The circular convolution involves periodic signals and it can be efficiently realized in the frequency domain via the discrete Fourier transform (DFT). To use the circular convolution, a block wise transmission is needed. Thus, we gather the low data rate sequences in blocks of L symbols $b^{(k)}(\ell N)$, $\ell \in \{0, \dots, L-1\}$, for each sub-channel. Then, we consider the prototype pulse $g(n)$ to be a causal finite impulse response (FIR) filter, with a number of coefficients equal to $L_g = LN$. If the length is lower than L_g, we can extend the pulse length to L_g with zero-padding, without loss of generality. The CB-FMT transmitted signal can be written as

$$
\begin{aligned}
x(n) &= \sum_{k=0}^{M-1} \left[a^{(k)} \otimes g \right](n) \\
&= \sum_{k=0}^{M-1} \sum_{\ell=0}^{L-1} a^{(k)}(\ell N) g((n - \ell N)_{L_g}) W_M^{-nk}, \\
&\quad n \in \{0, \dots, L_g - 1\},
\end{aligned}
\tag{5.24}
$$

where $\otimes$ denotes the circular convolution operator and $g((n)_{L_g})$ denotes the cyclic (periodic) repetition of the prototype pulse $g(n)$ with a period equal to L_g, i.e. $g((n)_{L_g}) = g(mod(n, L_g))$ where $mod(\cdot, \cdot)$ is the integer modulo operator. $W_M^{-nk} = e^{j2\pi nk/M}$ is the complex exponential function and j is the imaginary unit.

Similarly to the synthesis stage, we can apply the circular convolution to the analysis filter bank. The k-th sub-channel output is therefore obtained as follows

$$
z^{(k)}(nN) = \sum_{\ell=0}^{L_g-1} y(\ell) W_M^{\ell k} h((nN - \ell)_{L_g}),
\tag{5.25}
$$

$$
k \in \{0, \dots, M-1\}, \quad n \in \{0, \dots, L-1\},
$$

where $h((n)_{L_g})$ denotes the periodic repetition of the prototype analysis pulse $h(n)$ with period L_g.

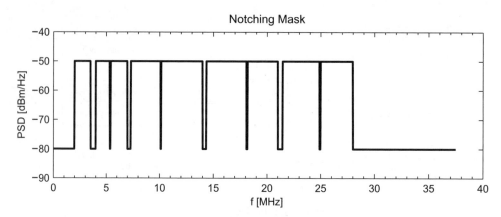

Figure 5.26 Notching mask.

Each sub-channel conveys a block of L data symbols over a time period equal to LNT. Therefore, the transmission rate equals $R = M/(NT)$ symbols/sec. Both critically sampled and non-critically sampled solutions exist respectively when $M = N$ and when $M < N$. A simple FD pulse design that renders the system orthogonal can be followed [60].

5.3.5 Coexistence and Notching

Wideband PLC systems currently operate in the band 2–28 MHz. To increase throughput wider bands up to 100 MHz and beyond are nowadays investigated. This translates into the important issue of coexistence with other systems and interoperability among different PLC technologies. Coexistence is firstly allowed by means of electromagnetic compatibility, while interoperability is reached through the use of media access control (MAC) and physical layer (PHY) mechanisms. Furthermore, coexistence has to be supported not only among PLC systems but also with other technologies such as radio amateur systems, radio broadcast systems, and even DSL technology that operate in the same or nearby spectrum.

Joint PHY and MAC algorithms that target spectrum sharing with a detect-and-avoid approach can allow coexistence. Interoperability, and more importantly compatibility, with existing standards are allowed by defining PHY mechanisms that grant communication both with existing devices and new devices. For instance we can envision new systems that support existing transmission technologies in the lower spectrum, e.g. Pulse-shaped OFDM, while operating with another MC scheme at higher frequencies. An optimized and efficient implementation of such an overlay hybrid MC PHY is possible once it is recognized that the various MC schemes are all based on an FB architecture as discussed in the previous sections.

The coexistence with other technologies implies that the PLC PHY shall be robust to interference and shall also generate low radiations in certain portions of the spectrum. The main source of interference in the 0–100 MHz frequency band are the radio amateur signals, and the AM and FM broadcasters. A robust approach to allow the coexistence is to notch the emitted signal at certain frequencies. As an example, in Figure 5.26 we plot a notching power spectral density (PSD) mask similar to that used by the HomePlug AV system in the 2–28 MHz band [46].

In order to optimize spectral efficiency, a detect-and-avoid approach can be used where the narrowband interferer is identified and the associated spectrum is notched. In contrast, a static approach would consist in notching an entire portion of the spectrum corresponding for instance to the AM broadcast radio spectrum. Especially with the former approach it is important to obtain a high notching selectivity that we can define as $1/L$ where L is the number of sub-channels that need to be switched off to fulfill the required PSD mask. Clearly, it is a function of the sub-channel spacing and the prototype pulse. A high notch selectivity implies lower losses in data rate since only few sub-channels need to be switched off. The MC schemes that deploy frequency confined sub-channel pulses can obtain high notch selectivity and consequently lower loss in data rate [38], [45], [48]. In the next section we show the achievable rate in the presence and the absence of notching for FMT modulation and for pulse-shaped OFDM assuming the notching mask of Figure 5.26.

5.3.6 Bit Loading

MC modulation is a practical approach to solve the resource allocation problem, i.e. bit and power loading across the sub-channels such that channel capacity is achieved [23]. Assuming to model the system as a set of parallel Gaussian channels the water filling algorithm provides the optimal power allocation and consequently the number of bits that can be loaded on a given sub-channel can be determined. Current norms impose a power spectral density constraint only. Consequently, commercial systems operate with a constant PSD, i.e. they uniformly distribute the power across the sub-channels. Under the assumption of Gaussian noise and interference, an achievable rate for the exponentially modulated FB is

$$R = \frac{1}{NT} \sum_{k=0}^{M-1} \log_2 \left(1 + \frac{\text{SINR}^{(k)}}{\Gamma} \right) \quad [\text{bit/s}],$$

where Γ represents a gap factor accounting for practical modulation and channel coding, and $\text{SINR}^{(k)}$ is the signal over noise-plus-interference ratio on sub-channel k. The parameter Γ can be computed as

$$\Gamma = \frac{1}{3} \left(Q^{-1} \left(\frac{P_e}{4 - \frac{4}{\sqrt{M_{\text{QAM}}}}} \right) \right)^2$$

for uncoded QAM constellations, where P_e is the target symbol error probability, M_{QAM} represents the constellation size, and $Q^{-1}(x)$ is the inverse function of the normalized Gaussian complementary distribution. Although Γ slightly varies with the constellation size, we assume it equal to 9 dB when showing numerical results. It is obtained for uncoded QAM of size 1024, and $P_e = 10^{-6}$. The number of bits that can be loaded to channel k is given by $b^{(k)} = \log_2(1 + \text{SINR}^{(k)}/\Gamma)$. In practical implementations, the number of bits is rounded to an integer number to take into account that M-QAM constellations are used.

It should be noted that the interference in the computation of the SINR includes both the narrowband interference, the ISI and the ICI (self interference). The self interference is the result of the distortions caused by the frequency selective and time variant PLC channel that destroy the FB orthogonality. For instance, in OFDM the ICI is significant when the CP is shorter than the channel length. In FMT the ICI is negligible but the ISI is present. As a result the achievable rate depends not only on the channel impulse response, and the background noise, but also on the particular prototype pulse used, the number of sub-channels M, and the interpolation factor N (equal to $M + \mu$ in OFDM). These parameters can be chosen to maximize achievable rate. In particular the CP length has not to be necessarily equal to the channel duration to maximize achievable rate. Further, the channel impulse response realizations may vary and therefore it is reasonable to consider the adaptation of the CP length to the specific channel realization. This translates into a more general resource allocation problem that is described in [61]–[63]. Another aspect that is herein not considered, is the effect of coupling and impedance adaptation. In PLC the line impedance shows a high frequency selective behavior which renders the adaptation problem challenging. In [64], a study about the effect of the optimal receiver impedance adaptation to the achievable rate is done.

In the following we report numerical results for Pulse-shaped OFDM and FMT as an example of FB modulation. We assume a sampling frequency for the low pass complex signal equal to $1/T = 37.5$ MHz with an effective transmission bandwidth in 2–28 MHz. A typical PLC indoor channel impulse response is considered and it corresponds to the worst channel realization obtained from the statistical channel model considered in [61], [62]. The signal-to-noise ratio at zero frequency is 20 dB or 40 dB. The average path loss profile in the transmission band matches typical measured profiles with an approximately linear decay of about 0.2 dB/MHz. We report in Figure 5.27 and Figure 5.28 the achievable rate as a function of the total number of system tones that is equal to $M = \{96, 192, 384, 768, 1536\}$. The achievable rate for the unmasked case (figures labeled with A) are obtained by switching

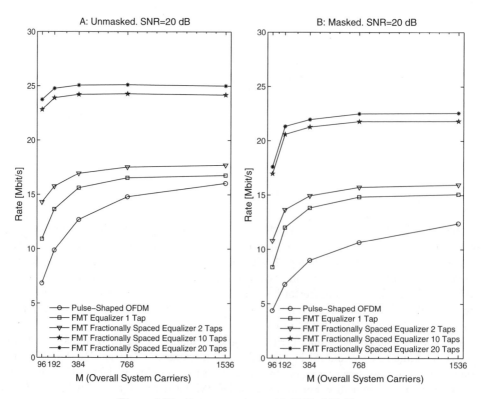

Figure 5.27   Rate comparison with SNR of 20 dB.

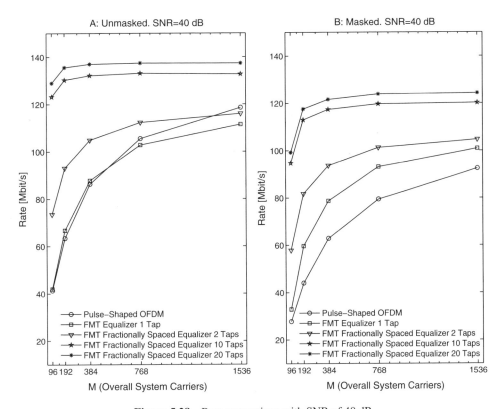

Figure 5.28 Rate comparison with SNR of 40 dB.

off only the tones at the band edges to obtain an effective transmission band in 2–28 MHz. The achievable rate for the masked case (figures labeled with B) are obtained by switching off extra tones such that the mask in Figure 5.26 is satisfied. We point out that pure OFDM experiences a high rate penalty to fulfill the mask. Pulse-shaped OFDM has better spectral containment if it deploys a raised cosine window with roll-off $\alpha = \mu = CP/2$, as herein considered. Therefore, it exhibits higher achievable rate. Significant improvements in sub-channel containment are obtained with FMT that uses a conventional truncated RRC pulse with roll-off 0.2 and length $12N$ coefficients. The interpolation factor N and the CP length have been obtained via rate maximization as described in [62].

Now looking at Figure 5.27 and Figure 5.28, we can see that FMT achieves significant higher rate than Pulse-shaped OFDM both for the masked and the unmasked case. The rate gain is a function of the affordable receiver complexity. In Pulse-shaped OFDM single tap sub-channel equalization is deployed. In FMT an identical equalization scheme can be used (labeled with Equalizer 1 Tap). Since FMT exhibits sub-channel ISI, significant improvements can be obtained with more complex equalization. Herein, we have considered MMSE fractionally spaced sub-channel equalization with a number of taps equal to 2, 10, and 20. More than 20 taps do not yield significant improvements. Other more complex equalization schemes can be considered [37]. For the masked case, even single tap equalization in FMT outperforms pulse-shaped OFDM for all the values of SNR and number of tones herein considered.

An increase in the number of tones increases rate in both systems since the added redundancy, e.g. the CP length, weights less. If the number of tones goes to infinity all MC systems converge to the same performance since the sub-channel pulses tend to become a delta function in the frequency domain. However, it should be noted that the higher the number of tones, the longer the symbol period is. Therefore, the number of tones has to be limited such that the symbol duration is shorter than the channel coherence time to avoid the interference from channel time variations [41].

For practical number of tone values, pulse-shaped OFDM has worse performance than FMT since it uses more redundancy, it suffers for an SNR loss, and requires to notch a higher number of tones to fulfill the imposed PSD mask. Interestingly, FMT achieves the maximum rate with a smaller number of tones provided that equalization is performed (equal to 384 in the scenario considered). In turn, this suggests to a trade-off between the FB parameters and the equalization scheme to lower the overall implementation complexity.

5.4   Current and Voltage Modulations

There are scenarios in which the typical transmitter consisting of a digital signal generator coupled to the power line is not satisfactory. In such cases signals must be generated by some other means. Unfortunately, generating the signal in that way usually constrains the available waveforms so that many modulation methods do not work. In this section we discuss scenarios in which those transmitters may be employed and how information is imparted to the transmitted signal and retrieved at the receiver.

5.4.1   VLF/ULF PLC

When discussing power line communications it must be remembered that the power distribution network was designed to transmit high voltage signals at relatively low frequencies (e.g. 50 and 60 Hz) efficiently. It was not designed to conduct higher frequency signals efficiently. Paired transmission lines, for example, have high inductances. High frequency signals experience significant attenuation with distance because of the resulting impedance. Although transformers are not strictly functional low-pass filters for purposes of power line communication [65], the entire distribution system, consisting of the transmission lines, transformers, and the various loads, functions as a low-pass filter over distances greater than just a couple of kilometers [66]. Thus, in cases where communication through multiple distribution transformers and over long distances is required it is advantageous to maintain a low carrier frequency. This is particularly true if repeaters are to be avoided, which is often the case given the added expense.

VLF and ULF band power line communications are often used in these cases. Not surprisingly, since VLF and ULF band signals are closer to the system design frequency of 50 or 60 Hz, these signals behave better on the power line channel. In particular, these lower frequency signals have little attenuation with distance [67] and can effectively penetrate distribution transformers. However, with reduced carrier frequency comes reduced bandwidth and reduced data rate. Most extant VLF/ULF systems are for AMR/AMI applications, though, which have significantly reduced data rate requirements over other communication systems. So, for certain markets VLF/ULF PLC is an attractive option.

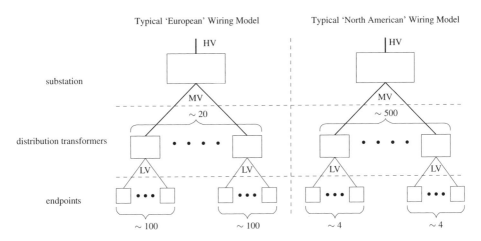

Figure 5.29 Illustrations of the differences between the 'European' wiring model (left) and the 'North American' wiring model (right) for power networks. In both scenarios a substation services approximately 2000 endpoints. In the European model each distribution transformer services on the order of 100 endpoints, while in the North American model each distribution transformer services only a few endpoints, typically far less than 10.

Power distribution networks vary widely across the globe, but for purposes of this discussion can be subdivided into two basic wiring categories. The model predominant in Europe and other countries outside of North America is one in which medium-voltage (MV) to low-voltage (LV) transformers service tens or even hundreds of endpoints (see Figure 5.29). We shall henceforth refer to this as the 'European' wiring model. In North America, however, each LV-MV transformer typically services only four endpoints, although this may be as few as one. This seemingly subtle difference has important implications in power line communication design.

In networks conforming to the European wiring model it is economical to implement power line communications on the LV network. The European wiring model is conducive to PLC because endpoints are typically very close together and served by a single distribution transformer. Neither signal attenuation due to distance nor the low frequency cutoff of the service transformer are an issue in this environment. Thus, a single router can be placed on the low side of a distribution transformer and effectively service a large number of endpoints.

In networks conforming to the North American wiring model the matter is different. In this model there are very few endpoints per service transformer. In rural areas the meter density is also very low, possibly as low as 1 endpoint per square mile. If a router is placed at every distribution transformer, the number of routers required per endpoint will be excessively high and cost prohibitive. Furthermore, because of the enormous distances between endpoints, signal attenuation with distance is a significant problem for higher frequency PLC. In order to maintain the required signal strength it may also be necessary to install many signal boosters and repeaters, thus further increasing cost.

VLF/ULF PLC provides a cost effective solution in many North American wiring models. In such a system a router can be placed somewhere on the MV network, often inside the substation itself, and modems are placed inside the electric meters on the LV network. Because the signal is affected very little by distance effects and penetrates the distribution transformers a link is maintained without the need for signal boosters or repeaters.

Arbitrary waveform generators are usually connected to the power network using a coupling capacitor to block the 50 or 60 Hz mains signals from the amplifier while simultaneously allowing the transmitter's relatively high frequency signals to pass to the power line unimpeded. Larger capacitors are required as the transmission frequency decreases. For VLF and ULF band signals the capacitor required may be too large to fit inside a typical residential electric meter. Thus capacitive coupling is often not an option for VLF/ULF band PLC.

One alternative is to connect directly to the power network using a switched load transmitter. We discuss two such transmitters in the following sections. As will be shown, these transmitters have the advantage of generating VLF and ULF band signals without requiring large coupling capacitors or transformers. However, they are at a disadvantage in that they are limited by the possible signals that may be transmitted. Arbitrary waveform generation, for example, is not possible with these transmitters. Modulation and demodulation, therefore, must be designed with the transmitter signal constraints in mind. While the resulting communications link may not have a high data rate compared to higher band power line communication systems, the data rate is sufficiently high for most AMI/AMR applications.

5.4.2   OOK with a Switched-load Transmitter

The circuit in Figure 5.30 illustrates an alternative to capacitive coupling of an arbitrary waveform generator to the power line. The box to the left models the power network using its Thévenin equivalent, with a purely resistive source impedance. Although the network impedance is predominantly inductive, a real resistance is used here to simplify the mathematics to follow. The concepts illustrated in this derivation can be generalized for an inductive source resistance. $v_s(t)$ is the mains voltage and R_s is the source impedance of the entire power distribution system as seen by the transmitter. The transmitter is connected directly to the power network and consists of a switch and load resistance, R_L. The switch, when activated,

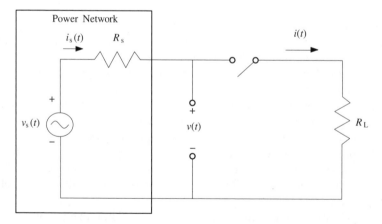

Figure 5.30   This schematic illustrates a simplified circuit for transmitting data on a power network by switching a load in and out of the network.

connects the load to the network. Prudent operation of this switch is used to create current in the distribution network, $i_s(t)$, that can be sensed by an appropriate receiver. We shall derive the equation for a single pulse generated by this circuit in what follows.

Let us model the mains signal as a cosine wave with amplitude V and frequency ω_0:

$$v_s(t) = V \cos(\omega_0 t). \tag{5.26}$$

The load is connected to the power network at some time $t \in \left[0, \frac{\pi}{2\omega_0}\right]$ which is before the first zero crossing. The switch is then disconnected from the power network at the first zero crossing, $t = \frac{\pi}{2\omega_0}$. The switch disconnects here because, often, it is the case that the physical implementation of the switch is such that it switches off when the current through it is zero. The mains current due to this switching operation is

$$i_s(t; \tau) = \frac{V}{R_s + R_L} \cos(\omega_0 t) \left[u(t - \tau) - u\left(t - \frac{\pi}{2\omega_0}\right) \right] \tag{5.27}$$

where

$$u(t) = \begin{cases} 1, & t > 0, \\ \frac{1}{2}, & t = 0, \\ 0, & t < 0. \end{cases} \tag{5.28}$$

This signal is illustrated graphically in the first half-cycle of the mains in Figure 5.31.

This is the communications pulse for what may be called a switched-load system and is the basis for an OOK modulation method to be discussed below. The simplicity of the circuit is attractive for low-cost transmitters that are to be manufactured in bulk and deployed in quantities as high as tens of thousands of endpoints for each utility. Functional examples include the TWACS AMR system ([68]–[70]).

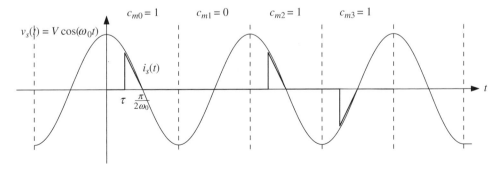

Figure 5.31 An example of a current generated with the circuit in Figure 5.30 using on-off keying.

The energy of each pulse is given by

$$
\begin{aligned}
E_p &= \int_{-\infty}^{\infty} i_s^2(t;\tau)(R_L + R_s)\, dt \\
&= \frac{V^2}{R_s + R_L} \int_{\tau}^{\frac{\pi}{2\omega_0}} \cos^2(\omega_0 t)\, dt \\
&= \frac{V^2}{R_s + R_L} \frac{1}{4\omega_0} \left[\pi - \sin 2\omega_0\tau - 2\tau\omega_0 \right].
\end{aligned}
\tag{5.29}
$$

We define a new parameter, β, which is the length of the pulse as a fraction of a quarter cycle:

$$
\beta = \frac{\frac{\pi}{2\omega_0} - \tau}{\frac{\pi}{2\omega_0}}.
\tag{5.30}
$$

The switch on time as a function of this parameter is found by solving for τ giving

$$
\tau = \frac{\pi}{2\omega_0} - \beta \frac{\pi}{2\omega_0}.
\tag{5.31}
$$

Substitution of (5.31) into (5.29) yields

$$
E_p = \frac{V^2}{R_s + R_L} \frac{1}{4\omega_0} (\beta\pi - \sin \beta\pi).
\tag{5.32}
$$

If these pulses are transmitted with probability $\frac{1}{2}$ in each mains half-cycle (as we will see they are in at least one modulation mode) then the average transmitted power is

$$
P = \frac{1}{2} \frac{E_p}{\frac{\pi}{\omega_0}} = \frac{E_p \omega_0}{2\pi}.
\tag{5.33}
$$

Substituting (5.32) into this expression yields

$$
P = \frac{1}{8\pi} \frac{V^2}{R_s + R_L} (\beta\pi - \sin \beta\pi).
\tag{5.34}
$$

As an example, consider the case in which the switched-load transmitter is used on network with $V = 240\sqrt{2}$ and with a combined source and load impedance of $R_s + R_L = 2\,\Omega$. If the transmitter duty cycle is $\beta = \frac{1}{3}$ then the average transmitter power is

$$
P = \frac{1}{8\pi} \frac{(240)^2 2}{2} \left(\frac{\pi}{3} - \sin \frac{\pi}{3} \right) = 415 \text{ W}.
\tag{5.35}
$$

This illustrates that the switched-load transmitter, in addition to being small and easy to implement, is also capable of generating considerable power. One would be hard pressed to fit a 415 W amplifier inside a residential electric meter. With such enormous transmit power the communication link is all but guaranteed to have a high SNR. However, it is also important to point out that much of this power is being dissipated as heat inside the load resistor. Thus the meter must be capable of dissipating large amounts of heat. The load resistor itself must also be capable of handling this enormous power requirement.

This is clearly a baseband signal, although it has spectral energy at high frequencies as well. The exact shape of the signal will vary with the mains voltage, which is not a perfect sinusoid. It will also be affected by variations in the network source impedance. If the network was modeled with a more inductive source impedance as would be more accurate, for example, the resulting pulse shape would be more rounded. Furthermore, the spectrum of the signal will vary from network to network, and even over time on the same network. It is important that for a given installation the spectrum be properly parameterized and the correct balance determined so that the signal spectrum satisfies any EMC requirements.

Figure 5.31 demonstrates a method of signal generation in which information is imparted to the transmitted signal by timing the switch-on and switch-off times for the switched-load circuit. Switching events occur at intervals the length of a mains half-cycle so that pulses may be generated at every mains zero-crossing. Since this interval length is $T = \frac{\pi}{\omega_0}$ a transmitted signal (current) of length K half-cycles has the form

$$s_m(t) = \sum_{k=0}^{K-1} c_{mk} \frac{V_s}{R_s + R_L} \cos \omega_0 t \left[u\left(t - \tau - \frac{k\pi}{\omega_0}\right) - u\left(t - \frac{k\pi}{\omega_0} - \frac{\pi}{2\omega_0}\right) \right] \quad (5.36)$$

where $c_{mk} \in \{0, 1\}$ indicates whether or not the switching event occurred in the k-th half-cycle. If $c_{mk} = 1$ then switching occurred in the k-th half-cycle, whereas if $c_{mk} = 0$ the switch stayed disconnected from the power network. The signal $s_m(t)$ can be created by setting the K coefficients, c_{mk}. There are therefore 2^K unique signals possible in this way and each one can be assigned a unique binary sequence. The set of all signals assigned to a transmitter in this way is the symbol constellation.

The signal expression (5.36) can be simplified by delaying $i_s(t)$, given by (5.27), by $\frac{k\pi}{\omega_0}$. This gives

$$i_s\left(t - \frac{k\pi}{\omega_0}; \tau\right) = \frac{V}{R_s + R_L} \cos\left(\omega_0 \left(t - \frac{k\pi}{\omega_0}\right)\right) \quad (5.37)$$

$$\times \left[u\left(t - \frac{k\pi}{\omega_0} - \tau\right) - u\left(t - \frac{k\pi}{\omega_0} - \frac{\pi}{2\omega_0}\right) \right]$$

$$= \frac{V}{R_s + R_L} \cos(\omega_0 t - k\pi) \left[u\left(t - \frac{k\pi}{\omega_0} - \tau\right) - u\left(t - \frac{k\pi}{\omega_0} - \frac{\pi}{2\omega_0}\right) \right] \quad (5.38)$$

$$= \frac{V}{R_s + R_L} (-1)^k \cos(\omega_0 t) \left[u\left(t - \frac{k\pi}{\omega_0} - \tau\right) - u\left(t - \frac{k\pi}{\omega_0} - \frac{\pi}{2\omega_0}\right) \right].$$

This equation can be used to simplify (5.36) to

$$s_m(t) = \sum_{k=0}^{K-1} c_{mk}(-1)^k i_s \left(t - \frac{k\pi}{\omega_0}; \tau \right). \tag{5.39}$$

If we define the normalized transmitted current

$$\phi_k(t) = \frac{1}{\sqrt{E_g}} i_s \left(t - \frac{k\pi}{\omega_0}; \tau \right) \tag{5.40}$$

where

$$E_g = \int_{\tau}^{\frac{\pi}{2\omega_0}} i_s^2(t; \tau) \, dt \tag{5.41}$$

then it is easy to show that

$$\int_{-\infty}^{\infty} \phi_n(t)\phi_m(t) \, dt = \delta_{mn}. \tag{5.42}$$

That is, $\phi_k(t)$ for $k = 0, 1, \ldots, K - 1$ define an orthonormal set of functions. Equation (5.39) can then be written in terms of these orthonormal functions as

$$s_m(t) = \sqrt{E_g} \sum_{k=0}^{K-1} (-1)^k c_{mk} \phi_k(t). \tag{5.43}$$

We model reception using an AWGN channel in periodic interference similar to what is observed on the power line in the VLF and ULF band. For a transmitted $s_m(t)$ in the VLF or ULF band the received signal is

$$r(t) = As_m(t) + \Re \left\{ \sum_{m=1}^{\infty} \alpha_m e^{jm\omega_0 t} \right\} + n(t). \tag{5.44}$$

Here A is the channel gain and $n(t)$ is AWGN. It can be shown that in the VLF and ULF band noise can be subdivided into periodic and non-periodic components [71]. The second term on the right side of (5.44) models the periodic component. α_m are deterministic and unknown complex valued constants representing the phasor of the m-th mains harmonic.

Substitution of (5.39) into (5.44) gives the received signal

$$r(t) = A\sqrt{E_g} \sum_{k=0}^{K-1} (-1)^k c_{mk} \phi_k(t) + \Re \left\{ \sum_{m=1}^{\infty} \alpha_m e^{jm\omega_0 t} \right\} + n(t). \tag{5.45}$$

Following the receiver design outlined in [4, Ch. 4] we form K receiver statistics by integrating the received signal with each of the $\phi_k(t)$:

$$
\begin{aligned}
r_k &= \int_0^{\frac{K\pi}{\omega_0}} r(t)\phi_k(t)\,dt \\
&= A\sqrt{E_g}(-1)^k c_{mk} + p_k + n_k
\end{aligned}
\tag{5.46}
$$

where the first term on the right follows from (5.42). The third term is a function of the non-periodic noise:

$$
n_k(t) = \int_0^{\frac{K\pi}{\omega_0}} n(t)\phi_k(t)\,dt.
\tag{5.47}
$$

From [72] these are independent, identically distributed real normal random variable. The second term is a function of the periodic noise:

$$
p_k = \int_0^{\frac{K\pi}{\omega_0}} \Re\left\{ \sum_{m=1}^{\infty} \alpha_m e^{jm\omega_0 t} \right\} \phi_k(t)\,dt.
\tag{5.48}
$$

A little bit of algebra will reveal their relatively simple structure. By rearranging terms this becomes

$$
p_k = \Re\left\{ \sum_{m=1}^{\infty} \alpha_m \int_0^{\frac{K\pi}{\omega_0}} e^{jm\omega_0 t} \phi_k(t)\,dt \right\}.
\tag{5.49}
$$

The integral at the center of this equation, though, is simply a Fourier transform. If we let $\Phi_k(\omega)$ denote the Fourier transform of $\phi_k(t)$ then this becomes

$$
p_k = \Re\left\{ \sum_{m=1}^{\infty} \alpha_m \Phi_k(m\omega_0) \right\}.
\tag{5.50}
$$

But from (5.40) we know that

$$
\Phi_k(\omega) = \frac{1}{\sqrt{E_g}} I_s(\omega) e^{-\frac{jk\pi\omega}{\omega_0}}
\tag{5.51}
$$

where $I_s(\omega)$ is the Fourier transform of $i_s(t;\tau)$. Substituting,

$$
\begin{aligned}
p_k &= \frac{1}{\sqrt{E_g}} \Re\left\{ \sum_{m=1}^{\infty} \alpha_m I_s(m\omega_0) e^{-\frac{jkm\pi\omega_0}{\omega_0}} \right\} \\
&= \frac{1}{\sqrt{E_g}} \Re\left\{ \sum_{m=1}^{\infty} \alpha_m I_s(m\omega_0)(-1)^{km} \right\}.
\end{aligned}
\tag{5.52}
$$

From this equation it is clear that p_k takes on one of two values, depending on whether k is even or odd. That is

$$p_k = \begin{cases} \dfrac{1}{\sqrt{E_g}}\Re\left\{\displaystyle\sum_{m=1}^{\infty} \alpha_m I_s(m\omega_0)\right\}, & k \text{ is even,} \\[4mm] -\dfrac{1}{\sqrt{E_g}}\Re\left\{\displaystyle\sum_{m=1}^{\infty} \alpha_m I_s(m\omega_0)(-1)^m\right\}, & k \text{ is odd.} \end{cases} \tag{5.53}$$

The periodic noise component can therefore be broken down into two components as

$$p_k = p_1 + p_2(-1)^k \tag{5.54}$$

where, as we shall show below, the values of p_1 and p_2 are not important. This shows that the periodic component of the noise lies in a subspace band spanned by the vector of all ones and the vector of alternating 1 and -1. The k-th receiver statistic in (5.46) now becomes

$$r_k = A\sqrt{E_g}(-1)^k c_{mk} + p_1 + p_2(-1)^k + n_k. \tag{5.55}$$

For equiprobable binary signaling, in which two signals, $s_1(t)$ and $s_2(t)$, are transmitted and the receiver must decide which of the two was most likely, the optimal AWGN receiver decision statistic is

$$\gamma = \int_0^{\frac{k\pi}{\omega_0}} r(t)\left(s_1(t) - s_2(t)\right) \, dt. \tag{5.56}$$

Applying this receiver to the transmitted signal presently under consideration, we substitute (5.43) into the above to get

$$\begin{aligned} \gamma &= \int_0^{\frac{k\pi}{\omega_0}} r(t)\sqrt{E_g}\sum_{k=0}^{K-1}(c_{1k} - c_{2k})(-1)^k \phi_k(t)\, dt \\ &= \sqrt{E_g}\sum_{k=0}^{K-1}(c_{1k} - c_{2k})(-1)^k \int_0^{\frac{k\pi}{\omega_0}} r(t)\phi_k(t)\, dt \\ &= \sqrt{E_g}\sum_{k=0}^{K-1}(c_{1k} - c_{2k})(-1)^k r_k. \end{aligned} \tag{5.57}$$

If we define the weight vector

$$w_k = (c_{1k} - c_{2k})(-1)^k \tag{5.58}$$

then the decision statistic can be written as

$$\gamma = \sqrt{E_g} \sum_{k=0}^{K-1} w_k r_k. \tag{5.59}$$

Substituting our expression for the receiver statistic in (5.55) into the above yields

$$\gamma = AE_g \sum_{k=0}^{K-1} c_{mk}(c_{1k} - c_{2k}) + p_1 \sum_{k=0}^{K-1} w_k + p_2 \sum_{k=0}^{K-1} w_k(-1)^k + \sum_{k=0}^{K-1} w_k n_k. \tag{5.60}$$

The second and third terms on the right in this equation can be made to go away so that the receiver statistic looks like the binary signaling statistic for AWGN if we select a weight vector that is orthogonal to the vector of all 1 and the vector of alternating 1 and -1. Since the vector space in which these vectors reside is K-dimensional and the subspace spanned by the periodic component of the noise is two-dimensional, the available subspace is $K - 2$ dimensional. That is, there are $K - 2$ possible weight vectors. Weight vectors may be selected from the K-dimensional Hadamard matrix. For example, if $K = 8$ this matrix is

$$\mathbf{W} = \begin{bmatrix} 1 & 1 & 1 & 1 & 1 & 1 & 1 & 1 \\ 1 & -1 & 1 & -1 & 1 & -1 & 1 & -1 \\ 1 & 1 & -1 & -1 & 1 & 1 & -1 & -1 \\ 1 & -1 & -1 & 1 & 1 & -1 & -1 & 1 \\ 1 & 1 & 1 & 1 & -1 & -1 & -1 & -1 \\ 1 & -1 & 1 & -1 & -1 & 1 & -1 & 1 \\ 1 & 1 & -1 & -1 & -1 & -1 & 1 & 1 \\ 1 & -1 & -1 & 1 & -1 & 1 & 1 & -1 \end{bmatrix}. \tag{5.61}$$

The rows of this matrix are orthogonal and the first two rows are the vectors spanning the subspace in which periodic noise resides. Thus any of the last six rows can be used as a weight vector. Once a weight vector is selected, code words may be selected by noting from the definition of the weight vector in (5.58) that

$$c_{1k} - c_{2k} = (-1)^k w_k. \tag{5.62}$$

Thus, if $w_k(-1)^k = 1$ then $c_{1k} = 1$ and $c_{2k} = 0$. If $w_k(-1)^k = -1$ then $c_{1k} = 0$ and $c_{2k} = 1$. In this way the receiver decision statistic in (5.60) becomes

$$\gamma = AE_g \sum_{k=0}^{K-1} c_{mk}(c_{1k} - c_{2k}) + \sum_{k=0}^{K-1} w_k n_k. \tag{5.63}$$

For $m = 1$

$$\gamma|_{m=1} = AE_g \sum_{k=0}^{K-1} c_{1k}(c_{1k} - c_{2k}) + \sum_{k=0}^{K-1} w_k n_k$$

$$= AE_g \sum_{k=0}^{K-1} c_{1k}^2 - c_{1k}c_{2k} + \sum_{k=0}^{K-1} w_k n_k \qquad (5.64)$$

$$= AE_g \frac{K}{2} + \sum_{k=0}^{K-1} w_k n_k$$

where the last step follows from the fact that exactly $\frac{K}{2}$ of the elements of c_{1k} must be 1 and that neither c_{1k} nor c_{2k} can both be 1 for the same k. It is also a corollary of the above that K must be even. Similarly, for $m = 2$

$$\gamma|_{m=2} = -AE_g \frac{K}{2} + \sum_{k=0}^{K-1} w_k n_k. \qquad (5.65)$$

The receiver decision is therefore

$$\gamma \underset{m=1}{\overset{m=2}{\lessgtr}} 0. \qquad (5.66)$$

Because the decision threshold for γ is zero, the $\sqrt{E_g}$ term in (5.59) may safely be ignored. This is orthogonal signaling and therefore has the same performance as any other binary orthogonal signaling communication system.

5.4.3   OOK Using a Resonating Transmitter

The circuit in Figure 5.30 generates current signals that span a wide range of frequencies. In practice, only a relatively narrow band may actually be sensed by the receiver. Energy generated outside that band is essentially wasted. It may also introduce unwanted interference in neighboring bands. In this section we discuss a transmitter introduced in [73] that generates narrowband signals using a technique similar to that in Section 5.4.2. In addition to the favorable spectral properties of the transmitted signal, the transmitter also dissipates less heat than the resistive switched-load transmitter.

The transmitter of Figure 5.30 is modified as shown in Figure 5.32 by replacing the load resistor with an inductor (L) and capacitor (C) connected in series. The resistance R_L models the equivalent series resistance of the inductor and capacitor. Also included is a resistor (R_D) that we will refer to as the drain resistor for reasons that will become obvious farther down. A switch is used to connect the inductor-capacitor pair to the network to either the distribution network or the drain resistor. Prudent operation of this switch is used to create current in the distribution network, $i_s(t)$, that can be sensed by an appropriate receiver. We shall derive the equation for a single pulse generated by this circuit in what follows.

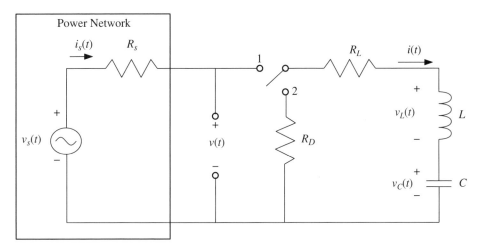

Figure 5.32 An illustration of a simplified circuit for transmitting data in a power network using a resonating switched load.

As we did in Section 5.4.2, we begin by assuming the mains voltage is a perfect sinusoid:

$$v_s(t) = V \cos(\omega_0 t). \tag{5.67}$$

We will use Laplace transforms to solve for the current $i_s(t)$ in Figure 5.32. The Laplace transform of the mains voltage is

$$V_s(s) = \frac{sV}{s^2 + \omega_0^2}. \tag{5.68}$$

Assume that the switch in Figure 5.32 is initially in position 2 and has been for a long time. At $t = 0$ the switch is moved to position 1. Notice that although in Section 5.4.2 the switch-on time could be any time in the first quarter cycle, we are now restricting ourselves to switching on when the mains voltage is at its peak. Laplace transforms are used to describe the resulting current:

$$I_s(s) = \frac{sC}{1 + sRC + s^2LC} V_s(s), \tag{5.69}$$

where $R = R_s + R_L$ is the total series resistance while the inductor-capacitor pair is connected to the network. Substitution of (5.68) into (5.69) yields the Laplace transform of the current

$$I_s(s) = \frac{s^2 VC}{(s^2 + \omega_0^2)(1 + sRC + s^2LC)}. \tag{5.70}$$

By partial fractions expansion this becomes

$$I_s(s) = \frac{\beta_1 s + \beta_0}{s^2 + \omega_0^2} + \frac{\alpha_1 s + \alpha_0}{1 + sRC + s^2 LC}. \tag{5.71}$$

From a table of Laplace transforms the time domain equivalent of the first term is

$$\frac{\beta_1 s + \beta_0}{s^2 + \omega_0^2} \xrightarrow{L^{-1}} \Re \left\{ I_L e^{j\omega_0 t} \right\} \tag{5.72}$$

where

$$I_L = \beta_1 - \frac{j}{\omega_0} \beta_0. \tag{5.73}$$

This is the forced response of the circuit, which is simply the current that would result if the switch were in position 1 for a long period of time. It is a function of the impedance of the inductor, capacitor and resistor at ω_0 and can be made small by making this impedance large. I_L is thus the current phasor at ω_0. We will not derive expressions for β_0 and β_1 here.

The natural response is given by the second term on the right in (5.71)

$$\frac{\alpha_1 s + \alpha_0}{1 + sRC + s^2 LC} \xrightarrow{L^{-1}} \Re \left\{ I_c e^{-\lambda_r t} e^{j\omega_c t} \right\} \tag{5.74}$$

where

$$\lambda_r = \frac{R}{2L} \tag{5.75}$$

$$\omega_c = \sqrt{\frac{1}{LC} - \lambda_r^2} \tag{5.76}$$

$$I_c = \frac{1}{LC} \left(\alpha_1 + \frac{j}{\omega_c} (\alpha_1 \lambda_r - \alpha_0) \right). \tag{5.77}$$

This is the natural response of the circuit. We will not derive expressions for α_1, α_0, or I_c here. See [73] for more details. An illustration of this signal is shown in Figure 5.33. When the switch is thrown at $t = 0$ the circuit begins to resonate at ω_c and is dampened with a time constant of $\frac{R}{2L}$. The inductor and capacitor can be set to the desired resonant frequency. Increasing the inductor size decreases the rate of dampening. These parameters should be set to maximize the overall signal energy.

We can now write an expression for the total current

$$i_s(t; 0) = \Re \left\{ I_L e^{j\omega_0 t} \right\} + \Re \left\{ I_c e^{-\lambda_r t} e^{j\omega_c t} \right\}. \tag{5.78}$$

The first term on the right is, for our purposes, a nuisance signal. This signal will come to dominate the signal in time if the signal is allowed to continue long enough. It has the same

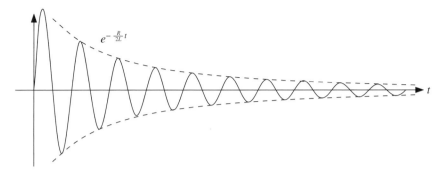

Figure 5.33 When the resonator in Figure 5.32 is switched in to the power network at the mains peak it resonates briefly as a dampened sinusoid. By switching it out and connecting the resonator to the resistor R_D during that period the capacitor charge is discharged and the above waveform can be repeated in another mains half-cycle.

frequency, ω_0, as the mains signal, and is a poor choice for a communications signal given its low peak power. The frequency, ω_c, of the second component, however, is variable, determined by L and C using (5.76). It can be used for narrowband communications provided the inductor and capacitor can be selected so that ω_c has the desired frequency and $|I_c| \gg |I_L|$. This is done by selecting inductors and capacitors such that the impedance of the circuit at ω_0 is low while maximizing I_c.

Ignoring the forced response, then, if the switch is moved to position 1 at time $t = \tau$ and position 2 at time $t = T_g + \tau$, then the current is

$$
\begin{aligned}
i_s(t; \tau) &\approx \Re \left\{ I_c(\phi) e^{-\lambda_r(t-\tau)} e^{j\omega_c(t-\tau)} \right\} \left(u(t-\tau) - u(t - \tau - T_g) \right) \\
&= \Re \left\{ g_l(t-\tau) e^{j2\pi f_c(t-\tau)} \right\},
\end{aligned}
\tag{5.79}
$$

where we have introduced the complex envelope

$$
g_l(t) = I_c e^{-\lambda_r t} (u(t) - u(t - T_g)).
\tag{5.80}
$$

This transmitter offers improved spectral efficiency over the resistive switched-load transmitter described in Section 5.4.2, which, along with the data rate improvements described below, confers an advantage to it in VLF/ULF power line communication links. This is demonstrated by the plot in Figure 5.34. Here, the power spectral density of representative signals generated by each transmitter are compared. The resistive load transmitter provides very little control of the distribution of signal energy in frequency. The reactive load transmitter, by comparison, concentrates signal energy in a comparatively narrow band while generating less energy overall. In this particular example, the reactive load transmitter actually has a higher energy at its resonant frequency than the resistive load transmitter.

The signal can be repeated indefinitely, provided the circuit components are reset after each switching event. This is done using the resistor R_D. After the switch is moved from position 1 to position 2, some residual charge will reside on the capacitor. The above derivation presumed that there is no charge when the load is first connected. Before the load can be connected again,

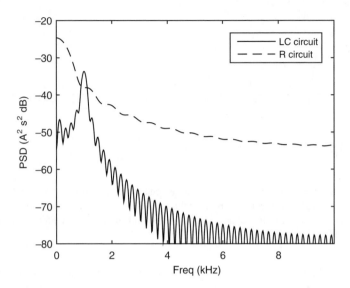

Figure 5.34 We illustrate here the difference between signals generated by a resistive switched-load circuit and by a reactive switched-load circuit. The reactive transmitter in this example has an inductance of 3.1 mH, a capacitance of 8.0 μF and a series resistance of 2 Ω. The resistive transmitter has a series resistance of 2 Ω.

the capacitor must be discharged by connecting it to the drain resistor. The residual charge is then dissipated as heat in the resistor. R_D is selected so that the resulting RLC circuit is critically damped, and the time for the charge to be dissipated is therefore minimized.

As with OOK with the resistive switched-load, the resonator may be switched in each half cycle as show in Figure 5.35. A signal constellation can be defined in which the m-th signal is

$$s_m(t) = \sum_{k=0}^{K-1} c_{mk}(-1)^k i_s\left(t - \frac{k\pi}{\omega_0}; 0\right) \tag{5.81}$$

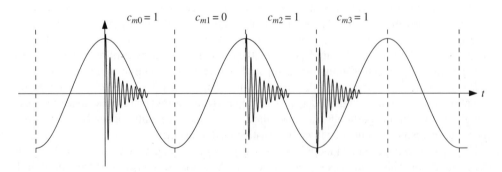

Figure 5.35 Communications signals can be generated by switching the resonating load into the network during some half-cycles and leaving it switched out during others.

where $c_{mk} \in \{0, 1\}$. Such a signal can be created by monitoring the voltage at the transmitter, $v(t)$. At times t_k the switch is moved to position 1 if a 1 bit is to be transmitted. This corresponds to $c_{mk} = 1$. After T_g seconds the switch is moved to position 2 and any remaining charge is dissipated in the drain resistor. If a 0 bit is to be transmitted, then the switch remains in position 2 at time t_k. This corresponds to $c_{mk} = 0$. Note that each signal comprises K such switching events and the symbol time is thus $T_s = KT_p$.

Note that (5.81) is functionally equivalent to (5.39). The only difference is that (5.81) has complex-valued functions as a result of the remixing operation necessitated by the transmitted signal residing at passband. Demodulation is therefore the same as in Section 5.4.2.

5.4.4 PSK Using a Resonating Transmitter

Both OOK modulation methods described in Sections 5.4.2 and 5.4.3 are limited to 1 symbol per half cycle. For a symbol length of K half-cycles only 1 bit can be represented. Although the SNR at the receiver may be high due to the extraordinary transmit power both circuits have, it is difficult to increase the data rate accordingly because of this limitation. However, the modulation method using the resonating transmitter described above in Section 5.4.3 and illustrated graphically in Figure 5.35 can be modified slightly to allow increased spectral efficiencies. Let us amend that technique so that the switch is thrown at every interval, although with a slight temporal modification. This is illustrated in Figure 5.36 where the m-th symbol is generated by switching from position 2 to position 1 in Figure 5.32 at K times given by

$$t_k(m) = kT_p + t_{mk} \tag{5.82}$$

and from position 1 back to position 2 at times $t_k(m) + T_g$. As with OOK, m is used to indicate signal index within a signal constellation. The m-th signal using this modulation method is

$$s_m(t) = \sum_{k=0}^{K-1} i_s(t; kT_p + t_{mk}). \tag{5.83}$$

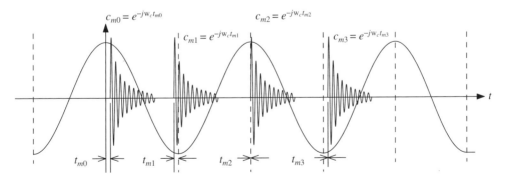

Figure 5.36 An illustration of how controlled switching of a resonant load can be used to generate communications signals. The phase of each pulse can be modified by delaying (or advancing) the switch-in time.

By (5.79)

$$i_s(t; kT_p + t_{mk}) = \Re \left\{ g_l(t - kT_p - t_{mk}) e^{j\omega_c(t - kT_p - t_{mk})} \right\}. \tag{5.84}$$

However, the switching delay must be small: $t_{mk} \ll T_p$. For small changes in the switch-on time the complex envelope changes very little allowing us to make the approximation

$$i_s(t; kT_p + t_{mk}) \approx \Re \{ g_l(t - kT_p) e^{j\omega_c(t - kT_p - t_{km})} \}. \tag{5.85}$$

t_{km} remains in the phase term, though, because even slight variations in the switching time can cause significant changes in the phase. The complex envelope of (5.83) is

$$s_{ml}(t) \approx \sum_{k=0}^{K-1} g_l(t - kT_p) e^{-jk\omega_c T_p} e^{-j\omega_c t_{km}}. \tag{5.86}$$

This can be written as

$$s_{ml}(t) \approx \sqrt{E_g} \sum_{k=0}^{K-1} c_{mk} \phi_k(t) \tag{5.87}$$

where we have defined

$$E_g = \int_0^{T_p} |g_l(t)|^2 \, dt \tag{5.88}$$

$$\phi_k(t) = E_g^{-1/2} g_l(t - kT_p) e^{-jk\omega_c T_p} \tag{5.89}$$

$$c_{mk} = e^{-j\omega_c t_{km}}. \tag{5.90}$$

Note that (5.87) is identical to (5.43) in form. However, the coefficients c_{mk} in (5.87) are any complex exponential, whereas in (5.43) they are strictly binary. This increases the possible number of bits that may be mapped to each symbol and thus may be used to increase the data rate. This method of modulation most closely resembles PSK in which the phase of a signal is used to convey digital information. Prudent selection of t_{km} can be used to construct a symbol constellation similar to a PSK constellation. Using, for example,

$$t_{km} = \frac{2\pi c'_{km}}{M\omega_c} \tag{5.91}$$

where c'_{mk} are integers between 0 and $M - 1$, will result in modulation resembling M-PSK.

To derive the receiver we use the same channel model as (5.44). After demixing at center frequency ω_c the received baseband signal is

$$r_l(t) = A s_{ml}(t) + \sum_{m=1}^{\infty} \alpha_m e^{j(m\omega_0 - \omega_c)t} + n_l(t). \tag{5.92}$$

Using the baseband modulation equation (5.87) we get

$$r_l(t) = A \sum_{k=0}^{K-1} c_{mk} \sqrt{E_g} \phi_k(t) + \sum_{m=1}^{\infty} \alpha_m e^{j(m\omega_0 - \omega_c)t} + n_l(t). \tag{5.93}$$

$n_l(t)$ is the complex envelope of $n(t)$. As before we generate statistics r_k using the orthonormal set of vectors $\phi_k(t)$:

$$r_{kl} = \int_0^{T_s} r_l(t) \phi_k^*(t) \, dt. \tag{5.94}$$

The resulting statistics are

$$r_{kl} = A\sqrt{E_g} c_{mk} + p_{lk} + n_{lk} \tag{5.95}$$

where

$$p_{lk} = \sum_{m=1}^{\infty} \alpha_m \int_0^{T_s} \phi_k^*(t) e^{j(m\omega_0 - \omega_c)t} \, dt \tag{5.96}$$

are the coefficients representing the periodic component of the noise and

$$n_{lk} = \int_0^{T_s} \phi_k^*(t) n_l(t) \, dt \tag{5.97}$$

are independent, identically distributed complex normal random variables. It can be shown [73] that the periodic noise coefficients, p_{lk}, are the same for all k even and the same for all k odd, thus satisfying an equation similar to (5.54). Equation (5.95) can thus be rewritten as

$$r_{kl} = A\sqrt{E_g} c_{mk} + p_{l1} + p_{l2}(-1)^k + n_{lk}. \tag{5.98}$$

For OOK receivers codewords are selected using a weight vector. But since for this modulation methods c_{mk} can take on values of complex roots of unity our choice of codeword is less constrained. The coefficients may be selected using

$$c_{mk} = s_m w_k \tag{5.99}$$

where s_m is a complex root of unity and $w_k \in \{-1, 1\}$. w_k may be selected using the procedure outlined in Section 5.4.2 surrounding (5.61). A new statistic can be formed by computing the inner product of r_{lk} with w_k using

$$\hat{s}_m = \sum_{k=0}^{K-1} r_{lk} w_k$$

$$= A\sqrt{E_g} s_m \sum_{k=0}^{K-1} w_k^2 + p_{l1} \sum_{k=0}^{K-1} w_k + p_{l2} \sum_{k=0}^{K-1} w_k(-1)^k + \sum_{k=0}^{K-1} w_k n_{lk}.$$

(5.100)

By our selection of w_k as rows of the Hadamard matrix (such as in (5.61)) the second and third terms on the right go to zero. Furthermore

$$\sum_{k=0}^{K-1} w_k^2 = K.$$

(5.101)

Thus

$$\hat{s}_m = AK\sqrt{E_g} s_m + \sum_{k=0}^{K-1} w_k n_{lk}.$$

(5.102)

This is a biased estimate of s_m. After accounting for the bias the transmitted symbol is estimated using

$$\hat{m} = \arg\min \left| AK\sqrt{E_g} s_m - \hat{s}_m \right|.$$

(5.103)

5.5 Ultra-wideband Modulation

Another kind of spread spectrum transmission technique is ultra-wideband (UWB) modulation. UWB has attracted considerable attention for wireless communications. Interestingly, also the commercial broadband PLC systems that deploy the 2–28 MHz band can be classified as UWB systems, according to the FCC definition, because the ratio between the used bandwidth and the central carrier is larger than 0.2. UWB modulation is generally considered synonymous of impulsive modulation, referred here as I-UWB. The basic idea behind impulse modulation is to convey information by mapping an information bit stream into a sequence of short duration pulses that occupy a large bandwidth [74]. No carrier modulation is required. Pulses (referred to as monocycles) are followed by a guard time in order to cope (at least in part) with the channel time dispersion. If the guard time is sufficiently long the inter-symbol interference is negligible at the receiver side such that detection simplifies to a matched filter receiver that basically correlates the received signal with a template waveform. The monocycle can be appropriately designed to shape the spectrum occupied by the transmission system and in particular to avoid the low frequencies where we typically experience higher levels of man-made background noise. Further, the power spectral density of the transmitted signal can be kept at low levels so that notching for coexistence purposes is not required. Impulse modulation can be combined with CDMA, allowing high data rate multi-user transmission that enjoys simple implementation, and robust performance in the presence of severe channel frequency selectivity, multiple access interference, and impulse noise [75]. Although not much research has been carried out yet, impulse modulation can also be an attractive technique for moderate rate applications (applications with large monocycle bandwidth with respect to data rate, e.g. 80 MHz band with data rate of 1 Mbps), that require simple modulation and demodulation architectures. These simple transceivers can be used for power line command/control systems and sensor networks as those used for in-home and industrial automation, or for in-vehicle device connectivity. Other applications that can be envisioned are energy monitoring and control in smart grids.

5.5.1 I-UWB transmitter

In impulsive UWB (I-UWB), the transmission takes place in frames. The frame has a period equal to T_f. Thus, the transmission rate is equal to $1/T_f$ symbols/s. For every frame, a short duration pulse is transmitted followed by a guard time. A single pulse maps one information symbol. The transmitted signal can be written as

$$x(t) = \sum_k b_k g_{tx}(t - kT_f), \tag{5.104}$$

where b_k is the information symbol transmitted in the k-th frame, and $g_{tx}(t)$ is the monocycle impulse response. The information symbols can be real or complex, as a function of the adopted constellation. To keep the complexity low, a Binary I-UWB is considered. Thus, a single symbol maps one bit and it assumes a value that belongs to the alphabet $\{-1, 1\}$. In other implementation forms, higher order modulation can be used, furthermore, the pulse can also be spread in time through a code as in CDMA [75].

In the wireless scenario, a common choice is to shape the monocycle as one of the derivatives of the Gaussian pulse. This pulse is suitable for the PLC scenario too. The derivatives of the Gaussian pulse have no frequency components in the lower frequency range, where the PSD of the background noise is higher. Gaussian pulses are only an example of a possible monocycle shape, i.e. custom pulse shapes can be designed.

5.5.1.1 Gaussian Pulse Shape Design

The Gaussian pulse is defined as

$$g_0(t) = \frac{K_0}{\sqrt{2T_0}} e^{-\frac{\pi}{2}\left(\frac{t}{T_0}\right)^2}, \tag{5.105}$$

where T_0 is a parameter that sets the transmission bandwidth and K_0 determines the PSD peak of the transmission pulse. In frequency domain, the p-th derivative of the Gaussian pulse reads

$$G_p(f) = \mathcal{F}\left[\frac{d^p}{dt^p} g(t)\right](f) = K_0 (i2\pi f)^p e^{-2\pi T_0^2 f^2}, \tag{5.106}$$

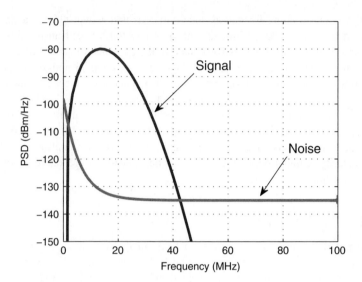

Figure 5.37 PSD of the I-UWB transmitted signal and the background noise. As monocycle, a second order derivative of the Gaussian pulse is considered.

where $\mathcal{F}[\cdot]$ denotes the Fourier transform operator. The transmission bandwidth of the monocycle is defined as $B = f_h - f_l$. f_l and $f_h > f_l$ are the two frequencies for which (5.106) falls 10 dB below its maximum.

Figure 5.37 shows the PSD of the I-UWB transmitted signal when a second order derivative of Gaussian pulse is used and the peak of the PSD is set a low level equal to -80 dBm/Hz. In the low frequency range, there are no spectral components of the monocycle. In the spectrum of interest, the SNR margin is still greater than 30 dB (and up to 45 dB) when considering a typical background noise PSD as shown in the figure.

5.5.2 I-UWB receiver

At the receiver input, the signal reads as

$$y(t) = x * g_{ch} + n(t), \tag{5.107}$$

where $g_{ch}(t)$ is channel impulse response and $n(t)$ is the background colored noise. The received signal in (5.107) is filtered with an analog front-end filter. At the front-end filter output, the signal reads

$$u(t) = \sum_{k} b_k g_{tx} * g_{ch} * g_{fe}(t - kT_f) + d(t), \tag{5.108}$$

where $g_{fe}(t)$ is the front-end filter impulse response and $d(t) = n * g_{fe}(t)$ is the filtered background colored noise. The equivalent impulse response between the transmitter and the receiver is referred as $g_{eq}(t) = g_{tx} * g_{ch} * g_{fe}(t)$. The front-end filter is set to $g_{fe}(t) = g_{tx}(-t)$, i.e. it is matched with the monocycle.

The output of the front-end filter is sampled with period $T_c = T_f/M$, where M denotes the number of samples/frame. The discrete-time signal in (5.108) is filtered with a receiver filter, denoted as $g_{rx}(mT_c)$. In the following, several discrete-time receivers are described [76].

5.5.2.1 Filter Receiver

The matched filter receiver is the simpler solution. The front-end filter is matched to the monocycle, i.e. $g_{fe}(t) = g_{tx}(-t)$ and the receiver filter is matched to the channel impulse response, i.e. $g_{rx}(mT_c) = g_{ch}(-mT_c)$. The decision signal is given by

$$\Lambda(k) = \sum_m u(mT_c)g_{rx}(kT_f - mT_c). \tag{5.109}$$

To discriminate the transmitted symbol (-1 or 1 for binary modulation), a threshold decision is made on the signal in (5.109). Strictly, the detected symbol in the k-th frame is given by

$$\hat{b}(kT_f) = \text{sign}\{\Lambda(k)\}. \tag{5.110}$$

This receiver is referred as matched filter (MF) receiver and it is optimal when the background noise is white.

5.5.2.2 Equivalent-matched Filter Receiver

The MF receiver is optimal in the wireless scenario, where the background noise is white. Contrarily, in PLC, the background noise can be considered Gaussian but it is colored. Therefore, to improve the MF receiver, the receiver filter is matched with the equivalent impulse response between the transmitter and the front-end output, i.e. $g_{rx}(mT_c) = g_{eq}(-mT_c)$. The front-end is matched to the monocycle. The decision process is equal to the MF receiver and it exploits (5.110). Since, the derivatives of the Gaussian pulse have a frequency response that approximates the shape of the inverse of the noise PSD in the lower frequency range, the front-end filter approximates a whitening filter in the lower frequency range.

This receiver is referred as equivalent-matched filter (E-MF) receiver.

5.5.2.3 Noise-matched Filter Receiver

To obtain the optimal filter, the colored noise correlation must be known. The discrete-time correlation of the background noise is defined as $R_n(mT_c) = E[n(mT_c + t)n(t)]$, where $E[\cdot]$ is the expectation operator. Then, the convolutional inverse of the noise correlation is denoted as $R_n^{-1}(mT_c)$. This function is defined s.t. $R_n^{-1} * R_n(mT_c) = \delta(mT_c)$.

The receiver filter is defined as $g_{rx}(mT_c) = R_n^{-1} * g_{ch}^-(-mT_c)$, where $g_{ch}^-(mT_c) = g_{ch}(-mT_c)$. Symbol decision exploits the metric in (5.109).

This receiver is referred as noise-matched filter (N-MF) receiver.

5.5.2.4 Frequency-domain Implementation of the N-MF Receiver

The previous three receivers are implemented in time domain. A frequency domain implementation is possible under two assumptions. First, the noise is assumed to be independent frame by frame. Second, the frame duration is greater than the channel impulse response duration. A more general derivation of the optimal FD receiver can be found in [75]. A vector notation is firstly introduced. At the output of the front-end filter, the samples related to the k-th frame are denoted as

$$\mathbf{u}_k = \left[u(kMT_c), \ldots, u((M-1+kM)T_c) \right]^{\mathrm{T}}. \tag{5.111}$$

Furthermore, the noise samples of the k-th frame are denoted with the vector $\mathbf{d}_k$. Under the previous assumptions, the frames can be processed independently. An M-point discrete Fourier transform (DFT) is computed on the $\mathbf{u}_k$ vector. In frequency domain, the frame is given by

$$\mathbf{U}_k = b_k \mathbf{G}_{eq} + \mathbf{D}_k, \tag{5.112}$$

where b_k is the symbol transmitted in the k-th frame, and $\mathbf{U}_k$, $\mathbf{D}_k$ and $\mathbf{G}_{eq}$ are $M \times 1$ vectors of the M-point DFT of $\mathbf{u}_k$, $\mathbf{d}_k$ and the equivalent impulse response vector $\mathbf{g}_{eq} = \left[g_{eq}(0), \ldots, g_{eq}(M-1) \right]$, respectively. The frequency spacing is equal to $f_n = n/MT_c$.

To obtain the decision metric, (5.112) is maximized. The metric reads

$$\Lambda(k) = b_k \, \mathbf{G}_{eq}^{\mathrm{H}} \mathbf{K}^{-1} \mathbf{U}_k, \tag{5.113}$$

where $\{\cdot\}^{\mathrm{H}}$ indicates the Hermitian transpose and $\mathbf{K}_k^{-1} = \left(E \left[\mathbf{D}_k \mathbf{D}_k^{\mathrm{H}} \right] \right)^{-1} = \mathbf{K}^{-1}$ is a matrix that takes into account the noise statistic. This matrix is constant for each frame under the stationary noise assumption. To decide the k-th transmitted symbol, the sign of (5.113) is evaluated, according to (5.110).

This receiver is referred as frequency domain (FD) receiver.

5.5.2.5   Comparison of the Receivers

In Figure 5.38, a comparison between the I-UWB receivers above described is shown. It is assumed to transmit over an MV channel whose impulse response was measured in a real life network [77]. The noise PSD is the one shown in Figure 5.37. As monocycle, a second order derivative of Gaussian pulse is chosen, with $B = 20$ MHz, $T_f = 5$ μs and $1/T_c = 50$ MHz. The peaks of the pulse PSD is varied between -100 and -80 dB. The N-MF receiver and its frequency domain implementation exhibit the best performance which is more than 4 dB better than the other receivers for bit-error-rate below 10^{-3}.

In general, I-UWB exhibits good performance with simple receiver structures. Immunity to impulse noise can be achieved with channel coding and spreading the pulse in time [75]. It is inherently immune to narrowband interference. Furthermore, it has been shown that it can coexist with both narrowband PLC systems and broadband PLC systems [78]. This is due to the low PSD of I-UWB and the high processing gain deriving from spectrum spreading.

5.6   Impulse Noise Mitigation

The PLC channel is considered a harsh environment for data communication, due to distortion of transmitted PLC signals and due to superposed interference seen at the PLC receiver side.

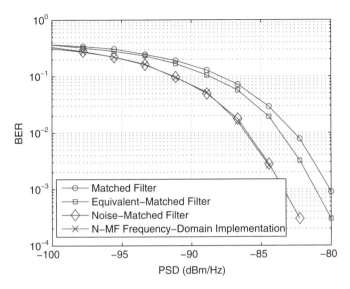

Figure 5.38 BER performance of different receiver schemes for the MV scenario as a function of the peak PSD value.

This section discusses methods to cope with noise spikes that occur on PLC channels, or so-called impulse noise. We start with a brief review of impulse-noise modeling and then present impulse-noise mitigation methods based on data transmission and detection.

5.6.1 Preliminaries on Noise

Noise in PLC channels including impulse (or impulsive) noise is discussed in some detail in Section 2.7. Impulse noise is considered as sequences of short rises of noise amplitudes. It is typically classified into impulse noise synchronous to the AC mains cycle (in AC grids), impulse noise with an underlying periodicity shorter than the AC mains cycle, and isolated impulse noise without a recurrence pattern. The sources of impulse noise are emissions from devices (i.e. electrical loads) connected to the power grid and ingress from wireless signals. For the modeling of impulse noise it is important to differentiate whether the model applies to the source, i.e. the noise transmitter, or the sink, i.e. the communication receiver. The former permits a separation of the features characteristic for a noise source and the effects from the power grid through which the noise signal travels to the receiver. It is thus suited for more deterministic channel emulation approaches, see Section 2.8. The latter is often in the form of statistical models, which are more immediately applicable for detector design and analysis.

Statistical modeling approaches for impulse noise commonly try to capture the distributions of noise amplitude, impulse inter-arrival time or impulse distance, and impulse width. Figure 5.39 illustrates the definition of these quantities according to [79]. Here, an impulse is associated with a pulse shape that is scaled with an amplitude. Different models differentiating narrowband and broadband PLC (and their associated frequency bands), application environments (indoor and outdoor) and the different noise types are available in the literature, e.g. [79]–[85].

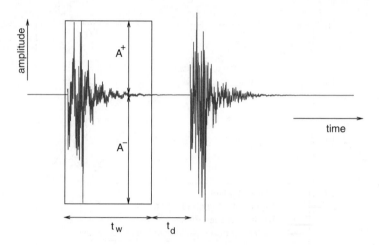

Figure 5.39   Illustration of impulsive noise and its parameters: amplitude $A = \max(A^+, A^-)$, impulse distance t_d and impulse width t_w, according to [79, Figure 1].

For noise-mitigation methods, the marginal distribution for the amplitude of impulse-noise samples is often focused on. This is relevant for transmission with 'perfect interleaving' and disregards the differences between the type of impulse noise. References [83], [84][2] suggest that the instantaneous impulse-noise amplitude experienced at the receiver at a certain point in time, which results from the superposition of multiple noise processes filtered through the PLC channel, is well modeled by Middelton's class-A distribution

$$p_a(x) = \sum_{i=0}^{\infty} p_i \mathcal{N}(x, 0, \sigma_i^2), \tag{5.114}$$

where

$$\mathcal{N}(x, 0, \sigma_i^2) = \frac{1}{\sqrt{2\pi\sigma_i^2}} \exp\left(-\frac{x^2}{2\sigma_i^2}\right), \quad p_i = \frac{e^{-A}A^i}{i!}, \quad \sigma_i^2 = \sigma_a^2 \left(\frac{i/A + \Gamma}{1 + \Gamma}\right). \tag{5.115}$$

The noise according to the probability density function (pdf) in (5.114) can be interpreted as drawing with probability p_i a random variable from a zero-mean Gaussian distribution with variance σ_i^2. Therefore, $p_0\sigma_0^2$ and $\sum_{i=1}^{\infty} p_i\sigma_i^2$ are considered as the power of the background and impulse noise components, respectively. Furthermore, $\sigma_a^2 = \sum_{i=0}^{\infty} p_i\sigma_i^2$ is the total noise power,

---

[2] While [83] considers general impulse noise, [84] is specifically concerned with asynchronous impulse noise.

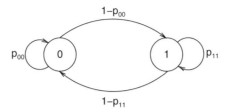

Figure 5.40 Gilbert-Elliott model, used for modeling the discrete-time impulse noise process.

the parameter $\Gamma = p_0\sigma_0^2 / \sum_{i=1}^{\infty} p_i\sigma_i^2$ is referred to as the background-to-impulse-noise power ratio, and A is the impulsive index.

An approximation of this distribution is the so-called two-term Gaussian mixture model given by

$$p_a(x) = (1 - \epsilon)\mathcal{N}(x, 0, \sigma_0^2) + \epsilon\mathcal{N}(x, 0, \sigma_1^2), \tag{5.116}$$

where impulses occur with probability ϵ.

The impulse inter-arrival time has been modeled as exponentially distributed random variable with a mean value of the order of milliseconds [83]. Impulse widths of the order of tens of microseconds have been reported in [79], [83].

Substituting the real-valued Gaussian pdf $\mathcal{N}(x, 0, \sigma^2)$ with its counterpart for a circularly symmetric complex Gaussian distributed variable with variance σ^2, $\mathcal{N}_c(x, 0, \sigma^2) = \frac{1}{\pi\sigma^2}\exp(-|x|^2/\sigma^2)$, in (5.114) and (5.116) leads to the impulse-noise distribution in the equivalent complex baseband, which will be needed for transmission using in-phase and quadrature components such as quadrature-amplitude modulations (QAM) [86].

Instead of considering measurement fitted models for the impulse inter-arrival time and impulse width, e.g. as exponentially and two-term Gaussian mixture distributed random variables as in [83], the Gilbert-Elliott model is an often used discrete-time model to capture the temporal characteristics of impulse-noise samples when developing mitigation techniques. This model is based on the notion of a 'good' state (state '0') representing background noise and a 'bad' state (state '1') representing impulse noise, and the transition between states is governed by the state transition matrix (see Figure 5.40)

$$T = \begin{bmatrix} p_{00} & 1 - p_{00} \\ 1 - p_{11} & p_{11} \end{bmatrix}. \tag{5.117}$$

The probability of an impulse-noise sample is given by

$$\epsilon = \frac{1 - p_{00}}{2 - p_{00} - p_{11}}, \tag{5.118}$$

which would be used in (5.116) for the marginal noise pdf. The run-length of bad states, which corresponds to impulse-noise width, is geometrically distributed with mean $1/(1 - p_{11})$.

Similarly, the run-length of good states, which corresponds to impulse inter-arrival time, is geometrically distributed with mean $1/(1 - p_{00})$. If we denote the sampling interval by T_s, then the inter-arrival time is approximately exponentially distributed with parameter $p_{00}T_s$. The Gilbert-Elliott model emphasizes the notion of noise bursts associated with impulse noise. Small $(1 - p_{11})$ will lead to long noise bursts affecting many successive received samples.

5.6.2 Transmission Methods

We now move on to describe methods to deal with impulse noise. We start with transmission strategies, which can be applied to mitigate the effect of impulse noise.

Retransmission: Due to the large amplitude and burst-character of impulse-noise events, an entire transmitted packet may be lost, i.e. not decodable, at the receiver side. In this case, retransmission can be applied. That is, the receiver would notify the transmitter to retransmit the lost packet. This approach treats the impulse-noise channel as a (packet-)erasure channel. Denoting the erasure probability e, then the throughput is $(1 - e)$ packets per transmission, which equals the capacity of the channel [87, Ch. 9]. Retransmission requires receiver feedback. Interestingly, the channel capacity is independent of whether there is feedback or not [87, Ch. 50].

Interleaving: A transmission strategy that attempts to mitigate the effect of impulse noise without the need for feedback is interleaving. The purpose is to break up long impulse bursts into several shorter impulses. This can help when interleaving extends over several codewords of an error-correction code that is applied for reliable transmission, as the maximal number of affected symbols per codeword is reduced. Furthermore, convolutional codes are vulnerable to burst errors and interleaving is beneficial even when applied within a codeword. The benefit of interleaving for decoders that do not attempt to make use of the bursti-ness in the noise process has been demonstrated in [88] considering channel cutoff rate and bit-error rate of convolutional coded transmission. Examples for the appli-cation of strong interleaving in PLC are the IEEE 1901.2 and G3-PLC standards [89].

While interleaving is an effective means to cope with burst noise using conven-tional decoding methods (which ignore the burstiness of the noise), we note that it is also considered a 'poor way to protect data against burst errors, in terms of amount of redundancy required' [87, Exercise 11.4]. Intuitively, it makes sense that knowing that after an impulse-noise sample another one is likely to come helps the receiver to estimate the location of impulse-noise events and to adjust its processing accordingly. This corresponds to state estimation for the Markov chain in Figure 5.40. With interleaving, on the other hand, there is no state information and the receiver has to operate on a (poor) average channel.

Application-layer coding: A transmission strategy that is adjusted to noise bursts is to shift error-correction coding from transmission symbols to data packets, or in other words, from the physical layer to the application layer. This strategy has been proposed and studied in [90] (see also [91]) for video distribution over PLC.

It is argued that due to burst errors, redundancy for error-correction is better spent at the application layer to correct the relatively few affected packets rather than the physical layer to correct the many affected symbols during a noise burst. In particular, erasure codes are applied. The quasi-standard codes for such purposes are Reed-Solomon codes, which have also widely been applied in PLC as outer codes in concatenated coding schemes (to correct burst errors after inner decoding). In [90], it is suggested to use so-called Raptor codes, which are an instance of digital Fountain codes [92], [93], [87, Ch. 50]. Raptor codes are encoded in a rateless fashion, i.e. as many encoded packets as needed are generated. This allows Raptor codes to adjust to channel (i.e. noise) conditions instantaneously, and makes them particularly useful in a broadcast communication scenario, where the same message is transmitted over multiple different channels simultaneously. Furthermore, Raptor codes have lower encoding and decoding complexity compared to Reed-Solomon codes. The use of Raptor codes requires infrequent feedback for the receiver to communicate to the transmitter that decoding has been successful.

Adaptation: Due to the 50 Hz or 60 Hz mains cycle in AC power grids, the PLC channel often exhibits periodicity with the mains or double the mains frequency. This can also cause impulse noise to occur (in part) periodically, already at the noise source or at the receiver because of the periodically varying channel frequency response. Such periodicity would not be described by a Markov model as in Section 5.6.1, but by a more deterministic filtering model as presented in Section 2.7. Noise periodicity can be learnt by the receiver and fed back to the transmitter, which in turn can adapt its modulation parameters according to the time (and frequency) variation of the noise, or more generally, the signal-to-noise power ratio (SNR). For example, an average of 10% improvement in data rate at the medium-access control (MAC) layer due to adaptation is reported in [94], considering measured channels, and [95] presents significant improvements in error rate when adapting modulation and power considering a time-varying continuous noise model. Furthermore, the work in [96] determines the capacity of the channel with periodic impulse noise modeled through a cyclostationary Gaussian process [81] or a linear periodically time-varying filtered Gaussian process [85] (see Section 2.7). It is shown that this capacity is higher than what can be achieved with time-frequency partitioning á la [95], [97] and that it can be approached using pre- and post-processing as applied to multiple-input multiple-output channels with stationary noise.

5.6.3   Detection Methods

We now consider signal processing at the receiver side to deal with impairments due to impulse noise. To start our discussion, let us consider the simple discrete-time transmission model

$$r_k = s_k + n_k, \tag{5.119}$$

where s_k is the transmitted symbol, n_k is the additive noise sample, and r_k is the received sample at discrete-time k. We assume real-valued baseband signals, and in particular, binary phase-shift keying (BPSK). If n_k was zero-mean Gaussian noise with variance σ_n^2, then the optimal receiver evaluates the log-likelihood ratio (LLR)

$$
\lambda^{GN}(r_k) = \log\left[\frac{1}{\sqrt{2\pi\sigma_n^2}}\exp\left(-\frac{(r_k-1)^2}{2\sigma_n^2}\right)\right] - \log\left[\frac{1}{\sqrt{2\pi\sigma_n^2}}\exp\left(-\frac{(r_k+1)^2}{2\sigma_n^2}\right)\right] \tag{5.120}
$$

$$
= 2r_k/\sigma_n^2.
$$

In the case of impulse noise according to (5.114), the receiver LLR would be approximated as

$$
\lambda^{MN}(r_k) \approx \log\left[\sum_{i=0}^{S-1} p_i \frac{1}{\sqrt{2\pi\sigma_i^2}}\exp\left(-\frac{(r_k-1)^2}{2\sigma_i^2}\right)\right]
$$
$$
- \log\left[\sum_{i=0}^{S-1} p_i \frac{1}{\sqrt{2\pi\sigma_i^2}}\exp\left(-\frac{(r_k+1)^2}{2\sigma_i^2}\right)\right] \tag{5.121}
$$

where S (instead of infinity) noise states are assumed. The special case $S = 2$ would correspond to the two-term noise model in (5.116).

The difference between the LLRs for the two detectors (5.120) and (5.121) is shown in Figure 5.41. We assume the parameters $\sigma_n^2 = \sigma_a^2 = 1$, i.e. the SNR is 0 dB, and $A = 0.01$, which means that $1 - e^{-A} \approx 1\%$ of samples are affected by impulse noise, and $\Gamma = 0.1$. We note that the LLR $\lambda^{GN}(r_k)$ is linear in the received sample r_k, i.e. it scales with the amplitude of the received sample. This means that the larger $|r_k|$ the more reliable a decision variable is assumed to be. In the case of $\lambda^{MN}(r_k)$, there are peaks around $r_k = \pm 1$, which indicates high reliability if the received sample is close to an actual signal point, i.e. $+1$ or -1. Large received sample amplitudes are discounted, as they are likely due to the occurrence of impulse noise, which makes the received sample highly unreliable.

Of course, the detector that uses $\lambda^{MN}(r_k)$ needs to know (i.e. estimate) the parameters A, Γ and S, and it relies on the assumption that the Middleton class-A distribution is a faithful representation of the actual impulse-noise distribution. However, the basic insight gleaned from Figure 5.41, i.e. curtailing large received samples in the case of impulse noise, can be used to develop simpler detectors. These detectors apply different nonlinearities to the received sample to mitigate impulse noise. The perhaps most practical ones are the soft-limiter (or clipping) and the blanking nonlinearities, which produce

$$
y_k = \begin{cases} r_k, & |r_k| \le T, \\ \frac{r_k}{|r_k|}T, & |r_k| > T, \end{cases} \tag{5.122}
$$

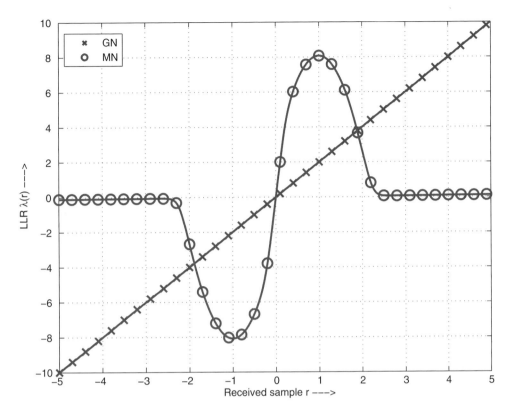

Figure 5.41 LLR $\lambda(r)$ from (5.120) (GN) and (5.121) (MN) versus received sample r.

and

$$y_k = \begin{cases} r_k, & |r_k| \le T, \\ 0, & |r_k| > T, \end{cases} \tag{5.123}$$

respectively. The parameter T in (5.122) and (5.123) is a threshold, which needs to be adjusted according to some criterion. The corresponding LLRs for BPSK would be $\lambda(r_k) = 2y_k/\sigma_n^2$. Figure 5.42 shows $\lambda^{SL}(r_k)$ and $\lambda^{BL}(r_k)$ as a function of the received sample r_k. $T = 2$ has been chosen for this figure. For a comparison, we included the LLRs from Figure 5.41. We observe that the soft-limiter and blanking nonlinearities mimic the effect of the impulse-noise LLR (5.121). The parameter T enables us to adjust to what extent amplitudes of the received sample are clipped or blanked.

The two clipping and blanking nonlinearities can also be combined into one nonlinearity which applies two different thresholds [98]. Furthermore, several other nonlinearities have been proposed and studied in, e.g. [88], [99]–[101].

An example for bit-error rate (BER) results for the impulse-noise channel is shown in Figure 5.43. Two-term Gaussian noise with $\epsilon = 0.1$ and $\sigma_1^2/\sigma_0^2 = 100$ is assumed, which is a close approximation to Middleton class-A noise with $A \approx 0.1$ and $\Gamma \approx 0.1$. Coded BPSK

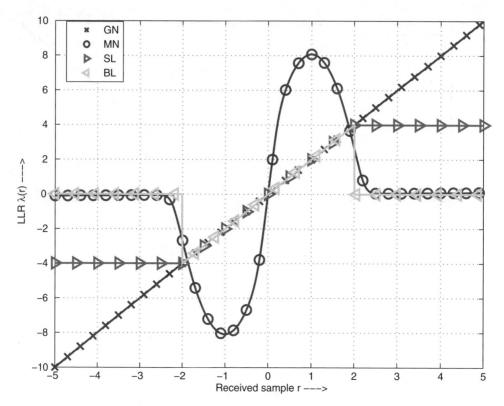

Figure 5.42 LLR $\lambda(r)$ versus received sample r for soft-limiter (SL) and blanking (BL) nonlinearities. For comparison, LLR for Gaussian noise (GN) and Middleton class-A noise (MN).

transmission using a rate-1/2, memory-4 convolutional code and maximum-likelihood detection (MLD) using $\lambda^{MN}(r_k)$ and Euclidean-distance based detection (EDD) using $\lambda^{GN}(r_k)$ have been considered. Both simulated BER and analytical BER approximations are shown, and the SNR is defined as $SNR_0 = 1/\sigma_0^2$. We observe the substantial gains by applying an LLR metric that is matched to the impulse noise, rather than the conventional EDD metric based on the assumption of Gaussian noise. We also note that the shape of the MLD BER curve is similar to the superposition of two shifted and scaled Gaussian-Q functions. This is due to the effect of the two noise terms, i.e. the background-noise and impulse-noise components. More details can be found in [88].

While MLD uses a decoding (or detection) metric which is based on the noise distribution, it does not exploit the fact that impulse noise occurs in bursts. In fact, the notion of noise states, S in the approximation of the Middleton class-A noise (5.114) and two for the two-term Gaussian mixture noise in (5.116), suggests a detector which estimates the current state and then applies a Gaussian-noise based metric (as in (5.120)) with the appropriate noise variance. The key to estimate the noise state is the memory (i.e. burst-structure) of the discrete-time noise process, which can be represented by, for example, the Gilbert-Elliott model in Figure 5.40. References [102], [103] present algorithms that perform noise-state estimation and data detection in an

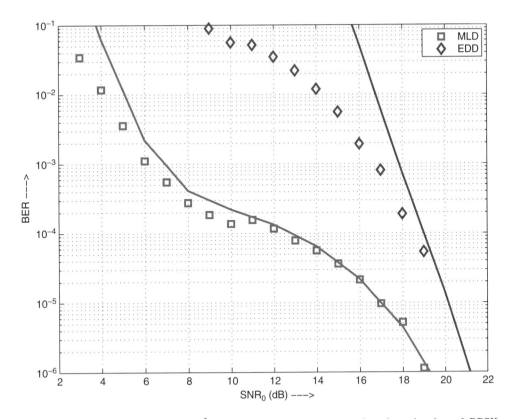

Figure 5.43 BER versus $SNR_0 = 1/\sigma_0^2$ for transmission over a two-term impulse noise channel. BPSK and rate-1/2 memory-4 convolutional code. Maximum-likelihood detection (MLD) using $\lambda^{MN}(r_k)$ and Euclidean-distance based detection (EDD) using $\lambda^{GN}(r_k)$. Markers correspond to simulated results and lines are analytical approximations from [88].

iterative fashion. In [103] this is demonstrated to approach the ultimate performance limit of knowing the noise state, if the channel memory that can be defined as [104]

$$\mu = p_{00} + p_{11} - 1 \tag{5.124}$$

is close to one.

Once the noise state is known, also the blanking nonlinearity in (5.123) could be applied. This approach is pursued in [105], where erasure marking is integrated into Viterbi decoding for convolutional codes. If the impulse noise variance is much larger than that of the background noise, little is lost by erasing the received samples deemed to be hit by impulse noise. Figure 5.44 illustrates the achievable performance assuming perfect noise state estimation for the same coded transmission example as in Figure 5.43. We present BER curves for the cases of MLD with known noise state (MLD-KS), i.e. the Euclidean distance metric with appropriate noise variance is used, and erasure-marking decoding with known noise stated (EMD-KS), i.e. the blanking nonlinearity is applied for the impulse-noise state. As a reference, the MLD curve from Figure 5.43 is included. We can see the significant performance gains achievable by

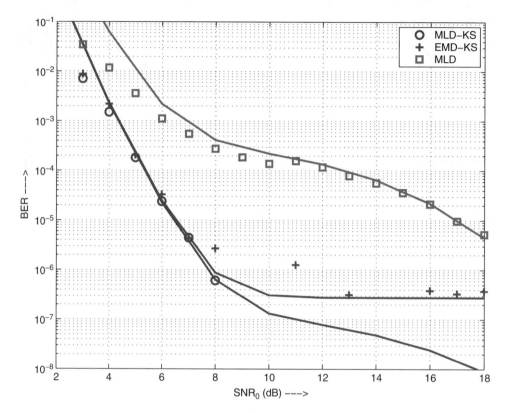

Figure 5.44 BER versus $SNR_0 = 1/\sigma_0^2$ for transmission over a two-term impulse noise channel. BPSK and rate-1/2 memory-4 convolutional code. Results for the cases maximum-likelihood detection with known noise state (MLD-KS), erasure-marking decoding with known noise state (EMD-KS) and MLD using $\lambda^{MN}(r_k)$. Markers correspond to simulated results and lines are analytical approximations from [88].

noise-state estimation, and we also note the similar BER performance of MLD-KS and EMD-KS for large BER ranges. For higher SNR_0, when the impulse noise becomes dominating, erasure marking is suffering from an error floor which depends on the minimum distance of the code, cf. [88].

Finally, we note that similar statements on the performance for different detection methods have been obtained in [106] for a partitioned Markov chain model for bursty impulse noise. This model was presented in [107] to describe the statistical dependence of errors in binary bursty channels. It splits the two good and bad states of the Gilbert-Elliott model in Figure 5.40 into multiple good and bad states, and it has been applied in [79] to provide a good fit to measurements of impulse-noise realizations in PLC.

5.6.4 Mitigation Methods for Multicarrier Transmission

While single-carrier transmission has long been used for narrowband PLC, multicarrier modulation and in particular orthogonal frequency-division multiplexing (OFDM) have been adopted for many recently developed PLC systems for multimedia and smart grid communications, see Chapters 8 and 9. Noise models and transmission methods discussed in Sections 5.6.1 and 5.6.2 apply to multicarrier transmission in the same way they apply for single-carrier transmission. On the detection side, however, there are some differences. Soft-limiter and blanking in (5.122) and (5.123), respectively, can be applied to the received samples from OFDM transmission without changes, as this happens before the time-to-frequency-domain transformation at OFDM receivers [98]. But methods based on the noise distribution and iterative detection need to take this transformation into account.

To see this, let us consider the sequence of time-domain received samples r_k and define the frequency-domain decision variable

$$R_l = \frac{1}{\sqrt{K}} \sum_{k=0}^{K-1} r_k e^{-j\frac{2\pi kl}{K}}, \quad l = 0, 1, \ldots, K-1, \tag{5.125}$$

for the l-th subcarrier in an OFDM system with K subcarriers. Under the assumption that complex-baseband noise samples n_k are independent and identically distributed according to the complex-valued variant of the two-term distribution in (5.116), then the noise component of R_l has the distribution [108]

$$p_A(x) = \sum_{i=0}^{K-1} \binom{K}{i} \epsilon^i (1-\epsilon)^{K-i} \mathcal{N}_c(x, 0, \sigma_i^2), \quad \sigma_i^2 = \sigma_0^2 + i\frac{\sigma_1^2}{K}. \tag{5.126}$$

Hence, we have a K-term noise distribution. We further note that this distribution is independent of the subcarrier index l. However, even if time-domain impulse noise occurs as independent spikes, still all subcarriers will be affected jointly, which can be seen from the discrete Fourier transform (DFT) equation (5.125). Hence, the marginal noise distribution (5.126) can only be applied for detection or decoding if interleaving over many OFDM symbols is applied. This is assumed in [109] to show that OFDM is generally more sensitive to impulse noise than single-carrier transmission, when decoding is based on noise distribution. Furthermore, noise-state estimation as discussed in the previous section would not be based on the frequency-domain variable R_l, but operate on the time-domain sample r_k before noise is 'spread' via the DFT.

This leads to the application of impulse noise cancellation for multicarrier transmission, for which the general receiver structure is illustrated in Figure 5.45. The variables in this figure are vectors of length K, corresponding to the number of OFDM subcarriers. The estimate for

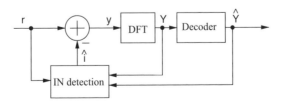

Figure 5.45   Block diagram of receiver structure for impulse noise (IN) cancellation in OFDM systems.

the impulse-noise vector $\hat{i}^3$ is subtracted from the received vector r to obtain a noise-reduced received vector y.

A fairly simple such noise-cancellation method has been proposed in [110]. The decoder output $\hat{Y}$, i.e. the received vector reconstructed based on decoding decisions and channel estimation, is subtracted from the input Y and then the 'inverse' blanking nonlinearity (5.123) is applied to the time-domain difference signal

$$d = F^H(Y - \hat{Y}), \tag{5.127}$$

where F is the $K \times K$ DFT matrix, i.e.

$$\hat{i}_k = \begin{cases} 0, & |d_k| < T, \\ d_k, & |d_k| \geq T, \end{cases} \quad k = 0, 1, \ldots, K - 1. \tag{5.128}$$

Decoding and noise cancellation is repeated in a number of iterations and significant error-rate gains are demonstrated for uncoded transmission.

Another class of algorithms to estimate the impulse noise makes use of *a priori* known OFDM subcarrier symbols. These are often considered pilot symbols or null-subcarriers, where the latter are used in PLC to notch certain frequency bands. The time-domain impulse-noise signal contributes to the signals received in those subcarriers and thus can be estimated after the known data signal is subtracted. The underlying principle is that we expect only few of the K received samples per OFDM symbol to be affected by impulse noise. To elaborate on this, let us separate impulse and background noise into the two length-K vectors i and n. Furthermore, let us assume that i is m-sparse, i.e. only at most m out of the K components of i are non-zero. Then, these non-zero elements, and there position in the length-K vector i can be estimated from the $2m < K$ measurements

$$w = F_1 i, \tag{5.129}$$

where F_1 is a $2m \times K$ sub-DFT matrix, using a discrete version of the Prony method [111, Ch. 2]. Examples of methods that apply this principle can be found in [112], [113].

An important continuation of this line of work has been presented in [114]. It builds on recently popularized convex optimization methods to reconstruct sparse signals from rank-deficient measurements (as in (5.129)), known as compressive sensing [111]. In the context of our problem, these methods provide higher flexibility in terms of where pilot and null-subcarriers can be located and better robustness to background noise (i.e. the vector n, which has been ignored in (5.129)) compared to the Prony-type methods. For example, Figure 5.46 shows the broadcast tone mask defined in the IEEE 1901 standard for the OFDM physical layer in North America [115]. We observe that the use of certain OFDM subcarriers is disabled, which is done to avoid interference with wireless systems operating in for example amateur radio bands. These null-subcarriers can be used for impulse-noise estimation, assuming that there is not interference from wireless systems at the OFDM receiver [116]. That is, denoting

---

[3] Vectors and matrices are written in bold font.

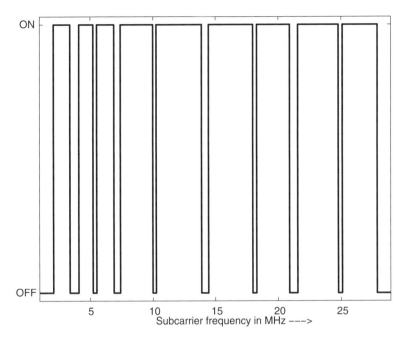

Figure 5.46 Broadcast tone mask for the OFDM physical layer of IEEE 1901 for North America [115, Table 13-29].

the number of data subcarriers by N, then the $(K - N) \times K$ sub-DFT F_1 whose rows correspond to the positions of $K - N$ null-subcarriers is used for impulse-noise detection [114], [116].

While [114] assumed that impulse-noise samples arrive statistically independently, the work [116] included the burst-structure of impulse-noise samples in compressive-sensing based impulse-noise cancellation. The impulse-noise vector i is represented as the block-vector

$$i = [\underbrace{i_1, \ldots, i_\delta}_{i^T[1]}, \ldots, \underbrace{i_{K-\delta+1}, \ldots, i_K}_{i^T[p]}]^{\mathrm{T}}, \qquad (5.130)$$

where the block-size δ is associated with the expected burst length. The vector i is thus assumed to be block-sparse [117]. Then, the following detection algorithm is applied.

(a) Form the measurement vector

$$w = F_1 r. \qquad (5.131)$$

(b) Solve

$$\tilde{i} = \min_i \sum_{j=1}^{p} \|i[j]\|_2 \qquad (5.132)$$
$$\text{s.t.} \quad \|w - F_1 i\|_2 \leq \varepsilon$$

where ε is adjusted such that $\|F_1 n\|_2 \leq \varepsilon$ with some probability, and thus it depends on the background noise variance σ_0^2.

(c) Estimate the support of i as

$$\hat{I} = \{j : |\tilde{i}_j|^2 > \sigma_0^2\}. \tag{5.133}$$

(d) Let $\hat{m} = |\hat{I}|$ and $l[j]$ be the position of j in $\hat{I}$, i.e. $1 \leq l[j] \leq \hat{m}$. Create the $(K - N) \times \hat{m}$ selection matrix S with elements $s_{jl[j]} = 1$ for $j \in \hat{I}$ and zero otherwise, and letting $A = F_1 S$, obtain the least-squares estimate

$$\hat{i} = S(A^H A)^{-1} A^H w. \tag{5.134}$$

Figure 5.47 shows a sample result obtained with the algorithm described above under the following assumptions. The tone mask as in Figure 5.46 is used, that is $K = 1224$ OFDM

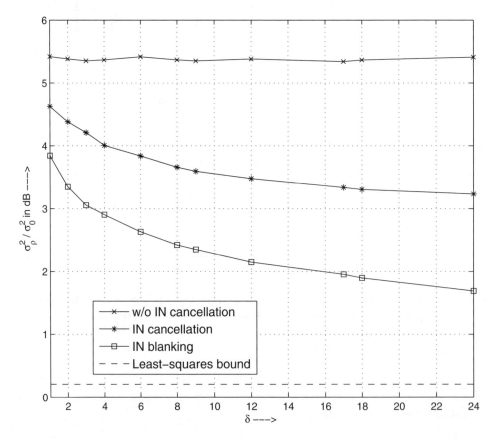

Figure 5.47 Relative empirical residual interference-plus-noise variance σ_p^2/σ_0^2 versus detection block-size δ for impulse-noise (IN) suppression using compressive sensing. 'IN blanking' assumes erasing received samples hit by IN. 'Least-squares bound' assumes that the location of IN samples is known perfectly.

subcarriers are located in the transmission band, of which 917 would be active. Another 31 subcarriers, spaced uniformly across active subcarriers, are nulled so that $N = 886$ subcarriers are active (i.e. 3% less than in IEEE 1901). This is applied for better stability of the least-squares step in (5.134). One impulse-noise burst of $m = 30$ samples affects the received vector r. The location of this burst is chosen uniformly at random, and the impulse-plus-background noise variance is 100 times stronger than that of the only background noise, i.e. $\sigma_1^2/\sigma_0^2 = 20$ dB. For each impulse-noise realization and its subsequent cancellation, the residual interference-plus-noise signal

$$\rho = i + n - \hat{i} \tag{5.135}$$

is measured. Figure 5.47 shows the relative empirical residual noise variance σ_ρ^2/σ_0^2 based on 500 impulse-noise realizations as function of the block-size δ. Without impulse-noise (IN) cancellation, $\sigma_\rho^2/\sigma_0^2 = \frac{100m+K-m}{K} \approx 5.34$ dB, i.e. IN deteriorates the average SNR by more than 5 dB. With IN cancellation this is reduced to almost 3 dB, for sufficiently large block-size δ. Instead of subtracting the estimated impulse-noise vector $\hat{i}$, one could also use the estimated support set $\hat{\mathcal{I}}$ from (5.133) and blank the corresponding received samples (similar to the blanking nonlinearity in (5.123)). The corresponding curve (IN blanking) in Figure 5.47 suggests a better noise suppression through blanking. We note however that blanking also removes part of the useful signal. Finally, Figure 5.47 shows a performance bound (Least-squares bound) assuming that the location of the IN burst is known perfectly and least-squares estimation as in (5.134) is applied. This reduces the residual noise variance almost to the variance of the background noise. This suggests that estimation of the IN location is the more critical part for IN cancellation.

A number of extensions and variations of sparsity-based detection cancellation of impulse noise have been investigated. The application to narrowband PLC (PRIME systems) has been discussed in [118], and a Bayesian-learning approach that does not make assumptions about noise parameters is presented in [119].

5.7 MIMO Transmission

Multiple-input multiple-output (MIMO) systems arise when the availability of several transmitters and receivers offers a network of transmission channels. This can lead to a capacity increase compared to the case of a channel spanned by a single transmitter and receiver, which has resulted in the incorporation of MIMO techniques into many wireless and also some wireline—including PLC—standards. This section explores some of the MIMO backgrounds and begins with the MIMO channel and notation definition in Section 5.7.1. The capacity increase offered by MIMO systems is laid out in Section 5.7.2, and can be exploited in terms of spatial multiplexing in Section 5.7.3 or in terms of spatial diversity in Section 5.7.4. The estimation of MIMO channels and some important trade-offs in the sounding of such channels are covered in Section 5.7.5. Aspects of broadband MIMO systems are touched on in Section 5.7.6, followed by an overview over PLC-specific MIMO issues in Section 5.7.7.

Figure 5.48 Generic multiple-input multiple-output (MIMO) system with K transmit signals $s_k[n]$, $n = 1, 2, \dots, K$ and M receive signals $r_m[n]$, $m = 1, 2, \dots, M$; for simplicity, the channel is assumed to be noise free.

5.7.1 MIMO Channel and Definitions

A multiple-input multiple-output (MIMO) transmission system is characterized by an arrangement with K transmit and M receive nodes, as shown in Figure 5.48. In isolation, the channel between the k-th transmitter sending a signal $s_k[n]$ and the m-th receiver measuring $r_m[n]$ is described by an impulse response $h_{mk}[n]$, such that

$$r_m[n] = \sum_{v=0}^{N} h_{mk}[v]s_k[n - v], \tag{5.136}$$

where it is assumed that the channel order does not exceed N. Therefore, the overall MIMO channel comprises of KM channel impulse responses, which can be written in matrix form as

$$\mathbf{H}[n] = \begin{bmatrix} h_{11}[n] & h_{12}[n] & \cdots & h_{1K}[n] \\ h_{21}[n] & h_{22}[n] & & \vdots \\ \vdots & & \ddots & \vdots \\ h_{M1}[n] & \cdots & \cdots & h_{MK}[n] \end{bmatrix}, \tag{5.137}$$

whereby $\mathbf{H}[n] \in \mathbb{C}^{M \times K}$ is a matrix of finite impulse response (FIR) filters. Its z-transform

$$H(z) = \sum_{n=0}^{N} \mathbf{H}[n]z^{-n} \tag{5.138}$$

is a polynomial with matrix-valued coefficients, or polynomial matrix [40].

To condense the notation of a MIMO system, the transmit and receive signals can be written in vectorial form,

$$\mathbf{s}[n] = \begin{bmatrix} s_1[n] & s_2[n] & \cdots & s_K[n] \end{bmatrix}^{\mathrm{T}}, \tag{5.139}$$

$$\mathbf{r}[n] = \begin{bmatrix} r_1[n] & r_2[n] & \cdots & r_M[n] \end{bmatrix}^{\mathrm{T}}, \tag{5.140}$$

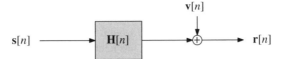

Figure 5.49 Generic multiple-input multiple-output (MIMO) system with input vector $\mathbf{s}[n] \in \mathbb{C}^K$, MIMO channel $\mathbf{H}[n] \in \mathbb{C}^{M \times K}$, noise vector $\mathbf{v}[n] \in \mathbb{C}^M$ and receive signal vector $\mathbf{r}[n] \in \mathbb{C}^M$.

such that $\mathbf{s}[n] \in \mathbb{C}^K$ and $\mathbf{r}[n] \in \mathbb{C}^M$. The notation $\{\cdot\}^T$ indicates transpose, whereby later also the Hermitian or complex conjugate transpose operator $\{\cdot\}^H$ will be used. With (5.137), this leads to a compact description

$$\mathbf{r}[n] = \sum_{v=0}^{L} \mathbf{H}[v]\mathbf{s}[n - v] + \mathbf{v}[n], \tag{5.141}$$

which also include additive channel noise in a vector $\mathbf{v}[n] \in \mathbb{C}^M$ that is structured akin to (5.140). This situation is also depicted in Figure 5.49.

The signals $\mathbf{s}[n]$, $\mathbf{v}[n]$ and $\mathbf{r}[n]$ are generally stochastic and defined by their statistics. Assuming all signals to be zero mean, the covariance matrix of the input is characterized by taking expectations $\mathcal{E}\{\cdot\}$, such that the covariance of the transmit signal is $\mathbf{R}_{ss} = \mathcal{E}\{\mathbf{s}[n]\mathbf{s}^H[n]\} \in \mathbb{C}^{K \times K}$, the noise covariance matrix $\mathbf{R}_{vv} = \mathcal{E}\{\mathbf{v}[n]\mathbf{s}^H[v]\} \in \mathbb{C}^{M \times M}$, and the covariance matrix of the receive signal vector $\mathbf{R}_{rr} = \mathcal{E}\{\mathbf{v}[n]\mathbf{v}^H[n]\} \in \mathbb{C}^{M \times M}$. All covariance matrices only reflect the spatial correlation between elements, and do not take into account any temporal correlations that may exist. Generally, independence between the transmit signal and the channel noise is assumed. The channel matrix $\mathbf{H}[n]$ can be deterministic, e.g. if measured and known, or can be treated as a statistical quantity if results over an ensemble of channel realizations is required.

Most MIMO techniques have been defined for the narrowband case, where the impulse responses contained in $\mathbf{H}[n]$ have order $L = 0$ and simplify to complex gains. Section 5.7.6 will show some approaches that allow to either turn a broadband channel into a system of narrowband channels, e.g. by means of OFDM, or by generalization of narrowband techniques to the broadband case. In the following, we assume the simplification $\mathbf{H}[n] = \mathbf{H}$.

The transition of the narrowband case means that the channel description according to Figure 5.49 and (5.141) simplifies to

$$\mathbf{r}[n] = \mathbf{H}\mathbf{s}[n] + \mathbf{v}[n]. \tag{5.142}$$

Based on (5.142) and with a deterministic $\mathbf{H}$ that acts as a constant w.r.t. the expectation operator, the covariance matrix of the received signal can be written as

$$\mathbf{R}_{rr} = \mathbf{H}\mathbf{R}_{ss}\mathbf{H}^H + \mathbf{R}_{vv}. \tag{5.143}$$

Often the noise $\mathbf{v}[n]$ is assumed to be independently and identically distributed, with $\mathcal{E}\{v_m[n]v_m^*[n]\} = \sigma_v^2, k = 1, 2, \ldots M$, such that $\mathbf{R}_{vv} = \sigma_v^2\mathbf{I}_M$ with $\mathbf{I}_M$-th $M \times M$ identity matrix. These statistics will now be used to characterize the capacity of a MIMO system.

5.7.2 MIMO Capacity

The well known Shannon limit of error-free throughput over a channel [120], [121], $C = \log_2(1 + \gamma)$, where γ is the SNR at the receiver, has been extended to the MIMO case as [122]

$$C = \max_{\mathbf{R}_{ss}} \log_2 \det \left(\mathbf{I}_M + \frac{1}{\sigma_v^2} \mathbf{H} \mathbf{R}_{ss} \mathbf{H}^H \right), \tag{5.144}$$

where $\det(\cdot)$ is the determinant. The maximization in (5.144) is over the spatial shaping of $\mathbf{s}[n]$, whereby the transmit power is constrained to P_0, i.e. $\text{tr}\{\mathbf{R}_{ss}\} = P_0$ where $\text{tr}\{\cdot\}$ is the trace operator.

If the channel $\mathbf{H}$ is deterministic but unknown to the transmitter, the spatial shaping of $\mathbf{s}[n]$ is neglected, and equal power is assigned to the mutually uncorrelated transmit signals such that $\mathbf{R}_{ss} = \frac{P_0}{K} \mathbf{I}_K$ and subsequently (5.144) simplifies to

$$C = \log_2 \det \left(\mathbf{I}_M + \frac{P_0}{K\sigma_v^2} \mathbf{H} \mathbf{H}^H \right). \tag{5.145}$$

The channel's singular value decomposition (SVD) [123]

$$\mathbf{H} = \mathbf{U} \mathbf{\Sigma} \mathbf{V}^H \tag{5.146}$$

can be computed with unitary $\mathbf{U} \in \mathbb{C}^{M \times M}$ and $\mathbf{V} \in \mathbb{C}^{K \times K}$, and a diagonal, real valued and positive semidefinite $\mathbf{\Sigma} \in \mathbb{C}^{M \times K}$,

$$\mathbf{\Sigma} = \begin{cases} \begin{bmatrix} \mathbf{\Lambda} & \mathbf{0} \end{bmatrix}, & M < K, \\ \begin{bmatrix} \mathbf{\Lambda} \\ \mathbf{0} \end{bmatrix}, & M \geq K, \end{cases} \tag{5.147}$$

with $\mathbf{\Lambda} = \text{diag}\{\lambda_1 \ \lambda_2 \ \cdots \ \lambda_R\} \in \mathbb{R}^R$ containing the singular values and $R = \min(K, M)$. Equivalently, the eigenvalue decomposition (EVD) $\mathbf{H}\mathbf{H}^H = \mathbf{U}\mathbf{\Lambda}^2\mathbf{U}^H$ provides the same factorization. Introducing the spurious $\mathbf{I}_M = \mathbf{U}\mathbf{U}^H$, (5.145) evolves into

$$C = \log_2 \det \left(\mathbf{U}(\mathbf{I}_M + \frac{P_0}{K\sigma_v^2} \mathbf{\Sigma}\mathbf{\Sigma}^T)\mathbf{U}^H \right) \tag{5.148}$$

$$= \log_2 \det \left(\mathbf{I}_M + \frac{P_0}{K\sigma_v^2} \mathbf{\Sigma}\mathbf{\Sigma}^T \right) \tag{5.149}$$

$$= \log_2 \prod_{m=1}^{R} \left(1 + \frac{P_0 \lambda_m^2}{K\sigma_v^2} \right) = \sum_{m=1}^{R} \log_2 \left(1 + \frac{P_0 \lambda_m^2}{K\sigma_v^2} \right). \tag{5.150}$$

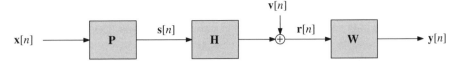

Figure 5.50 MIMO channel with precoding matrix **P** and equalization matrix **W**.

The step from (5.148) to (5.149) exploits the fact $\det(\mathbf{AB}) = \det(\mathbf{A})\det(\mathbf{B})$ for square $\mathbf{A}$ and $\mathbf{B}$ [123].

The maximum capacity for the case in (5.150) is achieved if the channel matrix $\mathbf{H}$ is orthogonal, such that $\mathbf{HH}^H = K\sigma_h^2\mathbf{I}_R$, and all singular values are identical, $\lambda_m^2 = K\sigma_h^2 \ \forall m$, with σ_h the gain of a channel coefficient contained in $\mathbf{H}$. Note that the MIMO capacity in (5.150) becomes

$$C = R\log_2\left(1 + \frac{P_0\sigma_h^2}{\sigma_v^2}\right), \tag{5.151}$$

which is equivalent to R times the capacity of a single-input single-output (SISO) system of same transmit power P_0 and noise power σ_v^2. Therefore, the channel capacity rises linearly with the number of transmitters or receivers, which ever is smaller [124]–[126].

If the deterministic channel $\mathbf{H}$ is known at the transmitter, the capacity can be maximized as demanded by (5.144) by an appropriate linear precoding matrix $\mathbf{P}$ and a linear equalizer $\mathbf{W}$, as shown in Figure 5.50.

To achieve optimality in a number of senses including capacity-optimality, $\mathbf{P}$ and $\mathbf{W}$ can be based on the SVD of the channel matrix in (5.146) by setting $\mathbf{P} = \mathbf{V\Gamma}$ and $\mathbf{W} = \mathbf{U}^H$ [127]. The dimension $\tilde{R}$ of the precoder input $\mathbf{x}[n] \in \mathbb{C}^{\tilde{R}}$ and equalizer output $\mathbf{y}[n] \in \mathbb{C}^{\tilde{R}}$ is the rank of the channel matrix $\mathbf{H}$, which is equivalent to the number of finite singular values in $\mathbf{\Lambda}$. The diagonal $\mathbf{\Gamma} = \mathrm{diag}\{\gamma_1\ \gamma_2\ \cdots\ \gamma_{\tilde{R}}\}$ controls how the transmit power P_0 is distributed across $\mathbf{x}[n]$, i.e. $\mathbf{\Gamma}$ is constrained by $\mathrm{trace}\{\mathbf{\Gamma}\} = \sqrt{P_0}$. The output

$$\mathbf{y}[n] = \mathbf{U}^H\mathbf{r}[n] = \mathbf{U}^H\mathbf{HV\Gamma}\mathbf{x}[n] + \mathbf{U}^H\mathbf{v}[n] = \mathbf{\Lambda\Gamma}\mathbf{x}[n] + \mathbf{U}^H\mathbf{v}[n] \tag{5.152}$$

shows that the precoder and equalizer according to Figure 5.50 have decoupled the channel. Note that due to the selection of unitary matrices in $\mathbf{P}$ and for $\mathbf{W}$ and their preservation of the Euclidean norm [123], the precoder permits precise control of the transmit power since $\|\mathbf{s}[n]\|_2 = \|\mathbf{\Gamma x}[n]\|_2$, and the equalizer does not amplify the channel noise, i.e. $\|\mathbf{U}^H\mathbf{v}[n]\|_2 = \|\mathbf{v}[n]\|_2$.

With the transmit power controlled by $\mathbf{\Gamma}$, the input covariance matrix becomes $\mathbf{R}_{xx} = \mathcal{E}\{\mathbf{x}[n]\mathbf{x}^H[n]\} = \mathbf{I}_{\tilde{R}}$, and the covariance matrix at the equalizer output is

$$\mathbf{R}_{yy} = \mathcal{E}\{\mathbf{y}[n]\mathbf{y}^H[n]\} = \mathbf{\Lambda\Gamma R}_{xx}\mathbf{\Gamma}^H\mathbf{\Lambda}^H + \mathbf{U}^H\mathbf{R}_{vv}\mathbf{U} \tag{5.153}$$

$$= \mathbf{\Gamma}^2\mathbf{\Lambda}^2 + \sigma_v^2\mathbf{I}_{\tilde{R}}. \tag{5.154}$$

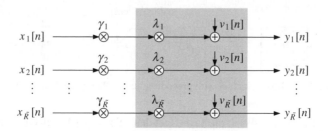

Figure 5.51 Decoupled MIMO channel equivalent to Figure 5.50. The shaded part represents the equivalent channel with gains λ_m, while γ_m are the transmit gains applied to the various channels $m = 1, 2 \ldots, \tilde{R}$.

This leads to the MIMO capacity formulation

$$C = \max_{\Gamma} \sum_{m=1}^{\tilde{R}} \tilde{R} \log_2 \left(1 + \frac{\gamma_m^2 \lambda_m^2}{\sigma_\nu^2} \right). \tag{5.155}$$

The capacity-optimal power allocation is provided by the waterpouring [128] or waterfilling algorithm [129], which maximizes (5.155) under the constraint trace$\{\Gamma\} = \sqrt{P_0}$.

The waterfilling algorithm is outlined in Figure 5.52 and orders the inverse subchannel SNRs calculated prior to power allocation, $\frac{\sigma_\nu^2}{\lambda_m^2}$, and determines a water level μ, such that the area under μ—i.e. the shaded area in Figure 5.52—is equal to the power budget P_0. The difference between the inverse subchannel SNR and the water level μ is the allocated transmit power, provided that the value is positive, i.e.

$$\gamma_m^2 = \max \left(\mu - \frac{\sigma_\nu^2}{\lambda_m^2}, 0 \right). \tag{5.156}$$

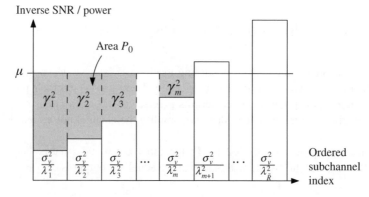

Figure 5.52 Visualization of the waterfilling algorithm [129], where the inverse SNRs are sorted in ascending order. The transmit power budget is under the waterlevel μ, which determines how many subchannels can be serviced, and at which transmit power γ_m^2.

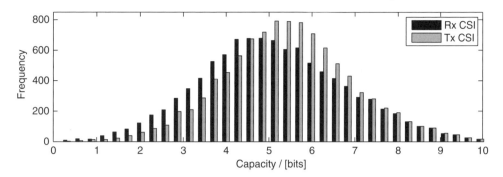

Figure 5.53 Distribution of the channel capacity C for an ensemble of 10^4 realizations of $\mathbf{H} \in \mathbb{C}^{2 \times 2}$ with zero mean unit variance complex Gaussian entries, at an SNR of 10 dB.

Subchannels whose inverse SNR is not reached by the water level μ are deselected from transmission and receive no transmit power. Note that this capacity-optimal power allocation assigns the largest transmit power to the subchannel with the highest SNR. This is contrary to the concept of equalization, where the largest power would be allocated to the weakest subchannel, i.e. transmit power would be wasted on the poorest modes of the MIMO channel.

In practical systems, the imbalance in SNR reached by waterfilling across different subchannels requires bit allocation or bit loading schemes, which offer different levels of optimality. Ideally bit loading should be performed such that subchannels achieve equal levels of BER. Since bit allocation through e.g. M-QAM schemes possesses a high granularity, often combined power and bit allocation is required to achieve optimality. In the literature, a number of methods such as incremental [130], [131] or greedy schemes [132]–[137] have been suggested.

To assess the capacity of a channel with random fluctuations, statements over an ensemble of channel realizations can be made to characterize the MIMO system. The distribution of the capacities according to (5.150) for Rx CSI and (5.155) for Tx CSI are obtained for an ensemble of 10^4 channel realizations, and shown in Figure 5.53 whereby the coefficients of $\mathbf{H} \in \mathbb{C}^{2 \times 2}$ are drawn from a complex Gaussian distribution with zero mean and unit variance with $P_0 = 10$ at an SNR of $\frac{P_0}{K\sigma_v^2} = 10$ dB. Unsurprisingly, the capacity reached by waterfilling in (5.155) for the case of a more globally available CSI will be higher than in the case of (5.150) where the channel is not known to the transmitter. From the type of distributions of measured or simulated capacities in Figure 5.53 often two important quantities are drawn: the ergodic capacity represents the mean $\mathcal{E}\{C\}$ of the distribution, while the outage capacity measures the tail of the distribution; a 10% outage capacity e.g. states the capacity value below which 10% of all measurements fall, i.e. it measures a type of confidence interval for the system performances and assesses the robustness of the MIMO transmission.

An example for the ergodic capacity of ensembles of MIMO channels of different dimension is depicted in Figure 5.54. Particularly for higher SNR values, it is clear that the channel capacity approximately doubles when proceeding from $K = M = 1$ via $K = M = 2$.

For non-square matrices with $\{K = 2, M = 1\}$ and $\{K = 1, M = 2\}$, the latter case achieves a high capacity, as the transmit power is the same, but with a higher number of receivers M, this configuration captures more of the dissipated power. The configuration $K = 2$ and $M \in \{3, 4\}$ are also shown for the relevance to PLC.

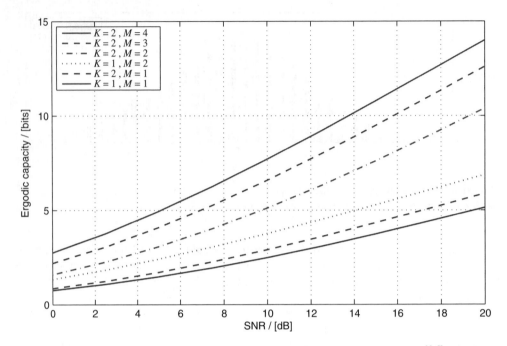

Figure 5.54 Ergodic capacity for ensembles of different MIMO channel matrices $\mathbf{H} \in \mathbb{C}^{M \times K}$, calculated for Rx CSI according to (5.150).

For the case of Rx CSI, it was stated that the maximum capacity is attainable with (5.151) if the channel matrix $\mathbf{H}$ is orthogonal. In practice, wireless systems often experience correlation because of the absence of local scatters or because the antenna separation of the MIMO system is insufficient. In PLC, where the live, neutral and common ground experience cross-coupling due to insufficient shielding, as well as similar impulse responses due to closely linked topologies, the consideration of the impact of spatial correlation of the matrix $\mathbf{H}$ onto the capacity is important. Here, we follow a simple model for a spatially correlated MIMO channel matrix [122], [138]–[140], which assumes a factorization of a channel by means of transmit and receiver spatial correlation matrices Ξ and Ψ, such that

$$\mathbf{H} = \Psi^{\frac{1}{2}} \tilde{\mathbf{H}} \Xi^{\frac{1}{2}}. \tag{5.157}$$

The channel matrix $\tilde{\mathbf{H}}$ is the same as the previously uncorrelated MIMO system matrix. The spatial correlation matrices capture correlation between adjacent nodes, whereby a particularly simple exponential correlation structure [141] is given by

$$\Xi = \begin{bmatrix} 1 & \rho_{\text{Tx}} & \cdots & \rho_{\text{Tx}}^{K-1} \\ \rho_{\text{Tx}}^{*} & 1 & & \vdots \\ \vdots & & \ddots & \vdots \\ \rho_{\text{Tx}}^{K-1,*} & \rho_{\text{Tx}}^{K-2,*} & \cdots & 1 \end{bmatrix}, \tag{5.158}$$

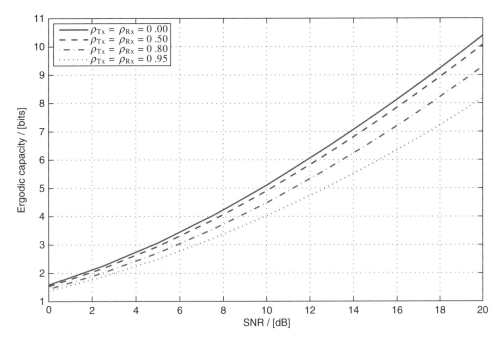

Figure 5.55 Ergodic capacity for ensembles of different spatially correlated MIMO channel matrices $\mathbf{H} \in \mathbb{C}^{2\times2}$, calculated for Rx CSI according to (5.150).

where ρ_{Tx} is the local correlation factor at the transmitter. A spatial receive correlation matrix Ψ can be defined with the same structure as in (5.158) but with a receive correlation factor ρ_{Rx}.

Simulation results for an ensemble of 10^4 MIMO channels $\mathbf{H} \in \mathbb{C}^{2\times2}$ using the factorization in (5.157) with spatial correlation factors in (5.158) are shown in Figure 5.55. As the spatial correlation increases, the MIMO capacity of the channel diminishes.

The capacity gain offered by a MIMO system can be exploited through spatial multiplexing or by utilizing the diversity gain afforded by the system. These will be explored in turns in the following two subsections.

5.7.3 Spatial Multiplexing

Maximizing the capacity in the case of Tx CSI was explored in Section 5.7.2, and utilized the SVD of the MIMO channel. This approach effectively implements spatial multiplexing in its most advanced form, and this section will only highlight some of the finer details of this technique, and its relation to beamforming and beamsteering.

The SVD of a channel matrix was calculated in (5.146), but can be expanded as

$$\mathbf{H} = \begin{bmatrix} \mathbf{U}_s & \mathbf{U}_s^\perp \end{bmatrix} \begin{bmatrix} \tilde{\Lambda} & \mathbf{0} \\ \mathbf{0} & \mathbf{0} \end{bmatrix} \begin{bmatrix} \mathbf{V}_s^{\mathrm{H}} \\ \mathbf{V}_s^{\perp\mathrm{H}} \end{bmatrix} \tag{5.159}$$

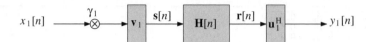

Figure 5.56   Single channel transmission over a MIMO system with a transmit steering vector $\mathbf{v}_1$ and a receive beamformer $\mathbf{u}_1^H$.

where $\tilde{\Lambda} \in \mathbb{R}^{\tilde{R} \times \tilde{R}}$ only contains the SVD's $\tilde{R}$ non-zero singular values, with rank$\{\mathbf{H}\} = \tilde{R} \le \min(M, K)$. The unitary matrices $\mathbf{U} \in \mathbb{C}^{M \times M}$ and $\mathbf{V}^H \in \mathbb{C}^{K \times K}$ in (5.146) are split into two subspaces in (5.159)—$\mathbf{U}_s = [u_1 \ \cdots \ u_{\tilde{R}}]$ and $\mathbf{V}_s = [v_1 \ \cdots \ v_{\tilde{R}}]$ span the signal subspace, from which $\mathbf{H}$ can be constructed entirely,

$$\mathbf{H} = \sum_{m=1}^{\tilde{R}} \lambda_m \mathbf{u}_m \mathbf{v}_m^H. \tag{5.160}$$

The nullspace of $\mathbf{H}$ is spanned by the $K - \tilde{R}$ columns in $\mathbf{V}_s^{\perp}$, i.e. any input to $\mathbf{H}$ formed by a linear combination $\mathbf{a} \in \mathbb{C}^K$ of the column vectors in $\mathbf{V}_s^{\perp}$ will lead to a zero vector output from the matrix, i.e. $\mathbf{Ha} = \underline{0}$. Therefore, to transmit across $\mathbf{H}$, successful transmit vectors have to lie within the space spanned by the columns of $\mathbf{V}_s \in \mathbb{C}^{\tilde{R} \times K}$; any transmission that includes the nullspace of $\mathbf{H}$ cannot be seen at the receiver and wastes transmit power.

If only one data stream was to be transmitted across $\mathbf{H}$, the optimum transmit processor would use a unit norm beamsteering vector $\mathbf{v}_1$ to map the data, weighted by a gain factor γ_1 to adjust the transmit power, into the K input ports of $\mathbf{H}$, as shown in Figure 5.56, with channel noise omitted for simplicity. In the receiver, a beamforming vector $\mathbf{u}_1^H$ will steer the sensitivity of the receiver in the direction of the principal mode of $\mathbf{H}$.

Note that the output

$$y[n] = \mathbf{u}_1^H \mathbf{H} \mathbf{v}_1 \gamma_1 x[n] = \mathbf{u}_1^H \sum_{m=1}^{\tilde{R}} \lambda_m \mathbf{u}_m \mathbf{v}_m^H \mathbf{v}_1 \gamma_1 x[n] = \lambda_1 \gamma_1 x[n] \tag{5.161}$$

extracts the input, weighted by the maximum singular value λ_1 as channel gain. Therefore the arrangement in Figure 5.56 implements the first transmit path of Figure 5.51. The vectors $\mathbf{v}_1$ and $\mathbf{u}_1^H$ in Figure 5.56 act like beamforming and beamsteering vectors [142] to direct the transmit power and receive sensitivity in the direction of the principal mode of the channel, characterized by the principal singular vector λ_1 of $\mathbf{H}$.

If several data streams are to be multiplexed across $\mathbf{H}$, it is possible to stack beamformers that are capable of exploiting the first principal modes of the MIMO channel matrix. Thus, it is possible to stack up to $\tilde{R}$ such beamformers as shown in Figure 5.57, whereby $\tilde{R}$ is the rank of $\mathbf{H}$. Since the modes extracted by the SVD are orthogonal, the SVD directly yields $\tilde{R}$ orthogonal transmit steering vectors $\mathbf{v}_m$, $m = 1, 2, \ldots, \tilde{R}$ forming the columns of $\mathbf{V}_s$ and receive steering vectors $\mathbf{u}_m$, $m = 1, 2, \ldots, \tilde{R}$ forming the columns of $\mathbf{U}_s$.

If all $\tilde{R}$ channels are used for multiplexing, the capacity-optimal arrangement shown in Figure 5.50 and its equivalent formulation in Figure 5.51 are implemented by the structure in Figure 5.57.

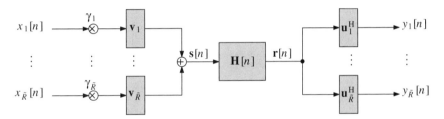

Figure 5.57 Stacking of orthogonal transmit and receive beamformers to multiplex up to $\tilde{R}$ data streams $x_m[n]$, $m = 1, 2, \ldots, \tilde{R}$ across the MIMO channel **H**.

5.7.4 Diversity

If several—ideally independent—links between transmitter and receiver exist, then transmitting the same signal across all links offers diversity. Simplistically, if ρ is the probability of a single channel having poor gain, e.g. by being in a fade, then the probability of none of $R > 1$ independent links offering a viable gain is $\rho^R < \rho$. To maximize what will be defined as the diversity gain of a MIMO system means to maximize the SNR at the output of the receiver, which is different from maximizing the data throughput via multiplexing in Section 5.7.3.

Receive diversity by acquiring a transmitted signal at M sensors—e.g. in a single-input multiple-output (SIMO) system with M receivers—can be exploited to maximize the SNR of a linearly combined signal. This scheme, termed maximum ratio combining (MRC) [143], [144], weighs the M contributions by the square root of the SNRs, γ_m, $m = 1, \ldots, M$, such that $r[n] = \sum_m \gamma_m r_m[n]$. Note that while the main contribution to $r[n]$ will come from MIMO subchannels with high SNR, poor subchannels are not discarded but are used to further enhance $r[n]$.

Transmit diversity arises from the availability of K transmitters, such as in a multiple-input single-output (MISO) system, and can be utilized, for example, by space-time block coding (STBC) [145]–[147]. Without transmit CSI, in STBC the transmitter will create K orthogonal transmit sequences. For the case $K = 2$, the Alamouti scheme [145] arises, where the orthogonal transmit matrix

$$\mathbf{S}_{2n} = \begin{bmatrix} \mathbf{s}_{2n} & \mathbf{s}_{2n+1} \end{bmatrix} = \begin{bmatrix} s[2n] & -s^*[2n+1] \\ s[2n+1] & s^*[2n] \end{bmatrix} \tag{5.162}$$

defines transmit vectors $\mathbf{s}[n]$ for two successive time slots n. If the received signal over the STBC period is stacked as

$$\begin{bmatrix} r[2n] \\ r^*[2n+1] \end{bmatrix} = \begin{bmatrix} h_{1,1} & h_{1,2} \\ h_{1,2}^* & -h_{1,1}^* \end{bmatrix} \begin{bmatrix} s[2n] \\ s[2n+1] \end{bmatrix} + \begin{bmatrix} v[2n] \\ v^*[2n+1] \end{bmatrix}$$

$$= \mathbf{H}_{\text{eff}}\mathbf{s}[2n] + \tilde{\mathbf{v}}[2n],$$

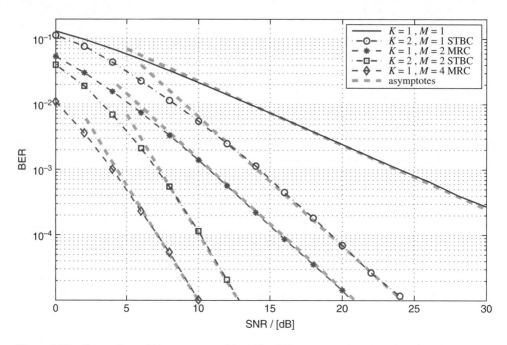

Figure 5.58 Comparison of bit error rates achieved by different transmit and receive diversity schemes on the basis of identical transmit power and channel energy σ_h^2.

then the orthogonality of $\mathbf{S}_{2n}$ is transferred onto the effective channel matrix $\mathbf{H}_{\mathrm{eff}}$. Due to its orthogonality, the SNR-optimal receiver is the matched filter, $\mathbf{H}_{\mathrm{eff}}^{\mathrm{H}}$,

$$\hat{\mathbf{s}}[2n] = \mathbf{H}_{\mathrm{eff}}^{\mathrm{H}}(\mathbf{H}_{\mathrm{eff}}\mathbf{s}[2n] + \tilde{\mathbf{v}}[2n]) = \beta\mathbf{s}[2n] + \mathbf{H}_{\mathrm{eff}}^{\mathrm{H}}\tilde{\mathbf{v}}[2n], \qquad (5.163)$$

where $\beta = |h_{1,1}|^2 + |h_{1,2}|^2$ is the diversity gain. When taking expectations over the channel with $\mathcal{E}\{|h_{i,j}|^2\} = \sigma_h^2$, the diversity gain becomes $\beta = 2\sigma_h^2$.

Transmit diversity schemes similar to Almouti exist for $K > 2$, and arise from the generation of a transmit matrix $\mathbf{S}$ similar to (5.162), which creates N transmit sequences from P successive symbols, leading to a code rate of $P/N \leq 1$. For complex symbol sequences $s[n] \in \mathbb{C}$, it is only possible to simultaneously achieve maximum diversity gain $\beta = K\sigma_h^2$, a code rate of one, and orthogonality of all transmit sequences for $K = 2$ transmitters. For real valued alphabets, $s[n] \in \mathbb{R}$, this limit rises to $K = 8$. It is however possible to bypass these limits by relaxing some constraints; for example, in the case of $K = 4$, the inclusion of transmit CSI in extended orthogonal STBC [148] can achieve not only a unity code rate and code orthogonality, but also a full diversity gain plus array gain of $\beta = (4 + \pi)\sigma_h^2$ [149].

A number of different transmit and receive diversity schemes are compared in Figure 5.58 w.r.t. their BER performance in a fading scenario.

It can be observed that an increase in transmitters K and/or receivers M will enhance the BER. For high SNR, the asymptotic behavior is typical of a fading scenario, with the gradient of the asymptotes in this log-log display indicating the system's diversity gain, which in all cases here is $\beta = KM$. As indicated earlier, receive diversity is preferable over transmit diversity, as the latter schemes—at identical transmit power—recover more dissipated energy and lead to a better SNR. This leads to a coding gain, indicated by a horizontal shift of BER curves in Figure 5.58, between the STBC and MRC schemes for $MK = \text{const.}$

5.7.5 Channel Estimation

Channel identification or estimation is generally performed using a pilot sequence for channel sounding. If the pilot sequences for the M transmit channels are held in a matrix $\mathbf{X} \in \mathbb{C}^{K \times L}$, where L is the length of the pilot sequences, then the system output $\mathbf{Y} \in \mathbb{C}^{M \times L}$ is

$$\mathbf{Y} = \mathbf{HX} + \mathbf{V}, \tag{5.164}$$

where $\mathbf{H} \in \mathbb{C}^{M \times K}$ is the MIMO channel matrix and $\mathbf{V} \in \mathbb{C}^{M \times L}$ models the additive Gaussian noise at the receiver. Denoting vectorized matrices by concatenating their column vectors as e.g. $\text{vec}\{\mathbf{X}\} \in \mathbb{C}^{KL}$, (5.164) can be rewritten as

$$\text{vec}\{\mathbf{Y}\} = \left(\mathbf{X}^{\mathrm{T}} \otimes \mathbf{I}_M\right) \text{vec}\{\mathbf{H}\} + \text{vec}\{\mathbf{V}\}. \tag{5.165}$$

In (5.165), the identity $\text{vec}\{\mathbf{ABC}\} = (\mathbf{C}^{\mathrm{T}} \otimes \mathbf{A})\text{vec}\{\mathbf{B}\}$ has been exploited, which uses the Kronecker product $\otimes$. Based on knowledge of second order statistics of $\text{vec}\{\mathbf{H}\} \in \mathcal{N}(\underline{0}, \mathbf{R}_H)$ and $\text{vec}\{\mathbf{V}\} \in \mathcal{N}(\underline{0}, \mathbf{R}_V)$ and the abbreviation $\tilde{\mathbf{X}} = \left(\mathbf{X}^{\mathrm{T}} \otimes \mathbf{I}_K\right)$, the Wiener-Hopf or minimum mean square error (MMSE) estimate $\hat{\mathbf{H}}_{\mathrm{MMSE}}$ for $\mathbf{H}$ is given by

$$\begin{aligned} \text{vec}\left\{\hat{\mathbf{H}}_{\mathrm{MMSE}}\right\} &= \left(\mathbf{R}_H^{-1} + \tilde{\mathbf{X}}^{\mathrm{H}}\mathbf{R}_V^{-1}\tilde{\mathbf{X}}\right)^{-1} \tilde{\mathbf{X}}^{\mathrm{H}}\mathbf{R}_V^{-1}\text{vec}\{\mathbf{Y}\} \\ &= \mathbf{R}_H\tilde{\mathbf{X}}^{\mathrm{H}}\left(\tilde{\mathbf{X}}\mathbf{R}_H\tilde{\mathbf{X}}^{\mathrm{H}} + \mathbf{R}_V\right)^{-1}\text{vec}\{\mathbf{Y}\}. \end{aligned} \tag{5.166}$$

The MMSE ξ_{MMSE} achieved by the solution in (5.166) is

$$\begin{aligned} \xi_{\mathrm{MMSE}} &= \mathcal{E}\left\{\|\text{vec}\{\mathbf{H}\} - \text{vec}\{\hat{\mathbf{H}}_{\mathrm{MMSE}}\}\|_2^2\right\} \\ &= \text{tr}\left\{\left(\mathbf{R}_H^{-1} + \tilde{\mathbf{X}}^{\mathrm{H}}\mathbf{R}_V^{-1}\mathbf{X}^{\mathrm{H}}\right)^{-1}\right\}, \end{aligned} \tag{5.167}$$

where $\text{tr}\{\cdot\}$ is the trace operator.

The increase in transmitters, K, and receivers, M, is desirable in terms of system capacity as demonstrated in Section 5.7.2, but comes with a drawback evident from the channel estimation simulations in Figure 5.59. Increasing the MK parameters to be estimated in $\mathbf{H}$ requires longer pilot sequences in order to achieve a pre-defined accuracy in terms of MMSE.

If the pilot sequence length L is kept constant, every quadrupling of the number of parameters MK in $\mathbf{H}$ leads to a deterioration in terms of MMSE by approximately 10 dB. Alternatively, to achieve the same MMSE accuracy using the same length L for the pilot sequences requires a 10 dB increase in SNR for every quadrupling of MK. In order to maintain the same MMSE accuracy at the same SNR, L will have to be increased by approximately an order of magnitude for every quadrupling of MK for the range of values shown in Figure 5.59. Therefore, the

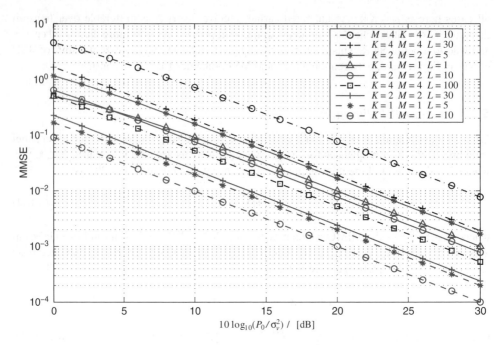

Figure 5.59 Median MMSE calculated over an ensemble of 10^4 pilot sequences for uncorrelated $\mathbf{R}_H$ and uncorrelated $\mathbf{R}_V$ for different number of transmitters K, receivers M, and pilot sequence lengths L.

capacity increase of MIMO systems is bought at the expense of bandwidth requirements during the channel sounding period.

The basic MMSE approach for the channel estimation in (5.166) can be enhanced by additionally adjusting the pilot sequence if the noise is correlated and $\mathbf{R}_V \neq \sigma_v^2 \mathbf{I}_{ML}$. Similarly, if the transmit or receive correlation matrices as defined in (5.157) are known or power constraints for the pilot sequences need to be taken into account, solutions in [150], [151] can be used to estimate both the optimum pilot sequences as well as the channel itself.

If $\mathbf{R}_H$ and $\mathbf{R}_V$ are not available or the sample size is insufficient for consistent estimates of the statistics to compute (5.166) reliably, it is possible to replace the Wiener-Hopf solution in (5.166) by iterative approaches. Two popular methods are the least mean squares (LMS) and the recursive least squares (RLS) solution [152], [153]. Both methods exploit the quadratic nature of the mean square error (MSE) cost function to employ gradient-based strategies. The LMS implements a stochastic gradient method, where a poor estimate of the cost function is compensated by averaging out the gradient noise in the iterative update procedure. The RLS minimizes the least squares error at every iteration, which leads to a more costly but also faster converging algorithm. Both LMS and RLS will approximate the MMSE solution in (5.166) given a sufficient set of samples and a suitable setting of parameters that govern convergence in both LMS and RLS. The previous comment about the increased requirement for pilot sequence length when M and K grow is echoed here: the data requirement for e.g. LMS rises linearly with the number of parameters that need to be identified [152].

5.7.6   Broadband MIMO

Most analyses and all techniques discussed in Section 5.7 so far have been concerned with the narrowband case. In the broadband case, where the channel between two points needs to be characterized in the time domain by an impulse response, the channel matrix now becomes a matrix of FIR filters, $\mathbf{H}[n]$ as outlined in (5.137). In the stationary case, its z-transform $H(z) = \sum_{n=0}^{N} \mathbf{H}[n]z^{-n}$ is a polynomial matrix. The matrix can be approximated by a narrowband system matrix $\mathbf{H}$ if the channel's delay spread is short compared to the symbol length, or if the channel only acts as a delay but is free of multi-path propagation. However, the latter part for a MIMO system forms a so-called key-hole channel, where both transmit and receive correlation matrices in (5.157) are rank one and no MIMO benefit can be derived. Therefore, below we now focus on the case of a broadband channel impulse response matrix $\mathbf{H}[n]$ and its transfer function $H(z)$.

Many optimal narrowband processing techniques—such as the utility of the SVD as applied to the decoupling of a MIMO channel in (5.146)—do not directly extend to the broadband case. In order to approximate a narrowband system, the inputs and outputs of a MIMO system can be P-times multiplexed by parallel-to-serial converters at its input and serial-to-parallel converters at its output. The resulting system has PK inputs and PM outputs, spanning in between a new MIMO system matrix $A(z) \in \mathbb{C}^{PM \times PK}$,

$$A(z) = \begin{bmatrix} H_0(z) & z^{-1}H_{P-1}(z) & \cdots & z^{-1}H_1(z) \\ \vdots & \ddots & \ddots & \vdots \\ H_{P-2}(z) & & H_0(z) & z^{-1}H_{P-1}(z) \\ H_{P-1}(z) & H_{P-2}(z) & \cdots & H_0(z) \end{bmatrix}. \tag{5.168}$$

This new MIMO system matrix $A(z)$ is block-pseudo-circulant and contains the polyphase components $H_p(z) = \sum_n \mathbf{H}[nP + p]z^{-n}$ [40], [154]. For $L \le P < \infty$, $A(z)$ reduces to order one. However, elements containing z^{-1} are now restricted to the upper right hand corner of $A(z) = \mathbf{A}_0 + z^{-1}\mathbf{A}_1$, where $\mathbf{A}_0$ is lower-left block-triangular as shown in Figure 5.60, and $\mathbf{A}_1$ is upper-right triangular. To eliminate the polynomial nature of $A(z)$, a guard interval can be inserted into the transmit vector and discarded from the receive vector. Thus, the effective channel matrix $\tilde{\mathbf{A}}_0$ now only contains components of $\mathbf{A}_0$ [155]–[157]; no longer containing polynomial components, standard narrowband techniques can now be applied to $\tilde{\mathbf{A}}_0$.

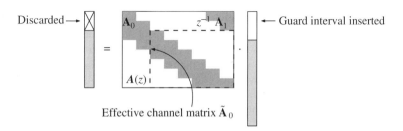

Figure 5.60   Multiplexed block-pseudo-circulant MIMO system matrix $A(z)$ can be made scalar for $P > L$ and by inserting a guard interval in the transmit vector, which is then discarded in the receiver, leading to a non-polynomial effective channel matrix $\tilde{\mathbf{A}}_0$.

If the effective channel matrix $\tilde{\mathbf{A}}_0$ is known at the transmitter, then optimal filter bank design techniques for precoding and equalization techniques can be applied to e.g. create spatio-temporally decoupled subchannels [155]–[157]. This can be followed by waterfilling in order to optimally allocate transmit power, as outlined in Figure 5.52.

If CSI is not available at the transmitter, but P can be selected to exceed the channel order L, then still a non-polynomial though now unknown effective channel matrix $\tilde{\mathbf{A}}_0$ arises. If additionally the guard interval is selected as a cyclic prefix of the transmit vector, then $\tilde{\mathbf{A}}_0$ becomes block-circulant and can be diagonalized by applying DFTs and inverse DFTs to each of the K groups of inputs and M groups of outputs of $A(z)$, which implements orthogonal frequency division multiplexing (OFDM, [158]). Within each frequency bin of an OFDM system, narrowband MIMO techniques can be applied—this can include both spatial multiplexing or diversity techniques such as space-time block coding. In addition to space-time block coding techniques, which would be restricted to operate within independent frequency bins, space-frequency block codes [159], [160] can additionally exploit diversity available across different subcarriers.

Filter bank-based multicarrier (FBMC) approaches [41], [161]–[164] have been gaining popularity in many communications fields including PLC (see also Section 5.3). Their aim is to divide the spectrum into narrower bands by means of filter banks that offer a higher selectivity than available in OFDM through the utility of the DFT. Also FBMC generally uses oversampled filter banks, where small guard intervals or spectral redundancy is inserted between frequency bands. As a result FBMC is more robust to synchronization mismatches than OFDM, but the subband signals still remain broadband. Applying narrowband MIMO techniques in combination with FBMC is therefore difficult.

Besides space-frequency block codes, a number of specific MIMO broadband diversity schemes have been considered. These include e.g. space-time block codes [165] for dispersive channels, named time-reversal STBC. Compared to STBC, the entries in the matrix of (5.162) are now symbol blocks, and Hermitian transposition of a component is accompanied by a time-reversal of the particular block. STBC approaches can also be combined with equalizers which can be constructed to exploit the specific structure of STBC codes [166], [167].

In order to extend the utility of optimal narrowband techniques such as EVD and SVD to the broadband case, a recent polynomial matrix eigenvalue decomposition (PEVD) [168], [169] can decompose a parahermitian matrix $R(z) \approx Q(z)\Lambda(z)\tilde{Q}(z)$, where $R(z) = \tilde{R}(z)$ with the parahermitian operation $\tilde{R}(z) = R^{\mathrm{H}}(z^{-1})$, implementing a Hermitian transposition and a time reversal. For the factorization, $Q(z)$ is a paraunitary matrix such that $Q(z)\tilde{Q}(z) = \mathbf{I}$. The matrix $\Lambda(z)$ remains polynomial but is diagonalized and spectrally majorized [40]. Two PEVDs can implement a polynomial SVD [168], which can be employed to decouple a MIMO system matrix $A(z)$. As the decomposition is achieved by means of paraunitary matrices representing lossless filter banks [40], similar to the SVD in (5.146), the polynomial SVD permits the extraction of linear precoders that do not alter the transmit power and linear equalizers that will provide no noise amplification. A number of communications-related applications of such broadband MIMO techniques have been reported in e.g. [154], [170]–[172].

5.7.7 PLC MIMO Aspects

The push to high data throughput in PLC has led to the exploration of MIMO techniques, whereby up to 2 transmitters and 4 receivers may be available over a PLC channel with three conductors (hot, neutral and ground wires) [173], [174] (see also Section 2.6). The capacity gain afforded by such a system has been exploited for both multiplexing [175], [176] and diversity via STBC [177]. Measurements of MIMO PLC channels have been undertaken in e.g. [178], and various degradations of the ideally expected MIMO capacity gain have been attributed to the spatial correlation of MIMO subchannels [179], [180], as well as the spatial correlation of noise [174], [181], [182]. Also, a MIMO channel simulator has been reported [183], and a real-world MIMO emulator been presented in [184].

With respect to PLC standardization efforts, HomePlug AV2 incorporates MIMO techniques with up to 2 transmit and 4 receive nodes, implementing spatial multiplexing [173], [185]. The broadband channel here is divided into narrowband subcarriers using OFDM, and the precoding and equalization matrices for each subcarrier are calculated by an SVD as outlined in (5.146). The HomePlug AV2 MIMO scheme is referred to as beamforming, where each beamformer implements one column of the singular matrices $\mathbf{U}$ and $\mathbf{V}^{\mathrm{H}}$ of (5.146).

5.8   Coding Techniques

Since the power line was never intended to be a communications channel, we cannot expect to find a communications-friendly channel. As already seen, not only the usual AWGN, but also narrowband interferers and impulse noise are present. As such, normal error control techniques are not sufficient, and we need adaptations, or in some instances, complete new schemes, to be able to handle the noise on the power line. Although not identical to the wireless channel, coding techniques from that scenario have found application in PLC.

Coding techniques need to make provision for these different noise types. In the subsection on coding techniques already employed in standards, we will see how ECC techniques make provision for these types of errors. Generally, some ECC code is used to correct errors caused by background noise, e.g. convolutional code, Reed-Solomon code, LDPC code or a combination of these. Frequency selective fading in PLC can lead to bursts of error, for this reason interleaving is applied to spread and separate adjacent bits. If designed properly, adjacent encoded bits are mapped to non-adjacent carriers during modulation. Depending on the severity of noise on the channel, information can also be repeated a number of times.

Note that we assume the reader to be familiar with terminology from the error correction field. If not, any good ECC books, such as [186], are recommended. The focus will be on the physical layer (PHY layer) and more specifically on the ECC being employed on this layer. We may touch on other aspects from the PHY layer, such as modulation, scrambling, etc., but this will not be the main focus of this section. References to the appropriate sections in this book with more details for these other aspects will be provided.

In most cases, a cyclic redundancy check (CRC) code is used on the MAC layer. This is an error detection code, and is normally used to cover and protect an entire packet as a second error handling mechanism, in addition to the ECC on the PHY layer. The ECC on the PHY layer will attempt to detect and correct any errors. However, in the event that the ECC cannot handle or correct the errors, the CRC code on the MAC layer should detect the errors and request that the packet be sent again.

First, we will look at coding techniques used in early products/protocols. Second, we look at coding techniques as proposed and used in PLC standards, some of which developed from the earlier protocols. Finally, a look at some interesting coding techniques proposed in research literature. This overview is by no means exhaustive, but should be considered as a good introduction into the varied coding techniques being applied in PLC.

5.8.1 Coding Techniques in Various Protocols

Here we will briefly discuss coding techniques in existing protocols used for narrowband (e.g. home automation) and wideband (e.g. multimedia), also discussed elsewhere in this book. One will notice that the techniques employed in the home automation protocols are rather simple, owing to the fact that these usually operate at low rates and as such the error events do not affect as many bits.

One of the very first home automation protocols was X10 (see Section 7.3.1). This protocol do not employ error correction, but makes use of repetition, i.e. sending the same information a number of times. In X10 the data packets are repeated twice. As will be seen later, even standards using more advanced error correction techniques, also include some form of repetition in their protocol.

In KNX (see Section 7.3.2 and [187]) a cross-checking method for error detection is used, with horizontal and vertical parity checks. This is essentially a very simple product code. Packets (or telegrams in KNX) are divided in 8-bit segments. A telegram then consists of different fields, each of 8-bit length, e.g. control, source address, destination address, user data, etc., as shown in Figure 5.61. A ninth parity bit is appended to each segment, ensuring that the segment has even parity horizontally. The final field in the telegram is the checksum, this consists of the parity that is added column-wise to ensure odd parity vertically. The segments are then serialized and sent over the channel. Using this method, single bit errors in the telegram can be corrected, as well as detecting a wide range of random errors.

Another home automation protocol is LONWorks (see Section 7.3.3), which uses a highly efficient, patented low-overhead forward error correction algorithm, able to detect and correct single bit errors. In addition, a CRC code is used to overcome packet errors.

Field	b_0	b_1	b_2	b_3	b_4	b_5	b_6	b_7	p
Control									
Source address, byte 1	0	1	0	1	0	1	1	0	0
Source address, byte 2	1	1	0	1	0	1	0	1	1
Target address, byte 1	1	0	1	0	1	0	1	0	0
Target address, byte 2	0	0	1	0	0	1	1	0	1
Routing/Length	1	1	0	1	1	1	1	1	1
User data, byte 1	0	1	0	0	0	0	0	0	1
User data, byte 2	0	1	1	1	0	0	0	0	1
Checksum, s	0	0	0	1	1	1	1	1	1

Odd parity

(Even parity, right side)

Figure 5.61 KNX telegram with horizontal and vertical parity checks.

The AMIS CX1-Profile protocol (see Section 7.5 and [188]) makes use of repetition and interleaving. The PHY block (consisting of preamble, PHY header and PHY data) is encoded using a binary repetition block code $(N, 1, N)_2$, where N is the number of hop frequencies being used in the transmission mode. The encoder thus has a rate of $1/N$, with N being chosen between 5 and 8 depending on the chosen transmission mode. If DBPSK is being used, then the PHY block is encoded as is using the $(N, 1, N)_2$ code. If DQPSK is being used, then the PHY block is first split into two streams, one containing odd indices and the other containing even indices. The two streams are then encoded independently and in parallel using the same repetition code. The data (one or two streams, depending on the chosen modulation) are then segmented into segments of length $2N^2$. After each segment is interleaved cyclically using $\text{mod}(2N^2)$ arithmetic, the data is sent to be modulated.

The last narrowband protocol we look at is DigitalSTROM (see Section 7.6 and [189]). In this case the system is designed such that downstream and upstream communication use different coding techniques. It was found for the downstream communication that a standard $(7, 4)$-Hamming code, extended by an additional parity bit, is sufficient. For upstream communication an $R = 1/2$ convolutional code is used, followed by a scrambler and an interleaver. A certain number of channel bits is transmitted within a mains half-period, and if the interleaver depth is set equal to this number, it helps to keep the length of error events below the constraint length of the convolutional code.

Next, we mention two protocols used for multimedia, but since these specifications have been incorporated into the IEEE 1901 standard (see Section 5.8.2.4) we omit the details here. HomePlug AV is a broadband specification for home networking that targets high bandwidth multimedia applications, like HDTV and VoIP. It makes use of turbo convolution codes with rates of 1/2 and 16/21. The IEEE 1901 FFT standard is backward compatible and thus uses the same FEC. Similarly, the HD-PLC specification from the HD-PLC Alliance was designed for high-speed in-home networking. It uses Reed-Solomon codes in combination with convolutional codes. The IEEE 1901 Wavelet standard is backward compatible with HD-PLC.

5.8.2   Coding Techniques in Standards

Here, we look at coding techniques that are presently being used in various standards. Specifically, we look at:

- PRIME,
- G3-PLC,
- ITU-T G.9960, and
- IEEE 1901.

5.8.2.1   PRIME

PRIME [190] is a specification used for narrowband PLC, popularly used in automatic metering. It has been standardized as ITU-T G.9904 PRIME (see Section 9.2.3 for more details).

A basic FEC scheme is employed. A convolutional code with bit interleaving is used, if enabled by higher layers. Should the channel be deemed good enough, the FEC is disabled.

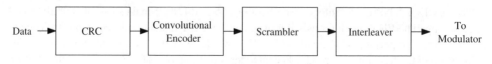

Figure 5.62   FEC encoder for PRIME.

Briefly, a cyclic redundancy check (CRC) is appended to the PHY header, as an error detection mechanism. No CRC is added to the payload, as this is performed by the MAC layer. If the FEC is enabled, then convolutional encoding is performed. Note here that the header is always encoded, irrespective of whether FEC is chosen or not. Scrambling is then always performed on both header and payload. Should the FEC be enabled, then interleaving is performed as a final step. The encoded bits are then modulated (using one of the available modulation schemes) and converted to OFDM symbols. The block diagram for the encoder is shown in Figure 5.62.

As stated already, the header always has FEC as On. However, the MAC layer will, based on errors from previous transmissions, decide whether to turn FEC On or Off for the payload. Based on this information it will also decide on the best modulation scheme to use. Note that the standard does not specify any target error rates for the selection of FEC or modulation schemes. This is left to the designer.

In more details: the uncoded header bits (and uncoded payload bits, if FEC is On) are encoded with a half-rate convolutional code with a constraint length of $K = 7$ and generator polynomials 1111001 and 1011011, as shown in Figure 5.63. The encoder state is set to zeros at the start of every PPDU transmission. Eight zeros are appended to the end of the uncoded header bits to flush the encoder and return it to the zero state. If FEC is On, then six zeros are appended to the uncoded payload for the same purpose.

The scrambler is used to randomize the bit stream. This helps to reduce the peak values after the IFFT, should a long series of ones or zeros occur in the header or payload after encoding. Essentially, the encoded bits are XOR'ed with a pseudo-noise sequence.

If FEC is On, then interleaving will be performed as a last step. Narrowband interference can cause deep fades in the spectrum, with groups of OFDM subcarriers being affected. This causes bursts of bit errors to occur. The interleaving ensures that these bursts of errors are spread out by reordering the bits in a predetermined manner, making it appear to be randomized. This aids the convolutional code, which is only able to handle random errors.

Refer to [190] for the specifics about the implementation of the scrambler and interleaver.

At the receiver all the encoding operations' counterparts are performed for decoding.

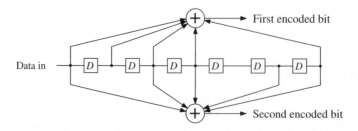

Figure 5.63   Convolutional encoder with $R = 1/2$ and $K = 7$ used for PRIME.

The following information bits per OFDM symbol are obtained in the payload, depending on which modulation is chosen and FEC being On or Off:

- DBPSK: 48 (On) and 96 (Off),
- DQPSK: 96 (On) and 192 (Off),
- D8PSK: 144 (On) and 288 (Off).

Since the header always uses DBPSK with FEC enabled, the information bits per OFDM symbol are 42.

In [191], a description of PRIME's physical specifications are given, followed by the modeling of the FEC and modulation components to ascertain the performance of PRIME when used on simulated channel models found in literature. Depending on the modulation used, results showed that an improvement of between 4 to 5 dB in SNR could be attained when using the FEC.

5.8.2.2 G3-PLC

G3-PLC [192] is a low-layer protocol aimed at the smart grid, promoted by various electricity distributors and vendors. It has been standardized as ITU G.9903 G3-PLC (see Section 9.2.2), and the standards ITU-T G.9902 G.hnem (see Section 9.2.1) and IEEE 1901.2 PLC (see Section 9.2.4) use it as a basis.

In this scheme convolutional and Reed-Solomon coding is used to combat errors caused by background noise and impulsive noise. A two-dimensional interleaving scheme is also used to provide diversity in both time and frequency, to combat bursts of errors. Additionally, a repetition coder can be used if required. The entire FEC encoder and decoder is shown in Figure 5.64.

Two different modes of operation are available: Normal and Robust. In Normal mode, the FEC consists of an RS encoder and convolutional encoder. In Robust mode, in addition to the RS and convolutional encoders, a repetition code is used that repeats each bit four times, for extra protection.

First, the frame control header (FCH) is protected by making use of a 5-bit cyclic redundancy check (CRC). Then, as with previous schemes, a scrambler is used to give the data a random distribution. The data is XOR'ed with a predetermined pseudo-noise sequence.

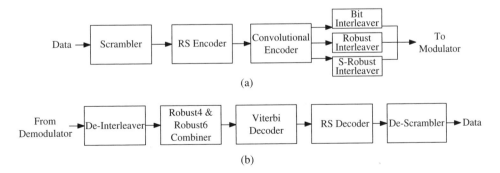

Figure 5.64 FEC blocks for G3-PLC showing (a) encoder, and (b) decoder.

After the scrambler, the data is encoded using one of two shortened, systematic RS codes. The parameters for the two codes are:

- $n = 255, k = 239, t = 8$, and
- $n = 255, k = 247, t = 4$

where n is the number of output symbols, k is the number of input symbols and t is the number of symbols that can be corrected. Both codes make use of Galois Field GF(2^8). The code generator polynomial is:

$$g(x) = \prod_{i=1}^{2t}(x - \alpha^i),$$

and the field generator polynomial is:

$$p(x) = x^8 + x^4 + x^3 + x^2 + 1.$$

Note that in Robust mode, the code with $t = 4$ is used.

Next, the output from the RS encoder is encoded using a half-rate convolutional encoder with a constraint length of $K = 7$ and generator polynomials 1111001 and 1011011. Note that this is the same convolutional encoder as used in PRIME (see Figure 5.63). To return the convolutional encoder back to the zero state, six zeros are inserted after the last bit of the data.

The final stage (in Normal mode) is interleaving. It is designed to provide diversity to combat burst errors that corrupt a few consecutive OFDM symbols, and protection against deep frequency fades that can affect a few adjacent frequencies for a number of OFDM symbols.

It achieves this by using a two-step process. Before interleaving starts, the bits are organized in a matrix where the columns represent different OFDM symbols and the rows represent different sub-carriers. First, each column is circularly shifted a different number of times to ensure that a corrupted OFDM symbols is spread over different symbols. Second, each row is circularly shifted a different number of times to ensure that a frequency fade does not disrupt an entire column. Refer to [192] for the specific implementation of the interleaver.

If Robust mode is enabled for the encoder, then the Repetition Coder will repeat the resulting packet from the interleaver four times. Note that the FCH is sent using the super Robust mode, which means that it will be repeated six times.

The coding rates achievable for different configurations are listed in Table 5.6.

Expected performance based on simulation results are found in [193]. When considering impulse noise, it was determined that there is about 3 dB gain between using the interleaver and not using the interleaver. Further results show performance for the uncoded case, convolutional encoder with interleaver, RS/convolutional encoder with interleaver and finally RS/convolutional encoder with interleaver/repetition. At a BER of 10^{-4}, adding the RS code provides a gain of approximately 1.3 dB. By adding the repetition coder, an additional gain of approximately 3 dB is attained.

A comparison of the physical layers of both PRIME and G3-PLC can be found in [194], with theoretical analysis and simulation results provided.

Table 5.6 RS block size and coding rates for various modulations.

Number of Symbols	DQPSK	DBPSK	Robust
12	37/53	10/26	—
20	73/89	28/44	—
32	127/143	55/71	—
40	163/179	73/89	13/21
52	217/233	100/116	20/28
56	235/251	109/125	22/32
112	—	235/251	54/62
252	—	—	133/141

The three components common to both PRIME and G3-PLC, i.e. convolutional encoding, bit interleaving and PSK modulation, are investigated in [195] in the presence of narrowband interference. Simple techniques, which do not require any changes in existing OFDM implementations, are suggested. It is shown that by changing the depth of the interleaver for specific orders of PSK modulation, the performance of the convolutional code can be improved. Further improvements can be obtained by employing signal nulling in the frequency domain. Doing this enables the convolutional decoder to operate in a soft-decision mode.

5.8.2.3 ITU-T G.9960

In this section we look at the coding used in ITU-T G.9960 [196]. This recommendation defines the physical layer specifications for the G.hn family of home networking technology, typically using power lines, phone lines and coaxial cables. The reader should refer to Section 8.8 for more details about other aspects of this standard.

The first step involves scrambling the information data, followed by two FEC and repetition encoders, one for the header and one for the payload. To complete the PHY frame, the encoded information is multiplexed and segmented into an integer number of symbol frames, in order to be sent as OFDM symbols.

All the data from the header and payload are scrambled with a pseudo-random sequence. This sequence is generated by a linear feedback shift register with the generator polynomial $p(x) = x^{23} + x^{18} + 1$, as shown in Figure 5.65. The shift register needs to be initialized with an initialization vector. A second initialization can be performed for the payload, and this

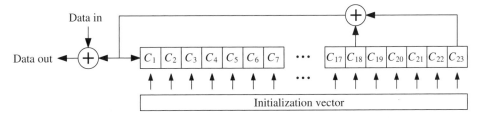

Figure 5.65 Scrambler used for ITU-T G.9960.

Table 5.7   FEC parameters for ITU-T G.9960.

	Mother Code	Code Rate, R	Information Size, k	Encoded Size, n	Puncturing Pattern, **pp**
Used for header	$(1/2)_H$	1/2	168	336	n/a
Used for payload	$(1/2)_S$	1/2	960	1920	n/a
	$(1/2)_L$	1/2	4320	8640	n/a
	$(2/3)_S$	2/3	960	1440	n/a
	$(2/3)_L$	2/3	4320	6480	n/a
	$(5/6)_S$	5/6	960	1152	n/a
	$(5/6)_L$	5/6	4320	5184	n/a
	$(5/6)_S$	16/18	960	1080	$\mathbf{pp}_{1152}^{(72)}$
	$(5/6)_L$	16/18	4320	4860	$\mathbf{pp}_{5184}^{(324)}$
	$(5/6)_S$	20/21	960	1008	$\mathbf{pp}_{1152}^{(144)}$
	$(5/6)_L$	20/21	4320	4536	$\mathbf{pp}_{5184}^{(648)}$

is indicated by the scrambler initialization value (a value of zero indicates that the second initialization did not take place).

After scrambling, the header and payload are sent to two different FEC encoders. The encoders consist of systematic quasi-cyclic low-density parity check (QC-LDPC) block codes, making use of three mother codes with rates $R = 1/2$, $R = 2/3$ and $R = 5/6$, which can then be adapted to codes with higher rates by making use of puncturing.

For each mother code, the rate of the code is determined by the $(n - k) \times n$-size parity-check matrix **H**, where k is the number of information bits and n is the number of coded bits. Each **H** is constructed using an array of circulant sub-matrices, by choosing certain parameters for the sub-matrices, different rates can be achieved. Seven different mother codes are used, depending on rate and the size of the information block (see Table 5.7 for details). Refer to [196] for the exact details of parameters and the **H**-matrices. As can be seen from Table 5.7, the header is always encoded with an $R = 1/2$ code, while the payload rate can be varied depending on channel conditions.

The encoding operation follows the usual block encoding principles: a group of information bits (length k) is put into the encoder and using the parity-check matrix the $n - k$ parity bits are calculated and appended to the information bits. The resulting codeword, say **x**, must satisfy the parity check equations such that $\mathbf{x}\mathbf{H}^T = \mathbf{0}$.

Next, if a higher rate code is to be used, the codeword is punctured according to the puncturing pattern, $\mathbf{pp}_T^{(i)}$, where i is the number of zeros in the pattern and T is the length of the pattern. The puncturing patterns used are:

$$\mathbf{pp}_{1152}^{(72)} = \left[\underbrace{1\,1\,\cdots\,1}_{720}\ \underbrace{0\,0\,\cdots\,0}_{36}\ \underbrace{1\,1\,\cdots\,1}_{360}\ \underbrace{0\,0\,\cdots\,0}_{36} \right],$$

$$\mathbf{pp}_{5184}^{(324)} = \left[\underbrace{1\,1\,\cdots\,1}_{3240}\ \underbrace{0\,0\,\cdots\,0}_{162}\ \underbrace{1\,1\,\cdots\,1}_{972}\ \underbrace{0\,0\,\cdots\,0}_{162}\ \underbrace{1\,1\,\cdots\,1}_{648} \right],$$

$$
\mathbf{pp}_{1152}^{(144)} = \left[\underbrace{1\ 1\ \cdots\ 1}_{720}\ \underbrace{0\ 0\ \cdots\ 0}_{48}\ \underbrace{1\ 1\ \cdots\ 1}_{240}\ \underbrace{0\ 0\ \cdots\ 0}_{96}\ \underbrace{1\ 1\ \cdots\ 1}_{48} \right],
$$

$$
\mathbf{pp}_{5184}^{(648)} = \left[\underbrace{0\ 0\ \cdots\ 0}_{216}\ \underbrace{1\ 1\ \cdots\ 1}_{4320}\ \underbrace{0\ 0\ \cdots\ 0}_{432}\ \underbrace{1\ 1\ \cdots\ 1}_{216} \right].
$$

A one in the puncturing pattern indicates that the corresponding bit will appear in the final codeword, and a zero indicates that the corresponding bit will be punctured out, i.e. it will not appear in the final codeword. Using the puncturing pattern together with a mother code results in code rates of $R = 16/18$ and $R = 20/21$, see Table 5.7 for details.

The final step after the FEC is the repetition encoders, which makes a certain number of copies of each codeword and then reorganizes the bits in predefined patterns.

The operation of the header encoder and payload encoder can be summarized as follows:

Header encoder: the header is encoded using the FEC as described above, always using $R = 1/2$ and $k = 168$. Then the header repetition encoder makes M copies of the encoded header, where M depends on the number of bits that the OFDM symbols carrying the header will be using. Two encoded header blocks are constructed by making use of shifted versions of the M copies of the encoded header. In the first block, the M copies of the encoded header codeword are concatenated together, with each subsequent copy being cyclically shifted by two bits. The second block is formed in the same way, except that the first copy is cyclically shifted by 168 bits (half the encoded codeword size) and subsequent copies again cyclically shifted by two bits.

Payload encoder: the payload is encoded using the FEC as described above, using any of the rates and input sizes available. For the payload there is the option of using the payload repetition encoder or not. In the *normal mode* of operation the repetition encoder is disabled. In *robust communication mode*, each codeword is repeated 2, 3, 4, 6 or 8 times. Each copy is divided into sections, each containing a number of bits that is dependent on the total number of bits to be loaded onto the OFDM symbol. The bits in each section appears in the same order as in the original codeword. The first copy out of the repetition encoder contains all the sections in the original order. Subsequent copies contain all the sections as well, but cyclically shifted. The shifts are defined by a cyclic section shift vector, which is determined by the number of repetitions chosen.

Based on the various PHY layer parameters, such as FEC block size, packet size and repetition numbers, [197] determines the throughput for the entire system.

5.8.2.4 IEEE 1901

The IEEE 1901 standard [198] is intended for broadband PLC, enabling high speed communications. It can make use of two different PHY layers, one based on the fast Fourier transform, and the other based on wavelets, see Section 8.5 for further details.

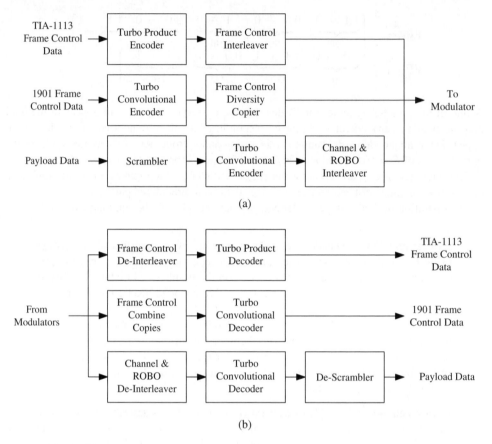

Figure 5.66 FEC blocks for IEEE 1901 FFT showing (a) encoder, and (b) decoder.

5.8.2.4.1 *FFT-OFDM*

The block diagram for the IEEE 1901 FFT-based encoder and decoder is shown in Figure 5.66. The encoder consists of separate encoders for the TIA-1113 header data, IEEE 1901 header data and for the payload data.

The *TIA-1113 frame control data* is first encoded using an $R = 1/4$ product code. This (100,25) product code with minimum Hamming distance 16, is derived from a (10,5) shortened extended Hamming code with minimum Hamming distance 4, using the generator matrix:

$$\mathbf{G} = \begin{bmatrix} 1 & 0 & 0 & 0 & 0 & 0 & 0 & 1 & 1 & 1 \\ 0 & 1 & 0 & 0 & 0 & 0 & 1 & 1 & 1 & 0 \\ 0 & 0 & 1 & 0 & 0 & 1 & 1 & 1 & 0 & 0 \\ 0 & 0 & 0 & 1 & 0 & 1 & 1 & 0 & 0 & 1 \\ 0 & 0 & 0 & 0 & 1 & 1 & 0 & 0 & 1 & 1 \end{bmatrix}$$

The product code is constructed using the following array:

$$\begin{bmatrix} i_{11} & i_{12} & i_{13} & i_{14} & i_{15} & p_r & p_r & p_r & p_r & p_r \\ i_{21} & i_{22} & i_{23} & i_{24} & i_{25} & p_r & p_r & p_r & p_r & p_r \\ i_{31} & i_{32} & i_{33} & i_{34} & i_{35} & p_r & p_r & p_r & p_r & p_r \\ i_{41} & i_{42} & i_{43} & i_{44} & i_{45} & p_r & p_r & p_r & p_r & p_r \\ i_{51} & i_{52} & i_{53} & i_{54} & i_{55} & p_r & p_r & p_r & p_r & p_r \\ p_c & p_c & p_c & p_c & p_c & p_p & p_p & p_p & p_p & p_p \\ p_c & p_c & p_c & p_c & p_c & p_p & p_p & p_p & p_p & p_p \\ p_c & p_c & p_c & p_c & p_c & p_p & p_p & p_p & p_p & p_p \\ p_c & p_c & p_c & p_c & p_c & p_p & p_p & p_p & p_p & p_p \\ p_c & p_c & p_c & p_c & p_c & p_p & p_p & p_p & p_p & p_p \end{bmatrix}$$

Information bits are placed in the i_{ij} positions, $1 \leq i,j \leq 5$, and the generator matrix is applied to the individual rows to produce the row parity, p_r. Next, the generator matrix is applied to the columns to produce the column parity, p_c, and the parity on parity, p_p. An interleaver is then used to spread the data over the carriers of the four frame control symbols.

The *IEEE 1901 frame control data* consists of 128 bits, which a turbo convolutional encoder (operating in $R = 1/2$ mode) encodes into 256 bits. The output of this encoder is randomized by using an interleaver. The final step is to apply a diversity copier to replicate the encoded and interleaved bits before being mapped onto the carriers. Both the turbo convolutional encoder and the interleaver are the same as used for the payload encoding, as discussed next.

The *payload data* is scrambled, turbo convolutional encoded and interleaved. The turbo encoder works with rates of $R = 1/2, R = 16/21$ or $R = 16/18$ (optional). When ROBO mode is used, only rate $R = 1/2$ is used, and the interleaver is followed by a ROBO interleaver. The payload encoder works on groups of octets called Physical Blocks of length 520 or 136.

As for ITU-T G.9960, the payload data is scrambled with a pseudo-random sequence, generated using the generator polynomial $p(x) = x^{10} + x^3 + 1$, as shown in Figure 5.67. The shift register is initialized with all ones before processing starts.

The scrambled data is encoded using an $R = 1/2$ turbo convolutional encoder. This encoder consists of two $R = 2/3$ recursive systematic convolutional (RSC) codes and one turbo interleaver. Puncturing is used to obtain additional rates of $R = 16/21$ and $R = 16/18$. The eight-state RSC encoders are shown in Figure 5.68. The input bits (u_1, u_2) are never punctured, and only the additional encoded bits (v_1, v_2) are considered for puncturing. The data is passed through the encoders twice, to ensure that tail-bitten termination is achieved. In the first pass, the encoders are initialized to the all-zeros state. The payload data is encoded as normal, except

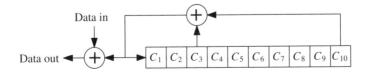

Figure 5.67   Scrambler used for IEEE 1901 FFT payload data.

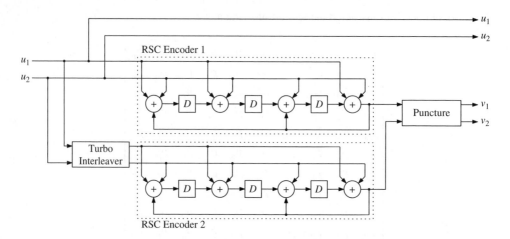

Figure 5.68 Turbo convolutional encoder used for IEEE 1901.

the outputs are not used on the first pass. The final state of the encoder then determines the starting state for the second pass, such that the starting state and the final state of the second pass will be the same. On the second pass, the payload data is again fed into the encoders, and the encoded data is passed on for puncturing.

When using the encoder in $R = 1/2$ mode, bits u_1, u_2, v_1 and v_2 are sent directly to the output. For the other two rates a puncturing pattern, $\mathbf{pp}_{v_i}$, is applied to the encoded bit v_i, where v_i is sent to the output if the puncturing pattern contains a one. For $R = 16/21$:

$$\mathbf{pp}_{v_1} = [1001001001001000],$$
$$\mathbf{pp}_{v_2} = [1001001001001000],$$

and for $R = 16/18$:

$$\mathbf{pp}_{v_1} = [1000000010000000],$$
$$\mathbf{pp}_{v_2} = [1000000010000000].$$

Finally, a channel interleaver is used to reorder the bits received from the turbo convolutional encoder.

Three robust modes of operation, or ROBO modes, are available that are used for beacon broadcast, multicast communication, session setup and management messages. The ROBO interleaver can be regarded as an interleaver and repetition encoder, as the data is copied several times and interleaved together with the original data. In any ROBO mode, the $R = 1/2$ turbo convolutional encoder is always used, together with the following modes and their respective parameters:

- Standard ROBO mode: Physical Blocks of size 520, and 4 repetitions,
- High-speed ROBO mode: Physical Blocks of size 520, and 2 repetitions,
- Mini ROBO mode: Physical Blocks of size 136, and 5 repetitions.

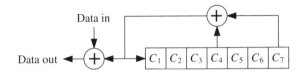

Figure 5.69   Scrambler used for IEEE 1901 Wavelet payload data.

Normally Standard ROBO mode is used, however if reliable communication is possible then the encoder can change to high-speed ROBO mode, doubling the communication rate. For small payloads that needs high reliability, mini ROBO mode is used with the smallest Physical Block size and the highest number of repetitions.

5.8.2.4.2   Wavelet-OFDM

For the Wavelet-ODFM we will focus only on the encoding of the payload data.

As before, the payload data is scrambled with a pseudo-random sequence, in this case generated by the generator polynomial $p(x) = x^7 + x^4 + 1$, as shown in Figure 5.69. The shift register is initialized with all ones before processing starts.

Next, the scrambled data is encoded with a Reed-Solomon encoder or alternatively with concatenated encoder consisting of a Reed-Solomon encoder followed by a convolutional encoder. The RS encoder uses the same field generator polynomial and code generator polynomial as for G3-PLC. Depending on the chosen parameters, coding rates of $R = 239/255$ (used for payloads), $R = 40/56$ (used for payloads making use of diversity-OFDM) and $R = 34/50$ (used for control frames) are obtainable. The $R = 1/2$, $K = 7$, convolutional encoder is also the same as for PRIME and G3-PLC (see Figure 5.63). It is reset to zero at the start of each data field and 6-bit tail is inserted at the end of each data field to force the encoder to the zero state. Different rates are obtained by applying puncturing according to the patterns in Table 5.8. The coding rates 1/2 to 7/8 are used for payloads, depending on channel estimation, and the rate 1/2 is used for payloads with diversity-OFDM, control frames and tone map indices.

Low-density parity-check convolutional codes (originally proposed in [199]) can optionally be used instead of the RS/convolutional encoder. Generally these convolution codes consist of shift registers for information bits, parity bits, a weight controller and adder, and weight multipliers; and are defined infinite check matrices that are periodically time varying. The chosen coding rates for the LDPC convolutional codes are $R = 1/2$, $R = 2/3$, $R = 3/4$ and

Table 5.8   Code rates and puncturing patterns for IEEE 1901 Wavelet convolutional encoder.

	Puncture Patterns	
Code Rate	First Encoded Bit	Second Encoded Bit
2/3	10	11
3/4	101	110
4/5	1000	1111
5/6	10101	11010
6/7	100101	111010
7/8	1000101	1111010

$R = 4/5$. Tail bits (the number of which depends on the coding rate and information size used) are included to force the encoder back to a known state.

Finally, the output from the convolutional encoder or the LDPC convolutional encoder is interleaved. In the case where only Reed-Solomon encoding is applied, the encoded data is not interleaved. The interleaver's block size is chosen to correspond with the number of bits in an OFDM symbol.

5.8.3 Other Coding Techniques

Finally, in this section we look at interesting coding techniques found in research literature. Some of the techniques discussed adds to the techniques already being used in the standards of the previous section.

We again consider the combined coding and modulation of permutation codes and M-FSK as discussed in Section 5.2.1. Where previously block permutation codes were considered, here we will look at combining permutation codes with convolutional codes to form permutation trellis codes. Some disadvantages of permutation block codes are that the construction of long block codes is a difficult mathematical problem, and a general decoding algorithm for permutations in this scenario is not known. To overcome this, distance-preserving mappings (DPMs) are used to map the binary outputs of a standard convolutional encoder to permutation symbols. This way the advantages of permutation codes when combined with M-FSK are retained, while making decoding easier by using the well known Viterbi algorithm. Figure 5.70 shows a block diagram of the proposed system. It is envisaged that these codes can be used in control or security applications where reliability is more important than the speed of communication.

The distance-preserving property of DPMs requires that the Hamming distance between one code word and another in the unconstrained set must have the same or larger Hamming distance between the corresponding code words in the other set which is mapped to. If an error correcting code is used as input to the mapping, then the distance preserving ensures that the resulting code after the mapping also possesses error correcting capabilities. In [200] and [201] it were shown that by making use of distance-preserving mappings, convolutional codes

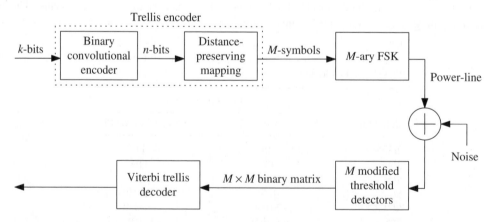

Figure 5.70 Block diagram of a system combining permutation coding and M-FSK modulation.

can be mapped to permutation codes, thereby forming permutation trellis codes. A binary convolutional code is used as a base code, with the outputs being mapped to a permutation code by using a DPM. The resulting permutation trellis code will then possess the same structure as the original base code, making Viterbi decoding possible.

We consider the M-FSK from Section 5.2.1 with the modified hard-decision threshold detector. The trellis encoder (the binary encoder combined with a DPM) outputs M symbols from a permutation codeword that belong to a particular transition in the trellis. The M-ary symbols are transmitted in time as the corresponding frequencies. With the modified threshold detectors, a transmitted codeword of length M thus leads to an $M \times M$ received binary matrix. The influence of the different noise types on the decoding process is the same as described in Section 5.2.1 and Figure 5.2.

Taking the effect of each noise type on the error metric into account, conditions can be placed on when decoding will be correct. Consider two paths in the trellis that differ in d positions. For background noise or impulse noise, $d-1$ errors can occur and the Viterbi decoder will still choose the correct path, when they merge. Narrowband noise is expected to last for long periods and will affect all sequences in a path. If two paths of length L are compared, narrowband noise can cause the error metric for the incorrect path to be decreased by L. If L is smaller than d, then correct decoding will take place irrespective of the narrowband noise. Should the narrowband noise occur in conjunction with other noise, then L plus the number of background and impulse noise errors must be less than d for correct decoding to occur.

The reader is referred to [201], [202], [203] and [204] for simulation results of permutation trellis codes.

The concept of permutation codes has also been used in other coding schemes for PLC.

In [205], M-FSK modulation is combined with a time-diversity permutation coding scheme. The setup is similar to the permutation trellis code where an $M \times M$ binary matrix is received. However, whenever a row or column is detected to contain ones (corresponding to narrowband noise and impulse noise, respectively), nulling is used to cancel the effect of the noise. A majority-logic decoding process with optional soft-decision outputs is used to correct additive errors from background noise. Furthermore, a time-diversity encoder is used to combat burst errors that can affect several codewords.

A rateless permutation coding scheme combined with OFDM is suggested in [206]. The permutation encoding and decoding is the same as in [205], here a fountain code is introduced to combat against disturbances of bursts of errors that affect several codewords. The fountain code provides time-diversity, and since it does not have a fixed rate, it can be adapted as channel conditions change, resulting in a more efficient use of resources.

Based on permutation codes, [207] suggests a more general code to deal with narrowband noise and signal fading. They show how a new parameter, the symbol equity, can influence the performance of the system. Equitable symbol weight codes are shown to be optimal in regards to this new parameter and simulation results confirm better performance for the codes compared to previous known codes.

As we have seen, LDPC codes are used in ITU-T G.9960, and are an important class of codes in other communication systems as well. QC-LDPC codes, with rates of 1/2, 1/3, 2/3, 3/4 and 4/5, are investigated in [208] and how different decoding algorithms affect the performance in PLC conditions. The iterative belief propagation algorithm is investigated, as well as an adaptation of the bit flipping algorithm where single bits are flipped per iteration instead of several bits. The authors also propose a hybrid decoding algorithm where the belief propagation algorithm is applied first, if it fails to find a correct codeword, the adapted bit flipping is applied. The investigation is extended [209] to specifically look at how QC-LDPC codes perform in impulsive noise conditions. QC-LDPC codes as inner codes and outer codes in a concatenated scheme together with RS codes or convolutional codes are compared. The authors conclude that LDPC codes as the outer code with RS codes as the inner code give the best performance.

A column extension method and optimization of degree distributions are proposed to create rate-compatible QC-LDPC codes [210]. It is shown that the new codes perform better than the current codes being used in ITU-T G.9960. In [211], short, optimized irregular LDPC codes for narrowband PLC are constructed using differential evolution to find degree distribution pairs that will improve the performance.

A comparison study between QC-LDPC codes and turbo codes are done in [212]. It is shown that with longer block lengths the LDPC codes can approximate the performance of turbo codes at low SNR values.

Carrying on with turbo codes, [213] proposes a turbo-coded single-carrier system that can operate in harsh impulsive noise conditions. Statistical knowledge of the impulse noise is generally needed for the turbo decoder to function optimally. The authors propose a clipping operation to be implemented in the trellis structure, therefore each received symbol can be considered as a symbol with additive memoryless clipping noise, of which the PDF can easily be obtained. A double binary turbo coded system together with OFDM is considered in [214].

Where the KNX protocol used a simple array code, [215] and [216] investigate a wider range of array codes, specifically looking at generalized array codes and row and column array codes. Also, where the bits were serially concatenated and sent in KNX, this investigation lets each row of bits be mapped onto one modulation symbol. Thus the number of columns will determine the order of the modulation scheme to be used. The performance of the codes are obtained by using computer simulations and compared with that of concolutional codes.

Similar to [206], [217] suggests using an LT-code as an outer code, together with irregular LDPC codes as the inner code for OFDM systems. Based on the decoding performed by the LDPC decoder, packets sent to the LT decoder can be marked as erased if they are found to be highly affected by impulse noise. Since the LT encoder introduces a small percentage of additional packets, the erased packets can be handled by the decoder and the original data recovered.

As a final topic, we discuss some research making use of network coding. Although network coding does not address error correction as was discussed in the rest of this section, and is normally implemented on one of the layers above PHY, it does contribute to the network's throughput and efficiency. It has already become a hot topic in other areas of communication, and it will surely be one in the coming years for PLC as well. In [218], a G.hn system is proposed that makes use of network coding on the physical layer and in [219], a cooperative scheme based on network coding is proposed for the link layer. Cooperative networking is discussed in detail in Section 6.5.

References

1. T. Schaub, Spread frequency shift keying, *IEEE Trans. Commun.*, 42(2/3/4), 1056–1064, Feb./Mar./Apr. 1994.

2. K. Dostert, Frequency hopping spread spectrum modulation for digital communications over electrical power lines, *IEEE J. Sel. Areas Commun.*, 8(4), 700–710, May 1990.

3. J. G. Proakis, *Digital Communications*. New York: McGraw-Hill, 1989.

4. J. G. Proakis and M. Salehi, *Digital Communications*, 5th ed. McGraw-Hill, 2008.

5. I. F. Blake, Permutation codes for discrete channels, *IEEE Trans. Inf. Theory*, 20(1), 138–140, Jan. 1974.

6. J. Häring and A. J. H. Vinck, Iterative decoding of codes over complex numbers for impulsive noise channels, *IEEE Trans. Inf. Theory*, 49(5), 1251–1260, May 2003.

7. J. O. Onunga and R. W. Donaldson, A simple packet retransmission strategy for the power-line throughput and delay enhancement on power line communication channels, *IEEE Trans. Power Delivery*, 8(3), 818–826, Jul. 1993.

8. R. C. Dixon, *Spread Spectrum Systems with Commercial Applications*, 3rd ed. Wiley-Interscience, 1994.

9. P. Baier, K. Dostert, and M. Pandit, A novel spread-spectrum receiver synchronization scheme using a SAW tapped delay line, *IEEE Trans. Commun.*, 30(5), 1037–1047, May 1982.

10. K. Dostert, Ein neues Spread-Spectrum Empfängerkonzept auf der Basis angezapfter Verzögerungsleitungen für akustische Oberflächenwellen, Ph.D. dissertation, University of Kaiserslautern, Germany, 1980.

11. H. Ochsner, Data transmission on low voltage power distribution lines using spread spectrum techniques, in *Proc. Canadian Commun. Power Conf.*, Montreal, Quebec, Oct. 15–17, 1980, 236–239.

12. P. van der Gracht and R. W. Donaldson, Communication using pseudonoise modulation on electric power distribution circuits, *IEEE Trans. Commun.*, 33(9), 964–974, Sep. 1985.

13. K. Dostert, A novel frequency hopping spread spectrum scheme for reliable power line communications, in *Proc. IEEE Int. Symp. Spread Spectrum Techniques and Applic.*, Yokohama, Japan, Nov. 29–Dec. 2, 1992, 183–186.

14. ——, *Power Line Kommunikation (in German)*. Franzis Verlag, 2000.

15. ——, *Powerline Communications*. Prentice Hall, 2001.

16. Signalling on low voltage electrical installations in the frequency range 3 kHz to 148.5 kHz, European Committee for Electrotechnical Standardization (CENELEC), Brussels, Belgium, Standard EN 50065-1, 1991.

17. ITRAN Communications Ltd., Spread spectrum communication system utilizing differential code shift keying, Sep. 2001, International Patent Application WO 01/67652 A1.

18. ——, Code shift keying transmitter for use in a spread spectrum communications system, Jun. 2001, International Patent Application WO 01/41383 A2.

19. ——, Receiver for use in a code shift keying spread spectrum communications system, Oct. 2001, International Patent Application WO 01/80506 A2.

20. T. Waldeck and K. Dostert, Comparison of modulation schemes with frequency agility for application in power line communication systems, in *Proc. IEEE Int. Symp. Spread Spectrum Techniques and Applic.*, 2, Mainz, Germany, Sep. 22–25, 1996, 821–825.

21. T. Waldeck, Einzel- und Mehrträgerverfahren für die störresistente Kommunikation auf Energieverteilnetzen (in German), Ph.D. dissertation, Logos Verlag, Berlin, Germany, 2000.

22. M. L. Doelz, E. T. Heald, and D. L. Martin, Binary data transmission techniques for linear systems, *Proc. IRE*, 45(5), 656–661, May 1957.

23. I. Kalet, The multitone channel, *IEEE Trans. Commun.*, 37(2), 119–124, Feb. 1989.

24. S. Weinstein and P. Ebert, Data transmission by frequency-division multiplexing using the discrete Fourier transform, *IEEE Trans. Commun. Technol.*, 19(5), 628–634, Oct. 1971.

25. F. Sjöberg, R. Nilsson, M. Isaksson, P. Ödling, and P. O. Börjesson, Asynchronous zipper, in *Proc. IEEE Int. Conf. Commun.*, 1, Vancouver, Canada, Jun. 6–10, 1999, 231–235.

26. G. Cherubini, E. Eleftheriou, and S. Ölçer, Filtered multitone modulation for very high-speed digital subscriber lines, *IEEE J. Sel. Areas Commun.*, 20(5), 1016–1028, Jun. 2002.

27. B. Saltzberg, Performance of an efficient parallel data transmission system, *IEEE Trans. Commun. Technol.*, 15(6), 805–811, Dec. 1967.

28. S. D. Sandberg and M. A. Tzannes, Overlapped discrete multitone modulation for high speed copper wire communications, *IEEE J. Sel. Areas Commun.*, 13(9), 1571–1585, Dec. 1995.

29. J. Tan and G. L. Stuber, Constant envelope multi-carrier modulation, in *Proc. IEEE Mil. Commun. Conf.*, 1, Anaheim, USA, Oct. 7–10, 2002, 607–611.

30. High Definition PLC Alliance. [Online]. Available: http://www.hd-plc.org

31. HomePlug Powerline Alliance. [Online]. Available: http://www.homeplug.org

32. Universal Powerline Association. [Online]. Available: http://www.upaplc.org

33. S. Galli and O. Logvinov, Recent developments in the standardization of power line communications within the IEEE, *IEEE Commun. Mag.*, 46(7), 64–71, Jul. 2008.

34. A. M. Tonello, Time domain and frequency domain implementations of FMT modulation architectures, in *Proc. IEEE Int. Conf. Acoustics, Speech and Sig. Proc.*, 4, Toulouse, France, May 14–19, 2006, 625–628.

35. S. Weiss and R. W. Stewart, Fast implementation of oversampled modulated filter banks, *Electron. Lett.*, 36(17), 1502–1503, Aug. 2000.

36. J. M. Cioffi, A multicarrier primer, *Amati Communications Corporation and Stanford University*, Nov. 1991, TIE1.4/91-157.

37. A. M. Tonello, Asynchronous multicarrier multiple access: Optimal and sub-optimal detection and decoding, *Bell Syst. Tech. J.*, 7(3), 191–217, Mar. 2003, Special issue: Wireless Radio Access Networks.

38. A. M. Tonello and F. Pecile, Efficient architectures for multiuser FMT systems and application to power line communications, *IEEE Trans. Commun.*, 57(5), 1275–1279, May 2009.

39. C. Siclet, P. Siohan, and D. Pinchon, Perfect reconstruction conditions and design of oversampled DFT-modulated transmultiplexers, *EURASIP J. Adv. Signal Process.*, 2006, article ID 15756.

40. P. P. Vaidyanathan, *Multirate Systems and Filter Banks*. Prentice Hall, 1993.

41. A. M. Tonello and F. Pecile, Analytical results about the robustness of FMT modulation with several prototype pulses in time-frequency selective fading channels, *IEEE Trans. Wireless Commun.*, 7(5), 1634–1645, May 2008.

42. N. Al-Dhahir and J. M. Cioffi, Optimum finite-length equalization for multicarrier transceivers, *IEEE Trans. Commun.*, 44(1), 56–64, Jan. 1996.

43. G. Arslan, B. L. Evans, and S. Kiaei, Equalization for discrete multitone transceivers to maximize bit rate, *IEEE Trans. Signal Process.*, 49(12), 3123–3135, Dec. 2001.

44. T. Pollet, M. Peeters, M. Moonen, and L. Vandendorpe, Equalization for DMT based broadband modems, *IEEE Commun. Mag.*, 38(5), 106–113, May 2000.

45. D. Umehara, H. Nishiyori, and Y. Morihiro, Performance evaluation of CMFB transmultiplexer for broadband power line communications under narrowband interference, in *Proc. IEEE Int. Symp. Power Line Commun. Applic.*, Orlando, USA, Mar. 26–29, 2006, 50–55.

46. K. H. Afkhamie, H. Latchman, L. Yonge, T. Davidson, and R. E. Newman, Joint optimization of transmit pulse shaping, guard interval length, and receiver side narrow-band interference mitigation in the HomePlugAV OFDM system, in *Proc. IEEE Workshop Signal Process. Adv. Wireless Commun.*, New York City, USA, Jun. 5–8, 2005, 996–1000.

47. A. J. Redfern, Receiver window design for multicarrier communication systems, *IEEE J. Sel. Areas Commun.*, 20(5), 1029–1036, Jun. 2002.

48. A. Skrzypczak, P. Siohan, and J.-P. Javaudin, Application of the OFDM/OQAM modulation to power line communications, in *Proc. IEEE Int. Symp. Power Line Commun. Applic.*, Pisa, Italy, Mar. 26–28, 2007, 71–76.

49. P. Siohan, C. Siclet, and N. Lacaille, Analysis and design of OFDM/OQAM systems based on filterbank theory, *IEEE Trans. Signal Process.*, 50(5), 1170–1183, May 2002.

50. B. Farhang-Boroujeny, Multicarrier modulation with blind detection capability using cosine modulated filter banks, *IEEE Trans. Commun.*, 51(12), 2057–2070, Dec. 2003.

51. T. Ihalainen, T. H. Stitz, M. Rinne, and M. Renfors, Channel equalization in filter bank based multicarrier modulation for wireless communications, *EURASIP J. Adv. Signal Process.*, 2007, article ID 49389.

52. P. Tan and N. C. Beaulieu, A comparison of DCT-based OFDM and DFT-based OFDM in frequency offset and fading channels, *IEEE Trans. Commun.*, 54(11), 2113–2125, Nov. 2006.

53. S. Hara and R. Prasad, Overview of multicarrier CDMA, *IEEE Commun. Mag.*, 35(12), 126–133, Dec. 1997.

54. M. Crussiere, J.-Y. Baudais, and J.-F. Hélard, Adaptive spread-spectrum multicarrier multiple-access over wirelines, *IEEE J. Sel. Areas Commun.*, 24(7), 1377–1388, Jul. 2006.

55. A. M. Tonello, A concatenated multitone multiple antenna air-interface for the asynchronous multiple access channel, *IEEE J. Sel. Areas Commun.*, 24(3), 457–469, Mar. 2006.

56. A. Tonello, Method and apparatus for filtered multitone modulation using circular convolution, 2008, patent n. UD2008A000099, PCT WO2009135886A1.

57. A. M. Tonello and M. Girotto, Cyclic block filtered multitone modulation, *EURASIP J. Adv. Signal Process.*, vol. 2014, 2014.

58. ——, Cyclic block FMT modulation for broadband power line communications, in *Proc. IEEE Int. Symp. Power Line Commun. Applic.*, Johannesburg, South Africa, Mar. 24–27, 2013, 247–251.

59. M. Girotto and A. M. Tonello, Improved spectrum agility in narrow-band PLC with cyclic block FMT modulation, in *Proc. IEEE Global Telecom. Conf.*, Austin, USA, Dec. 8–12, 2014, 2995–3000.

60. ——, Orthogonal design of cyclic block filtered multitone modulation, in *Proc. European Wireless Conf.*, Barcelona, Spain, May 14–16, 2014, 1–6.

61. A. M. Tonello, S. D'Alessandro, and L. Lampe, Bit, tone and cyclic prefix allocation in OFDM with application to in-home PLC, in *Proc. IFIP Wireless Days Conf.*, Dubai, United Arab Emirates, Nov. 24–27, 2008, 1–5.

62. F. Pecile and A. M. Tonello, On the design of filter bank systems in power line channels based on achievable rate, in *Proc. IEEE Int. Symp. Power Line Commun. Applic.*, Dresden, Germany, Mar. 29–Apr. 1, 2009, 228–232.

63. A. M. Tonello, S. D'Alessandro, and L. Lampe, Cyclic prefix design and allocation in bit-loaded OFDM over power line communication channels, *IEEE Trans. Commun.*, 58(11), 3265–3276, Nov. 2010.

64. M. Antoniali, A. M. Tonello, and F. Versolatto, A study on the optimal receiver impedance for SNR maximization in broadband PLC, *J. Electric. Comput. Eng.*, vol. 2013, 1–11, 2013.

65. W. C. Black and N. E. Badr, High-frequency characterization and modeling of distribution transformers, in *Proc. IEEE Int. Symp. Power Line Commun. Applic.*, Rio de Janeiro, Brazil, Mar. 28–31, 2010, 18–21.

66. B. Varadarajan, I. H. Kim, A. Dabak, D. Rieken, and G. Gregg, Empirical measurements of the low-frequency power-line communications channel in rural North America, in *Proc. IEEE Int. Symp. Power Line Commun. Applic.*, Udine, Italy, Apr. 3–6, 2011, 463–467.

67. D. W. Rieken and M. R. Walker, Distance effects in low-frequency power line communications, in *Proc. IEEE Int. Symp. Power Line Commun. Applic.*, Rio de Janeiro, Brazil, Mar. 28–31, 2010, 22–27.

68. S. T. Mak and D. L. Reed, TWACS, a new viable two-way automatic communication system for distribution networks. Part I: Outbound communication, *IEEE Trans. Power Apparatus Syst.*, 101(8), 2941–2949, Aug. 1982.

69. S. T. Mak and T. G. Moore, TWACS, a new viable two-way automatic communication system for distribution networks. Part II: inbound communications, *IEEE Trans. Power Apparatus Syst.*, 103(8), 2141–2147, Aug. 1984.

70. S. T. Mak and R. L. Maginnis, Power frequency communication on long feeders and high levels of harmonic distortion, *IEEE Trans. Power Delivery*, 10(4), 1731–1736, Oct. 1995.

71. D. W. Rieken, Periodic noise in very low frequency power-line communication, in *Proc. IEEE Int. Symp. Power Line Commun. Applic.*, Udine, Italy, Apr. 3–6, 2011, 295–300.

72. H. L. van Trees, *Detection, Estimation, and Modulation Theory, Part I*. Wiley-Interscience, 2001.

73. D. W. Rieken and M. R. Walker, Ultra low frequency power-line communications using a resonator circuit, *IEEE Trans. Smart Grid*, 2(1), 41–50, Mar. 2011.

74. M. Z. Win and R. A. Scholtz, Impulse radio: How it works, *IEEE Commun. Lett.*, 2(2), 36–38, Feb. 1998.

75. A. M. Tonello, Wide band impulse modulation and receiver algorithms for multiuser power line communications, *EURASIP J. Adv. Signal Process.*, 2007, article ID 96747.

76. F. Versolatto, A. M. Tonello, M. Girotto, and C. Tornelli, Performance of practical receiver schemes for impulsive UWB modulation on a real MV power line network, in *IEEE Int. Conf. Ultra-Wideband*, Bologna, Italy, Sep. 14–16, 2011, 610–614.

77. A. M. Tonello, F. Versolatto, and C. Tornelli, Analysis of impulsive UWB modulation on a real MV test network, in *Proc. IEEE Int. Symp. Power Line Commun. Applic.*, Udine, Italy, Apr. 3–6, 2011, 18–23.

78. A. M. Tonello, F. Versolatto, and M. Girotto, Multitechnology (I-UWB and OFDM) coexistent communications on the power delivery network, *IEEE Trans. Power Delivery*, 28(4), 2039–2047, Oct. 2013.

79. M. Zimmermann and K. Dostert, Analysis and modeling of impulsive noise in broad-band powerline communications, *IEEE Trans. Electromag. Compat.*, 44(1), 249–258, Feb. 2002.

80. T. Esmailian, F. R. Kschischang, and P. G. Gulak, In-building power lines as high-speed communication channels: Channel characterization and a test-channel ensemble, *Int. J. Commun. Syst.*, 16(5), 381–400, Jun. 2003, Special Issue: Powerline Communications and Applications.

81. M. Katayama, T. Yamazato, and H. Okada, A mathematical model of noise in narrowband power line communication systems, *IEEE J. Sel. Areas Commun.*, 24(7), 1267–1276, Jul. 2006.

82. J. A. Cortés, L. Diéz, F. J. Cañete, and J. J. Sánchez-Martínez, Analysis of the indoor broadband power line noise scenario, *IEEE Trans. Electromagn. Compat.*, 52(4), 849–858, Nov. 2010.

83. L. Di Bert, P. Caldera, D. Schwingshackl, and A. M. Tonello, On noise modeling for power line communications, in *Proc. IEEE Int. Symp. Power Line Commun. Applic.*, Udine, Italy, Apr. 3–6, 2011, 283–288.

84. M. Nassar, K. Gulati, Y. Mortazavi, and B. L. Evans, Statistical modeling of asynchronous impulsive noise in powerline communication networks, in *Proc. IEEE Global Telecom. Conf.*, Houston, USA, Dec. 5–9, 2011, 1–6.

85. M. Nassar, A. Dabak, I. H. Kim, T. Pande, and B. L. Evans, Cyclostationary noise modeling in narrowband powerline communication for smart grid applications, in *Proc. IEEE Int. Conf. Acoustics, Speech and Sig. Proc.*, Kyoto, Japan, Mar. 25–30, 2012, 3089–3092.

86. S. Miyamoto, M. Katayama, and N. Morinaga, Performance analysis of QAM systems under class A impulsive noise environment, *IEEE Trans. Electromagn. Compat.*, 37(2), 260–267, May 1995.

87. D. J. C. MacKay, *Information Theory, Inference, and Learning Algorithms*. Cambridge University Press, 2003.

88. J. Mitra and L. Lampe, Convolutionally coded transmission over markov-gaussian channels: Analysis and decoding metrics, *IEEE Trans. Commun.*, 58(7), 1939–1949, Jul. 2010.

89. IEEE standard for low-frequency (less than 500 kHz) narrowband power line communications for smart grid applications, IEEE Standards Association, IEEE Std. 1901.2-2013, Dec. 2013.

90. M. Luby, M. Watson, T. Gasiba, and T. Stockhammer, High-quality video distribution using power line communication and aplication layer forward error correction, in *Proc. IEEE Int. Symp. Power Line Commun. Applic.*, Pisa, Italy, Mar. 26–28, 2007, 431–436.

91. T. Stockhammer, Internet protocol television over PLC, in *Power Line Communications: Theory and Applications for Narrowband and Broadband Communications over Power Lines*, H. C. Ferreira, L. Lampe, J. Newbury, and T. G. Swart, Eds. John Wiley & Sons, Jun. 2010.

92. A. Shokrollahi, Raptor codes, *IEEE Trans. Inf. Theory*, 52(6), 2551–2567, Jun. 2006.

93. M. Luby, LT codes, in *Proc. IEEE Symp. Found. Comput. Sci.*, Vancouver, Canada, Nov. 16–19, 2002, 271–280.

94. S. Katar, B. Mashburn, K. Afkhamie, H. Latchman, and R. Newman, Channel adaptation based on cyclostationary noise characteristics in PLC systems, in *Proc. IEEE Int. Symp. Power Line Commun. Applic.*, Orlando, USA, Mar. 27–29, 2006, 16–21.

95. N. Sawada, T. Yamazato, and M. Katayama, Bit and power allocation for power-line communications under nonwhite and cyclostationary noise environment, in *Proc. IEEE Int. Symp. Power Line Commun. Applic.*, Dresden, Germany, Mar. 29–Apr. 1, 2009, 307–312.

96. N. Shlezinger and R. Dabora, On the capacity of narrowband PLC channels, *IEEE Trans. Commun.*, 63(4), 1191–1201, Apr. 2015.

97. M. A. Tunc, E. Perrins, and L. Lampe, Optimal LPTV-aware bit loading in broadband PLC, *IEEE Trans. Commun.*, 61(12), 5152–5162, Dec. 2013.

98. S. V. Zhidkov, Analysis and comparison of several simple impulsive noise mitigation schemes for OFDM receivers, *IEEE Trans. Commun.*, 56(1), 5–9, Jan. 2008.

99. J. Häring and A. J. H. Vinck, Performance bounds for optimum and suboptimum reception under class-A impulsive noise, *IEEE Trans. Commun.*, 50(7), 1130–1136, Jul. 2002.

100. J. Mitra and L. Lampe, Robust decoding for channels with impulse noise, in *Proc. IEEE Global Telecom. Conf.*, San Francisco, USA, Nov. 27–Dec. 1, 2006, 1–6.

101. D. Fertonani and G. Colavolpe, A simplified metric for soft-output detection in the presence of impulse noise, in *Proc. IEEE Int. Symp. Power Line Commun. Applic.*, Pisa, Italy, Mar. 26–28, 2007, 121–126.

102. ——, On reliable communications over channels impaired by bursty impulse noise, in *Proc. IEEE Int. Symp. Power Line Commun. Applic.*, Jeju Island, Korea, Apr. 2–4, 2008, 357–362.

103. J. Mitra and L. Lampe, On joint estimation and decoding for channels with noise memory, *IEEE Commun. Lett.*, 13(10), 730–732, Oct. 2009.

104. M. Mushkin and I. Bar-David, Capacity and coding for the Gilbert-Elliott channels, *IEEE Trans. Inf. Theory*, 35(6), 1277–1290, Nov. 1989.

105. T. Li, W.-H. Mow, and M. Siu, Joint erasure marking and Viterbi decoding algorithm for unknown impulsive noise channels, *IEEE Trans. Wireless Commun.*, 7(9), 3407–3416, Sep. 2008.

106. J. Mitra and L. Lampe, Coded narrowband transmission over noisy powerline channels, in *Proc. IEEE Int. Symp. Power Line Commun. Applic.*, Dresden, Germany, Mar. 29–Apr. 1, 2009, 143–148.

107. B. D. Fritchman, A binary channel characterization using partitioned Markov chains, *IEEE Trans. Inf. Theory*, 13(2), 221–227, Apr. 1967.

108. M. Ghosh, Analysis of the effect of impulse noise on multicarrier and single carrier QAM systems, *IEEE Trans. Commun.*, 44(2), 145–147, Feb. 1996.

109. R. Pighi, M. Franceschini, G. Ferrari, and R. Raheli, Fundamental performance limits for PLC systems impaired by impulse noise, in *Proc. IEEE Int. Symp. Power Line Commun. Applic.*, Orlando, USA, Mar. 27–29, 2006, 277–282.

110. J. Häring and A. J. H. Vinck, OFDM transmission corrupted by impulsive noise, in *Proc. Int. Symp. Power Line Commun. Applic.*, Limerick, Ireland, Apr. 5–7, 2000, 9–14.

111. S. Foucart and H. Rauhut, *A Mathematical Introduction to Compressive Sensing*. Birkhäuser, Basel, 2013.

112. F. Abdelkefi, P. Duhamel, and F. Alberge, Impulsive noise cancellation in multicarrier transmission, *IEEE Trans. Commun.*, 53(1), 94–106, Jan. 2005.

113. A. Mengi and A. J. H. Vinck, Impulsive noise error correction in 16-OFDM for narrowband power line communication, in *Proc. IEEE Int. Symp. Power Line Commun. Applic.*, Dresden, Germany, Mar. 29–Apr. 1, 2009, 31–35.

114. G. Caire, T. Y. Al-Naffouri, and A. K. Narayanan, Impulse noise cancellation in OFDM: An application of compressed sensing, in *Proc. IEEE Int. Symp. Inform. Theory*, Toronto, Canada, Jul. 6–11, 2008, 1293–1297.

115. IEEE standard for broadband over power line networks: Medium access control and physical layer specifications, IEEE Standards Association, IEEE Standard 1901-2010, Sep. 2010. [Online]. Available: http://grouper.ieee.org/groups/1901/

116. L. Lampe, Bursty impulse noise detection by compressed sensing, in *Proc. IEEE Int. Symp. Power Line Commun. Applic.*, Udine, Italy, Apr. 3–6, 2011, 29–34.

117. Y. C. Eldar, P. Kuppinger, and H. Bölcskei, Block-sparse signals: Uncertainty relations and efficient recovery, *IEEE Trans. Signal Process.*, 58(6), 3042–3054, Jun. 2010.

118. J. Matanza, S. Alexandres, and C. Rodriguez-Morcillo, Compressive sensing techniques applied to narrowband power line communications, in *Proc. IEEE Int. Conf. Signal Process. Comput. and Control*, Solan, India, Sep. 26–28, 2013, 1–6.

119. J. Lin, M. Nassar, and B. L. Evans, Impulsive noise mitigation in powerline communications using sparse Bayesian learning, *IEEE J. Sel. Areas Commun.*, 31(7), 1172–1183, Jul. 2013.

120. C. E. Shannon, A mathematical theory of communications (Part I), *Bell Syst. Tech. J.*, 27(3), 379–423, Jul. 1948.

121. ——, A mathematical theory of communications (Part II), *Bell Syst. Tech. J.*, 27(4), 623–656, Oct. 1948.

122. A. Paulraj, R. Nabar, and D. Gore, *Introduction to Space-Time Wireless Communications*. Cambridge University Press, 2003.

123. G. H. Golub and C. F. van Loan, *Matrix Computations*, 3rd ed. John Hopkins University Press, 1996.

124. G. J. Foschini and M. J. Gans, On limits of wireless communications in a fading environment when using multiple antennas, *Wireless Personal Commun.*, 6(3), 311–335, Mar. 1998.

125. I. E. Telatar, Capacity of multi-antenna Gaussian channels, *European Trans. Telecommun.*, 10(6), 585–595, Nov.–Dec. 1999.

126. D. Gesbert, M. Shafi, D.-S. Shiu, P. J. Smith, and A. Naguib, From theory to practice: An overview of MIMO space-time coded wireless systems, *IEEE J. Sel. Areas Commun.*, 21(3), 281–302, Apr. 2003.

127. M. Vu and A. Paulraj, MIMO wireless linear precoding, *IEEE Signal Process. Mag.*, 24(5), 86–105, Sep. 2007.

128. T. M. Cover and J. A. Thomas, *Elements of Information Theory*. John Wiley & Sons, Inc., 1991.

129. D. P. Palomar and J. R. Fonollosa, Practical algorithms for a family of waterfilling solutions, *IEEE Trans. Signal Process.*, 53(2), 686–695, Feb. 2005.

130. R. F. H. Fischer and J. B. Huber, A new loading algorithm for discrete multitone transmission, in *Proc. IEEE Global Telecom. Conf.*, 1, London, United Kingdom, Nov. 18–22, 1996, 724–728.

131. A. M. Wyglinski, F. Labeau, and P. Kabal, Bit loading with BER-constraint for multicarrier systems, *IEEE Trans. Wireless Commun.*, 4(4), 1383–1387, Jul. 2005.

132. A. Fasano, G. Di Blasio, E. Baccarelli, and M. Biagi, Optimal discrete bit loading for DMT based constrained multicarrier systems, in *Proc. IEEE Int. Symp. Inform. Theory*, Lausanne, Switzerland, Jun. 30–Jul. 5, 2002, 243.

133. X. Zhang and B. Ottersten, Power allocation and bit loading for spatial multiplexing in MIMO systems, in *Proc. IEEE Int. Conf. Acoustics, Speech and Sig. Proc.*, 5, Hong Kong, China, Apr. 6–10, 2003, 53–56.

134. A. Goldsmith, S. A. Jafar, N. Jindal, and S. Vishwanath, Capacity limits of MIMO channels, *IEEE J. Sel. Areas Commun.*, 21(5), 684–702, Jun. 2003.

135. G. Kulkarni, S. Adlakha, and M. Srivastava, Subcarrier allocation and bit loading algorithms for OFDMA-based wireless networks, *IEEE Trans. Mobile Computing*, 4(6), 652–662, Nov.–Dec. 2005.

136. N. Papandreou and T. Antonakopoulos, Bit and power allocation in constrained multi-carrier systems: The single-user case, *EURASIP J. on Adv. in Signal Process.*, vol. 2008, 1–14, Jul. 2007.

137. W. Al-Hanafy and S. Weiss, Discrete rate maximisation power allocation with enhanced bit error ratio, *IET Commun.*, 6(9), 1019–1024, Jun. 2012.

138. D. P. Palomar and M. A. Lagunas, Joint transmit-receive space-time equalization in spatially correlated MIMO channels: A beamforming approach, *IEEE J. Sel. Areas Commun.*, 21(5), 730–743, Jun. 2003.

139. J. Adeane, W. Q. Malik, I. J. Wassell, and D. J. Edwards, Simple correlated channel model for ultrawideband multiple-input multiple-output systems, *IET Microwaves, Antennas & Propagation*, 1(6), 1177–1181, Dec. 2007.

140. T. Kaiser, F. Zheng, and E. Dimitrov, An overview of ultra-wide-band systems with MIMO, *Proc. IEEE*, 97(2), 285–312, Feb. 2009.

141. X. Zhang, D. P. Palomar, and B. Ottersten, Statistically robust design of linear MIMO transceivers, *IEEE Trans. Signal Process.*, 56(8), 3678–3689, Aug. 2008.

142. H. L. Van Trees, *Optimum Array Processing*. New York: John Wiley & Sons, Inc., 2002.

143. J. Proakis, *Digital Communications*. McGraw-Hill, 1995.

144. H. V. Poor and G. W. Wornell, *Wireless Communications: Signal Processing Perspectives*. Upper Saddle River, NJ: Prentive-Hall, 1998.

145. S. Alamouti, A simple transmit diversity technique for wireless communications, *IEEE J. Sel. Areas Commun.*, 16(8), 1451–1458, Oct. 1998.

146. V. Tarokh, N. Seshadri, and A. R. Calderbank, Space-time codes for high data rate wireless communication: performance criterion and code construction, *IEEE Trans. Inf. Theory*, 44(2), 744–765, Mar. 1998.

147. V. Tarokh, H. Jafarkhani, and A. R. Calderbank, Space-time block codes from orthogonal designs, *IEEE Trans. Inf. Theory*, 45(5), 1456–1467, Jul. 1999.

148. J. Akhtar and D. Gesbert, Extended orthogonal block codes with partial feedback, *IEEE Trans. Wireless Commun.*, 3(6), 1959–1962, Nov. 2004.

149. M. N. Hussin and S. Weiss, Extended orthogonal space-time block coded transmission with quantised differential feedback, in *Proc. Int. Symp. Wireless Commun. Syst.*, York, United Kingdom, Sep. 19–22, 2010, 179–183.

150. E. Bjornson and B. Ottersten, A framework for training-based estimation in arbitrarily correlated Rician MIMO channels with Rician disturbance, *IEEE Trans. Signal Process.*, 58(3), 1807–1820, Mar. 2010.

151. T. Kong and Y. Hua, Optimal design of source and relay pilots for MIMO relay channel estimation, *IEEE Trans. Signal Process.*, 59(9), 4438–4446, Sep. 2011.

152. B. Widrow and S. D. Stearns, *Adaptive Signal Processing*. Prentice Hall, 1985.

153. S. Haykin, *Adaptive Filter Theory*, 3rd ed. Prentice Hall, 1996.

154. S. Weiss, C. H. Ta, and C. Liu, A Wiener filter approach to the design of filter bank based single-carrier precoding and equalisation, in *Proc. IEEE Int. Symp. Power Line Commun. Applic.*, Pisa, Italy, Mar. 26–28, 2007, 493–498.

155. A. Scaglione, G. B. Giannakis, and S. Barbarossa, Redundant filterbank precoders and equalizers. I. Unification and optimal designs, *IEEE Trans. Signal Process.*, 47(7), 1988–2006, Jul. 1999.

156. ——, Redundant filterbank precoders and equalizers. II. Blind channel estimation, synchronization, and direct equalization, *IEEE Trans. Signal Process.*, 47(7), 2007–2022, Jul. 1999.

157. A. Scaglione, S. Barbarossa, and G. B. Giannakis, Filterbank transceivers optimizing information rate in block transmission over dispersive channels, *IEEE Trans. Inf. Theory*, 45(4), 1019–1032, Apr. 1999.

158. L. Hanzo, M. Münster, B. J. Choi, and T. Keller, *OFDM and MC-CDMA for Broadband Multi-User Communications, WLANs, and Broadcasting*. Wiley-IEEE Press, 2003.

159. K. F. Lee and D. B. Williams, A space-frequency transmitter diversity technique for OFDM systems, in *Proc. IEEE Global Telecom. Conf.*, 3, San Francisco, USA, Nov. 27–Dec. 1, 2000, 1473–1477.

160. H. Bölcskei and A. J. Paulraj, Space-frequency coded broadband OFDM systems, in *Proc. IEEE Wireless Commun. Netw. Conf.*, 1, Chicago, USA, Sep. 23–28, 2000, 1–6.

161. G. Cherubini, E. Eleftheriou, and S. Ölçer, Filtered multitone modulation for very high-speed digital subscriber lines, *IEEE J. Sel. Areas Commun.*, 20(5), 1016–1028, Jun. 2002.

162. M. G. Bellanger, G. Bonnerot, and M. Coudreuse, Digital filtering by polyphase network: Application to sample rate alteration and filter banks, *IEEE Trans. Acoust., Speech, Signal Process.*, 24(4), 109–114, Apr. 1976.

163. M. Bellanger, M. Renfors, T. Ihalainen, and C. A. F. da Rocha, OFDM and FBMC transmission techniques: a compatible high performance proposal for broadband power line communications, in *Proc. IEEE Int. Symp. Power Line Commun. Applic.*, Rio de Janeiro, Brazil, Mar. 28–31, 2010, 154–159.

164. B. Farhang-Boroujeny, OFDM versus filter bank multicarrier, *IEEE Signal Process. Mag.*, 28(3), 92–112, May 2011.

165. S. Geirhofer, L. Tong, and A. Scaglione, Time-reversal space-time coding for doubly-selective channels, in *IEEE Wireless Commun. Netw. Conf.*, 3, Las Vegas, USA, Apr. 3–6, 2006, 1638–1643.

166. S. Bendoukha and S. Weiss, Blind CM equalisation for STBC over multipath fading, *IET Electron. Lett.*, 44(15), 922–923, Jul. 2008.

167. A. Daas, S. Bendoukha, and S. Weiss, Blind adaptive equalizer for broadband MIMO time reversal STBC based on PDF fitting, in *Proc. Asilomar Conf. on Signals, Systems and Computers*, Pacific Grove, USA, Nov. 1–4, 2009, 1380–1384.

168. J. G. McWhirter, P. D. Baxter, T. Cooper, S. Redif, and J. Foster, An EVD algorithm for para-Hermitian polynomial matrices, *IEEE Trans. Signal Process.*, 55(5), 2158–2169, May 2007.

169. S. Redif, S. Weiss, and J. G. McWhirter, Sequential matrix diagonalization algorithms for polynomial EVD of parahermitian matrices, *IEEE Trans. Signal Process.*, 63(1), 81–89, Jan. 2015.

170. C. H. Ta and S. Weiss, A design of precoding and equalisation for broadband MIMO systems, in *Proc. Asilomar Conf. on Signals, Systems and Computers*, Pacific Grove, USA, Nov. 4–7, 2007, 1616–1620.

171. W. Al-Hanafy, A. P. Millar, C. H. Ta, and S. Weiss, Broadband SVD and non-linear precoding applied to broadband MIMO channels, in *Proc. Asilomar Conf. on Signals, Systems and Computers*, Pacific Grove, USA, Oct. 26–29, 2008, 2053–2057.

172. N. Moret, A. Tonello, and S. Weiss, MIMO precoding for filter bank modulation systems based on PSVD, in *Proc. IEEE Veh. Technol. Conf.*, Yokohama, Japan, May 15–18, 2011, 1–5.

173. L. Yonge, J. Abad, K. Afkhamie, L. Guerrieri, S. Katar, H. Lioe, P. Pagani, R. Riva, D. M. Schneider, and A. Schwager, An overview of the HomePlug AV2 technology, *J. Electric. Comput. Eng.*, vol. 2013, 1–20, 2013.

174. A. Pittolo, A. M. Tonello, and F. Versolatto, Performance of MIMO PLC in measured channels affected by correlated noise, in *Proc. IEEE Int. Symp. Power Line Commun. Applic.*, Glasgow, Scotland, Mar. 30–Apr. 2, 2014, 261–265.

175. A. Canova, N. Benvenuto, and P. Bisaglia, Receivers for MIMO-PLC channels: Throughput comparison, in *Proc. IEEE Int. Symp. Power Line Commun. Applic.*, Rio de Janeiro, Brazil, Mar. 28–31, 2010, 114–119.

176. M. Biagi, MIMO self-interference mitigation effects on power line relay networks, *IEEE Commun. Lett.*, 15(8), 866–868, Aug. 2011.

177. Z. Quan and M. V. Ribeiro, A low cost STBC-OFDM system with improved reliability for power line communications, in *Proc. IEEE Int. Symp. Power Line Commun. Applic.*, Udine, Italy, Apr. 3–6, 2011, 261–266.

178. R. Hashmat, P. Pagani, A. Zeddam, and T. Chonavel, MIMO communications for inhome PLC networks: Measurements and results up to 100 MHz, in *Proc. IEEE Int. Symp. Power Line Commun. Applic.*, Rio de Janeiro, Brazil, Mar. 28–31, 2010, 120–124.

179. A. Tomasoni, R. Riva, and S. Bellini, Spatial correlation analysis and model for in-home MIMO power line channels, in *Proc. IEEE Int. Symp. Power Line Commun. Applic.*, Beijing, China, Mar. 27–30, 2012, 286–291.

180. B. Nikfar and A. J. H. Vinck, Combining techniques performance analysis in spatially correlated MIMO-PLC systems, in *Proc. IEEE Int. Symp. Power Line Commun. Applic.*, Johannesburg, South Africa, Mar. 24–27, 2013, 1–6.

181. R. Hashmat, P. Pagani, T. Chonavel, and A. Zeddam, A time-domain model of background noise for in-home MIMO PLC networks, *IEEE Trans. Power Delivery*, 27(4), 2082–2089, Oct. 2012.

182. B. Nikfar, T. Akbudak, and A. J. H. Vinck, MIMO capacity of class a impulsive noise channel for different levels of information availability at transmitter, in *Proc. IEEE Int. Symp. Power Line Commun. Applic.*, Glasgow, Scotland, Mar. 30–Apr. 2, 2014, 266–271.

183. F. Versolatto and A. M. Tonello, A MIMO PLC random channel generator and capacity analysis, in *Proc. IEEE Int. Symp. Power Line Commun. Applic.*, Udine, Italy, Apr. 3–6, 2011, 66–71.

184. N. Weling, A. Engelen, and S. Thiel, Broadband MIMO powerline channel emulator, in *Proc. IEEE Int. Symp. Power Line Commun. Applic.*, Glasgow, Scotland, Mar. 30–Apr. 2, 2014, 105–110.

185. H. A. Latchman, S. Katar, L. W. Yonge, and S. Gavette, *HomePlug AV and IEEE 1901: A Handbook for PLC Designers and Users*. Wiley-IEEE Press, 2013.

186. S. Lin and D. J. Costello Jr., *Error Control Coding: Fundamentals and Applications*. Englewood Cliffs, NJ: Prentice Hall Inc., 1983.

187. KNX Powerline PL 110, KNX Association, Tech. Rep., Jun. 2007. [Online]. Available: http://www.knx.org/media/docs/KNX-Tutor-files/Summary/KNX-Powerline-PL110.pdf

188. SIEMENS AMIS CX1-Profil (Compatibly/Consistently Extendable Transport Profile V.1) Layer 1-4 (in German), SIEMENS, Tech. Rep., Sep. 2011. [Online]. Available: http://quad-industry.com/titan_img/ecatalog/CX1-Profil_GERrKW110928.pdf

189. G. Dickmann, digitalSTROM®: A centralized PLC Topology for Home Automation and Energy Management, in *Proc. IEEE Int. Symp. Power Line Commun. Applic.*, Udine, Italy, Apr. 3–6, 2011, 352–357.

190. PRIME Specification revision v1.4, Specification for powerline intelligent metering evolution, Oct. 2014. [Online]. Available: http://www.prime-alliance.org/wp-content/uploads/2014/10/PRIME-Spec_v1.4-20141031.pdf

191. J. M. Domingo, S. Alexandres, and C. Rodriguez-Morcillo, PRIME performance in power line communication channel, in *Proc. IEEE Int. Symp. Power Line Commun. Applic.*, Udine, Italy, Apr. 3–6, 2011, 159–164.

192. PLC G3 physical layer specification, Électricité Réseau Distribution France, Specification.

193. K. Razazian, M. Umari, and A. Kamalizad, Error correction mechanism in the new G3-PLC specification for powerline communication, in *Proc. IEEE Int. Symp. Power Line Commun. Applic.*, Rio de Janeiro, Brazil, Mar. 28–31, 2010, 50–55.

194. M. Hoch, Comparison of PLC G3 and PRIME, in *Proc. IEEE Int. Symp. Power Line Commun. Applic.*, Udine, Italy, Apr. 3–6, 2011, 165–169.

195. T. Shongwe and A. J. H. Vinck, Interleaving and nulling to combat narrow-band interference in PLC standard technologies PLC G3 and PRIME, in *Proc. IEEE Int. Symp. Power Line Commun. Applic.*, Johannesburg, South Africa, Mar. 24–27, 2013, 258–262.

196. Unified high-speed wire-line based home networking transceivers – system architecture and physical layer specification, ITU-T, Recommendation G.9960, 2011.

197. S. Mudriievskyi and R. Lehnert, Performance evaluation of the G.hn PLC PHY layer, in *Proc. IEEE Int. Symp. Power Line Commun. Applic.*, Glasgow, Scotland, Mar. 30–Apr. 2, 2014, 296–300.

198. IEEE P1901, Standard for broadband over power line networks: Medium access control and physical layer specifications. [Online]. Available: http://grouper.ieee.org/groups/1901/index.html

199. A. J. Felström and K. S. Zigangirov, Time-varying periodic convolutional codes with low-density parity-check matrix, *IEEE Trans. Inf. Theory*, 45(6), 2181–2191, Sep. 1999.

200. H. C. Ferreira and A. J. H. Vinck, Interference cancellation with permutation trellis codes, in *Proc. IEEE Veh. Technol. Conf.*, Boston, MA, USA, Sep. 24–28, 2000, 2401–2407.

201. H. C. Ferreira, A. J. H. Vinck, T. G. Swart, and I. de Beer, Permutation trellis codes, *IEEE Trans. Commun.*, 53(11), 1782–1789, Nov. 2005.

202. T. G. Swart, I. de Beer, H. C. Ferreira, and A. J. H. Vinck, Simulation results for permutation trellis codes using M-ary FSK, in *Proc. IEEE Int. Symp. Power Line Commun. Applic.*, Vancouver, Canada, Apr. 6–8, 2005, 317–321.

203. T. G. Swart, A. J. H. Vinck, and H. C. Ferreira, Convolutional code search for optimum permutation trellis codes using M-ary FSK, in *Proc. IEEE Int. Symp. Power Line Commun. Applic.*, Pisa, Italy, Mar. 26–28, 2007, 441–446.

204. T. G. Swart, Distance-preserving mappings and trellis codes with permutation sequences, Ph.D. dissertation, University of Johannesburg, Johannesburg, South Africa, Aug. 2006.

205. L. Cheng and H. C. Ferreira, Time-diversity permutation coding scheme for narrow-band power-line channels, in *Proc. IEEE Int. Symp. Power Line Commun. Applic.*, Beijing, China, Mar. 27–30, 2012, 120–125.

206. L. Cheng, T. G. Swart, and H. C. Ferreira, Adaptive rateless permutation coding scheme for OFDM-based PLC, in *Proc. IEEE Int. Symp. Power Line Commun. Applic.*, Johannesburg, South Africa, Mar. 24–27, 2013, 242–246.

207. Y. M. Chee, H. M. Kiah, P. Purkayastha, and C. Wang, Importance of symbol equity in coded modulation for power line communications, *IEEE Trans. Commun.*, 61(10), 4381–4390, Oct. 2013.

208. N. Andreadou and F.-N. Pavlidou, QC-LDPC codes and their performance on power line communications channel, in *Proc. IEEE Int. Symp. Power Line Commun. Applic.*, Dresden, Germany, Mar. 29–Apr. 1, 2009, 244–249.

209. ——, Mitigation of impulsive noise effect on the PLC channel with QC-LDPC codes as the outer coding scheme, *IEEE Trans. Power Delivery*, 25(3), 1440–1449, Jul. 2010.

210. Z. Liu, K. Peng, W. Lei, C. Qian, and Z. Wang, Rate-compatible QC-LDPC codes design in powerline communication systems, in *Proc. IEEE Int. Symp. Power Line Commun. Applic.*, Beijing, China, Mar. 27–30, 2012, 126–131.

211. N. Andreadou and A. M. Tonello, Short LDPC codes for NB-PLC channel with a differential evolution construction method, in *Proc. IEEE Int. Symp. Power Line Commun. Applic.*, Johannesburg, South Africa, Mar. 24–27, 2013, 236–241.

212. G. Prasad, H. A. Latchman, Y. Lee, and W. A. Finamore, A comparative performance study of LDPC and Turbo codes for realistic PLC channels, in *Proc. IEEE Int. Symp. Power Line Commun. Applic.*, Glasgow, Scotland, Mar. 30–Apr. 2, 2014, 202–207.

213. D.-F. Tseng, T.-R. Tsai, and Y. S. Han, Robust turbo decoding in impulse noise channels, in *Proc. IEEE Int. Symp. Power Line Commun. Applic.*, Johannesburg, South Africa, Mar. 24–27, 2013, 230–325.

214. E. C. Kim, S. S. Il, J. Heo, and J. Y. Kim, Performance of double binary turbo coding for high speed PLC systems, *IEEE Trans. Consumer Electron.*, 56(3), 1211–1217, Aug. 2010.

215. N. Andreadou and F.-N. Pavlidou, Performance of array codes on power line communications channel, in *Proc. IEEE Int. Symp. Power Line Commun. Applic.*, Jeju Island, Korea, Apr. 2–4, 2008, 129–134.

216. ——, PLC channel: Impulsive noise modelling and its performance evaluation under different array coding schemes, *IEEE Trans. Power Delivery*, 24(2), 585–595, Apr. 2009.

217. N. Andreadou and A. M. Tonello, On the mitigation of impulsive noise in power-line communications with LT codes, *IEEE Trans. Power Delivery*, 28(3), 1483–1490, Jul. 2013.

218. H. Gacanin, Inter-domain bi-directional access in G.hn with network coding at the physical-layer, in *Proc. IEEE Int. Symp. Power Line Commun. Applic.*, Beijing, China, Mar. 27–30, 2012, 144–149.

219. J. Bilbao, A. Calvo, I. Armendariz, P. M. Crespo, and M. Médard, Reliable communications with network coding in narrowband powerline channel, in *Proc. IEEE Int. Symp. Power Line Commun. Applic.*, Glasgow, Scotland, Mar. 30–Apr. 2, 2014, 316–321.

6

Medium Access Control and Layers Above in PLC

J. A. Cortés, S. D'Alessandro, L. P. Do, L. Lampe, R. Lehnert, M. Noori, and A. M. Tonello

6.1 Introduction

Power lines provide a shared communication channel so that medium access control (MAC) and protocols that reside above the physical layer have to be implemented. In this chapter, such protocols are described.

MAC principles are summarized in Section 6.2. In Section 6.3, a power line communication (PLC) network organized in multiple cells is considered. In such a scenario, media access schemes and resource sharing strategies among the cells for broadband PLC are discussed. Considering the fact that different quality-of-service (QoS) features have to be supported, the use of hybrid media access control protocols integrating contention and contention-free access is eventually advocated for. This shows close similarities to broadband wireless communication systems that also operate over a shared medium.

The problem of allocating resources in a single cell multiple-user network is considered in Section 6.4. The achievable rate of the multiple-access and the broadcast channel is discussed considering both time division multiple access (TDMA) and frequency division multiple access (FDMA). The resource allocation problem is addressed for the specific PLC channel that exhibits a periodically time variant response with cyclostationary noise.

Finally, in Section 6.5 the usage of cooperative communication mechanisms is presented. Such mechanisms can be implemented through a tight connection with the physical layer. They include cooperative coding and relaying techniques.

Power Line Communications: Principles, Standards and Applications from Multimedia to Smart Grid, Second Edition.
Edited by Lutz Lampe, Andrea M. Tonello, and Theo G. Swart.
© 2016 John Wiley & Sons, Ltd. Published 2016 by John Wiley & Sons, Ltd.

6.2 MAC Layer Concepts

As discussed in previous chapters, the power line channel exhibits variability, as a consequence of network topology changes and load variations, noise and interference. Some parts of the network may be segmented in terms of signal propagation due to the presence of circuit breakers and transformers. In general, large networks may require the usage of repeaters to provide full coverage. Taking all of this into account, the allocation of channel capacity to the contending requests of the transmit stations on the same medium becomes a challenge. The momentarily available bandwidth becomes time dependent due to the time varying signal-to-noise-ratio experienced at the receiver and location dependent due to the locally induced disturbances and impedance changes. Due to the distance dependent attenuation of signals, not all stations can hear each other. This results in the so-called hidden station problem, which may complicate the resource allocation mechanism.

From the above description, we have to send communication data over a medium, whose capacity is variable over time and that depends on the distance between the modems. In a multicarrier OFDM-based system, we can allocate in the frequency domain the OFDM carriers (using an FDMA approach) and in the time domain the time slots (using a TDMA approach). N time slots constitute a frame. F carriers are available. So we have a two-dimensional block of $M \times N_{AU}$ allocation units (AU), see Figure 6.1. Each AU represents a (set of) carriers(s) and a (a set of) time slot(s). Due to the time-variant channel and the attenuation between different PLC-modems, the applicable modulation order of an AU varies. Therefore, the capacity of an AU varies accordingly.

To realize a transmission with a given data rate between two modems, we have to allocate i AUs, $1 \le i \le N_{AU}$. If the number of required AUs is $i \ge N_{AU}$, the transmission request cannot be satisfied. This may happen when a high data rate transmission cannot be established among far away modems, since low order modulation is deployed.

The allocation can be done for each frame. It is the task of the scheduler to allocate requests according to some rules and to fulfill certain conditions. Scheduling can be done by one entity. This is the central scheduler case. Ideally, the central scheduler knows all requests and the channel state. Therefore, an optimal resource allocation is possible. Some PLC systems implement a distributed scheduler. For example, the well-known Carrier Sense Multiple Access (CSMA) principle may be used. As it is uncoordinated, collisions may occur, which waste part of the precious channel capacity. As an improvement CSMA/CA (CSMA with collision

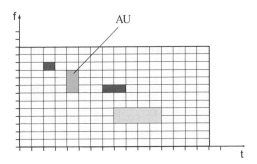

Figure 6.1 Two-dimensional capacity allocation.

avoidance) is usually implemented, as for instance in the broadband PLC standard IEEE P.1901. CSMA/CA cannot avoid collisions completely. A centralized scheduler can allocate more efficiently at the expense that all state information has to be made available.

PLC signals are attenuated along the link. Therefore, data cannot be transmitted over long distances. At a certain distance, the lowest modulation index is just sufficient to correctly decode the signal. For the modem behind, the signal cannot be decoded; it is just noise. We therefore have a *hidden station* case, which we already know from mobile communication systems. If we want to transmit over a larger distance, an intermediate modem has to repeat the signal. This means that the the received data chunk in a set of AUs has to be retransmitted and forwarded on another hop in the next frame and eventually in another set of AUs with possibly different modulation indices. As the channel is time variant, the achievable distance until we have to repeat the signal is also variable over time. That is why the topology of a network may change over time. This may occur due to local effects such as impedance changes but also more globally due to changes in the noise floor, e.g. daily variations.

The allocation of capacity to each request may be controlled by rules and goals:

- A flow may have a priority.
- Flows (of the same priority) shall be treated fair.
- Fairness may be expressed w.r.t. data rate, allocated AUs, or another parameter or combinations of all.
- Fairness can be seen locally (at the scheduler in a modem), end-to-end for each flow, or globally for all flows in the entire network.

The head elements in the queues of all active flows are considered to be scheduled in the next frame. The element may be a layer 3 IP packet, a segment from a convergence layer, etc. If just the head element is scheduled, we have a so-called *gated service*. In principle, queues may be emptied before considering other queues for scheduling. This so-called *exhaustive service* will be fair only w.r.t. the data rate at a single, local scheduler.

Very often the channel capacity is influenced by local effects, e.g. the noise floor comes from local disturbers. In this case an overall view for doing an optimal scheduling may not be required. A local or regional scheduler may perform near the optimum but with much less complexity. This allows to define PLC regions, which are managed by their own schedulers.

In the following sections, PLC regions are named *PLC cells*. As traffic varies statistically, transmission capacity may have to be temporarily moved to neighbor cells to cope with a momentarily high load. Mechanisms for implementing this are discussed and evaluated.

6.3 Protocols for Different Power Line Communications Applications and Domains

6.3.1 Transmission Resources Sharing Between Multiple PLC Cells

To cover a large area, a network may be configured into multiple smaller PLC network clusters, called PLC cells. In this section, the focus is given to the resource allocation among PLC cells whereas the problem of allocating resources within a cell is treated in Section 6.4. The resource allocation among PLC cells is also defined by standards, e.g. by HPAV or G3-PLC for co-existence of multiple logical networks or cells.

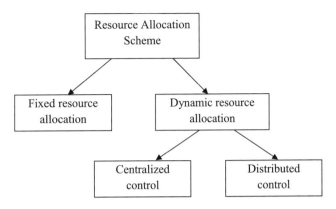

Figure 6.2   Classification of the resource allocation scheme for cellular network.

The cellular principle known from mobile networks is also investigated for PLC systems. In case a PLC network consists of several adjacent cells, in which parallel transmissions in different cells may interfere, a cell should not be allowed to utilize the whole transmission resources. To each cell should only be allocated a subset of the available transmission resources. In order to achieve this, the transmission resources are partitioned into a set of the so-called inter-cell resource units, or channels, to be allocated to each cell. Considering the interference conditions between the adjacent cells in a network and time variant of the channel conditions, a method to allocate the limited transmission resources into each cell is presented and analyzed in detail. Parts of this section are referred to the works in [1] and [2].

As known from mobile networks, e.g. GSM systems, the channel assignment techniques may be classified into two categories, fixed and dynamic allocation methods. Figure 6.2 presents the classifications of the channels allocation methods. Some studies have been conducted and solutions proposed for this problem in the PLC networks as well, e.g. in [3] and [4], available channel allocation solutions which can be applied to multiple cell PLC networks are analyzed.

6.3.1.1   Fixed Channel Allocation between PLC Cells

This method requires full knowledge of the network topology and the traffic demands from the cells in the entire network. In this method, the resources, fragmented into channels, are assigned to individual cells with the goal to operate over a long period. Since complete knowledge is required, the allocation is normally performed when the network is planned. The allocation can also be done by optimization algorithms [5], [6]. There are investigations that have proposed to share the transmission resources between the cells in PLC network in a statistical fashion; e.g. [3] has proposed to solve the channel allocation problem when planning the network.

One example of the solution for the Fixed Channel Allocation (FCA) is shown in Figure 6.3. The network depicted consists of four cells ($\mathbf{C} = \{C_i\}$, $i = 1, \ldots, 4$) and assumes that five channels ($\mathbf{F} = \{F_i\}$, $i = 1, \ldots, 5$) are available. As presented in this figure, transmissions in cell C_1 and cell C_2 may interfere with each other; therefore they are not assigned the same channels. Cell C_3 is located far from cell C_1, hence C_3 reuses channel F_1 which is already used in C_1. This allocation of channels in each cell may be optimal in the time of network

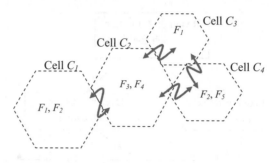

↯ Possible interference

Figure 6.3   PLC cells with fixed channel allocation scheme.

planning, but the fixed number of allocated channels leads to unfairness between the users in different cells when the traffic demand from users varies.

The advantage of this approach is that several calculation methods can be applied including exhaustive search or heuristic algorithms because the allocation algorithm does not have to run in real time. Therefore, the allocation for the planned traffic may be optimal. However, the traffic demand, as input for the problem solver, is a forecast from each cell and may differ from the actual traffic demand during run-time. Thus, the applied channel allocation solution is no longer optimal during run-time.

6.3.1.2   Dynamic Channel Allocation between PLC Cells

Dynamic Channel Allocation (DCA) is a method in which the channels are not permanently assigned to a certain cell, but are allocated to the cells according to the dynamic traffic demand from them. The channels are resource units that can be moved from one cell to another during run-time. There are several studies on the algorithms for dynamically solving the channel allocation problem, but most of them were originally proposed for wireless cellular system, e.g. for GSM in [7] or [8]. The dynamic channel allocation (DCA) can be implemented centralized or distributed. In the centralized approach, the requests for channels have to be transmitted from all cells to a common central controller. The controller estimates the overall network traffic demand, assigns channels to the cells, and informs the cells about the allocated channels. Using a distributed approach, all of the cells take part in the decision to determine the channel allocation in the network. For the master-slave model of a cell, each base station (BS) has to make decisions based on its own knowledge and the information from its neighbor cells.

6.3.1.2.1   *Review of Methods for Solving the Dynamic Channel Allocation Problems*
Several algorithms have been introduced for solving the DCA problem. For cellular GSM networks, a number of studies for the DCA have been conducted, e.g. as reviewed in [9] and [10]. In addition, the particular characteristics of the GSM network have been taken into account; for example, [11] proposes to dynamically distribute frequency bands between the cells by the so-called DFCA (Dynamic Frequency and Channel Assignment), while [12] and

[13] propose algorithms for solving DCA between the cells by means of the TDMA scheme for mobile communications systems.

Among some well-known DCA algorithms like the autonomous reuse partitioning scheme [14] and minimum interference [15], various solutions have also been proposed using other characteristics of the GSM systems. In [16], the authors use the geometric information in the strategy while solving the DCA with the bursty user mobility. The idea proposed in [17] and [18] is to solve the DCA with a distributed approach. The authors in [19] define and evaluate the DCA solutions by a cost function based on the provided services, the traffic load, and interference between the cells.

In other areas of telecommunications, some studies to solve DCA problems have also been conducted. The proposed solutions are based on the characteristics of the communication technologies, e.g. the solutions have been proposed by [20] for Ethernet Passive Optical Networks (EPONs).

Game theory has also been applied to solve the DCA problem. The methods of game theory and their applications in the communication areas have been reviewed and analyzed in [21]. It has been used, e.g. in [22] for power allocation in MIMO (Multiple Input Multiple Output) channels and for power control in CDMA (Code Division Multiple Access) systems in [23]. Generally, each game consists of several players that try to achieve certain objectives, called payoff [24]. In a problem which is defined as a game, the solution is conducted when the players decide to perform certain actions at the same time or consecutively. The payoff of the solution depends on the combination of the actions that the players performed. The authors in [25] use dynamic game theory and in [26] use cooperative games for spectrum sharing. [27] applies game theory to solve the DCA problem for the multiple cells wireless networks.

6.3.1.2.2 *Centralized Approach*

For the centralized approach, a central controller and its connections to the base stations in all cells are necessary, as depicted in Figure 6.4. This form of organization for the central controller is possible if the network is owned by one or a cooperating network service provider.

The information is periodically exchanged between the cells and the central controller. All the channel information, e.g. channel status, capacity of each channel, is stored in a

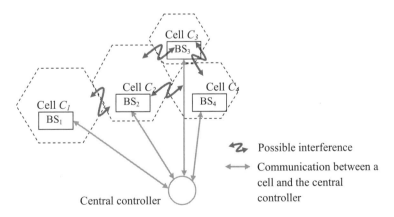

Figure 6.4 Centralized dynamic channel allocation.

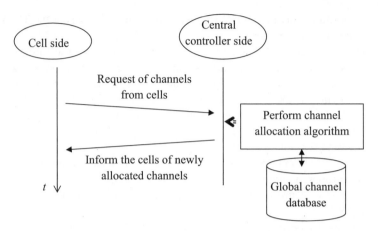

Figure 6.5   MSC of the process for centralized dynamic channel allocation with the central controller.

database of the central controller. The network structure, e.g. described by the interference between the cells is also stored in this unit. The cells regularly send their channel requests to the central controller. Based on the available channel information, the controller estimates the overall network traffic demand, calculating channels in each cell, and informs the cells of the new allocated channels. A message sequence chart (MSC) demonstrating the process of the centralized dynamic channel allocation in the central controller is presented in Figure 6.5. This method requires a relatively fast decision and response from the central controller [28].

6.3.1.2.3   *Distributed Approach*
A distributed approach becomes necessary when the centralized dynamic channel allocation is not applicable, e.g. due to large-scale networks with too many cells. In the situation of a PLC network, the cells may have different owners or service providers, meaning that organizing a common central station for the entire PLC network may not be possible. In the distributed approached, each base station has to participate in the decision for the allocated channels of its cell and its neighbor cells. This has to be performed without a common central station. In this case, the base stations are not connected to a common unit but only to its neighbor cells to exchange the channel information and to make decisions on the channel allocation. A common algorithm with a set of rules has to be applied to each cell allowing them to collaborate.

Let us consider an example network consisting of four cells as presented in Figure 6.6. The process for the channel allocation in this situation has to be made locally at each base station, as depicted in Figure 6.7. Cell C_2 is a neighbor of the cells C_1 and C_4, therefore, its channel decision will affect the channels used by the cells C_1 and C_4. If cell C_2 seizes channels F_3 and F_4, C_1 and C_4 should only use the remaining channels in order to avoid possible interference. However, the decisions of the cells C_1 and C_4 have to be made regarding the channels in use in the cell C_3, otherwise interference between cell C_1 or C_4 and C_3 may occur.

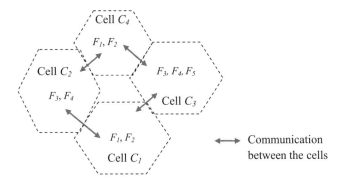

Figure 6.6 Multiple cells with distributed channel allocation scheme.

6.3.2 Inter-PLC Cell Transmission Resource Sharing

Due to interference between adjacent cells, both PLC specifications provided by the OPERA consortium [29] and IEEE-P1901 working group [30], propose to subdivide the transmission resources into smaller units to be distributed into the cells. However, these specifications are only aimed to deal with the co-existence between access cells and inhome cells, following a fixed allocation scheme. This does not address the dynamic traffic request from different cells in the overall PLC network. Previous studies have also proposed to allocate the transmission resources to the cells in the PLC network by utilizing an optimization problem solution. To solve these optimization problems, however, full knowledge about the traffic requirement throughout the entire network is needed. Therefore, the problem can only be solved during the phase of network design, e.g. using heuristic algorithms for channel allocation problems, see [3] and [31]. The results are therefore a fixed channel allocation solution, meaning that each cell is allocated a number of fixed channels and these allocated channels are not altered during

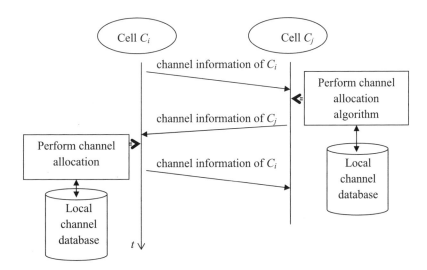

Figure 6.7 MCS of the distributed dynamic channel allocation scheme.

run-time. This kind of solution leads to unfairness between users in different cells, when the traffic demand from users varies or the cell structure is changed.

On the other hand, there are several studies to solve the channel allocation problem dynamically, but most of them were originally proposed for the GSM cellular system [9], [10], [13]. These available dynamic channel allocation methods, however, cannot be applied to the PLC directly, due to the differences between the systems. The main reason is that it is not feasible to organize a central unit for controlling the complete PLC network when the network is composed of several clusters operated by different service providers. Therefore, the re-organization of the PLC users between cells is not always possible, similar to the hand-over procedure in the case of the GSM system. Therefore, a novel negotiation strategy to be applied to every base station (BS) in the PLC network is proposed. The strategy allows all BSs to co-operate and to share the transmission resources dynamically during run-time and in a distributed fashion. This strategy allows every cell to maximize the reserved channels, while minimizing the channel interferences, and maintaining the fairness between users in different cells. The sharing methods have to take several network parameters into account, e.g. size and resource requirement from each cell, interferences between cells, as well as the number of available channels. In the investigations that follow, the reference networks used are based on the study of the electricity networks from the European Framework 7 project OPERA [29].

6.3.2.1   Interferences

Several interference sources are present in PLC, see [32]. Generally, when the communication signal is transmitted over the powerline, it can interfere with and be interfered by other signals, e.g. from radio signals or from the other PLC transmissions. Since the effect of the noise from the foreign systems is not avoidable, this investigation is interested in reducing the interferences caused by the PLC transmissions themselves. There are two kinds of interferences between the transmitting signals over the PLC networks, namely in-line interference and in-space interference [3], [32]. With the PLC cellular concept, the concern is to reduce the possibility of interferences between the transmissions in adjacent cells. Figure 6.8 demonstrates the potential of in-line and in-space interferences caused by the transmissions in two neighbor cells in a selected segment of a PLC network.

- In-line interferences occur when there is a power line used by both cells. E.g. in the line between two street cabinets in which BS1 and BS2 are installed. These two cells become neighbors. In this case, the transmission from a network device to one cell will be disturbed by the neighbor's if they use the same frequency band at the same time. It is a kind of collision between transmission signals. Therefore, these network devices from both the cells have to use different frequency bands or time slots for their respective transmissions.
- In-space interferences occur when two power lines carrying PLC signals are physically installed near each other. The cause of the interference is the electromagnetic field that one power line cable creates over the air of another cable. In this case, both of the power line cables passively work as antennas emitting and receiving the signal to and from the other. The in-space interference, which depends on the wiring of the network, causes fewer problems than the in-line interference when considering the actual locations of the neighboring cells [3], [32].

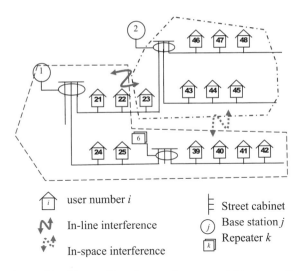

user number *i*   Street cabinet

In-line interference   Base station *j*

                                                   Repeater *k*

In-space interference

Figure 6.8   Types of interferences between neighboring cells.

The avoidance of the interference has to be organized by the master station when it does scheduling. If there is potential interference in the link to a station, the transmission in this link will be organized in a different time slot or frequency band of the interfering cell. The other links, which are not affected by any interference from the neighbor cells, can freely use all available time slots and frequency bands. In the worst case, if all the links in a cell are interfering with the neighbor cells, all the transmissions have to be scheduled in the time slots and frequency bands that are not used by the neighbor cells.

6.3.2.2   Channel Organization

We define a channel as a unit of transmission resources which can be allocated to a cell. The PLC systems are assumed to use Orthogonal Frequency Division Multiplexing (OFDM) modulation as specified in the PLC specifications [29], [30], [33]. Generally, a channel can be defined as a group of carriers, a repetition of a time slot in each time frame, or a hybrid of time slot and frequency carriers. Advantages and disadvantages of the channel organization are analyzed.

- A channel may be defined as repetition of time slots in each time frame, as presented in Figure 6.9. In this kind of channel organization, each time frame is divided into time slots. Each time slot may consist of several OFDM symbols. The advantage of this channel organization is that the entire frequency band is assigned to a cell at an instant of time to avoid frequency selective noise or notching in some specifically selected frequencies. Suppression of the signal in some carriers drop the transmission capacity of some specific frequency bands. However, when a cell has reserved some time slots that are not continuous in a time frame, the fragmentation of the transmission time increases the complexity of the later scheduling for users inside a cell. This may also reduce the efficiency of the resource utilization because multiple transmissions in different time slots of a user require additional

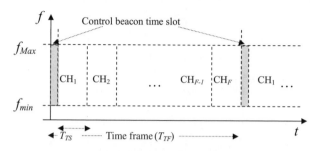

Figure 6.9   Channels as time slots.

overhead. This system also requires time slot synchronization between the neighbor cells. Based on the accuracy of the synchronization, a guard interval between the time slots may be necessary.

- A channel may be defined as group of carriers, as presented in Figure 6.10. In contrast to the previous case, this channel organization allows each cell to be reserved one or several groups of carriers in an entire time frame. With the utilization of the whole time frame, realization of sensitive time service which requires short transmission delay is feasible. The access to the entire time frame requires a lower calculation complexity for the scheduling, since each cell is allowed a full flexibility to access the medium during run-time. For supporting different PLC specifications, each cell can utilize a particular modulation scheme inside the reserved frequency channels. Main disadvantage of this channel organization is that it is too complex to include channel quality of each channel in all the cells into the inter-cell channel allocation process. This is because a channel may be scheduled in a cell in several intra-cell connections, each having a different channel quality.

- A channel is defined as group of time slots and carriers. This channel organization has the advantages and disadvantages of both previous channel organizations. It allows a full flexibility in both time and frequency dimensions for the resource allocation problem. However, it also increases the complexity of the scheduler inside each cell which may not applicable for the PLC system in reality.

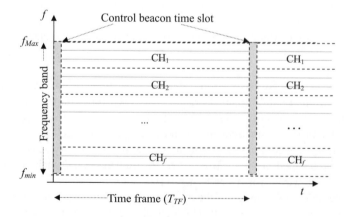

Figure 6.10   Channels as groups of carriers.

With the analysis above, to achieve the lowest scheduling complexity, a channel should be defined as a group of carriers. The organization of the channel as inter-cell resource units in the time frame is presented in Figure 6.10. If a channel consists of many carriers, there are only few channels remaining for sharing between the cells. This may lead to a low flexibility when allocating the channels to the cells. On the other hand, if there are a high number of channels, because a channel is a small group of carriers, the information which is used for indexing the channels and which has to be exchanged between the cells is large.

Considering a multi-cell PLC network which has $N_{\text{All}}^{\text{Carrier}}$ available carriers: if each channel is a group of $N_{1\text{Channel}}^{\text{Carrier}}$, the number of channels F is

$$F = \frac{N_{\text{All}}^{\text{Carrier}}}{N_{1\text{Channel}}^{\text{Carrier}}} \tag{6.1}$$

However, the proposed scheduling operates with channels as inter-cell resource; therefore it is independent from the divisions of the frequency and time domains.

The proposed inter-cell resource allocation provides the allocation of the transmission resources to each PLC cell. The task of channel allocation means to reserve channels to each cell in the network. Based on the exchange of the resource utilization and traffic requests with neighboring cells, a cell adjusts its transmission capacity reservation accordingly. Methods for transmitting information between neighboring cells and criteria for the adjustment of the transmission resources are proposed and analyzed. The performance of the proposed scheme is evaluated based on the network reference models.

In summary for this multi-cell PLC network, each PLC cell has one base station to provide communication services to all users and it is connected to the backbone network. The entire PLC network operates with limited transmission resources. When the cells are neighboring, the transmissions in a cell may produce interference to the transmissions of other cells. This means that neighboring cells cannot simultaneously utilize the same carriers. Another characteristic of the network is that the resource requirements from each cell are dynamically changing, e.g. users turn on or turn off during run-time. Therefore, a dynamic channel allocation protocol between the cells is necessary. This section proposes a channel allocation protocol to dynamically organize the limited transmission bandwidth between multiple PLC cells, which runs distributed fashion. The mechanisms are investigated to show their ability to improve the utilization of the overall transmission resources, while maintaining fairness between users and between the cells.

6.3.3   Distributed Inter-PLC Cell Resource Allocation Protocol

6.3.3.1   Overview of PLC Network Structures

Based on the distribution of customers in the power networks, a multi PLC-cell network can be organized in a chain, a ring, or a meshed structure as derived from European power grid topologies proposed in [34]. One meshed network can be fragmented into chain, ring, and star parts. When there are several in-home PLC cells activated near an access PLC cell, they can also form a star network structure. Therefore, three logical multi-cell PLC network structures, namely chain, ring, and star, as presented in Figure 6.11, are considered for the investigation

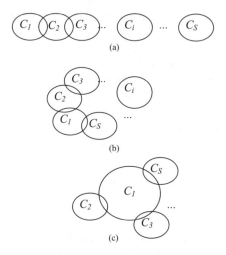

(a)

(b)

(c)

Figure 6.11   Three topologies of the investigated PLC networks: PLC cells in a (a) chain, (b) ring and (c) star structure.

of the resource allocation protocol. Each PLC network consists of S PLC cells, from C_1 to C_S. As previously mentioned, this section focus on the resource allocation among different cells. The resource allocation problem within a given cell is considered in Section 6.4.

6.3.3.2   Resource Unit Definition and Requirement

We define a channel as a group of carriers to be used for the transmission. A PLC system contains of several channels which are allocated to the PLC cells. Similar to the most of the other communication systems, the transmissions in the PLC networks are organized in the so-called time frames (TF) each has a constant duration T_{TF}. A time frame starts with a control beacon time slot, used for the synchronization between the PLC cells. There are F channels, from CH_1 to CH_F, for sharing between PLC cells in the available frequency band $[f_{min}; f_{max}]$. The capacity of a channel varies based on the location of the connection and the frequency selective noise from the environment.

6.3.3.3   Resource Utilization in a PLC Network

Considering the interference model in [35], a neighborhood is formed when the cells share the same power lines. The adjacent cells cannot use the same channels because of possible interference. When the cells are not neighboring, they are allowed to use the same channels. It is also assumed that the channel reuse inside a cell is not possible due the scheduler of this PLC cell, therefore all the network nodes in a cell have to use different channels for their transmissions comparing to its neighboring cell. Each PLC cell uses the allocated channels for further sharing between its users. It is assumed that all the allocated channels are utilized by users in the cell; therefore the resource utilization in a cell is defined as the number of allocated channels in that cell.

6.3.3.4   Description of the Transmission Resource Allocation Protocol

In order to adapt to the characteristics of the PLC network, a common set of actions, which form a so-called distributed inter PLC-cell resource allocation protocol assigned to every base station (BS), is proposed. The protocol is based on exchanging the channel information between neighboring cells. The exchanged information is processed by each BS individually. Based on the internal status and the information received, each BS may decide if it has to release or seize channels.

Using the principle presented in a previous study [36], the information to be exchanged between the cells is carried by the so-called channel announcement message (CA-Msg). This message is predefined and can be sent and received by all network nodes. A CA-Msg is generated and sent out from the base station of each cell regularly. The objectives of the protocol are realized by the functions inside the resource decision procedure. By executing this procedure, each BS tries to seize as many channels as possible to increase the resource utilization in the sub procedure called 'seize' procedure. The other procedure named 'release' is used for maintaining the fairness. It forces a cell to release channels if this cell has higher number of channels per user than its neighbors.

6.3.3.5   Communication Between the Base Stations

Transmission between the PLC cells is organized with the principle presented in the works presented in [36] and [2]. The specified message which is used for communicating among the PLC cells is simple to support for large PLC network. The message consists of basic status and channel information of the sending cell, as presented in Figure 6.12, in which:

- *AI*: for identifying if a PLC cell is access cell or in-home cell (1 Byte)
- *ID*: for identification of the sending base station (6 Bytes)
- *N*: for carrying the number of current active users (1 Byte)
- *H*: for the number of neighbor cells (1 Byte)
- Ch($\{f_i\}$): list of channels, in which each 2 bits for a channel f_i, the value of Ch(f_i) is:
  - 0: if this channel is free
  - 1: channel is allocated for this cell
  - 2: channel is used by one neighbor cell
  - 3: channel is used by more than one neighbor.

For informing all neighboring cells about its existence and channel utilization, each PLC cell sends its CA-Msg using a selected beacon time slot. When a cell decides to send out its

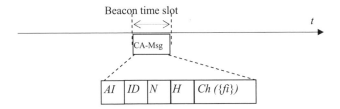

Figure 6.12   Channel Announcement Message (CA-Msg) format.

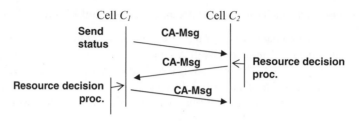

Figure 6.13   Exchange of CA-Msgs.

channel information, the BS of the cell generates and transmits its CA-Msg to its users during the data downlink transmission. All network elements in this cell send the CA-Msg the same next beacon time slot, as broadcasting of the CA-Msg from a cell to all possible neighboring cells. On the other hand, when a network node receives a CA-Msg from its neighbor, this message is forwarded to the BS in the data uplink transmission.

The transmission of the CA-Msg also requires transmission resources. With a short CA-Msg, one OFDM symbol is enough for carrying CA-Msg with the most robust modulation scheme. In accordance with the synchronization procedures in [37] and [29], it is also assumed that all PLC cells are synchronized, and the transmission of the CA-Msg at the same time from neighboring cells interfere each other. Therefore, each PLC cell chooses a beacon time slot to broadcast its CA-Mgs with a specific probability.

The estimation method of the probability in [36] to select a beacon time slot to send a CA-Msg from a BS is used. This probability of a cell C_j (with H_j neighboring cells) is calculated based on the set of the neighboring cells ($C_{nb}^{(j)}$),

$$p_j^{send} = \frac{1}{H_j} \cdot \sum_{C_k \in C_{nb}^{(j)}} \frac{1}{H_k + 1} \tag{6.2}$$

where C_k is a neighbor of C_j and itself has H_k neighbors.

6.3.4   Principles of Channel Reallocation Strategies

The resource allocation protocol defines the set of rules for the base stations. Following these rules, the allocated resources in a cell is dynamically adapt to the traffic changes in this cell and neighboring cells. When a channel is released from a cell and seized by another cell, this channel is virtually moved from a cell to the other cell in the network. The principle of the protocol presented in [36] is used, in which each base station executes 'resource decision procedure' to determine the number of reserved channels before sending out a CA-Msg. This procedure consists of two blocks, namely 'seize' and 'release' blocks. The illustration for the exchange CA-Msgs between two PLC cells and the flowchart of the resource decision procedure are presented in Figure 6.14. A cell which has higher number of channels per user than its neighbor releases channels by marking some of them as free channels. The released channel will be seized by its neighboring cell when the neighbor detects the free channel by receiving CA-Msg.

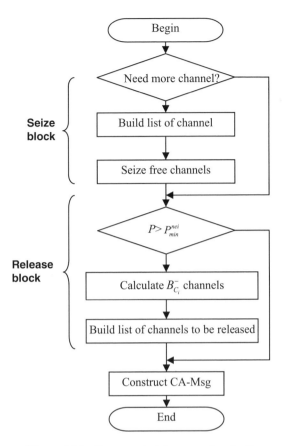

Figure 6.14 Resource decision procedure in C_i.

Each cell C_i has seized B_i channels, so the R-value of this cell is defined as the ratio of the number of allocated channels per number active users (N_i). R-value of a cell C_i is calculated by

$$R_i = \begin{cases} B_i/N_i, & \text{if } N_i > 0, \\ B_i, & \text{otherwise.} \end{cases} \tag{6.3}$$

The number of channels to be released (B_i^-) from a cell C_i is calculated based on its R-value (R_i) and the lowest R-value of its neighbor ($R_{\min}^{nei} = R_k$, R_k of the cell C_k). If the cell has R_i greater than $R_{\min}^{nei}$ ($R_i > R_{\min}^{nei}$), after releasing B_i^- channels from C_i and these channels are seized by C_k, then both of the cells should have the same R-value: $R_i^{\text{after\_release}} = R_k^{\text{after\_seize}}$. Therefore

$$\frac{B_i - B_i^-}{N_i} = \frac{B_k + B_k^-}{N_k} \tag{6.4}$$

where N_i and N_k are the numbers of active users in cell C_i and C_k, respectively. The number of channels to be released is an integer, therefore, B_i^- is rounded down by the floor function $\lfloor \cdot \rfloor$.

$$B_i^- = \left\lfloor \left| \frac{B_i N_k - B_k N_i}{N_i + N_k} \right| \right\rfloor \tag{6.5}$$

A channel f_t to be released by the cell C_i should be the channel which can be seized by the cell C_k. It means that, except of C_i, this channel is seized by no other neighbors of C_k. Therefore, this channel is found in the received CA-Msg from C_k in which $\mathrm{Ch}(f_t) = 2$.

6.3.5 The Evaluation Metrics

6.3.5.1 Throughput and Delivery Time of CA-Msg

The adaptation time of the channel reallocation to the change of the network depends on the delivery time of the CA-Msg between the cells. Therefore, the average delivery time of the CA-Msg between all the neighboring cells has to be evaluated. When C_j sends CA-Msg to C_i, who has H_i neighboring cells $\{C_k, k = 1, \ldots, H_i\}$ $(C_j \in \{C_k\})$, the receiving probability of CA-Msg at C_i from C_j is

$$p_{i,j}^{\mathrm{recv}} = p_j^{\mathrm{send}} \left(1 - p_i^{\mathrm{send}}\right) \prod_{k=1, k \neq j}^{H_i} \left(1 - p_k^{\mathrm{send}}\right) \tag{6.6}$$

The average duration for the delivery of one message from C_j to C_i is

$$D_{i,j} = 1/p_{i,j}^{\mathrm{recv}} \tag{6.7}$$

So, the average delivery time of CA-Msg in the network is

$$\overline{D_{Ch}} = \frac{1}{2n^{\mathrm{conn}}} \sum_{i=1}^{S} \sum_{j=1}^{S} c_{i,j} D_{i,j} \tag{6.8}$$

where $c_{i,j}$ indicates if there is a connection between cell C_i and cell C_j (1) or not (0); n^{conn} is the total number of end-to-end connections between two BSs of two cells and:

$$n^{conn} = \sum_{i=1}^{S-1} \sum_{j=i+1}^{S} c_{i,j} \tag{6.9}$$

6.3.5.2 Performance of the Allocation Protocol

6.3.5.2.1 Channel Reuse Factor

The channel reuse factor indicates the utilization gain of the resources in the network using the channel allocation protocol. The resource reuse factor is defined as the ratio of the sum of all the channels used in entire PLC network by the available number of channels (F). For the cell C_i, the number of the seized channels (B_i) is counted in the channel list (Ch($\{f_i\}$) in which the value of Ch(f_i) is equal to 1. The resource reuse factor is calculated for S PLC cells by

$$f_{CR} = \frac{1}{F} \sum_{k=1}^{S} B_i \qquad (6.10)$$

6.3.5.2.2 Average Number of Channels per Active User and the Fairness Index

The average number of channels per active user is simply calculated by the number of channels per active user in all S PLC cells in the network

$$\overline{R} = \frac{1}{S} \sum_{i=1}^{S} R_i \qquad (6.11)$$

The quantitative measurement of the fairness between the numbers of channels per active user in different PLC cells is derived from the Jain's fairness-index [38]. The value of the fairness-index is calculated by

$$F_{\text{Fair}} = \frac{\left(\sum_{i=1}^{S} R_i \right)^2}{S \cdot \sum_{i=1}^{S} (R_i)^2} \qquad (6.12)$$

The maximum fairness-index is 1 when all the PLC cells have the same R-value.

6.3.6 Numerical Results

It is assumed that the investigated PLC systems are based on the specification in [29]. In the physical layer, the system transmit using Orthogonal Frequency Division Multiplexing (OFDM) modulation scheme. The duration of an OFDM symbol is 71.2 µs. The duration of time frame $T_{TF} = 240$ ms, and it is assumed that the transmission frequency band is divided into $F = 80$ channels. For simulative evaluation, the YATS simulator [39] is used. The mean and confidence interval (with confidence level of 0.95) of the simulated result are calculated by means of 10 subruns, each lasts for 100 000 time frames.

6.3.6.1 Throughput and Delivery Time of CA-Msg

With 80 channels, the necessary length of a CA-Msg is 29 Bytes. It therefore requires one OFDM symbol for the transmission in accordance with the symbol capacity given in [29]. The ratio of this overhead for transmitting a CA-Msg and a time frame is small and negligible: 71.2 µs/240 ms $= 0.29 \times 10^{-3}$. The average delivery time of the CA-Msg is measured based on the simulation as well as calculated by means of (6.8). The throughput of CA-Msg in a PLC cell is the sum of all the receiving CA-Msg from other PLC cells inside a time frame which is calculated in (6.6). The average throughput of CA-Msgs in the entire PLC network ($\overline{p^{\text{recv}}}$) therefore can be calculated.

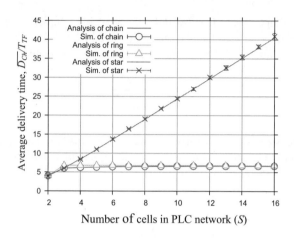

Figure 6.15   Average delivery time of CA-Msg.

The average delivery time and average throughput of the CA-Msg are evaluated for three aforementioned PLC network structures (chain, ring, and star structures) with different network sizes, as presented in Figure 6.15 and Figure 6.16. During simulation, each PLC cell determines its sending probability in a time frame based on the number of neighboring cells by using (6.2). Both analytical and simulation results are presented in the graphs. For each PLC network in accordance with a network structure, the calculated value is inside the confidence interval of the simulated value. By further calculation with higher number of PLC cells in a PLC network, the average delivery time of a CA-Msg in the chain and ring PLC networks converges to 6.75 time frames (TF) while that value in the star network linearly increases with the increasing of the number of PLC cells. The average throughput of the CA-Msg in the chain and ring networks converges to 0.269 (TF) and in the star network to 0.5. From the evaluated results, it can be seen that the reaction of reallocation of channels to change of traffic requests in the cells only slightly depends on size of the chain and the ring PLC networks. The reaction time however depends strongly on the number of the connections to a cell, as in the case of the star networks.

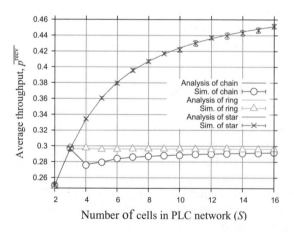

Figure 6.16   Average throughput of CA-Msg.

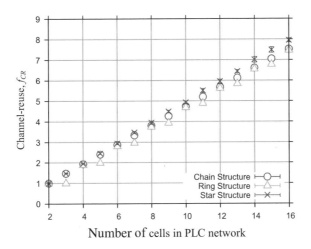

Figure 6.17   Channel-reuse factor.

6.3.6.2   Performance of the Allocation Protocol

For analyzing the performances of the protocol, each PLC cell is set to one base station and 20 users. Each user changes its status between active and inactive after a duration in accordance with a geometric distribution with a mean of $800TF$. Channel reuse factor, average number of channels per active user, and fairness-index are measured during simulation for different network structures and network sizes. The results are presented in Figures 6.17, 6.18, and 6.19, respectively.

As shown in Figure 6.17, the channel reuse factor depends on the number of PLC cells. With the same number of PLC cells, there is slightly different between channel reuse factors of different PLC network structures. In Figure 6.18, with high number of PLC cells in all three structures, the average number of channels per users are almost the same. It means that the average number of channels that each user can use only slightly depends on the network size.

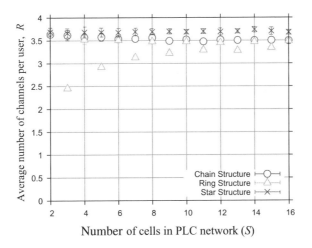

Figure 6.18   Number of channels per use.

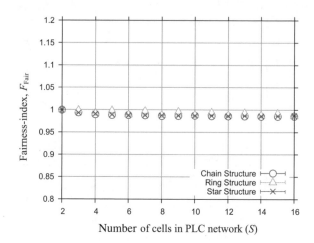

Figure 6.19   Fairness index.

When there are 16 cells in a PLC network in any one of the three structures, in average each user is reserved more than 3.5 channels for its transmissions. As presented in Figure 6.19, the fairness index is slightly decreased when the network is larger. However, the values of fairness index are almost 1 regardless the types of the network structures and the size of these PLC networks.

6.3.7 Remarks

The proposed channel allocation protocol allows for dynamic re-allocation of the channels between the PLC cells in the network during run-time. The objectives of the re-allocation of the channels are to maintain fairness between the users in different cells in terms of the number of channels per active user, while maximizing the channel reuse factor, and to avoid the usage of the same channels in the neighboring cells.

This protocol requires modifications in the base station of each cell. These modifications would allow a base station to exchange channel information with the other cells via the so-called channel announcement message (CA-Msg). One time slot with the duration of a OFDM symbol is reserved for sending this message between the base stations at the beginning of every time frame (TF). The transmission of the CA-Msg is based on the in-line interference characteristic of the PLC network: when two cells are interfering each other, they share at least the same electrical line.

The protocol is evaluated by means of simulation and analytic methods. In most of the referenced PLC networks, the protocol achieves near-perfect fairness between the users in different cells, the channel reuse factor is high and there are no occurrences of interfering channels.

Generally, the proposed protocol can be applied to both types of the PLC access cells and PLC in-home cells. Since it is designed to work in the distributed fashion, a cell needs only the channel information from its neighbor to locally determine its reserved channels. Based on the analysis of the results, the performance of the protocol strongly depends on the number of neighbors of a cell, but not on the size of the PLC network.

6.4 Multiple-user Resource Allocation[1]

The previous section has shown that the cellular principle used in mobile networks can be also applied to large PLC networks. In this context, transmission resources are firstly shared among the cells in which the coverage area is split and then allocated to the users located on them. While the previous section deals with the former problem, this section deals with the latter.

As explained in Chapter 5, a resource allocation problem in a single-user communication scenario aims at finding the optimal value of the system parameters under a given channel condition for achieving a target objective by satisfying a number of constraints. For example, in PS OFDM, many system parameters, namely, bits, sub-channel power, cyclic prefix duration, transmitter pulse shape, number of sub-channels and coding scheme, can be optimized in order to maximize the achievable rate under a power constraint. Furthermore, in time varying channels such as power line channels, the parameters optimization has to be done according to the channel condition, and this, in turn, increases the complexity.

In a multiple-user network, namely a network where one or more transmitters may wish to communicate with one or more receivers, the resource allocation problem becomes more complex inasmuch new elements such as cooperation, and feedback in presence of interference need to be considered to allocate the resources among the users. Accordingly, the capacity of a discrete memoryless channel in the general multiple-user case is unknown, in contrast to the single-user case. It is only for special cases that the capacity region for the multiple-user channels has been computed [40], [41]. Therefore, the performance achieved by a given resource allocation algorithm becomes complex, as well.

This section is organized as follows. In Section 6.4.1, we firstly briefly overview the achievable rate of two representative multi-user channels: the multiple-access channel, in which N_U users simultaneously communicate with one user, and the broadcast channel, in which one user sends data to N_U users simultaneously. For each channel, when possible, we report the achievable rate of FDMA and TDMA. In Section 6.4.2, we briefly describe the peculiarities of the main PLC application scenarios, namely, the indoor and outdoor, giving particular emphasis to the appropriate multi-user channel and access technique. In Section 6.4.3, we present the PHY layer system model that will be used in Sections 6.4.4 and 6.4.5 to solve practical resource allocation problems considering FDMA and TDMA contention free access techniques, respectively. Some observations regarding the hybrid use of contention free and contention based access techniques are made in Section 6.4.6. Finally, Section 6.4.7 is dedicated to an overview of the PLC literature on multi-user resource allocation.

6.4.1 An Information Theoretic Approach: the Multiple-user Gaussian Channels

In the rest of this section, unless otherwise stated, we consider the communication to be through Gaussian channels. The reason for this decision is twofold. Firstly, because it is the

---

[1] Portions of the material in this section are reprinted, with permission, from J. A. Cortés, L. Díez, F. J. Cañete, and J. T. Entrambasaguas, Analysis of DMT-FDMA as a multiple access scheme for broadband indoor power-line communications, *IEEE Transactions on Consumer Electronics*, vol. 52, no. 4, pp. 1184–1192, Nov. 2006, ©2006 IEEE, and from A. M. Tonello, J. A. Cortés, and S. D'Alessandro, Optimal time slot design in an OFDM-TDMA system over power-line time-variant channels, in *Proceedings IEEE International Symposium on Power Line Communications and Its Applications*, Dresden, Germany, Mar. 29–Apr. 1, 2009, pp. 41–46, ©2009 IEEE.

universal benchmarking communication channel. Having a reference bound is particularly important when no analytical expression for the capacity of the studied channel is known, as it occurs in PLC. Secondly, because it seems that noise in PLC channels can be assumed to be Gaussian provided that no impulsive noise is present [42]. The study can be extended to the case of channels that experience impulsive noise with statistics different from Gaussian [43]–[45]. Furthermore, we focus on the half-duplex transmission, namely, we consider the communication to happen in one direction. The emulation of the full-duplex case can be obtained by considering the bidirectional transmission over orthogonal channels, e.g. using different frequency bands (FDD) or time slots (TDD).

Before considering the multiple-user channels, we give an overview of the single-user Gaussian channel.

6.4.1.1 Single-user Gaussian Channel

Let $x(n)$, with $n \in \mathbb{Z}$, be a time-discrete baseband complex signal transmitted through a Gaussian channel, then the received signal is given by [40]

$$y(n) = x(n) + \eta(n), \tag{6.13}$$

where $\eta(n) = \eta_R(n) + j\eta_I(n)$ represents the noise whose components are drawn i.i.d. from a Gaussian distribution with variance $P_\eta/2$, i.e. $\eta_R(n), \eta_I(n) \sim N(0, P_\eta/2)$. Now, if we suppose the transmitted signal to have a constraint on the transmitted power and to be Gaussian white distributed, namely $x(n) = x_R(n) + jx_I(n)$ with $x_R(n), x_I(n) \sim N(0, P/2)$, then, the channel capacity is given by

$$C = C(\gamma) = \log_2(1 + \gamma) \quad \text{bits/use}, \tag{6.14}$$

where $\gamma = P/P_\eta$ represents the SNR. It can be shown that the previous definition can be extended to consider the case of transmission over bandlimited channels. Let us assume the transmission of a continuous time signal over a channel with impulse response $g_{ch}(t)$, with $t \in \mathbb{R}$. Let us further assume that the channel has bandwidth B, then from the sampling theorem it is known that the transmitted signal can be determined by its samples at rate $T = 1/B$. For each sample we can use the formulation (6.14) to compute the capacity for channel use, therefore, considering that $1/T$ samples are transmitted in one second, the capacity is given by

$$C = C(\gamma) = \frac{1}{T} \log_2(1 + \gamma) \quad \text{bits/s}, \tag{6.15}$$

where $\gamma = |H|^2 P/P_\eta$ and H is the channel gain.

Similarly, the previous definition can be extended to a set of M parallel Gaussian channels each having a bandwidth $B^{(k)}$, e.g. $B^{(k)} = 1/(MT)$ for $k = 0, \ldots, M-1$. In such a case, assuming $P^{(k)} \leq P$ to be the power of the signal transmitted on the k-th channel, and $H^{(k)}$ the k-th channel gain, the capacity is given by

$$C = \frac{1}{MT} \sum_{k=0}^{M-1} \log_2 \left(1 + \gamma^{(k)}\right) \quad \text{bits/s}, \tag{6.16}$$

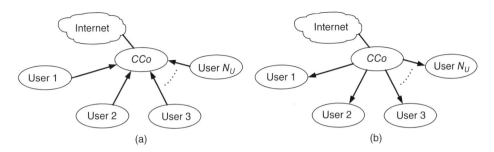

Figure 6.20 Example of (a) multiple-access channel, and (b) broadcast channel.

where $\gamma^{(k)} = |H^{(k)}|^2 P^{(k)}/P_\eta^{(k)}$. As we will see in the following, 6.16 can be used to compute a bound on the achievable rate of OFDM systems.

6.4.1.2 Multiple-access Channel

In a multiple-access channel, N_U users simultaneously communicate with one user. As an example, we can consider the indoor PLC network shown in Figure 6.20(a), where N_U nodes wish to send data to a CCo. For this channel, an achievable rate region has been found for an arbitrary number of users [40]. Considering the case with $N_U = 2$, and assuming $P^{(u)}$ to be the power constraint for user u, and Gaussian channels, then the bounds on the achievable rate region are given by

$$R^{(1)} \le C\left(\gamma^{(1)}\right),\tag{6.17}$$
$$R^{(2)} \le C\left(\gamma^{(2)}\right),$$
$$R^{(1)} + R^{(2)} \le C\left(\gamma^{(1)} + \gamma^{(2)}\right),$$

where $\gamma^{(u)} = P^{(u)}|H^{(u)}|^2/P_\eta$, $H^{(u)}$ is the channel gain between user u and the receiver, and P_η the noise power at the receiver. The capacity region, namely, the borders of the achievable rate region, is obtained when the transmitted signals are Gaussian distributed.

Figure 6.21(a) shows the capacity region. As it can be observed, this capacity region corresponds to a pentagon. Furthermore, as it is shown, the corners can be achieved considering a two-step decoding procedure where at the first step the signal of the first user is seen as noise and the signal of the second is decoded. Then at the second step, the signal of the first user can be decoded by subtracting the one of the second user. The procedure is called onion-peeling and can be extended to an arbitrary number of users. Figure 6.21(a) also shows two achievable rate regions attainable using the FDMA and the TDMA MAC techniques.

In the classical FDMA technique, each user occupies disjoint bands $B^{(u)}$ of the channel, so that $\sum_{u=1}^{N_U} B^{(u)} = B$. In this case, the achievable rate region can be computed using the capacity formula for bandlimited channels (6.15), and it is shown in Figure 6.21(a). As it is noticeable, the FDMA achievable rate intersects the capacity in one point. It is possible to show that this point is obtained when more bandwidth is allocated to the user with higher SNR. It is clear that FDMA can be easily implemented in networks where multicarrier modulation, e.g. OFDM, is

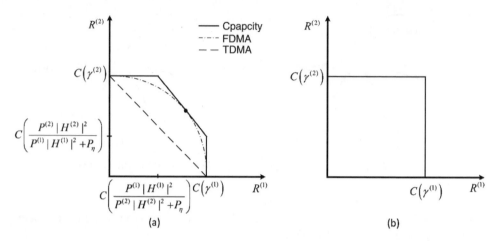

Figure 6.21   Achievable rate region for (a) the two users multiple-access channel, and (b) the two users orthogonal broadcast channel.

adopted at the PHY layer to exploit the frequency selectivity of the channel, as for example in the case of broadband PLC systems.

In TDMA, the time is divided in slots, and each user occupies the channel in one slot. In Figure 6.21(a), we show an achievable rate region using TDMA where each user achieves a rate $R^{(u)} = t^{(u)}C(\gamma^{(u)})$, with $t^{(u)} \in [0, 1]$. With this simple TDMA scheme, FDMA achieves higher rate than TDMA. It is possible to show that by optimally allocating resources among users (time slots duration and powers), TDMA and FDMA can achieve the same rate [40, cfr. Ch. 15], [41, §3.7].

It is interesting to note that the capacity region above described is also attainable adopting CDMA, i.e. by assigning different codes to the users and by decoding one user per time (onion peeling).

6.4.1.3   Broadcast Channel

The broadcast channel is essentially the dual of the multiple-access channel. In a broadcast channel (see Figure 6.20(b)), a node simultaneously communicates with N_U users. As an example, consider an in-home PLC network where a number of nodes want to download data coming from Internet (e.g. video streams) through a CCo.

The capacity of the broadcast channel has not been defined yet for a general number of users. However, there are cases where it is easy to proceed to its computation. The first presented case is the two-user Gaussian broadcast channel. For this channel, assuming $P_\eta^{(u)}$ the noise power experienced by user u, and $P = P^{(1)} + P^{(2)}$ to be the total power transmitted, the capacity region is given by

$$R^{(1)} < C\left(\gamma^{(1)}\right), \tag{6.18}$$

$$R^{(2)} < C\left(\frac{\gamma^{(2)}}{1 + P^{(1)}|H^{(2)}|^2/P_\eta^{(2)}}\right).$$

Secondly, the orthogonal broadcast channel, when signals are transmitted to the users over orthogonal channels, is considered. This situation is experienced for instance in satellite communications where different channels are used to send data to different users. The capacity region is the rectangle shown in Figure 6.21(b). Similarly to the multiple access channel, also in the case of a broadcast channel, when the channels are not orthogonal, e.g. in the PLC network example above, practical MAC access schemes as FDMA and TDMA can be adopted to orthogonalize the channels. Consequently, the corresponding theoretical achievable rates are the same of the ones illustrated for the multiple-access channel.

6.4.1.4   Observations on the Achievable Rate in Practical Implementations

In practical implementations, for both multiple-access and broadcast channels, the rate achieved with FDMA and TDMA techniques can result in much smaller values than the channel capacity. This can occur for various reasons:

- Non ideal conditions are usually present, e.g. interference caused by symbol misalignments both in time or in frequency, hardware impairments, etc.
- Feedback information needs to be exchanged among network nodes in order to compute the resource allocation: assign channels (namely, frequency bands in FDMA or time slots in TDMA) and power values to the network users.
- The transmitted symbols belong to finite size constellation (e.g. M-QAM) instead of belonging to a Gaussian distribution as assumed in the derivation of the previous capacity formulations.
- ISI (see Section 5.3) can be present due to channel frequency selectivity.

In the next sections, the resource allocation problem considering practical implementation of FDMA and TDMA schemes over typical PLC networks is presented and analyzed.

6.4.2   Multi-user Resource Allocation in PLC Scenarios

The characteristics of the multi-user resource allocation problem in the different LV PLC scenarios are quite varied. The following list summarizes the main characteristics of such scenarios:

- *Indoor PLC*:
  - *Broadband*: One of the most common deployments in this scenario consists of a set of different devices connected to a gateway that provides access to the Internet. Hence, the downlink is a broadcast channel and the uplink is a multiple-access one. The involved distances are short, which has two important implications. The first one is that no relays are generally needed (the reader is referred to Section 5.3.2.3 for details on relay transmission). The second one is that the differences among the propagation delays of the links are quite small. This considerably reduces the MAI when using an FDMA scheme, as it will be shown in Section 6.4.4.1. No network planning is generally needed in this scenario (except, probably, in offices or buildings). Hence, the resource allocation problem must be dynamically solved by the network devices. Because of it simplicity, most PLC systems in this category use a TDMA scheme with a contention free region, managed by a centralized coordinator, and a contention region.

– *Narrowband*: These are generally simple networks used for home automation purposes. They consists of a master device and a set of slave nodes. Hence, the downlink is a broadcast channel and the uplink is a multiple-access one. These networks require very low data rates and have nearly no QoS constraints. In Europe they use the CENELEC B (95–125 kHz) and C (125–140 kHz) bands. Moreover, only CSMA can be used in the latter. Cost efficiency rather than transmission efficiency is the leading design criterion.

- *Outdoor PLC* :
  – *Broadband*: The use of the LV network as a *last mile* medium is the most widespread application of this category. A broadband PLC access network is composed of the LV head end and the NTU of each end user. The average link length ranges from 150 m up to 800 m in western European networks [46]. This causes significant differences in the propagation delay from the NTUs to the LV head end, in contrast to the indoor case. In addition, repeaters are usually needed to compensate for the high attenuation values. The LV head end may be located in the transformer station or in another place where a convenient access to the backbone network is available either by PLC, DSL, wireless or any other technology. The downlink is a broadcast channel and the uplink is a multiple-access one. A LV distribution network can be segmented into several PLC access networks [47]. Each segment consists of a LV head end and a number of NTUs connected to it. As deeply explained in Section 6.3, network planning is an important task in this scenario. Two problems have to be solved: the placement of the LV head ends and the assignment of resources to them [48]. The latter can be accomplished by means of TDMA, FDMA or a combination of both of them. Broadband PLC access networks should provide high data rates and stringent QoS requirements.
  – *Narrowband*: This category encompasses the European systems that use the CENELEC A band (9–95 kHz) and the North American and Japanese ones that employ the band up to 500 kHz. It is being extensively used for Smart Grids applications, in particular for AMM. The last segment of AMM systems consists of a master node located at the transformer station and multiple end nodes located in the meters that act as slaves. Relaying is usually required. This task can be accomplished either by dedicated nodes or by end nodes [46]. As in the broadband case, significant differences in the propagation delay might exist. Nowadays, AMM are mostly restricted to remote metering, which require low data rates and modest QoS requisites.

- *In-vehicle PLC*: In this scenario, PLC is envisioned to be used to connect the large number of sensors, actuators and electronic control units installed in vehicles. There exist multiple functional domains inside a vehicle with very different traffic patterns (time-triggered and event-triggered), QoS requisites (response time, jitter, bandwidth, redundant communication channels for tolerating transmission errors, bit error rate, etc.) and data rate necessities, which range from tens of kbit/s up to a few Mbit/s [49]. Transmission distances are very short and priorities among the different nodes are an essential feature. The downlink is a broadcast channel and the uplink is a multiple-access one. However, they use to be quite asymmetric in some functional domains, e.g. the multimedia one, which is used for audio and video distribution. Both a TDMA scheme similar to the one currently employed in other PLC systems, with a contention free region managed by a centralized coordinator and a contention region, and an FDMA technique can be used to satisfy the different demands of each functional domain.

6.4.3   PHY Layer System Model

As discussed in Chapter 2, LV PLC channels are frequency selective and time variant. Measurements have shown that they can be modeled as LPTV systems having also the presence of additive cyclostationary colored noise [50], [51]. Therefore, in order to mitigate the frequency selectivity of the channel, the state-of-the-art systems employ OFDM at the PHY layer. Attractive features of OFDM are its straightforward implementation through DFT/IDFT operations, the interference mitigation offered by the use of the CP, the possibility to practically implement the water-filling principle by allocating the power across the sub-channels affected by different attenuations due to the channel frequency selectivity, and the possibility to easily implement the FDMA scheme by partitioning carriers among the network users realizing OFDMA. The reader is referred to Section 5.3.2.3 for a detailed treatment of OFDM.

In the following, we consider a network consisting of nodes located in the same cell (see Section 6.3) whose PHY layer is based on PS-OFDM (namely an OFDM that employs a Nyquist window at the transmitter) with M sub-channels, and a CP length of μ samples. The OFDM normalized symbol duration is equal to N samples and $N = M + \mu$. Assuming T equal to the sampling period, the symbol duration in seconds is equal to $T_0 = NT$. Denoting by $g_{ch}(n; i)$ the channel impulse response at time instant n to an impulse applied i time instants before, the received signal can be written as

$$y(n) = \sum_{i=0}^{\nu-1} x(n - i)g_{ch}(n; i) + \eta(n), \qquad (6.19)$$

where νT is the impulse response duration, $x(n)$ is the OFDM transmitted signal, and $\eta(n)$ is the cyclostationary additive noise. Both, the channel response and the cyclostationary noise have the periodicity of the mains signal (20 ms in Europe). At the receiver, after symbol synchronization, the CP is discarded and windowing is applied. As a further step, an M-point DFT is computed. The k-th sub-channel output can be written as

$$z^{(k)}(\ell N) = H^{(k)}(\ell N)a^{(k)}(\ell N) + I^{(k)}(\ell N) + W^{(k)}(\ell N), \qquad (6.20)$$

where $a^{(k)}(\ell N)$ is the ℓ-th data symbol transmitted on that sub-channel, $H^{(k)}(\ell N)$ is the effective channel transfer function, $W^{(k)}(\ell N)$ is the noise term and $I^{(k)}(\ell N)$ is the interference contribution. The latter is given by the ICI plus ISI terms in single-user scenarios, and also by a MAI term in multi-user scenarios. It must be emphasized that these distortion terms arise because of the loss of orthogonality due to an insufficient CP, to channel time variations, and to the multiple access, e.g. misalignment of symbols received by different transmitters. Provided that the number of sub-channels is sufficiently high, the interference term is expected to have a Gaussian distribution [52], [53] (which is strictly true when the information signals are Gaussian). Considering that transmissions are synchronized with the mains and that L OFDM symbols can be fitted into each mains cycle, the symbol index ℓ can be written as $\ell = m + Lr$, where $0 \le m \le L - 1$ and $r \in \mathbb{Z}$.

In the following, it is assumed that the transmitted signal needs to satisfy a constraint on the PSD mask. This is a practical assumption for indoor PLC systems to be compliant with EMC limits [54].

Now, assuming the transmission of the OFDM signal with a constant PSD, the power of all the terms in (6.20) is periodic and the SINR experienced in the k-th sub-channel at the m-th time instant can then be expressed as

$$\text{SINR}^{(k)}(mN) = \frac{P_U^{(k)}(mN)}{P_W^{(k)}(mN) + P_I^{(k)}(mN)}, \tag{6.21}$$

where

$$P_U^{(k)}(mN) = \left|H^{(k)}(mN)\right|^2 E\left[\left|a^{(k)}(mN)\right|^2\right],$$

$$P_W^{(k)}(mN) = E\left[\left|W^{(k)}(mN)\right|^2\right], \tag{6.22}$$

$$P_I^{(k)}(mN) = E\left[\left|I^{(k)}(mN)\right|^2\right].$$

It should be noted that the receiver can estimate the sub-channel SINR during the training phase via reception of known training symbols that are periodically sent by the transmitter. The estimate can be refined or updated using a data decision directed mode during data transmission [55]. As it will be clarified in the following, the previous observation is valid for TDMA scheme but for FDMA it is only true if the MAI term is present while estimating the SINR. This, in turn, will only occur if all the remaining users are transmitting while the desired user is estimating the SINR.

With the SINR definition (6.21), assuming the noise, the interference and the transmitted symbols to be i.i.d. and Gaussian distributed on M parallel Gaussian channels, the capacity can be computed using (6.16) as

$$C(mN) = \frac{1}{NT} \sum_{k \in \mathbb{K}_{ON}} \log_2\left(1 + \text{SINR}^{(k)}(mN)\right), \tag{6.23}$$

where $\mathbb{K}_{ON}$ is the set of sub-channel indices employed to transmit useful data in the considered cell (see Section 6.3), $\mathbb{K}_{ON} \subseteq \{0, \dots, M-1\}$.

It is interesting to note that under a PSD mask constraint and in absence of interference, the capacity 6.23 is maximized with a signal transmitted at the PSD limit level [56]. Furthermore, numerical results have shown that even in the presence of interference, the constant power allocation at the PSD limit level gives achievable rates close to the ones obtained using iterative water-filling algorithms [57]. Therefore, in the rest of the section, the OFDM signal is assumed to be transmitted at the PSD limit level.

The capacity formulation (6.23) assumes the number of bits loaded on each sub-channel to be real valued. This is not the case for real systems, where the number of bits loaded on each sub-channel is an integer value and furthermore the transmitted symbols, in general, belong to a M-QAM constellation. To overcome this problem, bit-loading algorithms are used in real systems [58], [59]. As an example, when showing numerical results, the following sub-channel bits assignment is adopted:

$$b^{(k)}(mN) = \left\{ \left\lfloor \log_2 \left(1 + \frac{\mathrm{SINR}^{(k)}(mN)}{\Gamma} \right) \right\rfloor \right\}, \tag{6.24}$$

where Γ is the gap factor that takes into account practical coding/modulation constraints [60], [61], and $\lfloor \lfloor \cdot \rfloor \rfloor$ denotes the rounding to the nearest available constellation.

6.4.4 *Frequency Division Multiple Access*

In the classical FDMA technique, the available bandwidth is divided into non-overlapping sub-bands that are assigned to the users. However, orthogonality among the signals of the different users can also be achieved with overlapped subbands. This strategy can be accomplished by using OFDM at the physical layer and by assigning a different set of carriers to each user. The resulting scheme is referred to as OFDMA and is the one considered in this section. The carrier allocation is usually accomplished by a central coordinator that dynamically modifies the assignment in order to guarantee the requested QoS.

The main advantage of OFDMA over TDMA in PLC scenarios is its higher efficiency. While in a Gaussian channel TDMA can achieve the same performance as OFDMA, provided that a proper time slot and power allocation scheme is employed, it may result in a waste of capacity in frequency-selective channels. The reason is that TDMA does not exploit the unused carriers in one link (because of their low SNR) that may experience acceptable SNR in other links. Moreover, since in OFDMA the protocol data units length is no longer limited by the time slots duration, the protocol overhead may be smaller.

On the other hand, OFDMA requires symbol and frequency synchronization to avoid MAI [62]. Additionally, the volume of signaling that the users must exchange with the CCo is higher because they have to inform the CCo of the amount of data per carrier, while in TDMA each user has to inform the CCo of the amount of data per time slot. In current broadband PLC systems, the number of carriers can be hundreds of times larger than the number of time slots per frame. These limitations exist in all multi-user scenarios. However, they are particularly severe in environments where multiple point-to-point connections are simultaneously carried out.

6.4.4.1 Carrier Allocation Techniques in OFDMA Networks

The first step in any resource allocation problem is to state the optimality criterion and the constraints that will rule the assignment process. Maximizing the aggregate data rate of all the users leads to an unfair resource sharing when the involved channel conditions are quite different. On the other hand, the max-min criterion penalizes the overall data rate and is not appropriate when the users request services with different data rates. To overcome this problem, the concept of balanced capacity was proposed in [63]. It is defined as the distribution of maximum simultaneously achievable bit rates that are proportional to the single-user rates. An alternative criterion that aims to provide a minimum data rate for each link and to allocate the remaining resources according to the quality of each link has been proposed in [64].

Fairness among real time and non-real time services is also considered in [65]. A power consumption constraint is usually included in the problem, whatever the optimization criterion is. However, a PSD constraint is also needed in PLC.

Another step in the carrier allocation problem is to decide whether the carriers will be assigned individually or on a group basis. The latter, usually referred to as tone-grouping, is more insensitive to MAI and reduces the computational complexity of the problem, but it is less flexible. In the presence of ICI or MAI, the SINR experienced by each carrier, and consequently the number of bits per carrier, depends on the assignation of the remaining carriers. To simplify the problem, the influence of both effects is discarded in practice. Hence, the obtained solution will be valid only if these impairments are really negligible.

Finally, the optimization problem has to be solved. The optimal carrier allocation strategy in a frequency-selective network is the solution to a non-linear optimization problem which is usually approximated by means of linear or integer programming [66]. In practice, the computational complexity of the latter scheme is so high that it can only be applied when tone-grouping is employed. On the other hand, the former uses to lead to solutions in which carriers have to be shared among users, as it is shown in the following example, leading to an hybrid TDMA-OFDMA scheme. In order to reduce the computational complexity, a reformulation of the problem into a convex optimization one is proposed in [67]. Other solutions have been proposed in [64] and [65].

As an example of a carrier allocation problem, let us consider the situation in which the optimality criterion is to maximize the aggregate bit rate of a network with N_L links. It should be noticed that, while in the broadcast and the multiple-access channel the number of links N_L equals the number of users N_U, this does not happen in scenarios with multiple point-to-point connections. The optimization is subject to the restriction that each link must achieve, at least, a given percentage, p (%), of the bit rate that it would attain in a single-user scenario. Let us denote by $b^{(u,k)}$ the number of bits that can be allocated to carrier k when used in link u. It can be computed using expression (6.24). M denotes the number of carriers, when they are assigned individually, and the number of group of carriers when tone-grouping is employed. The carriers allocated to each link can be obtained as the solution to the following optimization problem

$$\max \sum_{u=1}^{N_L} \sum_{k=0}^{M-1} b^{(u,k)} c^{(u,k)}, \qquad (6.25)$$

subject to

$$\sum_{k=0}^{M-1} b^{(u,k)} c^{(u,k)} \geq \frac{p}{100} \sum_{k=0}^{M-1} b^{(u,k)} \quad \text{for } u = 1, \dots, N_L, \qquad (6.26)$$

where $c^{(u,k)}$ denotes whether carrier k is assigned to link u, $c^{(u,k)} = 1$, or not, $c^{(u,k)} = 0$. To make each carrier, or group of carriers, to be assigned only to one link, the following constraint is also imposed

$$\sum_{u=1}^{N_L} c^{(u,k)} = 1 \quad \text{for } k = 0, \dots, M-1. \qquad (6.27)$$

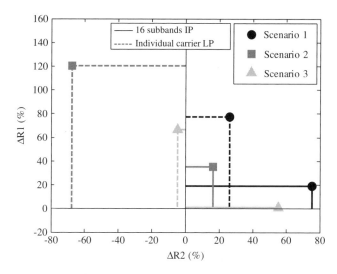

Figure 6.22 Relative bit rate imbalance in the two links of three networks when the 1024 carriers of the physical layer are assigned individually using LP and with tone-grouping (16 subbands) using IP.

In practice, condition (6.27) does not prevent $c^{(u,k)}$ to take non-integer values in the range [0,1] when the problem is solved by means of LP. This means that carrier k is assigned to link u a fraction of time given by $c^{(u,k)}$, leading to an hybrid TDMA-OFDMA. This is particularly common when the transmission characteristics of the involved links are quite similar. To obtain an OFDMA scheme the obtained $c^{(u,k)}$ must be rounded. This rounding process may cause the achieved solution to be far from the objective value.

To illustrate this end, Figure 6.22 depicts the relative bit rate imbalance obtained in the two links (labeled as R1 and R2) of three indoor PLC networks when the 1024 carriers of the underlying physical layer are allocated individually using LP and when they are grouped into 16 subbands that are assigned using IP. The imbalance is measured with respect to the objective value imposed in the carrier assignment process, which equals to $p = 40\%$ of the bit rate value achieved in each link in a single-user environment. The aggregate bit rates obtained in each case are quite similar, as shown in Table 6.1. However, the rounding of the $c^{(u,k)}$ coefficients provided by the LP may cause a severe imbalance in the bit rate achieved in each link. Moreover, it makes the obtained solution to violate the constraints imposed in the optimization process. This can be clearly observed in the Scenario 2, where the bit rate R2 is about 67% below the objective value.

Table 6.1 Aggregate values of the bit rates (Mbit/s) shown in Figure 6.22

Scenario	Individual Carrier (LP)	16 subbands (IP)
1	425.96	424.59
2	376.77	376.87
3	170.75	180.00

The carrier allocation process that has just been described assumes that the channel is time-invariant. However, both the response, and in particular the noise level, of broadband indoor PLC channels experience a periodical variation synchronized with the mains. Therefore, the carrier assignment should also be periodically varied. Nevertheless, the noise and the channel response changes tend to be correlated, at least in indoor networks. This allows accomplishing the carrier allocation using the channel state information obtained at any time instant within the mains cycle, and then performing a cyclic bit-loading in the carriers assigned to each user.

6.4.4.2 Multiple Access Interference in FDMA Networks

This section analyzes the MAI that occurs in OFDMA networks. In this scenario MAI is due to four causes: frequency selectivity of the channel, misalignment between the desired and interferer symbols at the input of the desired receiver, sampling frequency offsets between users and carrier frequency offsets between users [68]. The latter does not exist if DMT is used at the physical layer instead of OFDM. Similarly, no MAI due to symbol misalignment and sampling frequency offset exists in broadcast channels.

The frequency selectivity of the channel destroys the orthogonality of the carriers employed by each user, causing ICI and ISI, and of the carriers employed by different users, leading to MAI. Provided that symbols transmitted by the different users are perfectly aligned at the desired receiver, this type of MAI can be easily overcome by using a CP length longer than the channel impulse response. However, due to the long impulse responses of indoor power line channels, a shorter CP is usually selected and some residual MAI will exist.

In the multiple access channel and in scenarios with multiple point-to-point connections, MAI also occurs if the symbols transmitted by the different users are not aligned at the input of the desired receiver. Differences among the propagation delays of the links may cause symbol misalignment even if the transmissions of all the users are perfectly synchronous. Interestingly, a delayed and a time-advanced interferer symbol may cause significantly different MAI. As an example, let us consider the situation shown in Figure 6.23, where two consecutive symbols

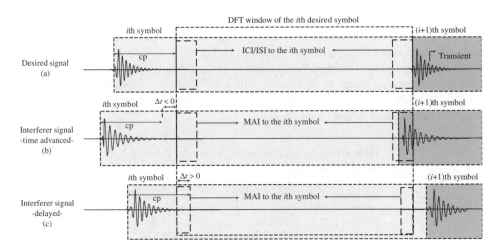

Figure 6.23   Illustration of symbol misalignment situations.

from a desired transmitter and from a delayed ($\Delta t > 0$) and a time-advanced interferers ($\Delta t < 0$) are depicted. The symbol synchronization strategy employed in the desired receiver leads to the DFT window shown in Figure 6.23(a). Figure 6.23 also shows the transient that occurs whenever an OFDM symbol traverses a frequency selective channel. As seen, while most of the distortion caused by a positive offset Δt is expected to be absorbed by the cyclic prefix, the situation is rather distinct for an offset in the opposite direction. Hence, MAI due to the transients in the interferer's symbols will be much greater in these circumstances. In a synchronized indoor PLC network, MAI due to symbol misalignment can be made very small with a proper selection of the OFDM parameters because no significant differences among the propagation delays of the links from a specific site are expected. On the other hand, significant delays might occur in outdoor PLC networks.

For a given symbol misalignment and sampling or carrier frequency error, the power of the resulting MAI depends on the difference between the attenuation of the interferer channel and the desired one; on the frequency selectivity of the interferer channel and on the frequency separation between the interferer and interfered carriers. Hence, assigning groups of tones instead of individual carriers is a common technique used to reduce this type of MAI. Nevertheless, the use of pulses with higher confinement than the rectangular one, both at the transmitter (PS-OFDM) and at the receiver (windowed OFDM), are the most powerful strategies to combat MAI in OFDMA systems. Other modulation schemes using pulses with higher spectral confinement, e.g. filtered multitone (FMT), may be used to deal with this problem (see Section 5.3.2.2) [69]. However, the following trade-off arises in all cases: using larger pulses, either at the transmitter or at the receiver, increases the SINR but reduces the symbol rate. The following subsections study the MAI due to symbol misalignment and to sampling frequency errors and the effect of pulse-shaping and windowing. The reader is referred to Section 5.3.2.2 for additional details on the latter techniques and their key parameters, α and β, and to [68] for the description of a procedure to estimate the MAI.

6.4.4.2.1 MAI Caused by Symbol Misalignment at the Receiver

To illustrate the effect of the MAI caused by symbol misalignment, Figure 6.24 depicts the aggregate bit rate in three indoor PLC networks with two point-to-point links. These networks have been categorized as *low MAI*, *intermediate MAI* and *high MAI*. The main difference between them is the attenuation of the channel from the desired transmitter to the desired receiver (desired channel) and of the channel from the interferer transmitter to the desired receiver (interferer channel). In the *low MAI* scenario, the desired channel has low attenuation, while the interferer channel is highly attenuated. Conversely, in the *high MAI* scenario, the desired channel is very attenuated and the interferer channel is lowly attenuated. In the *intermediate MAI*, both the desired and interferer channels experience similar attenuation. The physical layer employs 1024 carriers in the frequency band up to 30 MHz. Carriers are allocated using tone-grouping with 16 and 64 subbands. In order to obtain an upper bound of the MAI, adjacent subbands are assigned to different users (interleaved) and the misalignment between the desired and interferer symbols is about one half of the symbol length. The bit rate is computed as a function of the pulse-shaping and windowing parameters with $\alpha = \beta$, which was shown in [68] to maximize the performance in these circumstances. It is assumed that the noise and distortion terms are independent and Gaussian distributed.

Figure 6.24 also shows a reference curve labeled as 'bit rate loss asymptote'. It has been obtained by summing the single-user bit rates obtained by both users in the *low MAI* scenario. In

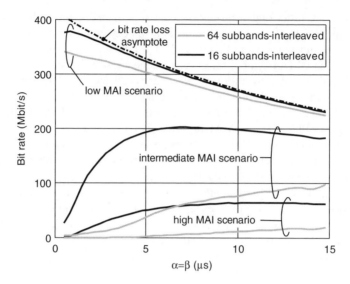

Figure 6.24 Aggregate bit rate in different MAI scenarios as a function of α and β for a symbol misalignment equal to $1/2$ of the symbol length. [68] © 2006 IEEE.

this case, the bit rate loss is almost due to the symbol rate reduction caused by the pulse shaping and windowing. It is worth noting that in the *low MAI* scenario pulse-shaping and windowing is nearly always counterproductive, since the MAI is very small. This fact is reinforced by noting that the bit rate for the 16 interleaved subbands scheme is nearly coincident with the bit rate asymptote. On the other hand, increasing α and β (in the considered range) in the *high MAI* network always improves the performance. In the *intermediate MAI* network, increasing both parameters improves the performance in first instance. However, when the reduction in the MAI power, which allows to use denser constellations, does not compensate for the reduction in the symbol rate, increasing α and β reduces the bit rate.

Figure 6.25 shows the performance in the *low MAI* and *high MAI* networks shown in Figure 6.24 as a function of the symbol misalignment, Δt (expressed as a percentage of the symbol length). Two physical layers are assessed. One uses OFDM with pulse-shaping and windowing with $\alpha = \beta = 8.3$ μs, which proved to be a good trade-off value for the three networks, as shown in Figure 6.24. The other physical layer employs OFDM with rectangular pulses and is denoted as *conventional OFDM*. Performance is measured in terms of the aggregate bit rate loss, where the sum of the single-user bit rates obtained with the conventional OFDM system in the considered scenario is taken as reference. Three carrier allocation schemes have been considered: one in which carriers are individually assigned using LP, and two sets of strategies that use tone-grouping with 64 subbands. In one of these strategies subbands are assigned using IP, while adjacent subbands are allocated to different users (interleaved) in the other. The latter is used to obtain a lower bound on the performance.

Figure 6.25 reveals that the performance degradation due to the frequency selectivity of the channel is negligible when the desired and the interferer symbols are aligned at the receiver side. On the other hand, the performance of the conventional OFDM system is very sensitive to small symbol misalignment and the network can be on outage in the *high MAI* scenario. The

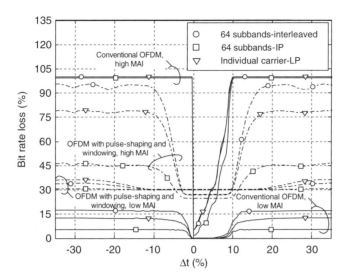

Figure 6.25 Bit rate loss for the *low MAI* and the *high MAI* scenarios as a function of the symbol misalignment. [68] © 2006 IEEE.

use of pulse-shaping and windowing causes a bit rate loss, due to the symbol rate reduction. In fact, it is always counterproductive in a *low MAI* scenario. However, it makes the performance much more resilient to symbol misalignment in the *high MAI* scenario. It is also important to note the impact of the carrier assignment in the performance degradation. As expected, the individual carrier allocation and the tone-grouping with interleaved subbands are the most sensitive strategies to the MAI.

6.4.4.2.2   MAI Caused by Sampling Frequency Offsets

Let us consider a network in which synchronization is accomplished by means of periodical synchronizations beacons sent at a proper rate. Users would adjust the phase of their clocks after the reception of the beacon, but the frequency of these clocks drifts between two consecutive synchronization beacons. This causes MAI even if the resulting symbol misalignment is kept less than one sample. In these circumstances, pulse-shaping at the transmitter is nearly always counterproductive, and only windowing reduces the MAI.

Figure 6.26(a) depicts the aggregate bit rate in the already presented MAI scenarios for a frequency mismatch of $\Delta f = 25$ ppm as a function of β. It can be seen that, while windowing is always counterproductive in the best MAI scenario, it improves the bit rate in the remaining cases. In fact, the behavior is quite similar to the one seen in the case of symbol asynchrony.

Figure 6.26(b) depicts the aggregate bit rate loss (with respect to the sum of the single-user bit rates attained with conventional OFDM system in the considered scenario) as a function of the frequency mismatch for a $\beta = 8$ μs. As seen, the performance of the conventional OFDM system is nearly unaffected by the frequency mismatch in the *low MAI* scenario. On the other hand, even small frequency offsets cause a severe performance degradation in the *high MAI* scenario. In this scenario, the bit rate loss of the windowed OFDM may also be considerable for some carrier assignment even when $\Delta f = 30$ ppm.

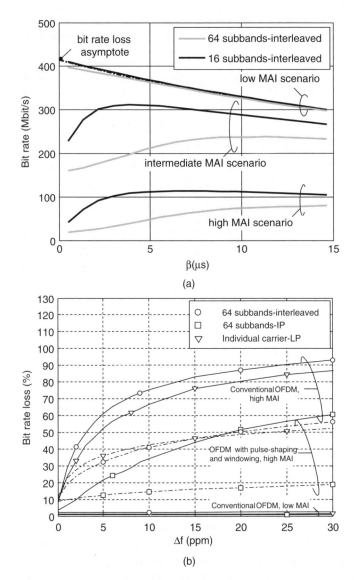

Figure 6.26 Aggregate bit rate caused by a sampling frequency offset in different scenarios as a function of β (a). Bit rate loss for the *high MAI* and the *low MAI* scenario as a function of the frequency mismatch (b). [68] © 2006 IEEE.

6.4.5 Time Division Multiple Access

6.4.5.1 Contention-free TDMA: Optimal Time-slot Design and Allocation Procedures

In this section, the resource allocation problem is analyzed for an indoor PLC system whose PHY layer is based on OFDM as specified in Section 6.4.3, and thus, a periodically time variant channel with cyclostationary noise are considered. The MAC layer consists of an

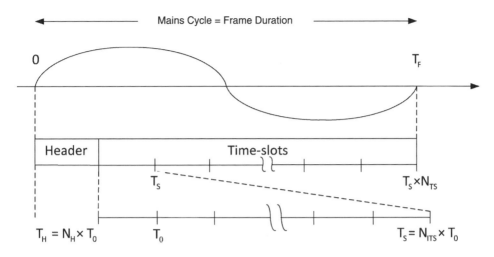

Figure 6.27 Frame structure. [72] © 2009 IEEE.

adaptive TDMA region used to provide high QoS. A similar scheme is described in many broadband PLC standards and industrial specifications, e.g. the IEEE P1901 and ITU-T G.hn [70] standards, the HomePlug AV HPAV and HomePlug AV2 [71] specifications. Similarly to HPAV, we assume to have a node in the network that acts as CCo and we focus on an orthogonal broadcast channel where the CCo sends data to the N_U network users (see Figure 6.20(b)). In particular, we consider the optimization of the TDMA region.

In such a scenario, the CCo is responsible for allocating resources by collecting information regarding the network state, i.e. number of users, channel conditions, QoS required from each user request, etc. It is worth noting that although we consider orthogonal broadcast channels, the algorithms presented in the following are also valid for the uplink case with opportune minor modifications.

As shown in Figure 6.27, we assume the MAC frame to have duration T_F equal to a mains cycle, i.e. $T_F = 20$ ms in Europe. The sub-frame consists of a header followed by a number of slots. The header and the slots have a duration equal to an integer number of OFDM symbols duration, i.e. they are equal to $T_H = N_H T_0$, and $T_S = N_{ITS} T_0$. Multiple MAC frames can be used to satisfy QoS constraints.

The MAC frame header carries the following information:

- Time slot duration,
- Scheduling for the time slots. The scheduling is a correspondence between each slot index and the physical address of the node whose slot has been reserved,
- Number of mains cycles where scheduling is valid.

Each slot carries also some OH information that is used by the PHY layer for synchronization and channel estimation algorithms.

6.4.5.1.1   MAC Procedures

The MAC protocol consists of three phases that we refer to as: network state learning, resource allocation, and data exchange. The first two steps have to be performed whenever a new node joins the network or the QoS required by an existing ones cannot be fulfilled with the current allocation, e.g. because of a significant channel variation has occurred.

During the *Network State Learning*, the CCo learns the network state, i.e. the links condition between itself and the nodes, the QoS required by the applications and so on. To accomplish it, the CCo may send training sequences to the users, like OFDM symbols in which each subchannel is modulated using QAM. All users estimate the SINR that they experience in each sub-channel and OFDM symbol in the TDMA sub-frame and compute bit-loading for all these symbols. The bit-loading map is denoted with $b^{(u,k)}(mN)$, and it provides the number of bits that can be transmitted in the k-th sub-channel to the u-th user if the CCo transmits during the m-th OFDM symbol of the MAC frame (cfr. eq. 6.24). The bit-loading map is then fed back to the CCo. It is worth noting that due to the fact that the channel is LPTV with period equal to the mains cycle, each user has to send only the information regarding one mains cycle, i.e. $b^{(u,k)}(mN)$, with $m = 0, \ldots, L - 1$. Furthermore, we point out that the bit-loading information per OFDM symbol is only required by the CCo to perform the time slot duration optimization, i.e. to find the optimal N_{ITS}. If a fixed time slot length is employed, only the bit-loading map per time slot is needed.

Once the CCo has received the bit-loading map from the nodes, it is able to determine the *Resource Allocation and the Scheduling*. To this end, it first computes the throughput for the transmission between itself and the $u - th$ user during the s-th time slot as

$$R_s^{(u)}(N_{ITS}) = \begin{cases} \frac{N_{ITS}-1}{T_F} \sum_{k \in \mathbb{K}_{ON}} \hat{b}_s^{(u,k)}, & N_{ITS} > 1, \\ 0, & \text{otherwise,} \end{cases} \qquad (6.28)$$

where $\hat{b}_s^{(u,k)}$ represents the number of bits loaded on sub-channel k during time slot s, and a scheduling procedure where a full OFDM symbol is used as PHY layer OH is assumed. This choice was described in [73]. In this case, we may send in the PHY header the bit map to be used in a slot as it is done in HPAV. A description of different scheduling procedures can be found in [72].

In 6.28, the number of bits loaded on each sub-channel is computed using 6.24 as

$$\hat{b}_s^{(u,k)} = \min_m \left\{ b^{(u,k)} \left(s N_{ITS} T_0 + m T_0 \right) \right\}, \ s = 0, \ldots, N_{TS} - 1, \ m = 0, \ldots, N_{ITS} - 1, \quad (6.29)$$

therefore assuming the bit-loading constant during each time slot.

Once the CCo has computed the rate that it achieves transmitting to each user in each time slot, namely $R_s^{(u)}(N_{ITS})$, it has to allocate the slots among the users and, further, it has to compute the optimal time slot duration. The problem can be formulated as an optimization problem, as follows,

$$\max_{c, N_{ITS}} \sum_{u=1}^{N_U} \sum_{s=0}^{N_{TS}-1} c^{(u,s)} R_s^{(u)}(N_{ITS}),$$

$$\text{s.t.} \sum_{u=1}^{N_U} c^{(u,s)} = 1, \quad s = 0, \ldots, N_{TS} - 1, \qquad (6.30)$$

$$\sum_{s=0}^{N_{TS}-1} c^{(u,s)} R_s^{(u)}(N_{ITS}) \geq \frac{p^{(u)}}{100} \sum_{s=0}^{N_{TS}-1} R_s^{(u)}(N_{ITS}), \quad u = 1, \ldots, N_U,$$

where $c^{(u,s)}$ denotes the binary coefficient equal to one if slot s is allocated to user u, and zero otherwise. The parameter $p^{(u)}$ is a weighting factor that denotes the percentage of data rate that the u-th user has to achieve with respect to the one that it would achieve in the corresponding single-user scenario. Problem (6.30) can be solved using integer programming once N_{ITS} is fixed. However, in some cases, integer programming is not able to provide a solution to (6.30) in a reasonable computation time. Moreover, it may also happen that the problem is not solvable satisfying the imposed constraints. In these circumstances, some constraints can be iteratively relaxed until integer programming gives a solution to the problem. Nevertheless, the problem may still be unsolvable in a reasonable time. To simplify the complexity, we propose to use LP [74, cfr. Ch. 4], once N_{ITS} is fixed. That is, for each value of N_{ITS}, the coefficients that give the slots allocation are returned via LP followed by rounding the coefficients $c^{(u,s)}$. Therefore, the optimal time slot duration is determined solving (6.30) for different values of time slot duration, i.e. N_{ITS}^{opt} is the number of OFDM symbols that maximizes the aggregate rate. Clearly, there are some cases where LP followed by rounding the coefficients may give a solution to (6.23) that does not satisfy all the constraints. In such cases heuristic solutions may be adopted.

It is interesting to note that since T_F and the SINR depend on the OFDM symbol duration and on the transmitter and receiver pulses, the optimization problem 6.30 may also be extended to take into account these parameters, which are neglected for reasons of simplicity.

Numerical Examples

To show numerical examples, we assume the following parameters. The transmission band is set equal to 0–37.5 MHz, and OFDM with rectangular transmit pulse and receiver window is used. The number of sub-channels is $M = 1536$ out of which 1066 are used, yielding a useful band in 2–28 MHz. The cyclic prefix duration is equal to 6.32 μs. Therefore, the OFDM symbol duration is 47.28 μs. The signal is transmitted with a power spectral density of −50 dBm/Hz. Hence, the number of OFDM symbols in the TDMA sub-frame is equal to 423. The header consists of 3 OFDM symbols. This results in 420 useful OFDM symbols in a frame. The slot duration can vary between a minimum of one OFDM symbol, up to 105 OFDM symbols for a number of nodes (users) N_U equal to 4. Bit-loading employing 2-PAM, 4, 8, 16, 64, 256 and 1024-QAM constellations is used. Furthermore, the SNR gap is set to $\Gamma = 9$ dB.

Figure 6.28 shows the aggregate rate for a four-user scenario obtained solving 6.30 for different time slot durations. The resulting single-user rates are also shown. Although not reported here, in [72], similar behaviors were found for two-user and three-user scenarios.

From Figure 6.28, it is observed that the aggregate rate has a maximum value. This behavior is explained by recalling that a fixed amount of OH per time slot is introduced (one OFDM symbol in our case). Therefore, the effect of the OH on the aggregate rate decreases by increasing the slot duration. This is true for a large set of time slot durations, and in particular, up to the point where the variation of the channel condition does not sensibly affect the bit-loading. After that, the aggregate rate starts to decrease. The quite flat behavior of the aggregate rate suggests to choose a 'global' optimum value of time slot duration for practical implementations. Consequently the resource allocation simplifies to the computation of the time slots assignment.

In [72], it is shown that in general, 50 OFDM symbols (~ 2.5μs) could be a good time slot value. This length leads to aggregate bit rate loss values smaller than 5%. In practice this loss can be much smaller (it could even be a gain), since employing a fixed time slot length

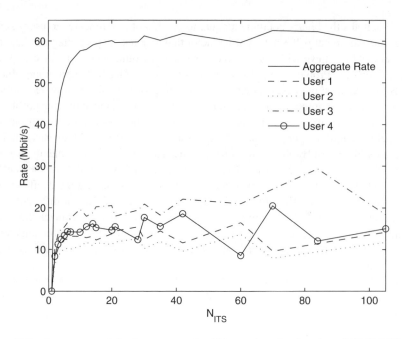

Figure 6.28   Aggregate and single-user rates for different time slot durations. [72] © 2009 IEEE.

considerably reduces the amount of signaling that the nodes have to transmit to the CCo. The result also agrees with the findings in [75], where a time slot duration of 1–2 ms was advocated.

6.4.6   Contention-based protocols with TDMA and FDMA

In the TDMA and OFDMA schemes discussed in the previous sections, a CCo schedules the access of each station to the shared resource, either time or frequency. These strategies are referred to as scheduled access. Their main advantages are the transmission efficiency and their ability to manage users with different QoS requisites. On the other hand, they have high signaling overhead and strict synchronization requirements, especially OFDMA. In contrast, a variety of random access strategies have been proposed. Among them, CSMA/CA is the most popular in PLC. This is a TDMA technique in which each user randomly attempts to transmit only after the channel has been sensed to be free. Its main advantage is its simplicity, since no CCo is needed, but idle slots and collisions, especially under heavy load conditions, lead to a reduced performance with respect to scheduled strategies.

In order to combine the ability of OFDMA to exploit the differences among the channel responses of the involved links, and the reduced signaling of the random access schemes, a generalized CSMA/CA for OFDMA systems has been proposed in [76]. In this strategy, a number of parallel subchannels is obtained by dividing the system carriers into subsets. Since users can sense the state of all the subchannels at each OFDM symbol, CSMA/CA can be employed in each subchannel. This leads to a two-dimensional CSMA/CA where users simultaneously contend in the time and frequency domains. When a user has data to be transmitted, it randomly occupies one subchannel among the idle ones.

In order to exploit the frequency selectivity of PLC channels, a slightly different version of the aforementioned OFDMA CSMA/CA technique is proposed in [77]. In this case, a CCo establishes the set of subchannels, from which each station chooses its own one. The CCo broadcasts a beacon with the number of users in each subchannel. After hearing the beacon, each user selects the subchannel that maximizes its achievable throughput. Hence, this scheme combines the lower collision probability of the generalized OFDMA CSMA/CA with the spectral efficiency of OFDMA, which is not fully exploited in [76]. This leads to an increased performance over the CSMA/CA used in conventional single-channel operation [77].

Following a similar idea, an opportunistic random access for OFDMA systems is proposed in [78]. In this strategy, subchannels are assigned to the user with better channel conditions at the end of the contention time. In addition, the backoff counter of each user, which determine transmission attempts, is updated taking into account the transmission characteristics of the sub-channels.

6.4.7   Related Literature

Other approaches have been presented in the literature to deal with the problem of multi-user resource allocation in PLC networks. In the following, a number of relevant articles on the topic is reported. They are gathered according to the channel access technique adopted, i.e. FDMA or contention free and contention based TDMA.

6.4.7.1   FDMA

Papers [64], [67], [77]–[83] treat the problem of multi-user resource allocation in indoor broadband PLC scenarios where the channel access is based on FDMA. More precisely, in [64], it is proposed a practical resource allocation algorithm that aims at allocating the channels among the network users to maximize the aggregate network rate under a PSD and a minimum data rate constraints, and according to the channel quality of each link. Both broadcast (downlink) and multiple-access (uplink) communications obtained through FDD are considered.

In [67], the authors formulate an optimization problem that aims at maximizing a weighted sum rate under a total power constraint. The resulting non-linear integer problem is solved by means of an approximation to a convex optimization problem. Numerical results show that the proposed method gives results close to the optimum. The algorithm is extended to two-hop PLC relay networks in [79].

In [80], an iterative algorithm that jointly computes the FDMA channels and powers allocation for the downlink case is proposed in order to maximize the aggregate network goodput [84] under a fairness, a total power and a PSD constraint. At each iteration, an outer cycle is responsible of allocating the channels among the network users, whereas an inner cycle computes the power allocation. In the same context, in [81], the authors propose an heuristic algorithm to partition channels among the network users with the objective of maximizing the gain given by the ratio between the FDMA and the TDMA bit rates under a fairness constraint. Numerical results show that, at PHY layer, FDMA performs better than TDMA and the proposed solution leads to higher gain w.r.t. the simpler channel allocation procedure proposed in [82]. Nevertheless, in [83], it is shown that FDMA does not always lead to throughput gains w.r.t. TDMA when considering the performance at the transport layer, i.e. under UDP traffic. The reason behind it is that although FDMA reaches higher PHY layer bit rate, it needs a greater quantity of OH to be implemented. Consequently, its use is beneficial only when the exploitation of frequency diversity accounts more than the loss caused by the extra OH.

An interesting hybrid scheme that makes use of CSMA/CA and FDMA for the uplink channel is proposed in [77]. According to the proposed scheme, each user contends only for the OFDM sub-channels that allow for high bit rate with a resulting decrease in collisions and thus an increase of the aggregate network throughput. The scheme is optimized in [78].

6.4.7.2   TDMA

The use of a contention free TDMA channel access scheme over indoor broadband PLC networks was firstly proposed in [85] to allow HomePlug AV devices to deliver HDTV streams without frame drops. According to the proposed scheme, the nodes can occupy a number of time slots in a persistent manner according to a request to send/clear to send protocol. The protocol is distributed, i.e. there is not a network coordinator. In [86], the authors proposed an opportunistic contention based protocol to deliver high QoS uplink traffic. The idea behind the proposed protocol is to exploit the cyclostationary behavior of the noise experienced at the receiver to divide a mains cycle into two kind of regions: the ones with high and low SNRs. Nodes with high priority traffic contend and occupy the channel during the high SNR regions, whereas nodes with low priority traffic contend for the channel during low SNR regions. Numerical results show good improvements in terms of throughput w.r.t. the prioritized CSMA/CA adopted by HomePlug AV devices.

The employment of TDMA was also proposed in [87] to improve the performance given by the CSMA/CA channel access protocol defined by the G3-PLC standard when used for smart-home/building applications.

6.5   Cooperative Power Line Communications

In this section, we present the application of cooperative transmission techniques in PLC. First, cooperative communications is briefly reviewed and then the state of the art on cooperative PLC is presented. It should be noted that here, by cooperative communications, we refer to the cooperative techniques at the physical layer. This is different from the conventional multi-hop transmission schemes where the network nodes convey a message from a source to a destination using consecutive transmission based on a routing table.

6.5.1   Introduction to Cooperative Communications

Cooperative communication [88] has been introduced as a paradigm shift from the conventional point-to-point communication. In a point-to-point system, each user communicates directly with the destination without incorporating any other user in the communication process.

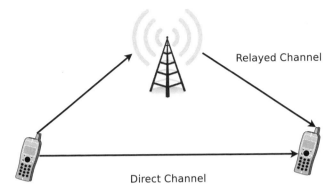

Figure 6.29   A simple three-node cooperative setup.

This approach may benefit more resourceful users, but hurts the weaker users by a lower quality of service. In a cooperative communication system, users cooperate with each other through sharing their resources, e.g. power and computation, in order to improve the system's performance measures for all users or save on the overall system's resources. In other words, a cooperative user takes a more altruistic approach and is in charge of transmitting other users' data in addition to its own data.

The simplest setup for a cooperative system is the three-node network depicted in Figure 6.29. In this setup, there exist one source, one destination and one intermediate or relay node. The relay node cooperates with the source node to assist it in the data communication with the destination node. In a three-node cooperative scenario, if the direct channel between the source and the destination is not reliable, the relay can help the source by delivering its message to the destination node through a different path. In other words, while in non-cooperative communication any damage to the direct link deteriorates the data communication, damage to the direct link can be handled using the alternative link through the relayed path in Figure 6.29. The simple three-node setup can be extended to the situations where multiple relays are involved in the data relaying. Relay nodes may operate in a parallel mode, i.e. all of them directly receive the signal from the source and directly communicate to the destination, or in a cascade mode where the data is delivered to the destination through multiple hops.

The first study on the cooperative communications can be traced back to [89] and [90] by Van der Meulen in the early 1970s, where the basic three-node setup was studied. Assuming that nodes cooperate with each other, the problem of efficient data transmission over such a channel was addressed. Another groundbreaking work on cooperative channels was presented by Cover and El Gamal in [91]. They investigated the information theoretic aspects of the three-node relay channel. Despite its early emergence in communications theory, cooperative communications stayed a relatively silent research field until recent years. During this quiet period, most studies, e.g. [92] and [93], were focused on the information theoretic aspects of the cooperative communications, and it was shown that cooperation can increase the system capacity. However, with an increasing need for improving coverage and reliability of wireless communications, this changed in early 2000s and cooperative communications gained considerable research interest and momentum. As a consequence, several relaying protocols were developed, cf. e.g. [88], [94]–[98].

Among the relaying strategies that have been developed for cooperative communication systems, two of them, namely amplify-and-forward (AF) and decode-and-forward (DF) [97], [98], have been the key strategies for users' cooperation. Consider the three-node setup presented in Figure 6.29. In AF, the source node transmits its message signal, which is received at both destination and relay nodes. After receiving the source signal, the relay node only amplifies the signal without further processing and forwards it to the destination node. For DF, however, the relay does more processing on its received signal and first decodes the source message. After re-encoding the source message, the relay transmits the coded signal to the destination node. The performance of AF and DF cooperative schemes along with different TDMA protocols was studied in [99] for fading channels showing the potential improvement offered by cooperative communications.

Considering the benefits of users' cooperation, especially extending the network coverage and boosting the system capacity, cooperative communications has been adopted in the elaboration of the new wireless standards, e.g. long term evolution (LTE) [100] and IEEE 802.16 [101]. While cooperative wireless communications have been widely studied, cooperative PLC is a relatively unexplored field. In the following, we discuss why cooperative communication is a promising approach to enhance the performance in PLC systems and review the state of the art in cooperative PLC.

6.5.2   Cooperative Power Line Communications

While using the existing power line infrastructure for data communication is cost-efficient, it faces several important challenges as well. One of the main challenges is the doubly-selective nature of the PLC channel meaning that the channel is selective over both frequency and time. The frequency selectivity is a result of the frequency characteristics of the power cables, loads connected to the network, and the mismatched power line junction points. Although the effect of the cables and junctions is almost time invariant, any temporal change in the impedance of the loads (abruptly when plugging an electric device or periodically with the mains cycle) alters the channel frequency response and results in the time selectivity. Another challenge in PLC networks is the significant signal attenuation mainly resulting from the lossy power cables and energy being absorbed at undesired load points. A direct result of having a lossy and frequency-selective channel is the low signal power at the receiver side and consequently poor data communication quality. To compensate for the signal power loss, one may suggest to increase the transmit power. However, this is not an option, mostly because of the electromagnetic compatibility (EMC) constraints. Hence, other solutions rather than increasing transmit power should be pursued.

Due to the possible rate and coverage improvement of the cooperative communication demonstrated for wireless systems, it seems to be a viable option to overcome the above mentioned challenges for PLC. In addition, the broadcast nature of the PLC channels further promotes the application of cooperative schemes. To explain this better, consider the PLC network depicted in Figure 6.30, where the source node u_s wants to send its data to the destination node u_d. The power cable connecting u_s and u_d traverses N other intermediate users, denoted by $u_1, u_2, \ldots, u_N$, on its way. When u_s transmits its signal toward u_d, some or all of the u_i's receive it due to the broadcast nature of the channel. The intermediate users can then apply a cooperative scheme, e.g. AF or DF, and use the overheard signal to help u_s in its data transmission.

Figure 6.30 Illustration of a PLC network with source u_s, destination u_d and N intermediate nodes.

Although cooperation between the users can potentially benefit both wireless and PLC systems, we should be aware of a substantial difference between these two communication media. In a wireless system, it is mainly true that the signals received at different cooperating nodes are statistically independent. For instance, the direct and relayed signals in Figure 6.29 often experience independent fading in a wireless setting. However, considering for example the PLC case illustrated in Figure 6.30, signals transmitted from u_s or any u_i and received at u_d travel over a common segment of the main power cable (horizontal line in the figure). This implies a dependency between the signals received from different transmitting nodes. A simple case to show this dependency is when a damage happens to the main power cable which hurts the signal quality received from all intermediate nodes. More generally, the channel frequency response between any two PLC network nodes can be written as the product of the frequency responses of channel segments along the path between the two nodes (cf. [102, Eq. 15]). Hence, signals going through the same segment will become dependent, which has implications for the diversity achievable with PLC as shown in [103]. Since PLC networks typically have a tree topology[2], the argument about the dependency between the signals of the cooperating nodes is legitimate also for PLC setups other than the one shown in Figure 6.30.

Depending on the flow of the information in the system, cooperative PLC systems are categorized into *one-way* and *two-way/multi-way* systems. In one-way cooperative PLC, there is a set of source nodes which send their data to a set of destination nodes via intermediate (relay) nodes. The sets of source and destination are disjoint in one-way networks and thus the information flow is only in one direction, i.e. from the source nodes to the destination nodes. The described PLC network scenario in Figure 6.30 is an example of a one-way cooperative PLC network. Different from this, source and destination sets are not disjoint in a two-way/multi-way system in which some or all users can simultaneously act as data source and destination. As a consequence, the information flows in two or more directions in the network.

In the sequel, we present the advances on one-way and two-way/multi-way PLC systems. Note that our study focuses on half-duplex PLC where each node is either in the transmit or receive mode at a time.

6.5.3 One-way Cooperative PLC Systems

The non-cooperative approach of *direct transmission* (DT) from the source to the destination node is often used as the simplest benchmark to evaluate the performance of cooperative PLC schemes. In addition to the direct transmission, one can also use the conventional multi-hop transmission based on a routing table in a one-way PLC system to deliver the data from the source to the destination. Note that the cooperation in this approach does not happen in the physical layer. Since only one user transmits at a time, according to the schedule acquired

---

[2] Note that in a tree graph, there is only one disjoint path between any two vertices of the graph.

from the routing table, this approach is called *single-node retransmission* [104]. Most of the studies on cooperative PLC, however, focus on the cooperation at the physical layer and several strategies have been proposed in the literature for physical-layer cooperation in one-way PLC systems. In the following, we review these strategies.

6.5.3.1 Single Frequency Networking (SFN)

The idea presented in [105] is one of the first studies addressing cooperative PLC at the physical layer. The suggested cooperative strategy builds on the application of single-frequency communication [106] and data flooding to enable data communication between users in a multi-hop fashion. While this approach is beneficial in terms of its robustness against network topology changes and reducing the routing overhead in the network, it suffers from several aspects. Specifically, the simultaneously transmitted signals from the users can be destructive and cancel each other out. Also, synchronization is critical for single-frequency communication. This necessitates the existence of a network-wide clock, which is however not unique to SFN but required for essentially all cooperative transmission schemes considered for PLC. Another aspect that needs to be considered for the scheme presented in [105] is the negative effect of flooding on the network traffic. Mechanisms to mitigate this are described in [107].

6.5.3.2 Distributed Space-time Block Coding (DSTBC)

An extension of the single-frequency transmission approach from [105] is a scheme based on space-time coding [108] proposed in [104], [109] and later studied further in [110]. Unlike single-frequency communication, where multiple retransmitted copies of a signal may superpose destructively, in this scheme, space-time coding is employed to improve the usefulness of the retransmissions made by the intermediate nodes. The basic idea behind the space-time coding is to smartly transmit multiple replicas of data, over both time and space (i.e. from different nodes) such that after signal combining at the destination side an overall SNR gain is achieved. For the simple case of a linear PLC network, similar to Figure 6.30, orthogonal space-time block codes (OSTBC) are studied first in [104]. While OSTBC provides a performance improvement over the single-frequency communication, which is called *simple transmission* in [104], it suffers from the rate loss and becomes inefficient when the number of users increases. This is due to the nonexistence of rate-one codes [104] for the situation when $N_u > 2$ intermediate users exist. To extend the application of space-time coding to larger networks, an approach based on the distributed space-time block codes (DSTBC) [111] is then proposed in [104]. Using DSTBC, a unique signature is assigned to each node that allows for the efficient combining of different users' signals at the destination node without significant rate loss.

Before proceeding to the performance evaluation of DSTBC, we introduce an important measure called *outage probability*. We here consider the probability of the event that the achievable data rate in the system drops below a certain rate threshold. More formally, if we denote the mutual information between the PLC channel input at u_s and the channel output at u_d by I, and the rate threshold by R, then the outage probability is

$$p_{out} = Pr\{I < R\} \qquad (6.31)$$

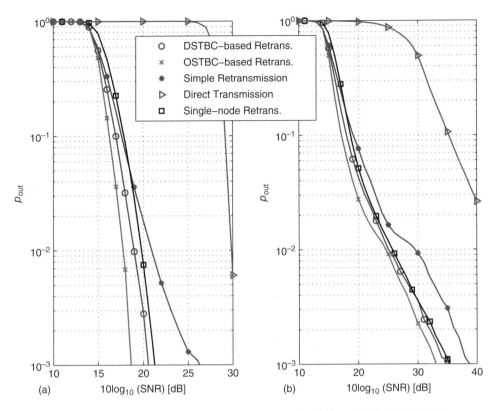

Figure 6.31 Outage probability versus SNR for $R = 1$ bits/sec. [104] © 2006 IEEE.

where the data rate is defined in bits per channel use. This is often also referred to as outage capacity, either with explicit optimization of I or by considering mutual information as a (e.g. signal constellation) constrained capacity.

To evaluate the performance of space-time coding for cooperative PLC, a linear PLC network with six intermediate nodes is considered. The source and destination nodes are 100 meters apart and the intermediate nodes are positioned randomly over the power cable between the source and the destination. The cable parameters are chosen based on a model suggested in [112]. The outage performance of the space-time coding along with other cooperative schemes is presented in Figure 6.31, considering a normalized target rate of $R = 1$ bit/(channel use) and the channel frequency responses at 1 MHz. Figure 6.31(a) is for the case when the network 'behind' the destination node is described by an impedance of $Z_d = 10^4$ Ω, and Figure 6.31(b) presents the results when the magnitude and phase of Z_d are uniformly drawn from [2, 200) Ω and [0, 2π) respectively. Although OSTBC is not viable for networks with more than two intermediate nodes, it is used in Figure 6.31 as an upper limit for the performance of DSTBC. It can be seen from the figure that DSTBC achieves an outage probability close to that of OSTBC and outperforms direct transmission, single-node transmission based on a routing table and simple transmission using single-frequency communication.

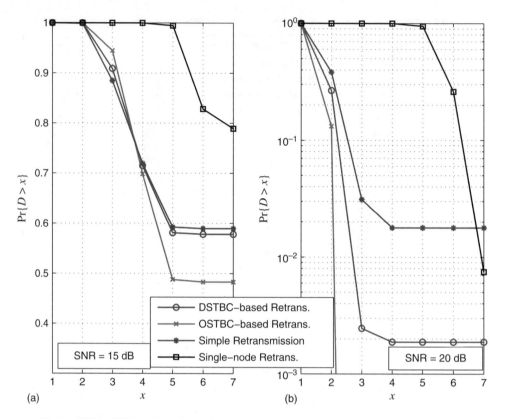

(a) x (b) x

Figure 6.32 CCDF of D for different schemes over two SNR values. [104] © 2006 IEEE.

In addition to the outage behavior, the delay characteristics of the aforementioned transmission strategies are also of interest. To study the delay behavior, the total number of transmissions by the network nodes that is required for delivering the data at the destination node is used as a measure. This parameter is called D and Figure 6.32 shows the complementary cumulative distribution function (CCDF) of D, denoted by $\Pr\{D > x\}$, for two different SNR values. We observe that DSTBC again outperforms simple and single-node retransmission and performs close to the OSTBC benchmark.

6.5.3.3 Cooperative Coding

Another approach for effective cooperative PLC is *cooperative coding* [103], [113], [114]. To explain the basic idea behind cooperative coding, consider a PLC network similar to the one in Figure 6.30. When u_s transmits its signal over the cable, all other nodes can fully or partially decode u_s's message. Considering the distance-dependent signal loss over the cable, u_1 is the most likely node to fully decode the source's message and u_d is the least likely one. All other intermediate nodes fall somewhere in between. To help u_d in fully decoding u_s's message, the intermediate nodes repeat u_s's message over the cable until the message is fully decoded at u_d. To save transmission energy and channel resources, instead of retransmitting the full source message, u_1 only sends as much of this message as needed by u_2 to fully decode it, after having received the original transmission. That is, combining the signal from first and second transmissions, u_2 is able to fully decode u_s's message. Then, u_2 partially retransmits the source message such that u_3 is capable of fully decoding u_s's message. This procedure continues until u_N transmits and u_d is able to fully retrieve the source message.

Here, we are interested in finding the achievable data rate of this cooperative coding. To this end, let us assume that u_s has k bits to send to the destination node, and for convenience, let u_s and u_d be referred to as u_0 and u_{N+1} respectively. Denoting the link capacity (in bits/(channel use)) for the line piece between u_i and u_j by $C_{i,j}$, it takes $n_0 = k/C_{0,1}$ transmissions by u_0 to fully receive the source message at u_1. Meanwhile, u_2 has received an information content of $n_0 C_{0,2}$ bits from the source message. Now, to make sure that u_s's message is decodable at u_2, u_1 needs to transmit only n_1 coded bits such that $k = n_0 C_{0,2} + n_1 C_{1,2}$. Similarly, u_2 transmits only the necessary amount of information over the line such that u_3 is capable of decoding the complete source message. Continuing this procedure, the relation between the link capacities and the number of transmissions by each node is formulated as follows

$$
\begin{bmatrix}
C_{0,1} & 0 & \cdots & 0 \\
C_{0,2} & C_{1,2} & 0 & \cdots \\
\vdots & & & \\
C_{0,N+1} & C_{1,N+1} & \cdots & C_{N,N+1}
\end{bmatrix}
\begin{bmatrix}
n_0 \\
n_1 \\
\vdots \\
n_N
\end{bmatrix}
=
\begin{bmatrix}
k \\
k \\
\vdots \\
k
\end{bmatrix}.
\tag{6.32}
$$

As can be seen from (6.32), information about the source message gradually accumulates at each node until the node is capable of fully decoding it. Since it takes $n_0 + n_1 + \cdots + n_N$ transmissions to convey the k-bit message from u_s to u_d, the overall data rate for the cooperative coding scheme is

$$
R_{\text{coop}} = k \left(\sum_{i=0}^{N} n_i \right)^{-1}.
\tag{6.33}
$$

Now, if we apply conventional DF, where the nodes do not exploit the overheard signals, to the same PLC setup, the achievable data rate is

$$
R_{\text{DF}} = \frac{1}{N+1} \min C_{i,i+1}, \quad i = 0, 1, \ldots, N.
\tag{6.34}
$$

Figure 6.33 depicts the comparison between the achievable data rate of cooperative coding and conventional DF where the source and destination are 1 km apart. The intermediate nodes are equally distanced between u_s and u_d and a binary modulation scheme is used for data transmission. Also, the signal attenuation over the channel is determined by an attenuation factor of δ dB/km and the signal's traveled distance. Here, the results are presented for three different values of $\delta = 40, 60, 100$ dB/km. As seen, cooperative coding significantly outperforms DF especially for larger values of N. Another interesting observation is that while the data rate of the cooperative coding is an increasing function of N, this does not hold for DF.

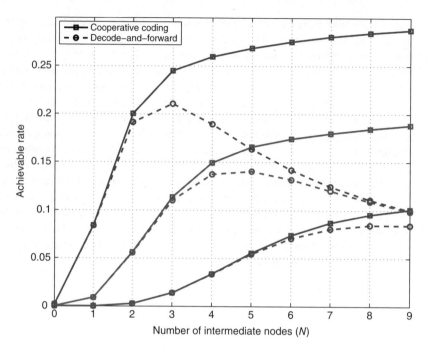

Figure 6.33   Achievable end-to-end rate (in bit/(channel use)) of cooperative coding and conventional DF for $\delta = 40, 60, 100$ dB/km (curves from top to bottom, numerical data according to [114, Figure 8]).

6.5.3.4   AF and DF Relaying

AF relaying has also been applied for cooperative one-way PLC systems. Cheng *et al.* [115], [116] study the achievable data rate in a dual-hop AF cooperative PLC system, i.e. $N = 1$ in Figure 6.30. Assuming a frequency-selective model for the channel, the capacity of such relay-aided system is analyzed through optimizing the power allocation over the frequency band. This analysis reveals that the application of AF relay-aided PLC is more beneficial in situations where the density of the load branches connected to the main power cable is high and the load impedances are low. In another study [117], the authors compare the performance of dual-hop AF and DF relaying over a medium-voltage electricity network when orthogonal frequency division multiplexing (OFDM) is used. While AF has the advantage of lower relaying complexity over DF, it is outperformed by DF in terms of the achievable data rate in the system. In [118], a new cooperative protocol based on AF is designed for large medium-voltage networks (in terms of the number of intermediate nodes). This protocol has a better cooperative gain compared to the conventional AF schemes and results in a notable increase in the achievable data rate of the system.

6.5.3.5   AF and DF Relaying for Indoor PLC

Cooperative schemes are also advantageous for indoor PLC systems. Here, the main purpose of user cooperation is to meet the high data-rate requirements, e.g. video gaming and high

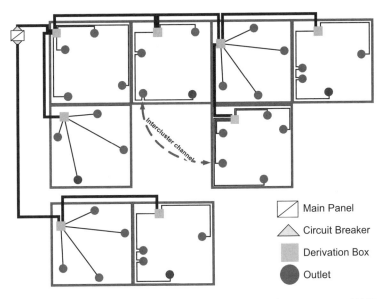

Main Panel

Circuit Breaker

Derivation Box

Outlet

Figure 6.34   Schematic of an in-home PLC system and its components [119].

definition televisions. Figure 6.34 illustrates an indoor PLC system, which consists of several clusters of outlets that are connected through their cluster head, which is commonly referred to as *derivation box* (DB). The DBs inside the house are then connected to each other as well as the main electricity panel (MP). The electricity comes from the step-down transformer to the house MP through the circuit breaker and then is distributed over the outlets through DBs. When two users (outlets) from different clusters communicate with each other, their signal passes through an *intercluster channel*. A signal transmitted over an intercluster channel experiences a significant attenuation due to: (i) traversing over long cable pieces, (ii) passing via two or more DBs and possibly MP, and iii) the effect of the loads connected to the outlets. To compensate the intercluster signal loss, a relay-aided cooperative scheme is applicable that improves the overall system performance including the data rate and energy efficiency.

One of the first studies on the application of cooperation for indoor PLC systems is reported in [79]. The authors argue that without proper resource allocation, using a relay can in fact degrade the data throughput. This comes as a result of taking away some parts of the resources from the direct link and allocating it to a weak relay channel. Considering this, a dual-hop relay-aided protocol based on OFDMA is then proposed, in which the subchannels and their assigned power are optimally allocated to the source and relay nodes. As concluded in [79], this carefully designed protocol significantly improves the data rate for the considered indoor PLC setup.

Another approach to exploit cooperative communications for indoor PLC systems is through opportunistic relaying [119]–[121] along with OFDM where the power allocation of the subchannels satisfies a power spectral density (PSD) mask constraint. Opportunistic relaying can be based on DF or AF, called opportunistic DF (ODF) and opportunistic AF (OAF) respectively. The concept of opportunistic cooperative PLC relies on using the relay whenever it is beneficial in terms of: i) improving the data rate while the PSD is kept below a predefined threshold, or ii) saving energy while meeting a target data rate and PSD threshold. When the cooperation through the relay is not beneficial, users communicate directly. The opportunistic relaying can be viewed as a combination of conventional DF and AF, where the relay is always involved in the data communication, and DT where there is no relay in the system. Considering a unit transmission time to convey the source message to the destination, the achievable data rate of ODF is found using the data rate of DT and conventional DF as follows [119]:

$$R_{\text{ODF}} = \max\{R_{\text{DT}}, R_{\text{DF}}(\tau)\} \tag{6.35}$$

where R_{DT} is the achievable data rate of direct transmission. Also, for $0 \leq \tau \leq 1$, $R_{\text{DF}}(\tau)$ denotes the achievable data rate of DF when source transmits to the relay and the destination during a fraction τ of the transmission time and the relay transmits to the destination for a fraction of $(1 - \tau)$ of the transmission time. Denoting the capacity of source-destination, source-relay, and relay-destination links by $C_{\text{s,d}}$, $C_{\text{s,r}}$, and $C_{\text{r,d}}$ respectively, we have

$$R_{\text{DT}} = C_{\text{s,d}},$$
$$R_{\text{DF}}(\tau) = \min\{\tau C_{\text{s,r}}, \tau C_{\text{s,d}} + (1 - \tau)C_{\text{r,d}}\}. \tag{6.36}$$

The achievable rate of OAF is described similarly in terms of the achievable rate of DT and AF. As can be seen from 6.36, the achievable rate of the opportunistic relaying depends on τ as well as the link capacities, which in turn are functions of the transmit powers of the source and the relay. Thus, to achieve the ultimate system performance, one should optimize τ and the power allocation between the source and the relay such that the data rate is improved or energy is conserved. For more details on solving the time and power allocation optimization problem, the interested reader is referred to [119].

In addition to the effect of the time and power allocation, relay placement also affects the system performance and should be considered as a design parameter. As argued in [119], however, the location of the relay is limited to the accessible points of the network. In other words, the relay can be placed only at outlets, DBs, or the MP. Considering this point, several relaying configuration are possible [119]:

- Outlet relay arrangement (ORA): The relay is connected to a randomly-chosen outlet.
- Main panel selection (MPS): The relay is placed in MP.
- Random derivation box (RDB): One of the DBs is randomly selected to put the relay.
- Source derivation box (SDB): The relay is located in the DB servicing the source cluster.
- Destination derivation box (DDB): The relay is placed inside the DB which feeds the destination outlet.
- Backbone derivation box (BDB): Let us denote the set of DBs on the path from the source to the destination by S. Note that both source and destination DBs belong to S. In this methods of relay placement, one of the DBs from S is randomly chosen to include the relay. This approach is different from RDB where also DBs not in S are included in the selection set.

The performance of the opportunistic relaying with optimized time and power allocation and different configurations for relay location is extensively investigated in [119] in terms of

Table 6.2 Achievable rates of different relaying schemes [119]

Configuration	R_{DT}	R_{DF}	R_{ODF}	U(ODF)	R_{AF}	R_{OAF}	U(OAF)
SDB	182.8	220.2	220.2	99.9	128.9	183.8	12.0
BDB	183.2	216.5	216.8	99.9	127.0	186.4	18.8
RDB	190.3	104.3	207.6	52.8	113.9	191.9	11.1
MPS	190.2	99.2	205.9	48.5	112.5	191.7	9.5
ORA	193.6	70.6	202.7	29.8	107.7	194.2	6.6
DDB	182.9	189.0	189.9	99.4	106.2	183.0	0.7

the data rate improvement and energy conservation. For this purpose, the authors consider both real-world channel model based on measurements as well as a statistical channel model based on the methodology from [102], [122]. Here, some of the results are presented in Table 6.2 and Figure 6.35. However, the interested reader is encouraged to see [119] for the complete results and details of the simulation setup.

Table 6.2 presents the comparison between the achievable data rates for different relaying schemes considering the described relay placement configurations. In this table, U(ODF) and U(OAF) represent the percentage of the relay usage in ODF and OAF during the transmission time, respectively. Also, all data rates are in Mbit/s. To obtain these results, it is assumed that the noise PSD is −110 dBm/Hz and data communication happens over the 1–28 MHz frequency band commonly used in indoor PLC standards. Also, the PSD mask constraint is set

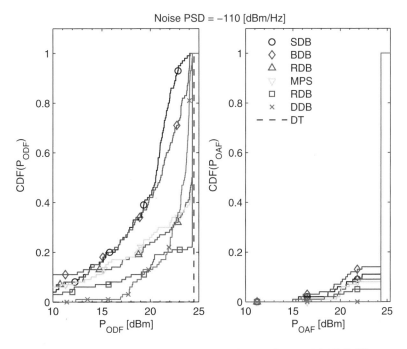

Figure 6.35 CDF of the transmit power for DT, ODF and OAF [119].

to −50 dBm/Hz and the PLC channel is generated based on the statistical model described in [119]. These results suggest that while ODF noticeably boosts the data rate, the performance improvement made by OAF over DT is marginal. It can be seen that ODF (OAF) outperforms both DT and DF (AF) by smartly combining them. The best performance by ODF is reached through SDB and BDB, where the relay is heavily involved in the data communication. Unlike ODF, OAF only little exploits the relay for data transmission and thus its performance is basically determined by the capacity of the direct link between the source and destination. Another observation is that the performance of DF and AF significantly depends on the relay's location and as pointed out in [79], they may actually be outperformed by DT under specific circumstances.

Figure 6.35 depicts the cumulative distribution function (CDF) of the transmit power, as a measure of energy efficiency, for direct transmission, ODF and OAF. To obtain the results in this figure, the target data rate is set to the capacity of the direct link, $C_{s,d}$. Then, the source and relay power as well as τ are optimized to provide maximum energy saving. Here, similar to the behavior seen in the data rate results, ODF provides a significant energy saving over direct transmission while OAF contributes a slight improvement.

6.5.4   Two-Way and Multi-Way Cooperative PLC Systems

After appreciating the advantages of using cooperative PLC to improve the system performance in a one-way PLC system, we go one step further and study the possibility of cooperative schemes for two-way (TW) [123] or multi-way (MW) [124] PLC scenarios. To gain a better understanding of the TW (MW) scenario, consider the setup presented in Figure 6.34 and assume that two (several) users want to exchange their data. Users can be under the same cluster or come from different clusters. Possible applications of this setup are video conferencing, file sharing and gaming. One approach to accomplish data communication for this scenario is to use TDMA and divide the transmission time to several slots. During each time slot, a specific user transmits its data with the help of the relay to the rest of the users based on a one-way relaying approach.

Another way to approach the data communication in the described PLC setup is to use two-way relaying (TWR) or multi-way relaying (MWR) based on network coding. Network coding [125] is an efficient technique to enhance the spectral efficiency and data throughput in communication systems. The essential notion of network coding is to allow data combining at the intermediate nodes instead of simply forwarding it. One simple approach for data combining at the intermediate nodes is to forward the sum of the received packets. Physical-layer network coding (PLNC) [126] is a variant of network coding that allows data mixing at the channel, instead of the data mixing at the intermediate nodes. This is achievable through permitting simultaneous transmissions by the users and proper processing of the interfering signals at the relay.

A visualization of how PLNC improves the data rate in a cooperative system is presented in Figure 6.36. The figure considers a communication setup where two users, u_1 and u_2, want to share their messages, x_1 and x_2. As it can be seen, when one-way relaying is used, users transmit their signals to the relay in separate time slots. After receiving each user's message, the relay then forwards the message to the other user in one time slot. This means that for the considered two-user setup, it takes four time slots for the users to exchange their messages.

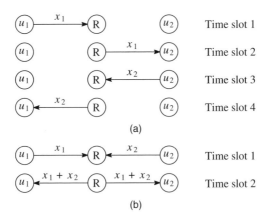

(a)

(b)

Figure 6.36 (a) Conventional one-way relaying with TDMA, and (b) two-way relaying with physical-layer network coding.

However, when TWR based on PLNC is used, both users transmit in the first time slot. For an additive channel, which is commonly assumed in communication systems, the relay receives the superposition of the users signal and broadcasts it in the second time slot. Now, after canceling its own signal, each user is able to obtain the message of the other user. Comparing the transmission time of one-way and two-way schemes, we observe that using TWR based on

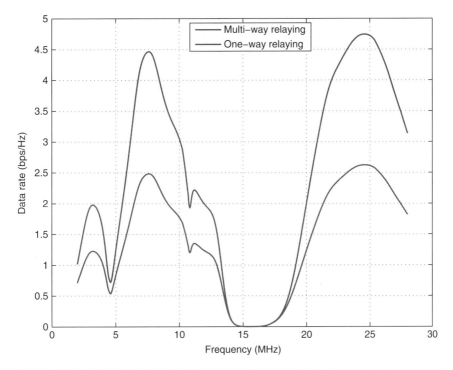

Figure 6.37 Achievable spectral efficiency over frequency for $K = 2$. [130] © 2013 IEEE.

PLNC cuts the channel usage in half which translates into improving the data rate. The same logic applies to MWR where more than two users share their data. For more information on the advantage of MWR over one-way relaying, see [127].

Application of TWR for cooperative PLC systems first appeared in [128]. In this work, the authors compare the achievable data rate of TWR and one-way relaying for both AF and DF relaying scenarios for an indoor PLC environment. Generally speaking, it is observed that depending on the distance between the source and destination, TWR provides a significant improvement over one-way relaying and direct transmission. Another study reported in [129] also considers the application of TWR for inter-domain communication based on the ITU-T G.hn standard where logical link control stack is proposed for efficient implementation of TWR.

MWR has also been employed to elevate the performance of PLC systems with more than two users. The first work to explore this idea is presented in [130] where $K \geq$ users want to share their data with the help of a relay in an indoor PLC setup similar to Figure 6.34. To overcome the frequency selectivity of the channel, OFDM has been employed, and AF relaying is applied to aid the users in their data communication. Then, the authors calculate the achievable common data rate and spectral efficiencies of TDMA one-way relaying and AF MWR considering the noise model presented in [112]. The spectral efficiency results for $K = 2$ and $K = 4$ are presented in Figure 6.37 and Figure 6.38 respectively. For the details

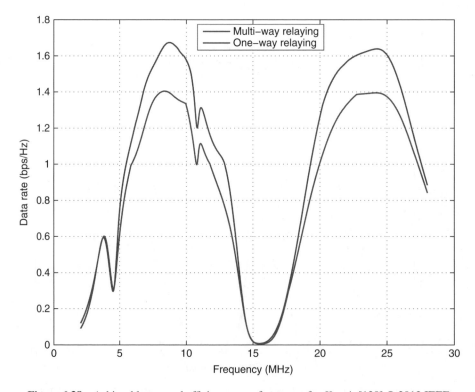

Figure 6.38   Achievable spectral efficiency over frequency for $K = 4$. [130] © 2013 IEEE.

of the simulation setup, including the length of the power cable and its parameters as well as the loads connected to the outlet, please refer to [130]. As it can be seen, MWR outperforms one-way relaying in both scenarios and provides better spectral efficiency. Furthermore, it is shown in [130] that the achievable data rate and spectral efficiency for MWR are proportional to $\frac{1}{(K-1)}$, while this is $\frac{1}{K}$ for one-way relaying. This explains why MWR has a more superior performance over one-way relaying for $K = 2$ compared to $K = 4$.

References

1. L. P. Do, Hierarchical resource allocation for powerline communications (PLC) networks, Ph.D. dissertation, Technische Universität Dresden, Dresden, Germany, 2012.
2. L. P. Do and R. Lehnert, Dynamic resource allocation protocol for large PLC networks, in *Proc. IEEE Int. Symp. Power Line Commun. Applic.*, Beijing, China, Mar. 27–30, 2012, 41–46.
3. A. Haidine and R. Lehnert, Analysis of the channel allocation problem in broadband power line communications access networks, in *Proc. IEEE Int. Symp. Power Line Commun. Applic.*, Pisa, Italy, Mar. 26–28, 2007, 192–197.
4. S. Galli, A. Kurobe, and M. Ohura, The inter-PHY protocol (IPP): A simple coexistence protocol for shared media, in *Proc. IEEE Int. Symp. Power Line Commun. Applic.*, Dresden, Germany, Mar. 29–Apr. 1, 2009, 194–200.
5. J. Chen, D. Seah, and W. Xu, Channel allocation for cellular networks using heristic methods, Cornell University, Tech. Rep. CS574, 2001. [Online]. Available: http://foulard.ece.cornell.edu/wxu/
6. S. H. Wong, Channel allocation for broadband fixed wireless access networks, Ph.D. dissertation, University of Cambridge, England, 2003.
7. Y. Aryropoulos, S. Jordan, and S. P. R. Kumar, Dynamic channel allocation in interference-limited cellular systems with uneven traffic distribution, *IEEE Trans. Veh. Technol.*, 48(1), 224–232, Jan. 1999.
8. S. Anand, A. Sridharan, and K. N. Sivarajan, Performance analysis of channelized cellular systems with dynamic channel allocation, *IEEE Trans. Veh. Technol.*, 52(4), 847–859, Jul. 2003.
9. F. D. Priscoli, N. P. Magnani, V. Palestini, and F. Sestini, Application of dynamic channel allocation strategies to the GSM cellular network, *IEEE J. Sel. Areas Commun.*, 15(8), 1558–1567, Oct. 1997.
10. L. Le Bris and W. Robion, Dynamic channel assignment in GSM networks, in *Proc. IEEE Veh. Technol. Conf.*, vol. 4, Amsterdam, Netherlands, Sep. 19–22, 1999, 2339–2342.
11. M. Salmenkaita, J. Gimenez, and P. Tapia, A practical DCA implementation for GSM networks: Dynamic frequency and channel assignment, in *Proc. IEEE Veh. Technol. Conf.*, 4, Rhodes, Greece, May 6–9, 2001, 2529–2533.
12. J.-H. Wen, W.-J. Chen, S.-Y. Lin, and K.-T. Huang, Performance evaluation of LIBTA/hybrid time-slot selection algorithm for cellular systems, *Intl. J. Commun. Syst.*, 14(6), 575–591, Aug. 2001.
13. A. Lozano and D. C. Cox, Distributed dynamic channel assignment in TDMA mobile communication systems, *IEEE Trans. Veh. Technol.*, 51(6), 1397–1406, Nov. 2002.
14. T. Kanai, Autonomous reuse partitioning in cellular systems, in *Proc. IEEE Veh. Technol. Conf.*, 2, Denver, USA, May 10–13, 1992, 782–785.
15. S. M. Shin, C.-H. Cho, and D. K. Sung, Interference-based channel assignment for DS-CDMA cellular systems, *IEEE Trans. Veh. Technol.*, 48(1), 233–239, Jan. 1999.
16. A. Baiocchi, F. D. Priscoli, F. Grilli, and F. Sestini, The geometric dynamic channel allocation as a practical strategy in mobile networks with bursty user mobility, *IEEE Trans. Veh. Technol.*, 44(1), 14–23, Feb. 1995.
17. A. Boukerche, S. Hong, and T. Jacob, Distributed dynamic channel allocation for mobile communication systems in *Proc. Intl. Symp. on Modeling, Analysis and Simulation of Computer and Telecommun. Syst.*, San Francisco, USA, Aug. 29–Sep. 1, 2000, 73–81.
18. N. Lilith and K. Dogancay, Distributed dynamic call admission control and channel allocation using SARSA, in *Proc. Asia-Pacific Conf. on Commun.*, Perth, Australia, Oct. 3–5, 2005, 376–380.
19. M. Bublin, M. Konegger, and P. Slanina, A cost-function-based dynamic channel allocation and its limits, *IEEE Trans. Veh. Technol.*, 56(4), 2286–2295, Jul. 2007.

20. R. Mastrodonato and G. Paltenghi, Analysis of a bandwidth allocation protocol for ethernet passive optical networks (EPONs), in *Proc. Int. Conf. Transparent Optical Netw.*, 1, Barcelona, Spain, Jul. 3–7, 2005, 241–244.

21. M. Felegyhazi and J. P. Hubaux, Game theory in wireless networks: A tutorial, École Polytechnique Fédérale de Lausanne, Switzerland, Tech. Rep., 2007. [Online]. Available: http://infoscience.epfl.ch/record/79715/files/

22. W. Zhong, Y. Xu, and Y. Cai, Capacity and game-theoretic power allocation for multiuser MIMO channels with channel estimation error, in *Proc. IEEE Int. Symp. Commun. Inform. Technol.*, Beijing, China, Oct. 12–14, 2005, 716–719.

23. F. Meshkati, M. Chiang, S. C. Schwartz, H. V. Poor, and N. B. Mandayam, A non-cooperative power control game for multi-carrier CDMA systems, in *IEEE Wireless Commun. Netw. Conf.*, vol. 1, New Orleans, USA, Mar. 13–17, 2005, 606–611.

24. M. B. Stinchcombe, Notes for a course in game theory, Massachusetts Institute of Technology Cambridge, The MIT Press, Tech. Rep., 2002.

25. D. Niyato and E. Hossain, Optimal price competition for spectrum sharing in cognitive radio: A dynamic game-theoretic approach, in *Proc. IEEE Global Telecom. Conf.*, Washington, USA, Nov. 26–30, 2007, 4625–4629.

26. J. E. Suris, L. A. DaSilva, Z. Han, and A. B. MacKenzie, Cooperative game theory for distributed spectrum sharing, in *Proc. IEEE Int. Conf. Commun.*, Glasgow, Scotland, Jun. 24–28, 2007, 5282–5287.

27. S. H. Wong and I. J. Wassell, Application of game theory for distributed dynamic channel allocation, in *Proc. IEEE Veh. Technol. Conf.*, 1, Alabama, USA, May 6–9, 2002, 404–408.

28. P. M. Papazoglou, D. A. Karras, and R. C. Papademetriou, A dynamic channel assignment simulation system for large scale cellular telecommunications, in *HERCMA Conf. Proc.*, Athens, Greece, Sep. 2005, 1–7.

29. OPERA specification Part 1: Technology, Part 2: System, The OPERA Consortium, Specification, Jan. 2006.

30. Merged of HomePlug and Panasonic as draft standard for broadband over power line networks: Medium access control and physical layer specifications, Sep. 2007, release 037.

31. A. Haidine and R. Lehnert, Solving the generalized base station placement problem in the planning of broadband power line communications access networks, in *Proc. IEEE Int. Symp. Power Line Commun. Applic.*, Jeju Island, South Korea, Apr. 2–4, 2008, 141–146.

32. K. Dostert, *Powerline Communications*. Prentice Hall, 2001.

33. Deliverable D28: Improved coexistence specification, The OPERA2 Consortium, Tech. Rep., Mar. 2007.

34. Deliverable D13: Reference guide on the design of an integrated PLC network, including the adaptations to allow the carriers' carrier model, The OPERA Consortium, Tech. Rep., Mar. 2007.

35. A. Haidine and R. Lehnert, The channel allocation problem in broadband power line communications access networks: analysis, modelling and solutions, *Int. J. Autonomous and Adaptive Commun. Syst.*, 3(4), 396–418, 2010.

36. L. P. Do and R. Lehnert, Distributed dynamic resource allocation for multi-cell PLC networks, in *Proc. IEEE Int. Symp. Power Line Commun. Applic.*, Dresden, Germany, Mar. 29–Apr. 1, 2009, 95–100.

37. S. Goldfisher and S. Tanabe, IEEE 1901 access system: An overview of its uniqueness and motivation, *IEEE Commun. Mag.*, 48(10), 150–157, Oct. 2010.

38. R. Jain, D. Chiu, and W. Hawe, A quantitative measure of fairness and discrimination for resource allocation in shared computer systems, Digital Equipment Corporation, Maynard, MA, Tech. Rep., 1984.

39. M. Baumann, YATS-simulator, users and programmers manual, Dresden University of Technology, Manual, 2008.

40. T. M. Cover and J. A. Thomas, *Elements of Information Theory*, 2nd ed. Wiley, Chichester, 2006.

41. G. Kramer, I. Marić, and R. D. Yates, Cooperative communications, *Foundations and Trends in Networking*, 1(3–4), 271–425, 2006.

42. J. A. Cortés, L. Díez, J. J. Cañete, and J. J. Sánchez-Martínez, Analysis of the indoor broadband power-line noise scenario, *IEEE Trans. Electromagn. Compat.*, 52(4), 849–858, Nov. 2010.

43. M. Zimmermann and K. Dostert, Analysis and modeling of impulsive noise in broad-band powerline communications, *IEEE Trans. Electromagn. Compat.*, 44(1), 249–258, Feb. 2002.

44. J. Häring and A. J. H. Vinck, OFDM transmission corrupted by impulsive noise, in *Proc. Int. Symp. Power Line Commun. Applic.*, Limerick, Ireland, Apr. 5–7, 2000, 9–14.

45. R. Pighi, M. Franceschini, G. Ferrari, and R. Raheli, Fundamental performance limits for PLC systems impaired by impulse noise, in *Proc. IEEE Int. Symp. Power Line Commun. Applic.*, Orlando, USA, Mar. 27–29, 2006, 277–282.

46. H. C. Ferreira, L. Lampe, J. Newbury, and T. G. Swart, Editors, *Power Line Communications: Theory and Applications for Narrowband and Broadband Communications over Power Lines*, 1st ed. Wiley, Chichester, 2010.

47. H. Hrasnica, A. Haidine, and R. Lehnert, *Broadband PowerLine Communications Networks: Network Design*. Wiley, Chichester, 2004.

48. A. Haidine and R. Lehnert, Analysis of the channel allocation problem in broadband power line communications access networks, in *Proc. IEEE Int. Symp. Power Line Commun. Applic.*, Pisa, Italy, Mar. 26–28, 2007, 192–197.

49. M. Thompson, The thick and the thin of car cabling, *IEEE Spectrum*, 33(2), 42–45, Feb. 1996.

50. F. J. Cañete, J. A. Cortés, L. Díez, and J. T. Entrambasaguas, Analysis of the cyclic short-term variation of indoor power line channels, *IEEE J. Sel. Areas Commun.*, 24(7), 1327–1338, Jul. 2006.

51. M. Katayama, T. Yamazato, and H. Okada, A mathematical model of noise in narrowband power line communication systems, *IEEE J. Sel. Areas Commun.*, 24(7), 1267–1276, Jul. 2006.

52. J. L. Seoane, S. K. Wilson, and S. Gelfand, Analysis of intertone and interblock interference in OFDM when the length of the cyclic prefix is shorter than the length of the impulse response of the channel, in *Proc. IEEE Global Telecom. Conf.*, vol. 1, Phoenix, USA, Nov. 3–8, 1997, 32–36.

53. J. A. Cortés, F. J. Ca nete, L. Díez, and L. M. Torres, On PLC channel models: an OFDM-based comparison, in *Proc. IEEE Int. Symp. Power Line Commun. Applic.*, Johannesburg, South Africa, Mar. 24–27, 2013, 333–338.

54. Final draft of EN-50561-1 standard. Power line communication apparatus used in low-voltage installations – radio disturbance characteristics – limits and methods of measurement – Part 1: Apparatus for in-home use, European Committee for Electrotechnical Standardization (CENELEC), Brussels, Belgium, Draft standard, 2012.

55. J. A. Cortés, A. M. Tonello, and L. Díez, Comparative analysis of pilot-based channel estimators for DMT systems over indoor power-line channels, in *Proc. IEEE Int. Symp. Power Line Commun. Applic.*, Pisa, Italy, Mar. 26–28, 2007, 372–377.

56. N. Papandreou and T. Antonakopoulos, Bit and power allocation in constrained multi-carrier systems: The single-user case, *EURASIP J. on Adv. in Signal Process.*, vol. 2008, pp. 1–14, Jul. 2007.

57. S. D'Alessandro, A. M. Tonello, and L. Lampe, On power allocation in adaptive cyclic prefix OFDM, in *Proc. IEEE Int. Symp. Power Line Commun. Applic.*, Rio de Janeiro, Brazil, Mar. 28–31, 2010, 183–188.

58. J. Campello, Optimal discrete bit loading for multicarrier modulation systems, in *Proc. IEEE Int. Symp. Inform. Theory*, Cambridge, USA, Aug. 16–21, 1998, 193.

59. ——, Practical bit loading for DMT, in *Proc. IEEE Int. Conf. Commun.*, vol. 2, Vancouver, Canada, Jun. 6–10, 1999, 801–805.

60. I. Kalet, The multitone channel, *IEEE Trans. Commun.*, 37(2), 119–124, Feb. 1989.

61. J. M. Cioffi, EE379C - Advanced Digital Communication, Lecture notes, 2008, ch. 4. [Online]. Available: http://www.stanford.edu/class/ee379c/

62. J.-J. van de Beek, P. O. Börjesson, M.-L. Boucheret, D. Landström, J. M. Arenas, P. Ödling, C. Östherg, M. Wahlqvist, and S. K. Wilson, A time and frequency synchronization scheme for multiuser OFDM, *IEEE J. Sel. Areas Commun.*, 17(11), 1900–1914, Nov. 1999.

63. T. Sartenaer, L. Vandendorpe, and J. Louveaux, Balanced capacity of wireline multiuser channels, *IEEE Trans. Commun.*, 53(12), 2029–2042, Dec. 2005.

64. N. Papandreou and T. Antonakopoulos, Resource allocation management for indoor power-line communications systems, *IEEE Trans. Power Delivery*, 22(2), 893–903, Apr. 2007.

65. Z. Xu, M. Zhai, and Y. Zhao, Optimal resource allocation based on resource factor for power-line communication systems, *IEEE Trans. Power Delivery*, 25(2), 657–666, Apr. 2010.

66. W. Yu and J. M. Cioffi, FDMA capacity of gaussian multiple-access channels with ISI, *IEEE Trans. Commun.*, 50(1), 102–111, Jan. 2002.

67. H. Zou, S. Jagannathan, and J. M. Cioffi, Multiuser OFDMA resource allocation algorithms for in-home power-line communications, in *Proc. IEEE Global Telecom. Conf.*, New Orleans, USA, Nov. 30–Dec. 4, 2008.

68. J. A. Cortés, L. Díez, F. J. Cañete, and J. T. Entrambasaguas, Analysis of DMT-FDMA as a multiple access scheme for broadband indoor power-line communications, *IEEE Trans. Consumer Electron.*, 52(4), 1184–1192, Nov. 2006.

69. A. M. Tonello and F. Pecile, Efficient architectures for multiuser FMT systems and application to power line communications, *IEEE Trans. Commun.*, 57(5), 1275–1279, May 2009.

70. M. M. Rahman, C. S. Hong, S. Lee, J. Lee, M. A. Razzaque, and J. H. Kim, Medium access control for power line communications: An overview of the IEEE 1901 and ITU-T G.hn standards, *IEEE Commun. Mag.*, 49(6), 183–191, Jun. 2011.

71. L. Yonge, J. Abad, K. Afkhamie, L. Guerrieri, S. Katar, H. Lioe, P. Pagani, R. Riva, D. M. Schneider, and A. Schwager, An overview of the HomePlug AV2 technology, *J. Electric. Comput. Eng.*, vol. 2013, 1–20, 2013.

72. A. M. Tonello, J. A. Cortés, and S. D'Alessandro, Optimal time slot design in an OFDM-TDMA system over power-line time-variant channels, in *Proc. IEEE Int. Symp. Power Line Commun. Applic.*, Dresden, Germany, Mar. 29–Apr. 1, 2009, 41–46.

73. S.-G. Yoon and S. Bahk, Rate adaptation scheme in power line communication, in *Proc. IEEE Int. Symp. Power Line Commun. Applic.*, Jeju Island, South Korea, Apr. 2–4, 2008, 111–116.

74. S. Boyd and L. Vandenberghe, *Convex Optimization*. Cambridge University Press, 2004.

75. S. Katar, B. Mashburn, K. Afkhamie, H. Latchman, and R. Newman, Channel adaptation based on cyclo-stationary noise characteristics in PLC systems, in *Proc. IEEE Int. Symp. Power Line Commun. Applic.*, Orlando, USA, Mar. 27–29, 2006, 16–21.

76. H. Kwon, H. Seo, S. Kim, and B. G. Lee, Generalized CSMA/CA for OFDMA systems: Protocol design, throughput analysis, and implementation issues, *IEEE Trans. Wireless Commun.*, 8(8), 4176–4187, Aug. 2009.

77. S.-G. Yoon, D. Kang, and S. Bahk, OFDMA CSMA/CA protocol for power line communication, in *Proc. IEEE Int. Symp. Power Line Commun. Applic.*, Rio de Janeiro, Brazil, Mar. 28–31, 2010, 297–302.

78. R. Dong, M. Ouzzif, and S. Saoudi, Opportunistic random-access scheme design for OFDMA-based indoor PLC networks, *IEEE Trans. Power Delivery*, 27(4), 2073–2081, Oct. 2012.

79. H. Zou, A. Chowdhery, S. Jagannathan, J. M. Cioffi, and J. Le Masson, Multi-user joint subchannel and power resource-allocation for powerline relay networks, in *Proc. IEEE Int. Conf. Commun.*, Dresden, Germany, Jun. 14–18, 2009.

80. M. Biagi and V. Polli, Iterative multiuser resource allocation for in-home power-line communications, in *Proc. IEEE Int. Symp. Power Line Commun. Applic.*, Udine, Italy, Apr. 3–6, 2011, 388–392.

81. P. Achaichia, M. Le Bot, and P. Siohan, Point-to-multipoint communication in power line networks: A novel FDM access method, in *Proc. IEEE Int. Conf. Commun.*, Ottawa, Canada, Jun. 10–15, 2012, 3424–3428.

82. T. Hayasaki, D. Umehara, S. Denno, and M. Morikura, A bit-loaded OFDMA for in-home power line communications, in *Proc. IEEE Int. Symp. Power Line Commun. Applic.*, Dresden, Germany, Mar. 29–Apr. 1, 2009, 171–176.

83. P. Achaichia, M. Le Bot, and P. Siohan, Frequency division multiplexing analysis for point-to-multipoint transmissions in power line networks, in *Proc. IEEE Int. Symp. Power Line Commun. Applic.*, Beijing, China, Mar. 27–30, 2012, 230–235.

84. M. Biagi, E. Baccarelli, N. Cordeschi, V. Polli, and T. Patriarca, Physical-layer goodput maximization for power line communications, in *Proc. IEEE IFIP Wireless Days*, Paris, France, Dec. 15–17, 2009.

85. Y.-J. Lin, H. A. Latchman, J. C. L. Liu, and R. Newman, Periodic contention-free multiple access for power line communication networks, in *Proc. IEEE Intl. Conf. Adv. Information Netw. Applic.*, 2, Taipei, Taiwan, Mar. 28–30, 2005, 315–318.

86. A. Chowdhery, S. Jagannathan, J. M. Cioffi, and M. Ouzzif, A polite cross-layer protocol for contention-based home power-line communications, in *Proc. IEEE Int. Conf. Commun.*, Dresden, Germany, Jun. 14–18, 2009.

87. L. D. Bert, S. D'Alessandro, and A. M. Tonello, Enhancements of G3-PLC technology for smart-home/building applications, *J. Electric. Comput. Eng.*, vol. 2013, 1–11, 2013.

88. A. Nosratinia, T. E. Hunter, and A. Hedayat, Cooperative communication in wireless networks, *IEEE Commun. Mag.*, 42(10), 74–80, Oct. 2004.

89. E. C. van der Meulen, Transmission of information in a T-terminal discrete memoryless channel, Ph.D. dissertation, University of California, USA, 1968.

90. ——, Three-terminal communication channels, *Adv. Appl. Prob.*, 3(1), 120–154, Spring 1971.

91. T. Cover and A. E. Gamal, Capacity theorems for the relay channel, *IEEE Trans. Inf. Theory*, 25(5), 572–584, Sep. 1979.

92. A. E. Gamal and M. Aref, The capacity of the semideterministic relay channel, *IEEE Trans. Inf. Theory*, 28(3), 536, May 1982.

93. R. Ahlswede and A. Kaspi, Optimal coding strategies for certain permuting channels, *IEEE Trans. Inf. Theory*, 33(3), 310–314, May 1987.

94. A. Sendonaris, E. Erkip, and B. Aazhang, User cooperation diversity. part I. System description, *IEEE Trans. Commun.*, 51(11), 1927–1938, Nov. 2003.

95. ——, User cooperation diversity. Part II. Implementation aspects and performance analysis, *IEEE Trans. Commun.*, 51(11), 1939–1948, Nov. 2003.

96. J. N. Laneman and G. W. Wornell, Distributed space-time-coded protocols for exploiting cooperative diversity in wireless networks, *IEEE Trans. Inf. Theory*, 49(10), 2415–2425, Oct. 2003.

97. J. N. Laneman, D. N. C. Tse, and G. W. Wornell, Cooperative diversity in wireless networks: Efficient protocols and outage behavior, *IEEE Trans. Inf. Theory*, 50(12), 3062–3080, Dec. 2004.

98. J. Boyer, D. D. Falconer, and H. Yanikomeroglu, Multihop diversity in wireless relaying channels, *IEEE Trans. Commun.*, 52(10), 1820–1830, Oct. 2004.

99. R. U. Nabar, H. Bolcskei, and F. W. Kneubuhler, Fading relay channels: performance limits and space-time signal design, *IEEE J. Sel. Areas Commun.*, 22(6), 1099–1109, Aug. 2004.

100. Q. Li, R. Q. Hu, Y. Qian, and G. Wu, Cooperative communications for wireless networks: techniques and applications in LTE-advanced systems, *IEEE Wireless Commun.*, 19(2), 22–29, Apr. 2012.

101. C. Eklund, R. B. Marks, K. L. Stanwood, and S. Wang, IEEE standard 802.16: a technical overview of the WirelessMAN™ air interface for broadband wireless access, *IEEE Commun. Mag.*, 40(6), 98–107, Jun. 2002.

102. A. M. Tonello and F. Versolatto, Bottom-up statistical PLC channel modeling – Part I: Random topology model and efficient transfer function computation, *IEEE Trans. Power Delivery*, 26(2), 891–898, Apr. 2011.

103. L. Lampe and A. J. H. Vinck, Cooperative multihop power line communications, in *Proc. IEEE Int. Symp. Power Line Commun. Applic.*, Beijing, China, Mar. 27–30, 2012, 1–6.

104. L. Lampe, R. Schober, and S. Yiu, Distributed space-time coding for multihop transmission in power line communication networks, *IEEE J. Sel. Areas Commun.*, 24(7), 1389–1400, Jul. 2006.

105. G. Bumiller, Single frequency network technology for medium access and network management, in *Proc. Int. Symp. Power Line Commun. Applic.*, Athens, Greece, Mar. 27–29, 2002.

106. M. Eriksson, Dynamic single frequency networks, *IEEE J. Sel. Areas Commun.*, 19(10), 1905–1914, Oct. 2001.

107. G. Bumiller, L. Lampe, and H. Hrasnica, Power line communication networks for large-scale control and automation systems, *IEEE Commun. Mag.*, 48(4), 106–113, Apr. 2010.

108. V. Tarokh, N. Seshadri, and A. R. Calderbank, Space-time codes for high data rate wireless communication: performance criterion and code construction, *IEEE Trans. Inf. Theory*, 44(2), 744–765, Mar. 1998.

109. L. Lampe, R. Schober, and S. Yiu, Multihop transmission in power line communication networks: analysis and distributed space-time coding, in *Proc. IEEE Workshop on Signal Process. Advances in Wireless Commun.*, New York, USA, Jun. 5–8, 2005, 1006–1012.

110. A. Papaioannou, G. D. Papadopoulos, and F.-N. Pavlidou, Hybrid ARQ combined with distributed packet space-time block coding for multicast power-line communications, *IEEE Trans. Power Delivery*, 23(4), 1911–1917, Oct. 2008.

111. S. Yiu, R. Schober, and L. Lampe, Distributed space-time block coding, *IEEE Trans. Commun.*, 54(7), 1195–1206, Jul. 2006.

112. T. Esmailian, F. R. Kschischang, and P. G. Gulak, In-building power lines as high-speed communication channels: Channel characterization and a test-channel ensemble, *Int. J. Commun. Syst.* 16(5), 381–400, Jun. 2003, Special Issue: Powerline Communications and Applications.

113. V. B. Balakirsky and A. J. H. Vinck, Potential performance of PLC systems composed of several communication links, in *Proc. IEEE Int. Symp. Power Line Commun. Applic.*, Vancouver, Canada, Apr. 6–8, 2005, 12–16.

114. L. Lampe and A. J. H. Vinck, On cooperative coding for narrow band PLC networks, *AEÜ Intl. J. Electron. and Commun.*, 65(8), 681–687, Aug. 2011.

115. X. Cheng, R. Cao, and L. Yang, On the system capacity of relay-aided powerline communications, in *Proc. IEEE Int. Symp. Power Line Commun. Applic.*, Udine, Italy, Apr. 3–6, 2011, 170–175.

116. ——, Relay-aided amplify-and-forward powerline communications, *IEEE Trans. Smart Grid*, 4(1), 265–272, Mar. 2013.

117. Y.-H. Kim, S. Choi, S.-C. Kim, and J.-H. Lee, Capacity of OFDM two-hop relaying systems for medium-voltage power-line access networks, *IEEE Trans. Power Delivery*, 27(2), 886–894, Apr. 2012.

118. K.-H. Kim, H.-B. Lee, Y.-H. Kim, J.-H. Lee, and S.-C. Kim, Cooperative multihop AF relay protocol for medium-voltage power-line-access network, *IEEE Trans. Power Delivery*, 27(1), 195–204, Jan. 2012.

119. S. D'Alessandro and A. M. Tonello, On rate improvements and power saving with opportunistic relaying in home power line networks, *EURASIP J. Advances Signal Process.*, vol. 2012, 1–17, Sep. 2012.

120. A. M. Tonello, F. Versolatto, and S. D'Alessandro, Opportunistic relaying in in-home PLC networks, in *Proc. IEEE Global Telecom. Conf.*, Miami, USA, Dec. 6–10, 2010, 1–5.

121. S. D'Alessandro, A. M. Tonello, and F. Versolatto, Power savings with opportunistic decode and forward over in-home PLC networks, in *Proc. IEEE Int. Symp. Power Line Commun. Applic.*, Udine, Italy, Apr. 3–6, 2011, 176–181.

122. A. M. Tonello and F. Versolatto, Bottom-up statistical PLC channel modeling – Part II: Inferring the statistics, *IEEE Trans. Power Delivery*, 25(4), 2356–2363, Oct. 2010.

123. W. Nam, S.-Y. Chung, and Y. H. Lee, Capacity of the Gaussian two-way relay channel to within $\frac{1}{2}$ bit, *IEEE Trans. Inf. Theory*, 56(11), 5488–5494, Nov. 2010.

124. D. Gunduz, A. Yener, A. Goldsmith, and H. V. Poor, The multi-way relay channel, in *Proc. IEEE Int. Symp. Inform. Theory*, Seoul, South Korea, Jun. 28–Jul. 3, 2009, 339–343.

125. R. Ahlswede, N. Cai, S.-Y. R. Li, and R. W. Yeung, Network information flow, *IEEE Trans. Inf. Theory*, 46(4), 1204–1216, Jul. 2000.

126. B. Nazer and M. Gastpar, Reliable physical layer network coding, *Proc. IEEE*, 99(3), 438–460, Mar. 2011.

127. M. Noori and M. Ardakani, On the achievable rates of symmetric Gaussian multi-way relay channels, *EURASIP J. Wireless Commun. Netw.*, vol. 2013, 1–8, Jan. 2013.

128. B. Tan and J. Thompson, Relay transmission protocols for in-door powerline communications networks, in *Proc. IEEE Int. Conf. Commun.*, Kyoto, Japan, Jun. 5–9, 2011, 1–5.

129. H. Gacanin, Inter-domain bi-directional access in G.hn with network coding at the physical-layer, in *Proc. IEEE Int. Symp. Power Line Commun. Applic.*, Beijing, China, Mar. 27–30, 2012, 144–149.

130. M. Noori and L. Lampe, Improving data rate in relay-aided power line communications using network coding, in *Proc. IEEE Global Telecom. Conf.*, Atlanta, USA, Dec. 9–13, 2013, 2975–2980.

7

PLC for Home and Industry Automation

G. Hallak and G. Bumiller

7.1   Introduction

The low 'media cost' for power line communication (PLC) technology renders it suitable for home and industry automation purposes. PLC uses an infrastructure already existing in every home and industrial facility, which eliminates the unnecessary expense and the difficulties of installing new wires for achieving high signal penetration. This enables a plug and play type use of PLC-enabled systems. In this chapter, an overview of PLC-based home and industry automation systems with interesting application examples is presented.

This chapter is organized as follows. Section 7.2 provides a brief discussion on the use of PLC for home and industry automation. Section 7.3 presents an overview of the most important protocols for home automation. First, the X10 narrowband PLC protocol for communication between electronic devices is introduced. Then, the KNX/EIB standard for in home and building automation bus systems is discussed. The standard is based on the OSI network communication protocol of the European Installation Bus (EIB). The KNX PL 110 is the standard for PLC as the physical communication method. Finally the LONWorks protocol for control networks and building automation is briefly described. Section 7.4 presents a sample application of PLC for industry automation, namely refrigeration container ships, based on the international standard ISO 10368:2006(E). The physical requirements of the system which include the transmission rate, modulation methods, and impedance specifications as well as the system components are discussed. Section 7.5 provides a summary of the communication protocol AMIS CX1 profile of the company Siemens AG for PLC between electricity meters, breakers and other grid elements and the data concentrator points at central stations in the low-voltage distribution grid. The AMIS CX1 profile describes the layers 1 to 4 and is designed as a multi hop master-slave system, where the data concentrator is always the master and all other nodes are slaves, which only answer on request. This chapter will be completed

Power Line Communications: Principles, Standards and Applications from Multimedia to Smart Grid, Second Edition.
Edited by Lutz Lampe, Andrea M. Tonello, and Theo G. Swart.
© 2016 John Wiley & Sons, Ltd. Published 2016 by John Wiley & Sons, Ltd.

by Section 7.6, which introduces the digitalSTROM technology as a PLC application for home automation in smart energy living. The idea of digitalSTROM is that the installed devices communicate through the existing power lines with separate in-home nodes over modulated up and downstream channels. Control can be done via physical switches or the Internet and smart phones. The digitalSTROM architecture, network components, installation and communication are described.

7.2   Home and Industry Automation Using PLC

Home and industry automation is a main part of modern life that helps to control and monitor the large variety of home and industry devices such as air-condition, refrigeration or lighting systems. In addition to the functionality and comfort these systems provide, they also help to improve energy efficiency of appliances and other electrical devices. The fast development of communication technologies has spurred the integration of automation systems into cyber-physical networks which enables managing these systems remotely, but on the other hand calls for improved measures to ensure security and also privacy.

PLC provides a large range of communication frequencies and systems are broadly classified into two categories. Narrowband PLC is used mainly for automation in a general sense, and broadband PLC enables home networking (multimedia) applications. A large number of home and industry automation solutions based on PLC are now available [1]. Figure 7.1 illustrates the

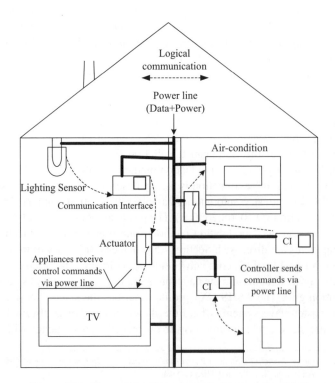

Figure 7.1   Illustration of home automation based on PLC.

connection of appliances, sensors and controllers via power lines in a home automation setting. Sensors like light, temperature and smoke sensors are deployed in every room of a modern home and their signals are sent through a communication interface such as Ethernet, RS232, etc. to the control unit, which is a central unit in a home automation system that records data and determines the required commands and sends them to actuator and regulators to switch appliances on and off. Appliances like air-condition units with thermostat and washing machines are also directly connected to actuators and switched off according to demand response notifications from the communication interface during peak times and cut off from the power board when unused. Central control units usually provide a friendly graphical user interface (GUI) for end users with a screen and keyboard to manage the system. The sensors, control unit(s), appliances and actuators are connected to each other via a communication channel, which in the case of PLC are the existing power lines.

Before the recent narrowband PLC standardizations such as IEEE 1901.2 and ITU-T G.990x, only relatively few companies developed and provided power line modem chip-sets and a limited number of applications reached quantities that allowed the development of individual chip-sets. The X10, KNX PL 110 and LONWorks systems are fairly representative of the used physical (PHY) layer technologies. In addition, three other systems, which are little known in the scientific community, will be discussed in this chapter. The first is an ISO standard for PLC on refrigeration container ships, which implements two PHY layers with frequency separation to simultaneously use the power lines. The second system is based on a published specification of a proprietary smart metering system, which is already in use by utilities. The selected approach for the same application is completely different from IEEE 1901.2 or ITU-T G.990x (see Chapter 9). The third system uses a PHY layer that enables very small form-factor modems without a coupling capacitor or transformer. A brief overview of these PLC systems for home and industry automation is presented in Table 7.1.

7.3 Popular Home Automation Protocols

Many popular home and building automation system protocols have been developed some time ago. In the following, we present a brief overview of important such protocols.

7.3.1 X10 Protocol

X10 is a narrowband PLC protocol for communication between electronic equipment for home automation. The X10 protocol delivers signals among transmitters and receivers over

Table 7.1 Overview of PLC home and industrial automation standards

Protocol	Rate (bit/sec)	Frequency Band (kHz)	Modulation Scheme
X10	60	95–125	Short 120 kHz pulses
KNX	1200	110	BFSK
LONWorks	5400/3600	132/86	BPSK
ISO 10368:2006(E)	1200/134400	53.9–56.1/130–400	FSK/BPSK
AMIS CX1-Profile	600–3000	39–90	DPSK
digitalSTROM	n.a.	10–120	Current on/off switching

the house electrical wiring. These signals are short radio frequency (RF) bursts that represent the transmitted information and control the electrical devices, such as lighting systems and audio/video equipment. X10 was developed in 1975 by Pico Electronics of Glenrothes, Scotland, and is also known as domotics network technology. It remains the most popular technology available for home automation systems because of the millions of installed units worldwide, and the low price of new units.

7.3.1.1   X10 Physical Layer Specification and Transmission

X10 transmissions are synchronized to the zero crossing point of the AC power line. The transmission bursts should be done as close to the zero crossing point as possible from negative to positive of the power signal, within 200 microseconds of the zero crossing point. Bursts are 120 kHz signals of 1 ms duration. The presence of a burst represents a '1', while the absence of a burst means '0'. Except for the Start Code (see below) binary information is encoded into burst pairs. That means, a binary '1' is transmitted as the presence of a pulse in one half cycle followed by no pulse in the next half cycle. A binary '0' is transmitted as the absence of a pulse, immediately followed by the presence of a pulse. In three-phase systems, the burst is sent three times, to reach the zero crossing point of each phase. Figure 7.2 shows the timing of the X10 signals in a 60 Hz system. Timing the signal at the zero crossings simplifies the receivers, by reading only from the power line for a short time after it detected a zero crossing point. Since the system only transmits one bit per cycle of the carrier, the raw signaling bit rate of the X10 system is 60 bps [2].

A complete code transmission includes eleven cycles of the power line. The first two cycles represent a Start Code. The next four cycles represent the House Code and the last five cycles represent either the Number Code (1 through 16) or a Function Code (On, Off, etc.). This

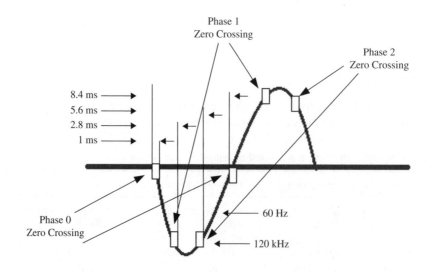

Figure 7.2   X10 transmission timing.

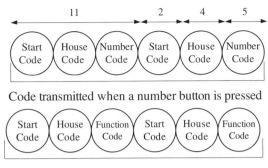

Code transmitted when a number button is pressed

Code transmitted when a function button is pressed

Figure 7.3 X10 coding and transmission. Numbers are mains cycles.

complete block, (Start Code, House Code, Key Code) should always be transmitted in groups of 2 with 3 power line cycles between each group of 2 codes, as shown in Figure 7.3.

7.3.1.2   X10 Limitations

The most common problem of X10 is the large attenuation of signals between the two live conductors in the 3-wire 120/240 volt system that is used in North America because of the high impedance of the distribution transformer winding between the live conductors. This problem could be overcome by installing a capacitor between the leg wires as a path for the X10 signals. Furthermore, a bare uninsulated wire is used for the ground connection. If the sender is connected to phase 1 and the receiver is connected to phase 2, the signal would sometimes be so poor that the X10 units would react intermittently. The X10 protocol is also slow and takes three quarters of a second to transmit an electronic device address and a command.

7.3.2   KNX/EIB PL 110 Standard

KNX/EIB is an open standard used in home and building automation bus systems. The standard is based on the OSI network communication protocol of EIB but amended with the physical layers, configuration modes and application experience of BatiBUS and European Home Systems (EHS). It is optimized for low-speed control applications like lighting systems. KNX/EIB is specified over various physical media, including power line (KNX PL 110), twisted pair, radio, infrared and Ethernet, and designed to be independent of any particular hardware platform [3].

7.3.2.1   KNX PL 110 Physical and Data Link Layer Specification

The KNX standard provides the possibility for developers to select between several physical layers, or to combine them. The KNX PL 110 enables communication over the power lines. The main communication characteristics are spread frequency-shift keying (FSK) signaling, asynchronous transmission of data packets and half duplex bi-directional communication. It

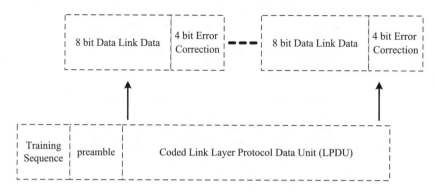

Figure 7.4   KNX telegram transmission.

uses a center frequency of 110 kHz and the rate is 1200 bit/s, which corresponds to a bit duration of 833 μs. The frequency for transmission a logical '0' is 105.2 kHz ± 100 ppm and the frequency for transmission a logical '1' is 115.2 kHz ± 100 ppm. The transmission starts at the mains zero crossing with a maximum level of 122 dBμV according to EN 50065-1 [3]. Each telegram starts with a 4-bit training sequence and a 16-bit preamble. The training sequence enables the receivers to adjust their reception to the network conditions. The preamble field has two purposes. First, it marks the start of the transmission and second, it controls the bus access. All frame information, except training sequence and preamble, is coded into 12-bit characters which allows to correct any two bits in the transmitted character as shown in Figure 7.4.

The Link layer Protocol Data Unit (LPDU) contains the following fields as shown in Figure 7.5.

- Control Field (CTRL): Contains information about the data link service, its priority (alarm messages, etc.), frame type (standard or extended) and whether the LPDU is a repeated one.
- Source Address (SA): The originator's unique address.
- Destination Address (DA): The unique address of destination node or the destination address of a group of nodes (multicast).
- Address Type (AT): Defines if the destination address belongs to a single node or a group of nodes.
- Network layer Protocol Control Information (NPCI): Controlled by network layer and contains the hop count information for routing.
- TPDU (Transport layer Protocol Data Unit): The payload from upper layer.
- Length (LG): Defines TPDU length.
- Check Octet (FCS): Helps ensure data consistency and reliable transmission.

Figure 7.5   KNX LPDU frame.

Training Sequence 4 bits	Preamble 16 bits	ACK/ NACK 8 bits	Error Correction 4 btis

Figure 7.6 KNX reply telegram.

The KNX PL110 protocol uses medium access control (MAC) mechanisms to avoid collisions, and an ACK/NACK telegram must be transmitted from the receiver to inform the telegram generator about the telegram delivery. The ACK telegram consists of a 20-bit training sequence and preamble, followed by a single character acknowledging or not acknowledging the received telegram, as shown in Figure 7.6. If the reply telegram is not sent, the telegram is repeated [4].

7.3.2.2 KNX PL 110 Topology and Addressing

The logical addressing of KNX-PL 110 is compatible with the KNX-TP1 standard for twisted pair transmission media. KNX organizes devices into areas and lines, with up to 8 areas, 16 lines per area and 256 devices per line. In larger installations, band-stop filters can be used for physical separation of areas [4].

7.3.2.3 KNX vs. X10

KNX system implementation is more expensive than using X10. The implementation is only worthwhile if several systems are to be connected with each other or if an installation needs to be more flexible so that it could be fast and effectively modified.

7.3.3 LONWorks

Local operating network (LONWorks) technology is an open solution and protocol for building and home automation based on a networking platform for controlling the required applications. A control network is any group of applications working in an end-to-end control system to monitor sensors, control actuators, communicate reliably, manage network operation, and provide full access to network data. Sensors and actuators are connected to a sub-panel, which is connected to the controller panel via a master/slave communication bus. In a LONWorks system, smart control units communicate with each other using the LONWorks protocol. LONWorks was developed by Echelon Corporation, and it is estimated that approximately 90 million devices have been installed by 2010. The LONWorks technology works over multiple media, including power line, twisted pair and wireless [5].

The power line modems PL 3120 operate at a primary frequency of 132 kHz in the C band of EN 50065-1 and can switch to a secondary frequency of 115 kHz, and PL 3150 devices use a primary frequency of 86 kHz in the A band of EN 50065-1 and can switch to a secondary frequency of 75 kHz. Furthermore, binary phase-shift keying (BPSK) modulation is applied, which provides data rates of 5.4 kbps in C band or 3.6 kbps in A band [5].

7.4 Power Line Communication Application for Refrigeration Containers Ships

Reefer container traffic through ports is increasing more than ever before. PLC enables monitoring and control of the refrigeration on board containers ships, remotely and in real time. The international standard ISO 10368:2006(E) (freight thermal containers-remote condition monitoring) regulates the manner in which a single central monitoring station exchanges information using power cables with a number of communication units and modems, which are fitted into refrigeration containers on board of ships. The purpose of this is to perform central monitoring of the climatic conditions in the individual refrigeration containers and when needed to perform changes in the otherwise local control of refrigeration power.

7.4.1 Physical Layer Specification

The standard ISO 10368:2006(E) [6] for remote monitoring of the temperature in refrigerated containers specifies two variants for power line transmission systems: low and high date rate transmission, which use different frequency bands as shown in Figure 7.7.

Low data-rate power line signals are sent in the frequency band from 53.9 kHz to 56.1 kHz. The messages are modulated by FSK, using a baud rate of 1200 symbols per second and a signal level of 6 V rms at a line impedance of 15 Ω. The receiver sensitivity is 1 mV rms. The carrier setup time is 10 ms and the out of band power spectrum shall not exceed 2 mV from 130 kHz to 400 kHz to protect the high data rate transmission system.

High data-rate power line signals use the frequency band from 140 kHz to 400 kHz. An anti-symmetric waveform is digitally synthesized using 32 chips of a 4.3008 MHz clock representing one raw data bit. The wave form is shown in Figure 7.8, and the power spectral density of the transmit signal is shown in Figure 7.9. Each bit is modulated by BPSK, resulting in raw data rate of 134.4 kbits/s. The modulation can use variable signal levels, with a maximum output power of 100 mW per 10 kHz for a line impedance of 18 Ω. Line impedance controlled voltage injection is used between the three phases and the ground. The receiver needs to be

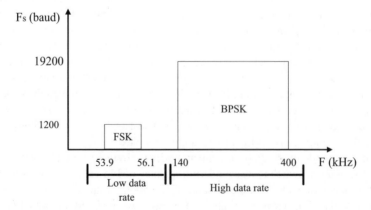

Figure 7.7   Frequency bands and modulation used for PLC according to ISO 10368:2006(E).

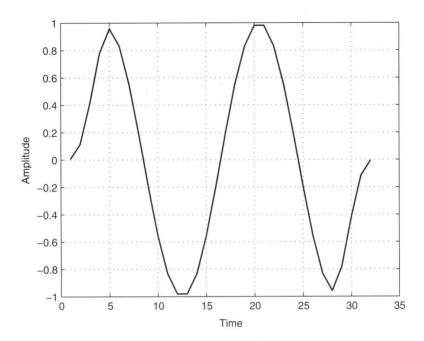

Figure 7.8   Anti-symmetric waveform with 32 chips.

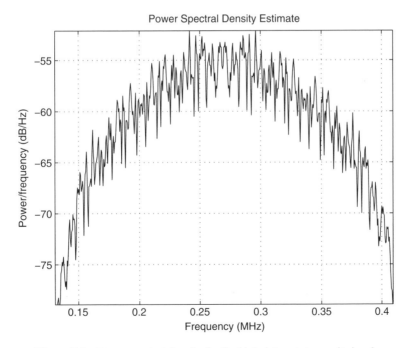

Figure 7.9   Power spectral density for the high data-rate transmit signal.

able to detect the signal for attenuations of 15 dB and operate with a 2 mV received signal amplitude [6].

The system has two independent physical layers, which are separated in frequency and used simultaneously on the same line. The impedance requirements for the receiving and the transmitting states for each device in the two modes are specified in ISO 10368:2006(E). In the low data-rate mode, the signal should be sent half duplex at 6 V rms into a line impedance of $15 + 15j\ \Omega$ and the signal is transmitted over lines with impedance less than 3 kΩ at 55 kHz at the receiver. In the high data-rate mode, the modem impedance for transmission is close to 18 Ω phase to phase and 21 Ω phase to ground. The modem impedance for reception is greater than 200 Ω. The detailed impedance specification are important to realize the coexistence on the same line for the frequency separated systems [7].

7.4.2   Data Link Protocol

In the low data-rate transmission mode, 8-bit messages are sent asynchronously with a start and a stop bit and without channel coding, similar to RS232. In the high data-rate transmission mode, eight data bits are encoded with a (32,8) block code for error correction. Due to the code rate of 1/4, the resulting data rate is 33.6 kbits/s. Unicast data and broadcast data packet exchange are used for transmission with specific data frames. The frame starts with a preamble of 5 bytes that contains the necessary sequences to synchronize communication between equipment on the power line. The data frames have different sizes depending on the transmission mode. Table 7.2 describes the different data frames that are used.

For unicast transmission, a stop-and-wait protocol is used to exchange packets between two connected nodes, which ensures that information is not lost due to dropped packets. Header and data frames are divided into small packets, with maximum payload of 5 bytes for each frame. The transmission of each packet is succeeded by an acknowledgement (ACK). If the sender receives a negative acknowledgment (NACK) or timeout because of dropped ACK, the last packet will be retransmitted for several times until ACK is received. As a result, the next packet will be transmitted if and only if an acknowledgment is received. Therefore, this method is done without using any segment numbers and the only two different types of intermediate frames are sufficient in this case to detect duplicated transmission.

In the case of broadcast transmission, the receiving units cannot confirm the receipt or request a retransmission. Therefore the frame header is repeated eight times to allow all receiving units to initiate a broadcast packet reception. The intermediate data frames follow and are repeated four times for improved reliability.

7.4.3   System Components

The refrigeration container ship system according to ISO 10368:2006(E) consists of five main components, which can be configured as shown in Figures 7.10 (Configuration A) and 7.11 (Configuration B).

A single remote condition monitoring system consists of one master monitoring unit (MMU) to control the entire remote condition monitoring system and one multiple data rate central control unit (MDCU) which establishes the connection between the MMU and the three-phase power line bus. The MDCU can include one high data rate central control unit (HDCU) and

Table 7.2   Data frames description

Packet Exchange	Frame Specification	Total Size Encoded/ Unencoded	Coding
Unicast header/first frame	|preamble (5) |long header (4) destination address high order (4) |destination address low order (4) |source address low order (4) |source address high order (4) |	25 Bytes/5 Bytes	(32,8) ECC and EDC
Unicast rest header frame and three bytes data	|preamble (5) |short header (1) |number of data bytes filed at high order (4) |number of data bytes filed at low order (4) |data1 (4) |... |data3 (4) |	26 Bytes/6 Bytes	(32,8) ECC and EDC, except the short header
Broadcast header /first frame	|preamble(5) |long header (5) |sequence is the repetition count of consecutive broadcast header and ranges in value from 00 to 07 |	13 bytes/2 Bytes	(32,8) ECC and EDC
Broadcast rest header frame and first byte data	|preamble (5) |short header (1) |source address low order (4) |source address high order (4) |number of data high order (4) |number of data low order (4) |data1(4)	26 Bytes/6 Bytes	(32,8) ECC and EDC, except the short header
Unicast/Broadcast data intermediate frames and five bytes data	|preamble (5) |short header (1) |data1 (4) |...|data5 (4) |	26 Bytes/6 Bytes	(32,8) ECC and EDC, except the short header
Unicast/Broadcast data last frame (four bytes data and one byte checksum)	|preamble (5) |short header (1) |data1 (4) |...|data4 (4) |XOR |	10–26 Bytes/2–6 Bytes	(32,8) ECC and EDC, except the short header
General response/ long acknowledgment frame	|preamble (5) |short header (1) (NOT Ack) |long header (4) (LONGAck) |source address high order (4) |source address low order (4)	18 Bytes/4 Bytes	(32,8) ECC and EDC, except the short header
First frame acknowledgment	|preamble (5) |short header (1) (NOT Ack) |long header (4) (LONGAck) |	10 Bytes/2 Bytes	(32,8) ECC and EDC, except the short header
Control transfer response/handshake to transmit data	|preamble (5) |long header (4) (LONGAck) |	9 Bytes/1 Bytes	(32,8) ECC and EDC

one low data rate central unit (LDCU). The center control unit (CCU) interface joins the HDCU and LDCU together. The MMU/MDCU interface represents a method to connect the MMU to the MDCU through a single port as shown in Configuration A in Figure 7.10. But certain expansion paths may require multiple connections as shown in Configuration B in Figure 7.11. Remote communications devices (RCDs) communicate data. The RCDs should

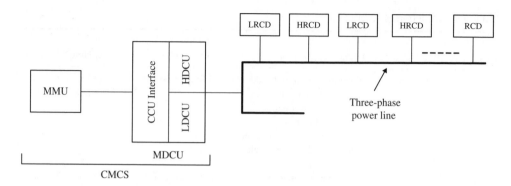

Figure 7.10   Remote condition monitoring system according to ISO 10368:2006(E), configuration A.

coexist on the same power line network and not interfere with simultaneous communications of the HDCU and the LDCU. An MMU, MDCU and operator interface devices compose a central monitoring and control system (CMCS) which monitors and controls one or more RCDs [6].

7.4.4   Communication Protocol

Each refrigeration container ship system applying ISO 10368:2006(E) has three interface areas that manage the communication between units as shown in Figure 7.12.

> **Communications between MMU and MDCU**: The communication between the MMU and the MDCU is handled by a full duplex Electronic Industries Association (EIA) RS232-C serial interface. The baud rate should be two times the baud rate of the fastest RCD in the system, typically 4800 baud.

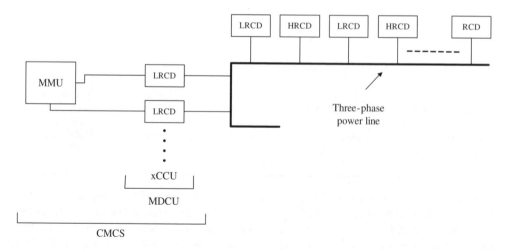

Figure 7.11   Remote condition monitoring system according to ISO 10368:2006(E), configuration B.

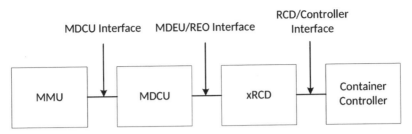

Figure 7.12 Communication interfaces for remote condition monitoring according to ISO 10368:2006(E).

Communications between MDCU and LRCD/HRCD: In the low data-rate case, the communication system consists of a single control station LDCU which initiates all transactions as a master device for multiple LRCDs as slave devices. There is no LRCD to LRCD communications. The PLC transmission has four components: a preamble to determine the start of a valid message and synchronize communications between devices on the power line, ten ASCII control characters reserved to provide control functions during communications, a prefix with 13 ASCII characters, and data portion that consists of the information to be exchanged between two devices (master to slave) on the PLC network. The data block structure has a 16-bit cyclic redundancy check (CRC) field.

In the high data-rate case, two basic types of communication are possible between an HDCU and an HRCD: data packet exchanges and control exchanges. Data packet exchanges transfer information from HDCU or HRCD to its peer layer in an HRCD or HDCU, and control exchanges are used by HDCU to give an HRCD temporary access to the power line network to perform a data packet exchange. If the slave has any data to send, it responds with an acknowledgment and the master in turn, sends a handshake. This allows the slave to start transmitting its data packet and after that returns control to the master.

The communication transfer diagram session between MDCU and LDCU/HDCU is shown in Figure 7.13. The communication commands between MDCU and LDCU/HDCU are described in Table 7.3.

Communications between RCD and container controller: The communication between the RCD and the container controller is not specified in the standard and it could be given for example by an RS-232 serial interface.

7.4.5 Remarks

This system shows a typical approach for industrial PLC systems. The complete system is handled as an island and end-to-end IP communication is often not required. A central control unit manages the traffic on the line and collisions are avoided. CSMA or slotted ALOHA as channel access are only used for the registration phase of new equipment. The predictability of the communication has a high importance. While IP traffic will be included more and more, a change to a decentralized medium access control and the loss of predictability of the communication will not be acceptable for this applications.

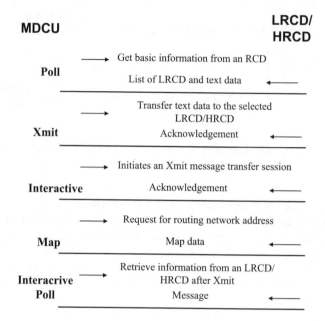

Figure 7.13   Communication session between MDCU and LRCD/HRCD.

In this special application an international standard is very important, because the trading companies cannot accept a restriction that their refrigerating containers can only use some special ships. Therefore, a significant number of components are installed worldwide. For each change, it has to be ensured that the compatibility with the existing nodes is not lost, which also limits the possibility to update these systems to state of the art technologies. The networks have to be installed very fast and run only for days or weeks during the transfer time from harbor to harbor. The installing effort for any other wired technology is not acceptable and wireless technologies have problems to run reliably in this environment of metal walls and containers.

Table 7.3   Communication command between MDCU and LRCD/HRCD

Message Type	Message Description
Poll	The control station queries a slave station to receive basic information.
Xmit1/2	Xmit1 data from the control station to a particular slave station using fast select, or selection with response in Xmit2.
Interactive1/2	Xmit1/2 then, after delay, Interactive1/2 poll the same slave station to receive response.
Map	Map slave stations on the network and uses a control exchange to log in new devices.
Interactive poll	The control station polls a particular slave station to receive information from a previous Xmit.

7.5 Windowed Frequency Hopping System AMIS CX1-Profile

The communication protocol AMIS CX1-Profile of the company Siemens AG is used for PLC over the low voltage distribution grid between final nodes like meters, breakers, control units and data concentrators at central stations. It is part of an integrated data management system for utilities and a classical CENELEC A-Band (9–95 kHz) application according to EN 50065-1. Very limited information about this proprietary protocol was available until recently, when Siemens AG decided to publish this specification [8] to enable other manufacturers to implement it. The plan to bring this specification to international standardization was announced, but it is not part of the recently approved IEEE 1901.2 and ITU-T G.990x standards (see Chapter 9).

The AMIS CX1-profile describes the layers 1 to 4 (i.e. physical to transport layer) of the OSI model. The protocol is closely linked to the applications and no transfer of IP packets is supported. Analyzing the specification [8] some similarities to a street lighting automation system with PLC of the company Zumtobel can be noted. Furthermore, although the system was developed more than one decade ago, there are some very interesting features and solutions and not supported or as efficiently solved in the current versions of IEEE 1901.2 or ITU-T G.990x standards.

7.5.1 Physical Layer

The packet structure of the PHY layer is shown in Figure 7.14. It starts with a preamble, followed by PHY-Header, MAC-Header and MAC-Data (which form PHY-Data), and PHY-Trailer. Preamble and PHY-Header have a fixed transmission mode with 600 bits/s. PHY-Data and PHY-Trailer can use 1 out of 16 transmission modes with data rates between 600 and 3000 bits/s. The information about the used transmission mode is transmitted within the PHY-Header. All transmission modes use the frequencies between 39 kHz and 90 kHz and fit into the CENELEC A-Band of EN 50065-1.

The requirements on the PHY-layer design are slightly different from those for orthogonal frequency-division multiplexing (OFDM) based designs used in IEEE 1901.2 and ITU-T G.990x. Of course robustness to frequency selective channels and impulsive and narrowband interference as well as electromagnetic compatibility are required. However, other design objectives are low requirements on the linearity of the analog-front-end, low peak-to-average-power ratio of the transmit signal to realize high power efficiency of the transmit amplifier and path.

To meet these requirements, the AMIS CX1-profile uses a wavelet based frequency hopping technique in combination with differential phase shift keying (DPSK), interleaving and a repetition code as high redundancy channel coding. The signal is generated using a constant

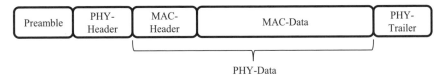

Figure 7.14 Packet structure of the physical layer of AMIS CX1.

sampling rate of 347.2 kHz (with a deviation of less than 25 ppm). Due to the low carrier frequencies, no frequency synchronization between transmitter and receiver is necessary. The basic transmission mode uses a $(8,1,8)_2$ repetition code, 8 hopping frequencies and a binary DPSK (DBPSK). Other transmission modes use $(5,1,5)_2$, $(6,1,6)_2$, $(7,1,7)_2$ or $(8,1,8)_2$ repetition codes, 5 to 8 hopping frequencies and either DBPSK or $\pi/4$-shifted quaternary DPSK (DQPSK) modulation. The number of coded bits is always equal to the number of hopping frequencies.

A block interleaver ensures a spreading of the coded bits over time and all hopping frequencies. The PHY-Trailer is used for bit stuffing to ensure completely filled blocks for the interleaver. As each bit of information is transmitted on every hopping frequency, a very high resistance against narrowband interference and frequency selective fading can be realized. The time spreading increases the resistance against impulsive interference. The used repetition code has a low code rate of 0.125 to 0.2 and as direct consequence a relatively low data rate is realized. Of course a stand alone repetition code is not state of the art anymore. An equivalent or even better resistance could be reached by a convolutional code at a code rate of 0.333 to 0.5, which would result in more than double data rate, without any other changes to the system.

The wavelets are defined as sine tones on the hopping frequencies with a cosine-roll-off amplitude window with roll-off-factor $\alpha = 1$. Due to the large roll-off-factor the spectral forming of the wavelets is very good compared to classical frequency hopping approaches. The wavelets are superposed with a 50% time-overlapping, which results in a nearly constant envelope.

Figure 7.15 explains the basic transmission mode in a time-frequency-amplitude diagram. The hopping frequencies of the basic transmission mode are written on the y-axis. The hopping frequencies are used in a fixed order and the duration of a single wavelet is 416 μs. Due to the 50% time-overlap, every 208 μs one wavelet is transmitted. After 1.666 ms all 8 hopping frequencies are used and it continues with the same frequency from the beginning. The inter-symbol-time on each individual hopping frequency is constant 1.666 ms.

The information of the coded bits is mapped on the phase differences of consecutive wavelets on the same hopping frequencies. Therefore, the modulation is a temporal DPSK and similar to the differential coding used by the OFDM system in IEEE 1901.2. The OFDM system would add the sequentially transmitted signals on the hopping frequencies to a single signal, which

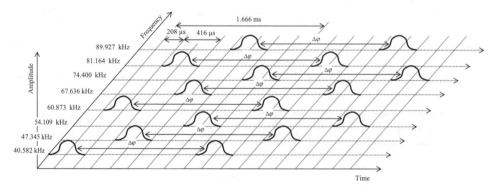

Figure 7.15   Signals in basic transmission mode of AMIS CX1.

would occupy all frequencies at the same time. This leads to the higher data rates in OFDM systems. The modulation scheme of AMIS CX1 uses each hopping frequency independent of the other hopping frequencies. Due to the high redundancy and the interleaving method, all information is transferred on each hopping frequency. Even if only one hopping frequency is transferred reliably, all information is delivered.

Compared to the typically used OFDM systems for the CENELEC A-band, this approach also has other advantages. The available transmit power can be concentrated on each frequency separately and the nearly constant envelope allows for more efficient transmit amplifier concepts. A notch in the access impedance and overload of the power amplifier would only affect a single hopping frequency, as compared to the complete OFDM symbol in other systems. The receiver can do individual matched filtering on each hopping frequency, which has a much better spectral suppression of out of band disturber than an FFT-based channel separation with a $\sin(x)/x$ spectral mask. Also, inter-frequency-modulation due to non-linearity of the channel has nearly no negative effect. Finally, impulsive noise bursts would only hit a single wavelet and not a complete OFDM symbol. Therefore, the AMIS CX1 PHY layer enables PLC over larger distances in the CENELEC A band compared to OFDM-based systems like IEEE 1901.2 or ITU-T G.9904 (PRIME).

7.5.2 Medium Access Control and Network Layer

The AMIS CX1 communication protocol is designed as multi-hop master-slave system, where the data concentrator is always the master and all other nodes are slaves, which only answer on request. A polling procedure is used to transfer spontaneous data from any slave to the master. All slaves are used as possible repeater and the relay function is realized by synchronous forwarding, which can also be understood as single frequency network (SFN) based flooding, cf. [9]. Due to the integration of the synchronous forwarding, MAC and networking cannot be separated any more. Hence, the system design can be understood as cross layer approach.

In principle two MAC-block formats exist: a data telegram with MAC-Header and MAC-Data, and a short telegram with only a MAC-Header. The MAC-Header has a fixed length of 80 bits, containing 8 bits for control information, 40 bits for addressing, 8 bits for data block length, 2 counters of 4 bits for synchronous forwarding and a 16 bit CRC for the header. The MAC-Data is protected by another 16 bit CRC.

The system is designed to allow up to 8 repetitions of a packet. The transmission mode and the number of repetitions are decided by the master dynamically and individually for each slave. The number of repetitions is added to both counters in the MAC-Header. If a slave receives a packet for which it is not the destination, and the second counter is not 0, then it prepares this packet for a retransmission. The second counter is decremented and the CRC for the MAC-Header re-calculated. Exactly 253.8 µs after the reception of the packet, all of these slaves retransmit this prepared packet synchronously. The method is called simultaneous forwarding in [8]. It can also be understood as an instance of SFN transmission.

Figure 7.16 illustrates this simultaneous forwarding for a single-packet request-response service in a tree topology with 200 participants (cross-markers). The master is located in the central node. The filled-square markers are transmitting and the square markers are receiving the packet in the depicted time slot. The master initiates the communication service by

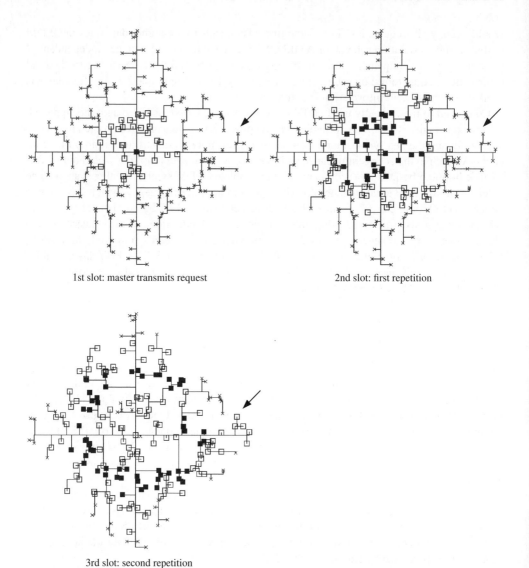

Figure 7.16 Illustration of simultaneous forwarding from master to destination slave (→).

transmitting a request. The messages flows in all directions, which is typically understood in networking as information flooding. Therefore, this simultaneous forwarding method can be characterized as an SFN-based flooding.

The destination node also uses this mechanism in the opposite direction for its response as illustrated in Figure 7.17. In the shown example, the master already receives the response with the second repetition and a third repetition actually is not required. However, if the third repetition is performed, the master gets a second chance to receive the response. The effort required for managing the network is independent of the number of repetitions. For each

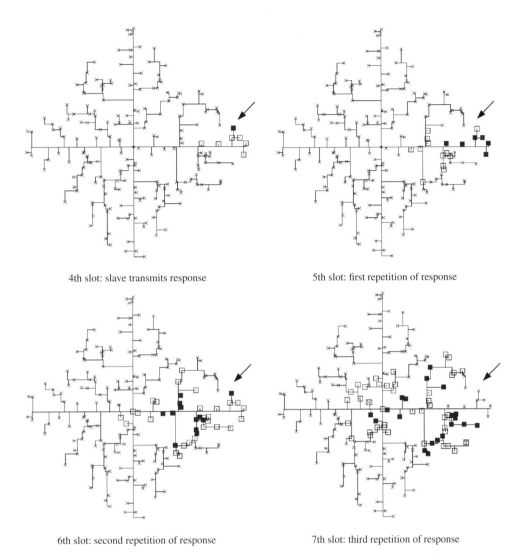

4th slot: slave transmits response

5th slot: first repetition of response

6th slot: second repetition of response

7th slot: third repetition of response

Figure 7.17 Illustration of response from slave to master in a simultaneous forwarding system.

downlink and uplink the same number of repetitions will be used, which however is an unnecessary restriction compared to the approach discussed in [9].

Compared to the routing mechanisms proposed in IEEE 1901.2 and ITU-T G.990x, simultaneous forwarding is much easier to implement. Repeaters do not need to be identified beforehand and thus, much faster adaptation to changes in the topology are possible. However, due to multiple simultaneous transmissions the overall energy consumption of the communication system will be higher and transmission opportunities are wasted in networks with a lot of repeater levels. On the other hand, especially in this type of network, routing based protocols are not efficient, because of the traffic required for network organization. The requirements of grid automation systems are hard to fulfill with the current routing approaches. SFN-based flooding for airfield lighting automation (see discussion in [10]) is in use in systems on several continents and has a proven track-record for safety critical automation.

7.5.3 Management Functions

The AMIS CX1 profile includes data transfer and management functions. A master-slave concept with central resource management is implemented. For efficient use and fair reduction of bandwidth in case of overload, a time credit technique is used. For automation systems a very reliable management information base and short detection times of interruptions is important and polling or keep-alive messages, which is equivalent to polling on a point-to-point link, are always implemented. On the negative side, the latency for spontaneous messages from nodes are fairly high in large networks due to the polling time.

Simultaneous forwarding requires synchronization between the transmitters. As several devices have already synchronization at the MAC-layer, it is efficient to integrate the absolute time synchronization of the complete system as a feature to this layer. Despite the low data rate, it is possible to synchronize all nodes in the network with a precision of < 1 ms. Precise time stamping is very helpful in automation systems, especially when alarm showers happen.

SFN-based flooding realizes an efficient broadcast in the system with very low resource consumption. This broadcast can use nodes as repeaters, which are not logged in already. Beacon transmission and login procedures are much easier. To enable configuration and software download a conformed broadcast is implemented. Several broadcast messages (i.e. firmware) are flooded through the network. Each node responds with a cumulative acknowledgement and lost messages are selectively retransmitted. For a system with low data rate, it is an efficient way to update the software of a large number of nodes in the field in a reasonable time.

7.5.4 Further Remarks

A recent deployment example of this technology is the AMIS smart meter system installed for Energie AG Oberösterreich in Austria. The current roll-out includes more than 100 000 smart meters. Due to regulations of the European Union, 80% of all households should be equipped with smart meters by 2020. Energie AG Oberösterreich is planning the installation of smart meters for 500 000 customers.

AMIS CX1 is a good example of an already available communication solution for (smart grid) system automation. However, due to its proprietary nature it has not been subject to a scientific discussion. New PLC developments often start from standardized and published solutions for wireless systems, such as IEEE 802.15-4. However, these solutions may not be a good fit for the power line channel and a lot of changes are necessary. The result is not as good and efficient as it could be using a direct design approach. A fundamental difference between AMIS CX1 and any OFDM-based PLC system is its robustness to non-linear channel effects. It is still an open question for CENELEC A-band modems, if it possible to ensure for any OFDM-system the required linearity in case of low and time-varying access impedances. The massive use of the new energy-saving power supplies compounds this problem.

Precise time synchronization, simultaneous forwarding and confirmed broadcasts are useful features of AMIS CX1, which are not fully available or not supported in IEEE 1901.2 and

ITU-T G.990x. Considering current chip designs for narrowband PLC, AMIS CX1 could be implemented as a software radio on most of them, which would give the utilities an exit strategy for the non-standardized system.

7.6 DigitalSTROM®

The digitalSTROM® technology is a PLC technology for 'smart energy living' [11]. The main idea of digitalSTROM® is that the installed technologies communicate through the existing power lines between the different in-home applications over modulated up and downstream channels. The system control and management can be done via normal switches, the internet and smart phones. The switch in digitalSTROM® is a 4 mm × 6 mm high-voltage (220 V) chip integrated in the digitalSTROM® terminal box. The integrated chip switches electricity on/off, therefore, digitalSTROM® does not need any capacitive coupling or galvanic isolation. Devices and home applications need only to install this chip.

7.6.1 The DigitalSTROM® Architecture and Components

DigitalSTROM® is a centralized network based on installed concentrators in the distribution panel. It operates as power meter for particular distribution circuits and communicates with in-home appliances over different modulated up and downstream channels. The basis of digitalSTROM® is a high-voltage chip called terminal block that can switch, dim and measure electricity in lighting systems, store data and communicate. To use the existing power lines to communicate from the distribution panel to digitalSTROM® Devices (dSD), a digitalSTROM® Meter (dSM) is installed in each power circuit in the distribution panel. Multiple dSMs (up to 62 units) are using a standardized protocol (dS485 bus interconnection) for communication. The installation of an optional digitalSTROM® Server (dSS) enables connection to a higher level system, the Internet or a local network. DigitalSTROM® Filters (dSF) are band-pass filters for CENELEC A band and are installed per phase between L and N upstream to optimize the network conditions for digitalSTROM® communication by adjusting the power signal to reduce the interference with other devices.

7.6.2 The DigitalSTROM® PLC Network Components and Installation

The digitalSTROM® installation uses the master-slave architecture in which a dSM creates the communication master for each circuit. The master communicates via the power line with the terminals (slaves) in the same circuit. For example, if the dSD registers a movement of the switch, it sends a telegram to the dSM. The dSM calculates what has to be done and sends the command via the power line. The system uses an asynchronous process for communication between meter and terminal in which the dSM transmits the information to the terminals in the common current. Upstream (from dSD to dSM), the terminals code their signals by modulating the current consumption. The dSMs communicate among themselves and with the dSS via RS485 interfaces and dS485 bus interconnection. The dSS has various functions. It links the dSMs and also connects to the Internet via TCP/IP, thus permitting access to all network functions through computers or smart phones. Finally, the dSF protects

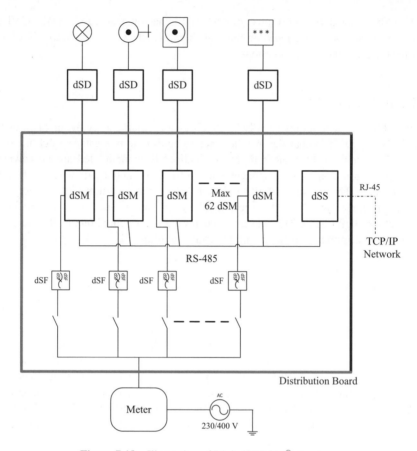

Figure 7.18   Illustration of DigitalSTROM® structure.

the network communication and is installed per phase between L and N upstream [12]. The digitalSTROM® components and topology are illustrated in Figure 7.18.

7.6.3   DigitalSTROM® Communication

For up and downstream communication, different signals are transmitted [12].

> **Downstream communication**: The downstream transmission technique of the digitalSTROM® is to cut off the distribution board from the supplying grid periodically and place the signal onto the mains voltage in close proximity to the mains zero crossings, which is different to other normal PLC systems that add the transmitted signal to the mains voltage, which reduces crosstalk into the neighboring circuits caused by low grid impedance. The cut off technique is simply done by a switch like a circuit breaker that opens for the period of the downstream signaling. The transmitter is realized using power semiconductors operated by a

pulse shaping control. The impulses are represented by a logical '1' and a logical '0'. A (7,4) Hamming code is implemented to detect and correct errors. The downstream transmitter could be considered as an optimal voltage source that does not influence the low network impedance, especially low-bit-rate PLC technologies. The signal attenuation should be small because of the low source impedance and the short length of cables inside homes. No interference is expected close to the zero-crossing because of the switched-mode power supplies.

Upstream communication: Upstream communication is based on the centralized architecture with a dSM for each circuit breaker. The transmitter is realized as current source. The dSM applies a current sensor to the received signal. Additional capacitors are added between the neutral and the phase conductors to keep the impedance of transmitter low and increase the signal level at the receiver. The upstream uses FSK modulation for CENELEC A or B bands. Interferences from home appliances and switching devices is expected. A convolutional code with an interleaver is implemented by keeping the events failure below the constraint-length of the convolutional code.

7.7   Conclusion

PLC systems enable communication for home and industry automation. In this chapter, we first reviewed some popular protocols and standards, which have been applied widely but are only a selection of those available in the market. We then introduced the international standard ISO 10368:2006(E) for refrigeration container ships, which applies two independent physical layers that are separated in frequency and used simultaneously on the same line. We discussed the AMIS CX1-Profile for PLC over voltage distribution grid, which uses a windowed frequency hopping system with a very robust PHY-Layer, low crest-factor and a high immunity to non-linear distortion. An SFN-based flooding is realized using precise time synchronization. We finally presented the digitalSTROM architecture which enables PLC modems with a very small size. The physical dimensions of the modem are even smaller than the capacitor of a classical coupling unit.

References

1. E. Mainardi and M. Bonfè, Powerline communication in home-building automation systems, in *Robotics and Automation in Construction*, (eds C. Balaguer and M. Abderrahim), InTech, 2008, ch. 4, 53–70.
2. J. Burroughs, X-10 home automation using the PIC16F877A, Microchip Technology Inc., Tech. Rep., 2002. [Online]. Available: http://ww1.microchip.com/downloads/en/AppNotes/00236a.pdf
3. KNX System Specifications, KNX Association, Tech. Rep., 2009. [Online]. Available: http://www.sti.uniurb.it/romanell/Domotica_e_Edifici_Intelligenti/110504-Lez10a-KNX-Architecture v3.0.pdf
4. KNX Powerline PL 110, KNX Association, Tech. Rep., Jun. 2007. [Online]. Available: http://www.knx.org/media/docs/KNX-Tutor-files/Summary/KNX-Powerline-PL110.pdf
5. Echelon Corporation. Introduction to the LONWorks system. United States of America. [Online]. Available: http://www.echelon.com/support/documentation/manuals/general/078-0183-01A.pdf
6. International Organization for Standardization, Freight thermal containers – Remote condition monitoring, ISO 10368:2006, ISO, Geneva, 2006.

7. G. Hallak and G. Bumiller, Coexistence analysis of impedance modulating transmitters, in *Proc. IEEE Int. Symp. Power Line Commun. Applic.*, Johannesburg, South Africa, Mar. 24–27, 2013, 279–284.
8. SIEMENS AMIS CX1-Profil (Compatibly/Consistently Extendable Transport Profile V.1) Layer 1-4 (in German), SIEMENS, Tech. Rep., Sep. 2011. [Online]. Available: http://www.quad-industry.com/titan_img/ecatalog/CX1-Profil_GERrKW110928.pdf
9. G. Bumiller, L. Lampe, and H. Hrasnica, Power line communication networks for large-scale control and automation systems, *IEEE Commun. Mag.*, 48(4), 106–113, Apr. 2010.
10. G. Bumiller and N. Pirschel, Airfield ground lighting automation system realized with power-line communication, in *Proc. Int. Symp. Power Line Commun. Applic.*, Kyoto, Japan, Mar. 26–28, 2003, 16–20.
11. digitalSTROM Alliance. [Online]. Available: http://www.digitalstrom.org
12. G. Dickmann, digitalSTROM®: A centralized PLC Topology for Home Automation and Energy Management, *Proc. IEEE Int. Symp. Power Line Commun. Applic.*, Udine, Italy, Apr. 3–6, 2011, 352–357.

8

Multimedia PLC Systems

S. Galli, H. Latchman, V. Oksman, G. Prasad, and L. W. Yonge

8.1 Introduction

A major driving force in the development of high speed power line communications has been
the seamless delivery of multimedia content throughout the home using the existing electrical
wiring within the home [1]. In order to provide multiple simultaneous channels of voice, high
fidelity stereo audio and standard as well as high definition video signals while coexisting with
traditional data traffic, multimedia PLC networks and systems have had to be designed to meet
stringent quality of service (QoS) specifications [2]. In this section we provide a description
of innovative PLC multimedia communication technologies that provide acceptable levels of
latency, jitter and reliability for the delivery of multimedia content via noisy and unreliable
power line channels.

8.2 QoS Requirements for Multimedia Traffic

The PLC channel is known to be plagued by various noise and interference sources that have
made reliable high speed communication a challenge. The HomePlug 1.0.1 [3] specification
provided the first physical layer (PHY) and Medium Access Control (MAC) protocols that
provided a 10 Mbps class of PLC LANs using existing in-building electrical wiring. The
HomePlug 1.0.1 chip also formed the basis of a 10 Mbps Broadband over Power Line (BPL)
access network using the incoming low-voltage distribution lines [4]. The HomePlug 1.0.1
system provided theoretical maximum application throughput of about 8 Mbps, with a typical
maximum in practice of 5–7 Mbps [5], [6], quite adequate for many data-centric LAN applications. In this way the PLC channel, formerly used only for the delivery of electrical power,
entered the digital age, now also providing a high speed digital pipe to the home with a 'no
new wires' multimedia communication network inside.

Though not optimized for multimedia communication, HomePlug 1.0.1 systems could also
support Voice over IP (VoIP), as well as IP Television (IPTV) using some variant of now

Power Line Communications: Principles, Standards and Applications from Multimedia to Smart Grid, Second Edition.
Edited by Lutz Lampe, Andrea M. Tonello, and Theo G. Swart.
© 2016 John Wiley & Sons, Ltd. Published 2016 by John Wiley & Sons, Ltd.

popular buffered streaming mechanisms [7]. These two multimedia applications – VoIP and IPTV – illustrate some of the challenges that have to be successfully met by PLC systems that target multimedia traffic. On one end of the spectrum VoIP traffic involves the transmission of relatively small packets (10–15 ms in duration) at almost symmetric bi-directional data rates of about 100 kbps but with the constraint that these packets must be delivered within a tight delay bound 10's of milliseconds to facilitate interactivity and naturalness in human communication. Voice applications are very tolerant to moderate packet loss rates and per packet delay variation (jitter), provided we are willing to accept marginal degradation in voice quality. On the other hand IPTV uses higher data rates (1–4 Mbps) with asymmetric bandwidth requirements and for optimal user experience should exhibit very low jitter and almost no video drop outs related to packet losses. Since IPTV is a broadcast or non-interactive service, we can buffer up to several hundred milliseconds of the multimedia data and ensure that the subsequent play out experience is acceptable for discriminating video enthusiasts and audiophile audiences. In all cases, it is desirable for the multimedia network to cover the entire home or building in a seamless manner.

The above issues associated with multimedia content delivery, so well illustrated by VoIP and IPTV, may be extended and scaled for other existing and emerging multimedia content, such as HDTV, Multi-channel high fidelity audio and video conferencing, multimedia gaming, etc.

8.2.1 Multimedia In-home Networking

Table 8.1 summarizes the typical QoS requirements for various multimedia content that could be delivered over PLC multimedia networks.

8.2.1.1 Multimedia Traffic Characteristics

The table shows a variety of multimedia content, ranging from digital video, audio, pictures and images as well as interactive services such as video conferencing, voice and gaming. Each of these groups of applications is briefly described next.

Table 8.1 Bandwidth and QoS requirements for Multimedia Applications

Application	Bandwidth [Mbps]	Latency [ms]	Jitter	Packet Loss Probability (PLP)
High Definition (HD) Video Streaming	11–25	100–300	0.5 μs to several msec	Quasi Error Free*
Standard Definition TV (SDTV) Video Streaming	2–6	100–300	0.5 μs to several msec	Quasi Error Free
IPTV	1–4	100–300	0.5 μs to several msec	Quasi Error Free
DVD Quality Video	6–8	100–300	0.5 μs to several msec	Quasi Error Free
Internet Video Conferencing	0.1–2	75–100	—	10^{-3}
Home Theater Audio (Multiple streams of audio)	4–6	100–300	0.5 μs to several μs	Quasi Error Free
Voice over IP (VoIP)	< 0.064	10–30	10–30 msec	10^{-2}
Network Gaming	< 0.1	10–30	10–30 msec	Quasi Error Free

* Less than one error in two hours

8.2.1.1.1   Digital Video

Digital video and audio form the bulk of the multimedia traffic encountered today. Early forms of digital video included digital formats for video streaming at a variety of speeds and resolutions by such applications as RealPlayer[1] or Microsoft Media Player[2]. Streaming protocols associated with these technologies provide for adaptation to the available bandwidth using a buffered playback mechanism. These methods are still used today for IPTV broadcasts over the Internet. Digital Video Disk (DVD) and HDTV formats ranging in data rates from 5–25 Mbps are now also available and can be played from a DVD player or HDTV player (such as BluRay) or played over a home network from a central media server. DVD and HDTV use compression techniques based on the MPEG-2 and MPEG-4 formats developed by the Motion Pictures Expert Group (MPEG).

8.2.1.1.2   Pictures and Images

Pictures and images form another popular media content that need to be transmitted over a home network. Digital pictures and images obtained via still digital scans or digital cameras with resolutions of several mega-pixels can also be stored on a central media server and slide shows can be generated on a screen by fetching and displaying the pictures in sequence. Each picture can be several megabytes in size and such slide shows will have a significant impact on the home network. Digital photographs and scanned images are often stored in the Joint Pictures Expert Group (JPEG) format or Graphics Interchange Format (GIF). In fact it is possible to generate motion video by using motion JPEG – effectively playing still frames at an appropriate frame rate.

8.2.1.1.3   Audio – Music and Speech

High fidelity stereophonic (or multi-speaker surround sound) music is another major component in home multimedia systems. Often the audio channels are synchronized to the video in DVD or HDTV presentations and these are transmitted simultaneously over the network channel, without significant delay variation or picture-sound synchronization problems. In broadcast applications, the video and audio can be buffered, decompressed and played out in the right sequence and rate to give optimal quality.

8.2.1.1.4   Interactive Voice and Video

Applications such as video conferencing and VoIP impose the additional demand of two-way spontaneous interactivity, with the attendant need for psychologically pleasing responsiveness. These applications are best provided in a channelized form in which a specific allocation of channel resources is reserved for each direction of the video and audio interchange. A home network designed to handle multiple channels of VoIP as well as video conferencing needs to consider these demands in conjunction with appropriate compression to conserve bandwidth and enhance interactivity.

8.2.1.1.5   Gaming and Simulations

Multiplayer network based games and simulators such as flight simulators also place great demands for interactivity and high video and audio quality on the underlying network. Some simulators and games have lifelike video or high quality graphics that must be loaded, displayed

---

[1] http://www.real.com
[2] http://www.microsoft.com/windows/windowsmedia

and rendered as quickly as the players make moves – thus creating the need for synchronization, interactivity and naturalness.

8.2.1.2   Quality of Service Parameters

For each major type of multimedia content, Table 8.1 also shows four major parameters, namely bandwidth, packet loss, latency and jitter, that are used to quantify quality of service.

8.2.1.2.1   *Bandwidth*
Multimedia traffic such as video and voice performs best when guaranteed bandwidth is available to allow the maximum expected bit rate of the stream to be supported without congestion. For example, traditional time division multiplexing (TDM) provides fixed allocations of bandwidth that supports standard telephony with uncompressed voice occupying 64 kbps and ISDN-based video that requires multiples of 64 kbps Basic Rate Interface channels. However in a shared network such as the Internet or a LAN based on Ethernet or Wireless or PLC technologies, bandwidth allocation for multimedia traffic becomes much more involved and sophisticated admission control and congestion management strategies are needed to ensure QoS. Nonetheless, it is certainly the case that services such as multiple channels of HDTV, IPTV and VoIP cannot be contemplated unless the overall available channel through put is adequate. Thus a major effort in multimedia PLC is to provide the largest possible raw data rate given the spectral and regulatory constraints. Currently available PLC technology provides a data rate of some 200 Mbps in the 2–30 MHz band, with usable application level bandwidth of about 90 Mbps after error and protocol management overheads are accounted for. Using a bandwidth up to 100 MHz and MIMO, the data rate goes up to around 500 Mbps.

8.2.1.2.2   *Packet Loss and Latency*
The delay experienced by a packet of data as it traverses the power line channel is composed of a number of contributory factors such as processing and propagation delays and retransmissions. It turns out that the overall delay is dominated by the ever-present need for re-transmissions in the PLC channel. Even after the use of powerful error control coding, the power line channel may still have FEC-corrected block error rates as high as 1 in 100 or worse. To get a block error rate of 10^{-10} (*quasi-error-free*) as needed to meet the QoS requirements for some multimedia applications, up to five retransmissions may be needed. Applications like HDTV, SDTV or music streaming allow for an initial delay of say 100–300 ms and this time can be used to accommodate the needed retransmissions. Interactive applications such as gaming, VoIP and video conferencing that cannot tolerate such long delays will need to drop or substitute packets and this may lead to some unavoidable quality degradation.

8.2.1.2.3   *Jitter*
Jitter is the term used to describe the variation in delay that may cause undesirable effects in multimedia traffic. Consider for example the case of HDTV streamlining in which an initial delay of 300 ms is used to accommodate needed packet retransmissions. Now if all packets are delayed (buffered and played out) with this identical delay of 300 ms relative to the start of the first packet being played, then the jitter will be precisely zero. In practice a jitter value for such applications of about 500 nanoseconds is desirable and multimedia PLC protocols should provide for this capability.

8.2.1.3   A PLC Solution for Multimedia Traffic

From the above it is clear that, while HomePlug 1.0.1 provided acceptable data rates (5–7 Mbps) and performance for data communication needs in connecting multiple computers and peripherals in a LAN setting, higher data rates and more stringent QoS controls are needed to support digital multimedia communication within the home [8].

For example, as seen earlier, a single stream of HDTV requires about 25 Mbps and a typical home scenario may have a number of simultaneous multimedia streams of voice, audio and video, along with traditional non-time sensitive network traffic. Moreover, in addition to bandwidth, multimedia applications also have prescribed latency, jitter and packet loss probability (PLP) requirements that must be met for optimal performance, as shown in Table 8.1.

It is worth noting in passing that there are several potential alternatives to PLC for multimedia communications. These include such technologies as 100 Mbps and even Gbps (gigabit per second) Ethernet, the popular IEEE 802.11x suite of wireless protocols (including the emerging IEEE 802.n standard [9]), the newer Ultra-Wide band standard [10] providing high bandwidth but limited reach and others like the Multimedia over Coax Alliance (MoCA) [11] standard that uses exiting coaxial video cabling and Phoneline networks (HomePNA) [12] that make use of telephone wiring. The attractiveness of the PLC solution lies in its ubiquity, with typically an average of some 44 outlets already available in homes (in the USA), as well as its reach in providing, in most cases, whole house coverage at acceptable data rates.

8.3   Optimizing PLC for Multimedia

The first challenge in developing a PLC system suitable for multimedia traffic is to design the PHY and MAC protocols, together with required cross layer MAC/PHY interactions, to ensure adequate bandwidth to support multiple simultaneous channels of voice video and data.

8.3.1   Overall Design Considerations for Multimedia PLC

8.3.1.1   Multipath Effects, Noise and Interference in the PLC Channel

By their very nature, in-building electrical wiring, consisting of an almost random interconnection of and variety of conductor types, exhibit terminal impedances that vary with frequency and time. This results in an undesirable multi-path effect that causes delay spread (averaging a few microseconds) as well as deep signal fades up to 70 dB at certain frequencies within the 2–30 MHz band currently used by PLC communications [13]–[15].

PHY design for the PLC channel must also contend with an ever-present and varied set of PLC noise sources such as halogen and fluorescent lamps, switching power supplies, brush motors, and dimmer switches, to name a few. In this regard it should be noted that a multimedia PLC design should recognize and exploit the known cyclic variation of noise within the power line cycle, with the best signal-to-noise ratio occurring near the zero crossings of the cycle.

Since currently available PLC devices operate in the 2–30 MHz range, care should be taken to avoid causing interference or being impacted by other legitimate signals, such as shortwave, citizen and amateur radio signals, while at the same time adaptively maximizing the capacity of the resulting PLC channel.

8.3.1.2   Multimedia PLC Design Choices

In recent years, a class of 200 Mbps PLC systems has emerged from the major players in the PLC chip development arena [16]–[18]. A common approach appropriate for the physical characteristics of the PLC channel outlined above is to use OFDM, with either Fast Fourier Transform (FFT) or Wavelet Transform as the underlying technology. It is beyond the scope of this section to discuss the differences and similarities between the FFT and Wavelet based approaches. Suffice it to say that both methods with appropriate design choices of basis functions, symbol lengths, guard intervals, modulation, error correcting code, and windowing selections can yield comparable levels of performance. In the rest of this section reference will be made to the FFT-based OFDM scheme used in HomePlug AV [18], [19].

The choice of OFDM is appropriate for PLC channels since the associated multicarrier approach allows the use of adaptive multi-level modulation schemes in the flat-fading sub-channels defined by OFDM from the 2–30 MHz frequency selective channel. In addition, the OFDM parameters can be so defined as to impose an operating spectral mask to avoid violating regulatory or interference limits.

It is generally agreed that for multimedia PLC systems, a TDMA approach is desirable to complement the CSMA/CA approach used in HomePlug 1.0.1 and other data centric protocols operating in noisy environments. Naturally the combined use of TDMA and CSMA/CA on the same medium introduces a great deal of complexity but this is the price one has to pay for QoS guarantees needed for multimedia traffic.

A basic need is for reliable communication of critical parameters of the hybrid TDMA/CDMA system to all nodes in the PLC network. To accomplish this, HomePlug AV defines a set of selectable low data-rate *robust modulation* (ROBO) schemes that use a high degree of time and frequency redundancy as well as low modulation order and powerful error correcting codes. ROBO mode is used to exchange critical information between transmitter and receiver, and then an adaptive high speed mode is negotiated between communicating pairs of nodes using adjustable *Tone Maps* to define the selected carriers, modulation and coding for inter-node communication. To maintain the performance of the network, every PLC packet contains a highly reliable *Frame Control* (FC) field that uses ROBO-like features to ensure that key PLC parameters are updated and reliably received.

8.4   Standards on Broadband PLC-Networking Technology

An important condition for mass deployment of broadband PLC technology and for multimedia applications in particular, is an international standard issued by a credible and globally recognized standards-setting body. Lack of international standardization initially led to the fragmentation of the home networking market. In fact, the PLC home networking market included mainly three industrial solutions: the HomePlug Powerline Alliance (HPA), High-Definition Power Line Communication (HD-PLC) Alliance, and the Universal Powerline Association (UPA). Furthermore, since the various PLC technologies available a dacade ago do not interoperate, the situation is very inconvenient to consumers, consumer electronics companies and service providers. Consumer confusion alone usually leads to higher return rates, which is a multi-billion dollar problem for consumer electronics companies.

The IEEE P1901 Corporate Standards Working Group [20] and the ITU-T Study Group 15 Question 18 initiated in 2005 standardization work to eliminate this fundamental barrier in development and deployment of home networking and access networks based on PLC technology. These two groups have finalized two BB-PLC standards, the IEEE 1901 [21], [22] and ITU-T G.996x (or G.hn) [23]–[28], and a stand-alone BB-PLC coexistence standard G.9972 (or G.cx) [29], [30]. These solutions include the physical (PHY) layer and the media access (MAC) sub-layer of the data link layer of the Open Systems Interconnection (OSI) Basic Reference Model. Upper layers of the protocol stack are generic. In the next two sections, the IEEE and ITU-T standards will be discussed in detail.

8.5   The IEEE 1901 Broadband Over Power Line Standard

The IEEE 1901 Working Group was established in 2005 to unify power line technologies with the goal of developing a standard for high-speed (> 100 Mb/s) communication devices using frequencies below 100 MHz and addressing both home networking and access applications [21], [22]. The standard was ratified in 2010 and defines two PLC technologies (an FFT-OFDM based PHY/MAC, and a Wavelet-OFDM based PHY/MAC) and a PLC coexistence protocol (the Inter-System Protocol or ISP). As per the scope of IEEE 1901, the standard will be usable by all classes of PLC devices, including those used for the first-mile/last-mile (< 1500 m to the premise) broadband services as well as devices used inside buildings for LANs and other data distribution (< 100 m between devices) applications.

The FFT-OFDM 1901 PHY specification facilitates backward compatibility with devices based on the HomePlug AV industry specification. Similarly, the Wavelet-OFDM 1901 PHY specification facilitates backwards compatibility with devices based on the HD-PLC Alliance industry specification.

A conceptual overview of the IEEE 1901 PHY and MAC standard is shown in Figure 8.1. The common MAC handles the two different PHYs via an intermediate layer called the Physical Layer Convergence Protocol (PLCP). There are two PLCPs: the O-PLCP that handles the

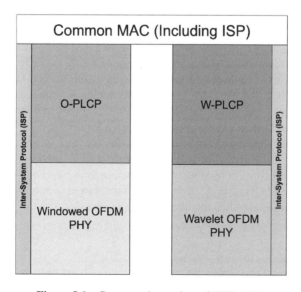

Figure 8.1   Conceptual overview of IEEE 1901.

interaction between the common MAC and the windowed FFT-OFDM PHY, and the W-PLCP that handles the interaction between the common MAC and the Wavelet-OFDM PHY. Example of functionalities present in each layer: Common MAC, frame formats, Addressing, SAP, SAR, Security, IPP/ISP, channel access, etc; W-PLCP and O-PLCP, channel adaptation, PPDU format, FEC, etc. Another key component of the standard is the presence of a mandatory Inter-System Protocol (ISP)[3] that will allow PLC devices based on the IEEE 1901 standards to share the medium efficiently and fairly regardless of the PHY differences; furthermore, the ISP also allows IEEE 1901 devices to coexist with devices based on the ITU-T G.9960 standard. The ISP is a new protocol that is unique to the PLC environment [30]. The ISP has also been standardized by the ITU-T as Recommendation G.9972 [29].

The decision of having a multiple-PHY solution in 1901 is not the consequence of a technical necessity. There are certainly some advantages in using either the Wavelet or the FFT-based OFDM PHYs, but those advantages alone do not really warrant the necessity of including both PHYs in the standard. Therefore, the multiple PHY nature of the IEEE 1901 standard descends more from a political necessity rather than a technical one. On the other hand, this decision will also enable continuity and a smooth migration from the currently deployed devices based on the HomePlug and Panasonic technologies to the IEEE 1901 devices. The ISP will also facilitate coexistence of IEEE 1901 and ITU-T G.996x devices avoiding the performance degradation due to the interference that devices based on these two non-interoperable standards will create. In fact, although the very concept of coexistence becomes moot once the industry aligns behind a common technology, including the ISP in the current standards and in the future next generation ones is a small price to pay in terms of complexity if a longer product life can be offered to PLC technologies based on the IEEE 1901 and ITU-T G.996x standards.

8.5.1   IEEE 1901 FFT-OFDM PHY

8.5.1.1   Overview

A block diagram of an IEEE 1901 FFT-OFDM transceiver is shown in Figure 8.2. On the transmitter side, the PHY layer receives its inputs from the Medium Access Control (MAC) layer. Three separate processing chains are shown because of the different error correction coding for HomePlug 1.0.1 Control Information, IEEE 1901 FFT-OFDM Control Information, and IEEE 1901 FFT-OFDM data. AV Control Information is processed by the AV Frame Control FEC Encoder block, which has an embedded FEC block and Diversity Copier while the IEEE 1901 FFT-OFDM data stream passes through a Scrambler, a Turbo FEC Encoder and a Channel Interleaver. The HomePlug 1.0.1 Frame Control (FC) information passes through a separate HomePlug 1.0.1 FEC. The outputs of the three FEC Encoders lead into a common OFDM Modulation structure, consisting of a Mapper, Inverse Fast Fourier Transform (IFFT) processor, Preamble and Cyclic prefix insertion, and windowed overlapping which eventually feeds the Analog Front End (AFE) module that couples the signal to the power line medium.

---

[3] Originally, the inter-technology coexistence mechanism introduced in IEEE 1901 was called Inter-PHY Protocol or IPP, and it handled coexistence only between the two IEEE 1901 PHYs. It was soon recognized by the IEEE 1901 Working Group that it was advantageous to extend coexistence also to other systems like the ITU-T G.hn and, thus, the original IPP was adapted becoming the ISP.

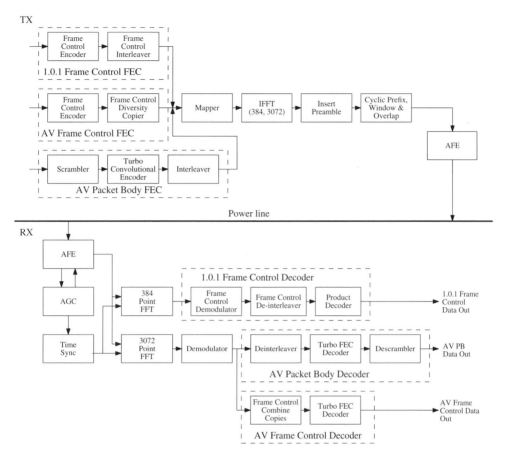

Figure 8.2   IEEE 1901 FFT-OFDM transceiver.

At the receiver, an AFE operates with an Automatic Gain Controller (AGC) and a time synchronization module to feed separate control and data information recovery circuits. The Frame Control is recovered by processing the received sampled stream through a 384-point FFT (for HomePlug 1.0.1 delimiters) and a 3072-point FFT (for IEEE 1901 FFT-OFDM), and through separate Frame Control Demodulators and Frame Control Decoders for the IEEE 1901 FFT-OFDM and HomePlug 1.0.1 modes. The sampled data stream (which contains only IEEE 1901 FFT-OFDM formatted symbols) is processed through a 3072-point FFT, a demodulator with Signal-to-Noise Ratio (SNR) estimation, a De-interleaver followed by a Turbo FEC decoder, and a De-scrambler to recover the data stream.

IEEE 1901 FFT-OFDM provides an order of magnitude throughput improvement over HomePlug 1.0, while also addressing key QoS issues. The bandwidth used has been extended and subcarrier spacing reduced in AV. Whereas HomePlug 1.0.1 uses the frequency range from 4.5 to 20.7 MHz quantized into 84 subcarriers, IEEE 1901 FFT-OFDM operates with 1155 carriers over the frequency range 1.8 to 30 MHz. While HomePlug 1.0.1 in its default configuration uses 76 active carriers in its bandwidth of operation, IEEE 1901 FFT-OFDM uses 917 in its default mode, after accounting for the masking of certain carriers to avoid interference [3], [18].

8.5.1.2 Carrier Modulation

In IEEE 1901 FFT-OFDM carriers can be modulated with BPSK, QPSK, 8-QAM, 16-QAM, 64-QAM, 256-QAM, or 1024-QAM, thus allowing the system to take full advantage of all possible ranges of SNRs that a particular subcarrier could encounter. IEEE 1901 FFT-OFDM also supports bit-loading with a rich mix of modulations, tailored for each channel such that each carrier communicates with the fastest modulation that the carrier's SNR can support.

8.5.1.3 Frame Control

The AV Frame Control (FC) field consists of 128 information bits that are encoded and modulated over one OFDM symbol. The AV Frame Control symbol has an IFFT interval of 40.96 μs and an effective (non-overlapped) guard interval (GI) of 18.32 μs. The long guard interval was chosen so that time-domain averaging can be used to increase the symbol signal to noise ratio (SNR). In addition, the frame control IFFT and guard interval is transmitted at 0.8 dB higher power that the payload, again to increase robustness. As the frame control duty cycle is low in typical traffic, this extra power does not result in any measurable effect on radiated emissions.

The 128 information bits are encoded at rate $\frac{1}{2}$ using the turbo convolutional code engine to produce 256 coded bits. These 256 bits are interleaved and then put through an outer repetition code that copies each bit as many times as is possible and the bits are mapped onto the frame control symbol with maximum spreading for time and frequency diversity.

The FC contains information needed by both the PHY and by the MAC. PHY related contents consist of delimiter type, Tone Map Identifier (TMI), and length of PHY Body. Delimiter type is needed for FC decoding, and TMI is required to demodulate the PHY Body, if present. The TMI is just a nine-bit index that indicates which Tone Map the transmitter used to modulate the OFDM symbols of the PHY Body. It is chosen by the receiver during channel adaptation and is sent along with the Tone Map to the transmitter. PHY Body length is needed by the PHY to know how many symbols to demodulate.

The PHY Body length information is also needed by the MAC for Virtual Carrier Sense (VCS). As in wireless communications, attenuation and noise are high on the PLC channel, so Physical Carrier Sense (PCS) is limited to synch detection. VCS allows stations to refrain from accessing the medium while another station is transmitting, particularly in CSMA/CA access mode.

8.5.1.4 Payload

The OFDM time domain signal, based on a 75 MHz system clock, is determined as follows. For Data Symbols, a set of data points from the mapping block is modulated onto the subcarrier waveforms using a 3072-point IFFT resulting in 3072 time samples (referred to as the IFFT interval). A fixed number of samples from the end of the IFFT are inserted as a cyclic prefix at the front of the IFFT interval, to create an extended OFDM symbol.

8.5.1.5   IEEE 1901 FFT-OFDM Enhancement of HomePlug AV 1.1

The IEEE 1901 FFT is an extension of HomePlug AV 1.1 specification. The extensions were made in such a way that the IEEE 1901 FFT-OFDM systems could co-exist with the already existing HomePlug AV system while providing higher performance. The following sections provide an overview of the technical specifications for these enhancements.

8.5.1.5.1   30–50 MHz Frequency Band

HomePlug AV Physical layer operates in the 1.8–30 MHz frequency band using only 917 of the available 1155 carriers in the 1.8–30 MHz band and the remaining carriers are masked (i.e. not used for transmitting data). In IEEE 1901 FFT-OFDM, the following two extensions were made to the frequency band:

(a) Frequency band was extended up to 50 MHz. The carrier spacing in the 30 MHz–50 MHz is the same as that in 1.8–30 MHz band (i.e. 24.414 kHz).
(b) The masked carriers in the 1.8–30 were also made available for transmitting data.

These two extensions enable IEEE 1901 FFT-OFDM systems using the 1.8 MHz–50 MHz band to support up to 1974 carriers (i.e. (50-1.8)*1000/24.414). These additional carriers along with a 16/18 code rate and a 1.6 μs Guard Interval enable IEEE 1901 FFT-OFDM systems to provide a peak PHY data rate of 500 Mbps (1974*12*(16/18)/(40.96+1.6)).

8.5.1.6   Additional Guard Intervals

HomePlug AV physical layer uses a 7.56 μs guard Interval on the first two OFDM symbols of the PPDU payload. IEEE 1901 FFT-OFDM allows the guard interval on the first two OFDM symbols to be either 7.56 μs or 19.56 μs. In-home networks are restricted to the 7.56 μs guard interval on the first two OFDM symbols, thus preserving compatibility with HomePlug AV.

HomePlug AV PHY limits the guard interval on the third and higher OFDM symbols in the PPDU payload to be either 5.56 μs or 7.56 μs. The receiver selects the guard interval to be used based on channel conditions and indicates it to the transmitter as part of channel adaptation. IEEE 1901 FFT-OFDM extends the number of Guard Intervals by allowing Extended Smaller Guard Intervals and Extended Larger Guard Intervals:

- Extended Smaller Guard Intervals enables support for {1.60, 3.92, 2.08 and 2.56} μs Guard Intervals. These enable IEEE 1901 FFT-OFDM stations to improve efficiency on channels with low delay spread channels.
- Extended Larger Guard Intervals enable support for {9.56, 11.56, 15.56 and 19.56} μs Guard Intervals. These are primarily intended for use on high delay spread channels (e.g. in Access networks).

Stations indicate their support for these extended Guard Intervals by setting Extended Smaller Guard Interval Support Flag (ESGISF) and Extended Larger Guard Interval Support Flag (ESGISF) in the Sound MPDU Frame Control. Receivers use this information to determine the Tone Map Guard Interval based on the channel conditions and the Guard Interval capabilities of the transmitter.

8.5.1.7 4096-QAM

The highest modulation supported by HomePlug AV is 1024-QAM. In IEEE 1901 FFT-OFDM, optional support for 4096-QAM was added to enhance the performance on high SNR channels.

Stations indicate their support for 4096-QAM by setting Extended Modulation Support (EMS) field in the Sound MPDU Frame Control. Receivers use this information to determine the modulation on each carrier of the Tone map based on channel conditions and receiver capabilities.

8.5.1.8 16/18 Code Rate

HomePlug AV supports FEC Code rates of 1/2 and 16/21. In IEEE 1901 FFT-OFDM, support for 16/18 code rate was added to enhance performance on higher SNR channels. 16/18 FEC Code uses a coding scheme similar to the Turbo Convolutional Encoder for the exiting rate 1/2 and 16/21 codes; however the puncturing pattern is modified to reduce the redundancy and increase the code rate.

Stations indicate their support for 16/18 code rate by setting the Extended FEC Rate Support (EFRS) field in the Sound MPDU Frame Control. Receivers use this information to determine the FEC code rate for the Tone map based on channel conditions and receiver capabilities.

8.5.2 IEEE 1901 Wavelet-OFDM PHY

Wavelet-OFDM is the second multichannel transmission technique contained in the IEEE 1901 standard. The fundamental characteristic of Wavelet-OFDM is that the usual FFT-based transform and the rectangular/raised-cosine windowing used in conventional OFDM is replaced with critically decimated Perfect Reconstruction Cosine Modulated Filter Banks which exhibit several desirable properties such as very low spectral leakage. One of the most interesting aspects of Wavelet-OFDM is that it is not necessary to introduce a guard interval between consecutive symbols. There is extensive literature on Wavelet-OFDM, see [31] and references therein.

The Wavelet-OFDM system places 512 evenly spaced carriers into the frequency band from DC to around 30 MHz. Of these 512 carriers, 338 of them (approximately 2 MHz to 28 MHz) are used to carry information. With the use of an optional band up to 50 MHz, maximum PHY rates on the order of half a Gbps can be achieved. Every carrier is loaded with real constellations such as M-PAM ($M = 2, 4, 8, 16, 32$). It is important to point out that the fact that Wavelet-OFDM employs real constellations does not mean that Wavelet-OFDM has lower spectral efficiency than conventional FFT-OFDM that employs 2D constellations such as QAM. In fact, the frequency resolution of Wavelet-OFDM is twice that of windowed OFDM because the use of non-rectangular windowing allows for a higher degree of spectral overlap. As a consequence, for the same total bandwidth and the same number of transform points K, Wavelet-OFDM uses K real carriers that employ PAM whereas OFDM uses $K/2$ complex carriers that employ QAM. Thus, OFDM and Wavelet-OFDM have the same spectral efficiency. Specified FECs include a mandatory concatenated Reed-Solomon/convolutional code scheme and an optional LDPC convolutional code that allows easy scalability to high data rates at reasonable complexity.

8.5.3   The MAC and the Two PLCPs

The IEEE 1901 MAC layer architecture used to coordinate the IEEE 1901 network is Master/ Slave. The Master (QoS Controller) authorizes and authenticates the slave stations in the network, and may assign time slots for transmissions using either CSMA-based or TDM-based access. Network stations can communicate directly with each other (as opposed to an access point that retransmits all traffic). This increases the efficiency of the network and reduces the load on the Master.

The MAC layer employs a hybrid access control based on TDMA and CSMA/CA by defining a Contention Free Period (CFP) and a Contention Period (CP) to accommodate data with different transmission requirements. The CFP is a portion of the total transmission cycle during which stations that have low-delay/low-jitter requirements are allowed exclusive use of the medium. All streams requiring transmission in the CFP are managed by a QoS Controller. The CFP starts with a beacon, which is periodically sent by the QoS Controller, and ends when all reserved streams are transported. The rest of the Beacon Cycle is used for CP. In the CFP, data streams that have a time allocated to them through a bandwidth reservation procedure managed by the QoS Controller are transported. Frequency division multiplexing (FDM) can also be supported to allow for co-existence between in-home and access networks. Fragmentation support, data bursting, group-ACK, and selective Repeat ARQ are also important features supported in IEEE 1901.

Intelligent TDMA is also defined in IEEE 1901. Intelligent TDMA is a dynamic bandwidth allocation mechanism that exploits information about the amount of traffic queued in each transmission station. This mechanism realizes stable transmission which can cope with errors and IP/VBR traffic. In each transmitted data packet, each station inserts the number of frames pending to be transmitted. Since traffic information is directly obtained from data packets, the QoS Controller can perform accurate real-time operation. An option for line cycle synchronization is also present for coping with the periodically time-varying channel and cyclostationary noise.

In the following section, only the details of the Physical Layer Convergence Protocol (PLCP) that describe the MAC-PHY interface for the IEEE 1901 FFT-OFDM PLCP will be given. For details on the IEEE 1901 Wavelet-OFDM PLCP, see [32].

8.5.4   IEEE 1901 FFT-OFDM Medium Access Control (MAC)

In order to support multimedia applications, the PLC Medium Access Control (MAC) protocol must operate in close coordination with the PHY services in order to provide bandwidth and QoS guarantees in the presence of limited and time-varying network resources in the PLC channel. The IEEE 1901 FFT-OFDM MAC that is described next, addresses these challenges.

8.5.4.1   Network Architecture

From the Medium Access Control perspective, an IEEE 1901 FFT-OFDM PLC network consists of a set of HomePlug stations connected to the AC power line, with stations in the same logical network cryptographically isolated using a 128-bit AES Network Encryption Key (NEK) thus forming an AV Logical Network (AVLN). Each AVLN is managed by a single controlling

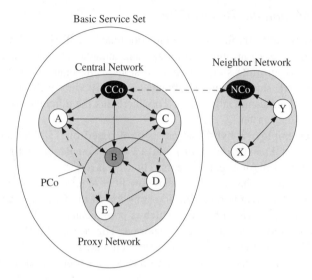

Figure 8.3 IEEE 1901 FFT-OFDM network architecture.

station called the Central Coordinator (CCo). The CCo performs network management functions such as

(a) association of new stations joining the AVLN;
(b) authentication of new stations joining the AVLN;
(c) admission control for TDMA sessions;
(d) scheduling TDMA and CSMA allocations during each Beacon period.

Figure 8.3 shows the organization of IEEE 1901 FFT-OFDM devices into different classes of networks. The CCo and the devices in the logical network that can directly communicate with it form the Central Network (CN). The attenuation and noise characteristics on the power line channel may give rise to situations where certain devices that belong to the same home network may not be able to communicate with the CCo. A Proxy Network (PN) is instantiated by the CCo in such scenarios to allow it to control the hidden stations (HSTAs) by relaying messages through a Proxy Coordinator (PCo). Note that direct, peer-peer communications are still enabled between HSTAs and devices in the CN with which their PN is associated. The PN approach improves coverage, although proxy networks are very rare due to the robust physical layer used by IEEE 1901 FFT-OFDM.

8.5.4.2 Network Modes of Operation

The operation mode of an AV network depends on whether Neighbor Networks can be detected. The CCo of each AVLN maintains an Interfering Network List (INL). The INL is the list of AVLNs whose beacons can be detected directly by the CCo. In neighbor network coordination, an AVLN is required to coordinate with only the AVLNs in its INL. Consequently, there is no chaining effect where an AVLN has to coordinate with AVLNs multiple hops away [33]. An AVLN can operate in one of two modes:

(a) Uncoordinated (Stand-alone) Mode, or
(b) Coordinated Mode.

An AVLN operates in Uncoordinated Mode if it cannot detect any Beacons reliably (i.e. when its INL is empty). This can happen because there are no existing networks in the vicinity of the AVLN or because the CCo cannot detect beacons of the existing networks. An AVLN operating in Uncoordinated Mode will generate and maintain its own Beacon Period timing.

An AVLN operates in Coordinated Mode if its CCo can detect beacons reliably from at least one existing AVLN. In Coordinated Mode, an AVLN shares the Beacon Period with all the AVLNs in its INL. The Beacon Region contains multiple beacon slots and each AVLN transmits its beacons in a designated beacon slot once every Beacon Period. Usually, two AVLNs in each others INL will share a common start time for their Beacon Periods.

8.5.4.3   MAC/PHY Cross Layer Design for Multimedia

IEEE 1901 FFT-OFDM uses a high level of MAC-PHY cross layer design to cope with the unique characteristics of power line channels while providing QoS guarantees for multimedia traffic. Important aspects of the IEEE 1901 FFT-OFDM MAC-PHY cross layer design are:

(a) Efficient MAC Framing to overcome impulse noise,
(b) AC Line cycle based channel adaptation [34], and
(c) Dynamic TDMA to handle changing channel conditions.

Impulse noise is the most common impairment over power lines [1], [2]. It is handled in IEEE 1901 FFT-OFDM by a combination of aggressive channel adaptation at the PHY level and efficient retransmission at the MAC level. Impulse noise power is typically much greater than the signal power level, and is broad spectrum, so it will severely disrupt one or more PHY symbols per event. Adapting the channel at the PHY layer to overcome this form of impairment would greatly degrade the data rate, and may simply not be feasible. Since there are often several undisrupted symbols between events, adapting to the highest rate that these good symbols will support provides better net throughput. However, this hinges on the ability to retransmit efficiently only the badly damaged portions of each data frame, which is realized by a novel 2-level MAC framing method with sub-frame level Selective Acknowledgments.

Noise on power lines varies with the AC line cycle; for example, noise levels at the zero crossings can be very low compared to noise at the peaks. Many impulse noise sources are synchronous with the line cycle (e.g. dimmers) causing even greater periodic noise effects [1], [2]. The achievable PHY rate at the cleanest portion of the line cycle may be more than 50% higher than that of the noisiest portion. MAC structures are used in IEEE 1901 FFT-OFDM to facilitate channel adaptation synchronized to the underlying AC line cycle, thus appropriately handling cyclostationary noise.

Power line channel characteristics also can change when electric appliances turn on and off. This may cause an allocation that once was sufficient to fail to satisfy the throughput

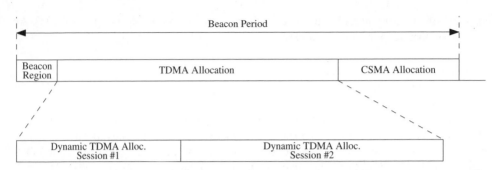

Figure 8.4   Beacon period structure.

requirements of a given application. To ensure that QoS guarantees are maintained under such conditions, IEEE 1901 FFT-OFDM TDMA allocations are dynamic. Short delivery deadlines (say 100 ms) require very rapid reallocation in order to meet the packet loss tolerance requirements.

8.5.4.4   Channel Access Control

8.5.4.4.1   Beacon Period Structure in Uncoordinated Mode

The Beacon Period structure consists of a Beacon Region followed by TDMA and CSMA Regions (Figure 8.4). The Beacon Region contains the Beacon transmitted by the CCo. Each Beacon consists of a Preamble, Frame Control, and 136-byte payload. Allocation information is broadcast in the Beacon payload using 'mini-ROBO,' one of the robust modulation methods described earlier.

TDMA allocations are provided for streams requiring QoS. They follow the connection setup procedure, during which the allocation requirements are negotiated. TDMA allocations in IEEE 1901 FFT-OFDM are dynamic. Sessions with TDMA allocation continually update the CCo with their allocation requirements enabling the CCo to update allocations rapidly as channel conditions or source rates change.

CSMA allocations in the Beacon Period are used by connectionless traffic and by connections that do not have strict QoS requirements. IEEE 1901 FFT-OFDM uses the same CSMA channel access mechanism as HomePlug 1.0.

Streams requiring QoS go through a connection establishment procedure to ensure that network and station resources are available to support the connection. Connection requests can be initiated either by the application or the Auto Connect in the Convergence Layer. Connection requests include a Connection Specification (CSPEC) that contains traffic characteristics and QoS requirements. Within the IEEE 1901 FFT-OFDM station, the 'Connection Manager' (CM) handles these requests.

Connection setup is divided into two stages. First, the Connection Manager initiating the connection communicates with the CM at the other end of the desired connection to determine if enough resources are available at the destination to handle the new connection. If either the source or the destination lacks resources, the connection fails, and the application is notified. Otherwise, the CM at the source communicates with the CCo, which performs Call Admission Control. If sufficient network resources are available, the CCo accepts the new connection and provides it with a Link Identifier (LID), which is used for provisioning of allocated resources within the AV network.

8.5.4.4.2   Beacon Period Structure in Coordinated Mode

The Beacon Period structure in Coordinated Mode is slightly more elaborate than that in Uncoordinated Mode. In Uncoordinated Mode, the CCo of the AVLN has complete control over the entire Beacon Period. Thus, it can decide autonomously the locations of the contention-free allocations and CSMA allocations. In Coordinated Mode, however, an AVLN first has to 'reserve' a portion of the Beacon Period from the AVLNs in its INL before it can schedule any contention-free allocations. All AVLNs in its INL will then refrain from transmitting in that time interval. A time interval is shared by all AVLNs for CSMA traffic.

8.5.4.4.3   Neighbor Network Coordination

The objective of Neighbor Network coordination is to ensure that each AVLN and its INL specify a consistent Beacon Period structure. Each CCo should find out the Beacon Period structures of all the AVLNs in its INL and compute a single 'unified' schedule of all interfering AVLNs. From this unified schedule of its INL, the CCo can choose a consistent Beacon Period structure, which it broadcasts in its beacon. A CCo exchanges messages (request, response, and confirm) with neighbor CCos (NCos) in its INL to request new Reserved Regions. The CCo first sends a request to all NCos specifying the time intervals that the CCo wants to use as its new Reserved Regions. Each NCo will send a response back to the CCo. If all the responses indicate that the request is accepted, then the CCo will send a positive confirm message to all NCos, update its Beacon Period structure to include the new Reserved Regions, and start using them.

8.5.4.5   Medium Activity

Medium activity in an IEEE 1901 FFT-OFDM network includes a series of MPDUs separated by inter-frame spaces. The most basic atomic of medium activity is a Long MPDU followed by a SACK. Due to the hostile nature of the power line channels, AV requires Long MPDUs to be acknowledged immediately, although up to four MPDUs may be sent in a burst followed by a single SACK. This not only saves overhead due to the SACK delimiters, but also the extra Response Gaps. Under hidden node conditions, this atomic is augmented by RTS/CTS. Activity in a TDMA allocation includes multiple SOF-SACK or RTS-CTS-SOF-SACK exchanges between source and destination of the LID to which the TDMA allocation is intended (refer to Figure 8.5). Medium Activity during CSMA allocations is similar to that used in HomePlug 1.0. After the end of a previous transmission, two priority resolution slots are used for priority contention. Priority contention ensures that only the stations with the highest priority in the network contend during the subsequent Contention Window. A modified Binary Exponential Back off algorithm is used to avoid collisions in the Contention Window. The station that gains access to the medium uses a SOF-SACK or RTS-CTS-SOF-SACK atomic for transmitting information (refer to Figure 8.6).

8.5.4.6   Channel Adaptation

Proper channel adaptation is crucial to successful use of the power line medium. Channel adaptation in IEEE 1901 FFT-OFDM is unique to a path (unidirectional transmitter-receiver

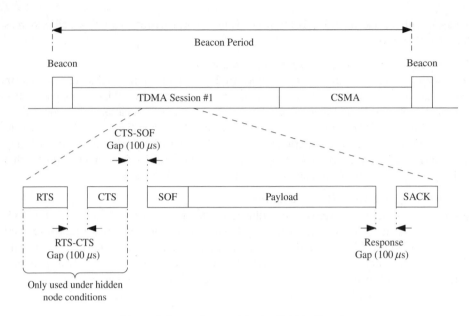

Figure 8.5   Medium activity in TDMA allocation.

pair). Further, IEEE 1901 FFT-OFDM tone maps are valid only for portions of the line cycle. Data rates achievable from one 'Tone Map Region' to another for a given path may vary by as much as 50%.

Sound MPDUs are used (by the receiver) to obtain an initial estimate of Signal to Noise Ratios (SNRs) of the OFDM sub-carriers. These SNR estimates are used to define initial adapted tone maps and tone map regions. Each tone map indicates the bit loading on each of the OFDM Carriers and the FEC Block rate. Receivers continually monitor the SNR and error rates on all received MPDUs, and constantly provide transmitters with updates on the tone

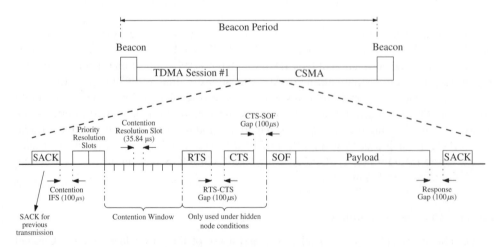

Figure 8.6   Medium activity in CSMA allocation.

maps and their regions of validity. This enables AV stations to operate close to the channel capacity and to react quickly to changing channel conditions.

8.5.4.7   Convergence Layer

The Convergence Layer (CL) forms the 'top half' of the ISO's OSI Data Link Layer. It is responsible for several functions that are not part of a traditional MAC but which are necessary to provide QoS services. These functions include packet classification and automatic connection establishment.

When a connection is established, a collection of rules is provided to the Convergence Layer. This rule collection (set of parameters) allows the Classifier to associate packets that flow across the Convergence Layer data Service Access Points (SAPs) with established connections. These rules are local to each device; they are not transmitted across the network. Examples of these rules include Source and Destination addresses, Protocol Types, port numbers, etc. The rule syntax is rich and allows specification of the priority in which rules are applied and Boolean combinations of individual rules during packet classification.

Some packets that would appropriately be provided with a level service arrive at the CL in a 'connection-less' state. Examples include packets received from legacy applications that start transmission without specifying QoS parameters, and packets bridged from another network. When the Classifier encounters a packet that does not associate with an existing connection, it routes the packet to the Auto-Connect function to possibly establish a connection. Auto-Connect will attempt to determine what level of QoS, if any, is appropriate for the stream to which the packet may belong. The techniques employed by the Auto-Connect function to identify connections include templates (e.g. associating traffic on a particular TCP port with a particular usage and inferring a level of QoS from that usage) and heuristics (i.e. attempting to identify a need for a particular level of QoS from the statistical behavior of traffic offered). Auto-Connect acts as a surrogate application for the connections it establishes, performing the activities that an application would normally perform.

8.5.5   Coexistence

Wavelet OFDM modulation is also supported by IEEE 1901 and their coexistence with the other standards are discussed in this section. IEEE 1901 specification uses an Inter-System Protocol (ISP) for coexistence between IEEE 1901 Access and in-home stations that use wavelet or FFT PHY. ISP can also be used for coexistence between IEEE 1901 and G.hn stations.

The ISP is a resource sharing mechanism that regulates access to the power line medium. ISP uses coexistence signals that enable IEEE 1901 stations to signal their presence and also determine the presence of other IEEE 1901 systems and G.hn systems. Coexistence signals are transmitted in a periodic manner during the ISP windows assigned to various systems.

A solution based on multiple non-interoperable PHYs with a common MAC is a common approach in standards, e.g. 802.11. However, the definition of two non-interoperable PHYs also leads to the necessity of handling the case when devices with different PHYs are in proximity and connected to the same shared medium. In its initial conception, the ISP handled only the coexistence of the two IEEE 1901 PHYs but now the ISP will also handle coexistence between 1901 and G.9960 ones.

8.5.5.1   ISP Waveform and the Network Status

IEEE 1901 access (AC) and in-home (IH) devices will indicate their presence and requirements by transmitting a set of simple signals. The ISP waveform included in the 1901 standard is called the Commonly Distributed Coordination Function (CDCF) waveform.

The CDCF signal is obtained by the repetition of R baseband windowed OFDM signals. Each OFDM symbol, formed by a set of all 'one' BPSK data, is modulated onto the carrier waveforms using a 512-point inverse fast Fourier transform (IFFT). The CDCF signal is defined as ($1 \leq n \leq 512R$)

$$S_I(n) = N_c W(n) \sum_{C_a} \cos\left(\frac{2\pi C_a n}{512} + \Phi(C_a)\right),$$

where N_c is a normalization factor, $W(n)$ is a windowing function, C_a is the carrier index and $\Phi(C_a)$ is a binary $\{0, \pi\}$ phase vector. Some of the carriers used in the above equation can be masked in order to meet the Transmit Spectrum Mask. Additional carriers may be masked by the equipments depending on local regulations. Samples of the base signal waveforms can be stored in memory and flushed directly to the DAC, thus allowing simple implementation by any PHY.

Several phase vectors can be defined to create a set of base signals, i.e. the common alphabet shared by all nodes. By defining multiple phase vectors, we can create a set of CDCF signals and this set will constitute the common 'alphabet' shared by all the non-interoperable devices. An obvious trade off with complexity arises when defining the cardinality of the set of CDCF signals belonging to the alphabet. However, the goal of ISP is to have as low complexity as possible so that the design choice is not to define a large alphabet of CDCF signals for data communication between non-interoperable devices but to define a sufficient number of CDCF signals for facilitating the detection of the network status. The exact number of phase vectors is still under discussion, but it will range between four and six.

CDCF signals will be transmitted in the ISP time-window, a region of time used by PLC devices for transmitting/detecting ISP signals. The ISP time-window occurs periodically every T_{isp} seconds and is further divided in F time sub-windows called fields. The presence/absence of ISP signals in a field conveys several kinds of information about the presence/absence of a device of a certain kind (AC, IH with FFT-OFDM PHY, IH with Wavelet PHY, etc.), bandwidth requirements (low, medium, high), re-synchronization requests, etc. Each field in the ISP window has a duration of around 250 μs, so there is a margin of around 85 μs at both ends of the ISP field. This allows handling imperfect zero crossing detection, load induced phase shifts of the mains signal, and other non idealities of the channel. The ISP window occurs every T_{isp} seconds (Allocation Period) at a fixed offset T_{off} relative to the underlying line cycle zero crossing. This is shown in Figure 8.7. Since there are two zero crossings in a cycle and there are often up to three phases in a building, there are actually six possible zero crossings instances. Proper synchronization techniques are also being defined to allow all devices in range of each other to synchronize to a common zero crossing instance.

When a device starts operating on the power line medium, it will first determine the correct location of the ISP window and then it will scan for ISP signals to determine the network status, i.e. what type of systems are present on the shared medium, what are their bandwidth requirements, etc. AC and IH devices will indicate their presence as well as

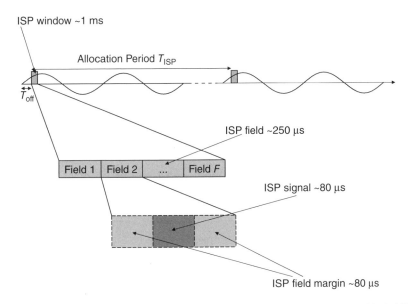

Figure 8.7   ISP time window, ISP fields, ISP field margins, and ISP signal window. [22] © 2008 IEEE.

other useful information by transmitting ISP signals in the appropriate ISP fields of the ISP window pertaining to their system. In particular, every system will use in exclusivity an ISP window every T_{isp} seconds. For example, all IH devices that use the OFDM PHY (IH-O) use simultaneously an ISP window, all AC devices use simultaneously the next ISP window, and then all In-Home devices that use the Wavelet-OFDM PHY (IH-W) use simultaneously the next one, and so on in a round robin fashion. This allows all devices to unequivocally determine the network status every $3T_{isp}$ seconds.

8.5.5.2   Support of Dynamic Bandwidth Allocation (DBA)

Depending on the status of the power line network, different resource allocations will be carried out. TDMA sharing between Wavelet and OFDM systems will be based on Allocation Periods. As shown in Figure 8.8(a), there will be N TDM Units (TDMUs) per Allocation Period, where an Allocation Period lasts T_{isp}. The duration of a TDMU is equal to two power line cycles, and each TDMU contains S TDMA time slots. Each TDMA slot will be exclusively assigned to either AC, IH-O, or to IH-W systems, and the allocation policy will be based on the network status. Fair sharing of resources will be accomplished by assigning a fair number of TDMA slots to each system that is present on the power line network. Sensible values for parameters N and S currently under discussion are: $3 \leq N \leq 10$, and $8 \leq S \leq 12$ and, as a consequence, T_{isp} has a value of a few hundred milliseconds. An example of three possible TDMA structures is given in Figure 8.8(b) for the case of $S = 12$ and for three different network statuses. With a period equal to T_{isp}, devices can update the network status and eventually change the utilized TDMA structure to ensure efficient DBA. The ISP window always occurs at the beginning of TDMU #0.

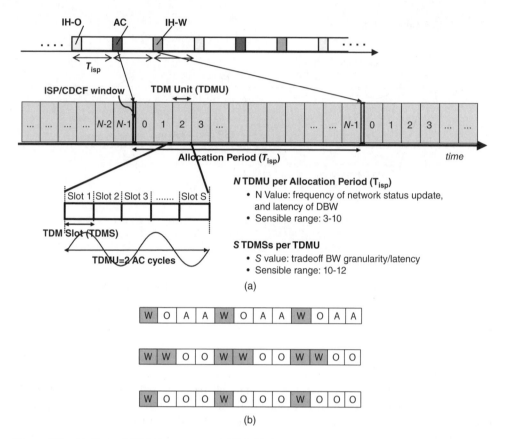

Figure 8.8 (a) General TDMA structure: N TDMUs in an Allocation Period, and S TDM Slots per TDMU (a TDMU is two line cycles long); (b) Illustration of three possible TDMUs for the case of $S = 12$: (Upper) the TDMSs are allocated 50% to the access system and 50% to the in-home systems (25% to Wavelet-OFDM systems and 25% to FFT-OFDM systems; (Center) the TDMSs are allocated 50% to Wavelet-OFDM systems and 50% to FFT-OFDM systems since no access system is present; (Lower) same as the center case but for a different network status, i.e. for the case when Wavelet systems have required reduced resources in the appropriate ISP field. [22] © 2008 IEEE.

The duration of a TDMS is either $40/S$ ms (50 Hz) or $33.33/S$ ms (60 Hz) and these values are equal to the minimum system latency that can be guaranteed by the network. For example, for the case $S = 12$, we have 3.33 ms (50 Hz) or 2.78 ms (60 Hz). Similar to the case of the ISP Fields, it is necessary to add a margin of some microseconds around the TDMS boundaries.

8.5.5.3 Support of TDMA Slot Reuse (TSR) Capability

The interference generated on shared power line networks is a random variable that depends on many factors, such as the transmitted power, the power line topology, wiring and grounding practices, the number of mains phases delivered to the premises, etc. PLC devices can interfere with other devices that are in close proximity, but also with devices that are located farther away, e.g. on another floor. In other cases, even within the same apartment, devices can cause very different levels of interference depending, for example, if they are located on the same phase of the alternating current mains or not.

Algorithms for TSR exploit this physical property of the power line channel by allowing devices, either in the same network or in different neighboring networks, to transmit *simultaneously* without causing interference to each other. Currently, no commercial PLC product has this capability. Usually, within the same network, nodes are either assigned orthogonal resources (e.g. different TDMA slots) or compete for resources (e.g. CSMA). Simulation results confirming the capability of achieving substantial TSR can be found in [30].

8.6 Performance Evaluation

Performance of the two-level framing approach, independent of the MPDU and channel access overheads, was presented in [35]. We summarize those results and then provide the results of detailed simulations based on measured PHY characteristics and known MAC parameters.

8.6.1 MAC Framing Performance

With two-Level Concatenation, only FEC PHY blocks (PBs) that are corrupted need to be retransmitted. If p is the probability of an FEC block error, then the expected number of FEC blocks delivered per MPDU is $(1 - p)$ times the number, N of FEC blocks. The efficiency with two-level Concatenation, η_{2L}, is given by

$$\eta_{2L} = (1 - p) \left(\frac{L_{fec} - L_{OH,2L}}{L_{fec}} \right) \left(\frac{L_{MSDU}}{L_{mf}} \right),$$

where $L_{OH,2L}$ is the overhead per FEC block, L_{fec} is the total length of an FEC block, L_{MSDU} is the size of an MSDU, and L_{mf} is the length of a MAC frame (MSDU plus framing overhead). These results show that IEEE 1901 FFT-OFDM MAC Framing efficiency is close to the theoretical limit, as the overheads at the MAC frame and PB levels are minimal.

8.6.2 Overall MAC Efficiency

Using considerable data on coverage, path rates, and error rates for the PHY obtained from field tests, we have run extensive simulations to predict the performance of the MAC. This section gives the results of these simulations.

MAC efficiency depends on many factors, including the size distribution of the MSDUs, source rate, PB error rate (PBER), frame control error rate, PHY data rate, PHY symbol size, etc. Generally, efficiency of the MAC is around 80%, usually a bit higher.

MAC efficiency is measured by dividing the total length of transmitted MSDUs by the total time allocated to transmit them. We do not include time that is allocated to other streams, but do consider the beacon overhead. Perhaps more important is the net data rate available to the application at the boundary to the MAC for a given PHY rate.

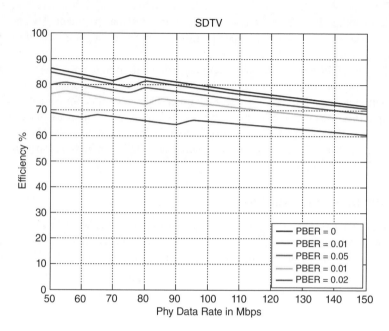

Figure 8.9   IEEE 1901 FFT-OFDM efficiency for 6-Mbps SDTV at various PB error rates.

Figure 8.9 shows MAC efficiency, including loss due to retransmissions, for SDTV, assuming 1378-byte MSDUs consisting of seven 188-byte MPEG Transport Frames with 40 bytes of UDP/IP header and 22 bytes of IEEE 802.3 overhead, including virtual LAN tag. Efficiencies for PB error rates ranging from 0 to 20% are shown. Discontinuities in the curves are due to MPDU and burst length boundary effects. Higher data rates result in shorter MPDUs for the same amount of data. With fixed delimiter overheads, shorter MPDUs result in lower efficiencies as data rate increases, particularly for applications with lower source rates. The corresponding result for HDTV, though not shown here show even better efficiencies.

8.7   HomePlug AV2

The HomePlug AV 2.0 Specification (AV2) adds new features to HomePlug AV/IEEE 1901 FFT-OFDM that provide a significant increase in throughput (up to 1.5 Gbps) and coverage. Field test measurements have shown that HomePlug AV2 provides a coverage performance of approximately 90 Mbps UDP network throughput (three equal UDP streams of 30 Mbps each) for 99% of networks with four or more devices assuming immediate repeating is implemented.

New physical layer features used to achieve this performance include Multiple Input Multiple Output (MIMO) schemes, extended frequency band, efficient notching and short delimiter. New MAC layer features that are included are delayed acknowledgement, immediate repeating and a power save mode. This section focuses on the key enhancements of HomePlug AV2, namely, MIMO, an extended frequency band, Short Delimiter, Delayed Acknowledgement, Effective Notching, and Immediate Repeating [4].

8.7.1   Extended Frequency Band

The HomePlug AV 1.1 Specification utilizes the frequency band from 1.8 to 30 MHz, and the IEEE 1901 standard increase that to 1.8 to 50 MHz. HomePlug AV2 extends the frequency band even further to 1.8–86.13 MHz [36].

One of the challenges with frequency bands above 30 MHz is that the regulations require a 25 to 30 dB reduction in transmit power spectral density (PSD) above 30 MHz. The performance gain provided by the 30–86.13 MHz band is generally quite high on medium to good channels due to the relatively wide channel bandwidth. However, this additional band does not provide much benefit on the poorest channels, e.g. the worst 5% of connections, due to the low PSD level allowed at the transmitter. However, this band does contribute to the coverage performance in two ways. Firstly, most power line channels fall in the good to medium category and the higher data rate provided on these channels enables them to reduce the time-on-wire for their traffic thus enabling more time-on-wire to be available for traffic on poorer channels. Also, when the higher frequency band is used with immediate repeating, discussed in Section 8.7.3, dramatic improvements in the performance can be seen even on poor paths by taking advantage of the higher data connections through a repeater.

8.7.1.1   Power Back-Off

Power Back-off is a feature introduced in the HomePlug AV2 to improve the performance on relatively good power line channels. Practical implementation of the transmitter-receiver system have a limited dynamic range in the analog-to-digital converters (ADC) and digital-to-analog converters (DAC) and thus the OFDM carriers in the high frequency band (i.e. above 30 MHz) suffer distortion because of the reduced PSD level when compared to that in the lower frequency band (i.e. below 30 MHz). This is due to the quantization noise in the ADC and DAC and the limited linearity of the line driver. To address this problem on good power line channels, the transmit PSD in the lower frequency band can be reduced so that there is lesser distortion in the OFDM carriers in the high frequency band.

8.7.2   Effective Notching

HomePlug AV 1.1 specified OFDM with windowing, to get the 30 dB deep notches that were required to be achieved to have no transmissions in the Amateur bands. This requirement was removed in HomePlug AV2 and alternate implementations were allowed. Alternate implementations such as adding fixed and/or programmable IIR or FIR filters or a combination of windowing and filters were allowed in HomePlug AV2. To support this, smaller guard intervals were added and other protocol changes were made to support additional OFDM carriers when supported by a transmitter.

8.7.3   Immediate Repeating

HomePlug AV2 supports repeating and routing of traffic to not only handle hidden nodes but also to improve coverage (i.e. performance on the worst channels). The repeating and routing function that is used by HomePlug AV2 is the same as in IEEE 1901 which was discussed in the previous section.

With HomePlug AV2 systems, hidden nodes are very rare. However, some links may not support the data rate required for some applications such as a 3D HD video stream and other high speed multimedia applications. In a network where there are multiple AV2 devices, the connection through a repeater provides a higher data rate than the direct path for the poor channels.

Immediate Repeating is a new feature in AV2 that enables high efficient repeating. Immediate Repeating provides a mechanism to use a repeater with a single channel access, and the acknowledgement does not involve the repeater. With this approach, latency is actually reduced with repeating, assuming the resulting data rate is higher, the obvious criteria for using repeating in the first place. Also the receiver has no retransmission responsibility for failed segments [4].

8.7.4   Short Delimiter and Delayed Acknowledgment

The Short Delimiter and Delayed Acknowledgment features were added to HomePlug AV2 to reduce the overheads in the transmitting payloads and increase the efficiency. The overheads result in relatively poor TCP efficiency in HomePlug AV 1.1 systems. One of the goals that was achieved with the introduction of these features was to get the TCP efficiency relatively close to that of UDP.

In order to send a payload data packet over a noisy channel, signaling is required to indicate to the receiver the beginning of the packet and for channel estimation so that the payload can be decoded. Additional signaling is required to acknowledge that the payload was received successfully. Inter frame spaces are required to be present between the transmission of the payload and the acknowledgment, for the receiver to decode, check the payload for successful reception and to encode the acknowledgement. This overhead is even more significant for TCP payload since the acknowledgement in this case has to be transmitted in the reverse direction.

8.7.4.1   Short Delimiter

The delimiter specified in AV 1.1 contains the preamble and frame control symbols and is used for the beginning of data PPDUs as well as for immediate acknowledgements. The length of the AV 1.1 delimiter is 110.5 microseconds and can represent a significant amount of overhead for each channel access. A new single OFDM symbol delimiter is specified in AV2 to reduce the overhead associated with delimiters by reducing the length to 55.5 microseconds.

One of the limitations of the Short Delimiter is that it cannot be easily detected asynchronously, which is necessary for CSMA channel access. Thus reception of the short delimiter requires that the receiver know the position in time where the Short Delimiter was transmitted. Thus the use of the Short Delimiter is limited to Selective Acknowledgement of CSMA and TDMA Long MPDUs, Reverse Start of Frame, and TDMA Start of Frame.

8.7.4.2   Delayed Acknowledgment

The processing time to decode the last OFDM symbol and encode the acknowledgement can be quite high, thus requiring a rather large Response Inter-Frame Space (RIFS). In AV 1.1, since the Preamble is a fixed signal, the preamble portions of the acknowledgement can be transmitted while the receiver is still decoding the last OFDM symbol and encoding the payload for the acknowledgement. With the Short Delimiter, the preamble is encoded in the same OFDM symbol as the payload for the acknowledgement, so the RIFS would need to be larger than for AV 1.1, eliminating much of the gain the Short Delimiter provides. Delayed acknowledgement solves this problem by acknowledging the segments ending in the last OFDM symbol in the acknowledgement transmission of the next PPDU. This permits practical implementations with a very small RIFS, reducing the RIFS overhead close to zero. AV2 also allows the option of delaying acknowledgement for segments ending in the second to last OFDM symbol to provide flexibility for implementers.

The combination of Short Delimiter and Delay Acknowledgment can provide a significant improvement in TCP and UDP efficiency.

8.8   ITU-T G.996x (G.hn)

The ITU-T started the 'G.hn' project in 2006 with a goal of developing a worldwide recommendation for a next generation unified HN transceiver capable of operating over all types of in-home wiring: phone lines, power lines, coax and Cat 5 cables and bit rates up to 1 Gbps. In December 2008, ITU-T consented on Recommendation G.9960, which is the G.hn foundation and specifies system architecture, most of the PHY and data-path related parts of the MAC. The technology targets residential houses and public places, such as small/home offices, multiple dwelling units or hotels. G.9960 did not originally address PLC Access and Smart Grid applications, but in mid 2009 a proposal for addressing Smart Grid applications was approved by the group. The G.996x family of standards underwent several amendments in the past few years, and the latest specifications can be found in [23]–[27].

G.9960 allows up to 250 nodes operating in the network. It defines several *Profiles* to address applications with significantly different implementation complexity. High-profile devices, such as residential gateways, are capable of providing very high throughput and sophisticated management functions. Low-profile devices, such as home automation or Smart Grid applications, have low throughput and only basic management functions, but can interoperate with higher profiles.

Past approaches emphasized transceiver optimization for a *single* medium only, i.e. either for power lines, or phone lines, or coax cables. The approach chosen for G.9960 is a single transceiver optimized for multiple media. Thus, G.9960 transceivers are parameterized so that relevant parameters can be set depending on the wiring type [28]. For example, a basic multicarrier scheme based on windowed OFDM has been chosen for all media, but some OFDM parameters, such as number of sub-carriers and sub-carrier spacing, are media dependent. Similarly, a three-section preamble is defined for all media, but durations of these sections change on a per media basis. A quasi-cyclic low-density parity-check (QC-LDPC) code has been chosen for forward error correction (FEC), but a particular set of coding rates and block sizes are defined for each type of media [22]. A parametrized approach also allows to some extent optimization on a per media basis to address channel characteristics of different wiring without sacrificing modularity, flexibility and cost.

8.8.1   *Overview of G.9960 Network Architecture*

A G.9960 network consists of one or more domains, as shown in Figure 8.10. In G.9960, a domain is constituted by all nodes that can directly communicate and/or interfere with

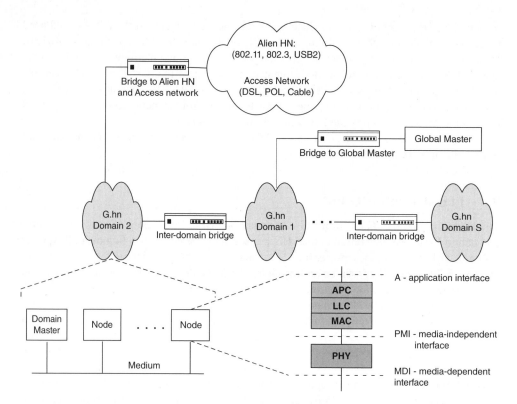

Figure 8.10 G.9960 network model, domain structure and protocol reference model of a node. [28] © 2009 IEEE.

each other. Thus, there is no interference between different domains of the same network, except crosstalk between closely routed wires. One of the nodes is a Domain Master (DM). It controls operation of all nodes in the domain, including admission to the domain, bandwidth reservation, resignation and other management operations. In case a DM fails, the DM function is passed to another node in the domain.

Since all nodes of the network that can communicate or directly interfere with each other are in the same domain, the DM can avoid interference between nodes by coordinating their transmission time. This is simpler and more efficient than coordinating transmissions in several domains sharing the medium. The latter is still necessary when the medium is shared between neighboring networks, such as in many deployments over power lines. The user can also establish multiple domains on the same medium, e.g. by enabling baseband and passband modes on power lines or using different RF channels on coax.

Domains of the same network are connected by inter-domain bridges. This allows nodes of any domain to 'see' any other node of any other domain in the network. Any domain may also be bridged to wireline or wireless alien networks, e.g. DSL, PLC Access, WLAN or other HN technologies.

Nodes of the same domain can communicate with other nodes directly or through one or more relays. In Centralized mode, nodes talk to each other via one dedicated relay node

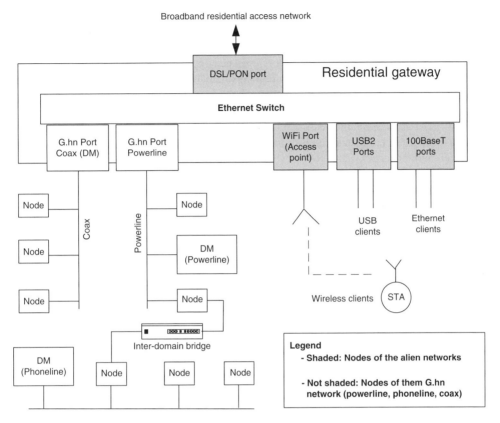

Figure 8.11   Example of HN topology associated with residential access. [28] © 2009 IEEE.

called the Domain Access Point. Nodes which are hidden from the DM are coordinated via DM-proxy nodes assigned by the DM.

An example of residential HN is presented in Figure 8.11. The network includes three domains: over coax, phone line and power line, each controlled by its DM. Alien networks are WLAN, USB2, Ethernet and residential access networks. A residential gateway bridges power line and coax domains and bridges the G.9960 network to alien networks. Each G.9960 node is configured to operate over the medium to which it is connected and it can communicate directly with any other node of its own domain and, via inter-domain bridges, with nodes of other domains. Communications with nodes of alien networks, including broadband access networks, are through the residential gateway.

G.9960 also envisions multi-port devices communicating over multiple media via separate ports. Since any device would be anyway plugged into a power outlet, designing a dual-port device (e.g. power line plus coax) appears to be a natural extension of a power line connection. Multi-port capability can increase data rate and coverage as data traffic may be split between media. From the application viewpoint, a multi-port device appears as a single entity while handling of network traffic over the available physical ports is done at the LLC layer. If enhanced by relaying capability, it can also serve as a PHY-layer inter-domain bridge.

Domains of the same network may require mutual coordination to avoid excessive crosstalk from one to another (due to closely routed wires), or when more than one domain is established on the medium in the same frequency band (this exceptional situation may happen if no other frequency band is available), or for performance optimization of connections routed via multiple domains. Coordination between domains is the responsibility of the Global Master (GM), seen in Figure 8.10. The GM collects statistics from domains and external management entities, derives appropriate parameters for each domain, such as transmit-power, timing, bandplan, etc., and communicates them to the DMs of the coordinated domains. Each DM imposes these parameters on all nodes of its domain.

G.9960 facilitates coexistence of networks sharing the same medium and frequency band (for instance, neighboring power line networks) by limiting their mutual interference. The following coexistence mechanisms allow operation of multiple networks with limited performance degradation:

- with neighboring G.9960 networks by mutual coordination of transmissions and resource sharing;
- with alien IH and access networks supporting the Inter-System Protocol (ISP), a coexistence mechanism currently under development in IEEE 1901 and ITU-T [30];
- with alien IH and access networks not supporting ISP via PSD shaping or sub-carrier masking, up-shifting of the spectrum to the passband or to a different RF channel (see Figure 8.12); additionally, a dual-mode device operating simultaneously as a G.9960 and an alien node can facilitate coexistence by coordinating G.9960 networks with the non-ISP neighboring alien networks (e.g. HomePlug AV, HD-PLC, UPA, etc.);
- with coax RF systems via a frequency agility mechanism: once an alien RF signal is detected the DM will move all nodes to another RF channel;
- with radio services by avoiding frequencies allocated to international amateur radio bands and switching off or reducing power of all interfering sub-carriers.

Details of coexistence protocols, including resource sharing policies, are currently under study.

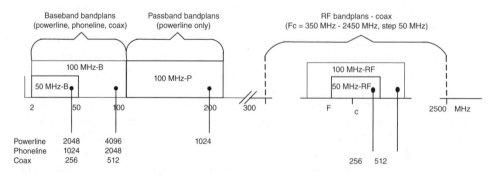

Figure 8.12 G.9960 bandplans. The number of the OFDM sub-carriers used for each medium and bandplan is also shown. [28] © 2009 IEEE.

8.8.2   Overview of the Physical Layer of ITU-T G.hn

8.8.2.1   Modulation and Spectrum Usage

G.9960 has adopted windowed OFDM with the following programmable set of parameters to address different types of wiring:

- number of sub-carriers, $N = 2^n$, $n = 8$ to 12;
- sub-carrier spacing as $F_{SC} = 2k \times 24.4140625$ kHz, $k = 0, 1, \ldots, 6$;
- center frequency F_C;
- window size.

The values of media-dependent parameters are selected taking into account channel characteristics of different media types. The following criteria were applied to simplify modulator design:

- all values for the sub-carrier spacing (F_{SC}) are power-of-2 multiples of a basic spacing;
- all values for the number of sub-carriers (N) are powers of two;
- all values of sampling frequency are dividers of a common reference frequency.

The set of values of sub-carrier spacing and sampling frequency includes those used by the 1901 OFDM PHY for PLC and by MoCA (Multimedia over Coax Alliance) for RF coax to simplify implementation of dual mode devices.

G.9960 defines operation in several frequency regions, referred to as bandplans, and includes baseband bandplans, passband bandplans and RF bandplans, see Figure 8.12. For each particular medium and bandplan, G.9960 defines only a single set of OFDM parameters so that overlapping bandplans use the same sub-carrier spacing. This rule, plus a unified per medium default preamble structure and PHY frame header, facilitates interoperability, i.e. allows all devices sharing the same frequency band to communicate with each other. The number of sub-carriers used in each bandplan depends on the type of the medium and varies from 256 to 4096 (Figure 8.12). There are also eight selectable values for the payload CP-length: $kN/32$, $k = 1, 2, \ldots, 8$. To address operation in baseband, in passband, and in RF, a generic passband OFDM modulator concatenated with an RF modulator is used. The passband part includes IDFT, cyclic extension, windowing, and frequency up-shift (for baseband operation, the frequency up-shift is set to the middle frequency of the band, associated with a sub-carrier index $N/2$. The RF modulator further up-shifts the spectrum to the RF band, between 0.3 GHz and 2.5 GHz.

Flexible bit loading in the range between 1 and 12 bits is defined on all sub-carriers. Gray-mapping is used for all constellation points of even-bit loadings and for almost all constellation points of odd-bit loadings. A particular bit loading for each sub-carrier of each connection can be negotiated between the transmitter and receiver, providing sufficient flexibility to adopt channels with wide range of frequency responses and noise PSDs.

8.8.2.2   Advanced FEC

The selected QC-LDPC code is a subset of the QC-LDPC codes defined in the IEEE 802.16e (WiMAX) with five code rates (1/2, 2/3, 5/6, 16/18 and 20/21) and two block sizes of 120

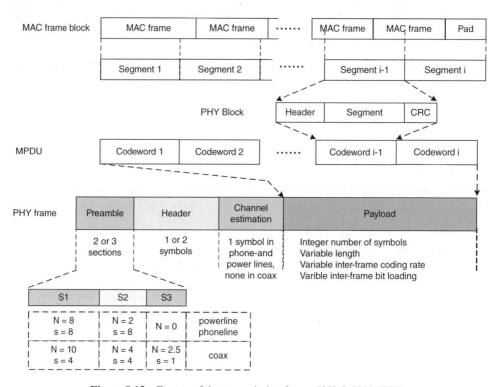

Figure 8.13   Format of the transmission frame. [28] © 2009 IEEE.

and 540 bytes. Three parity check matrices are used for code rates 1/2, 2/3 and 5/6, whereas the other two high code rates are obtained by puncturing the rate 5/6 code. The range of code rates and block sizes, together with bit loading capabilities, is designed to fit all types of media with their corresponding retransmission schemes: for media with frequent retransmissions, such as the power line, bit loading and FEC can be optimized to operate at Block Error Rates (BLERs) up to 10^{-2}, while for media with rare retransmissions the optimization can target operation with very low BLER, e.g. at 10^{-8}. The coding gain for the average BLER of 10^{-3} in the presence of Additive White Gaussian Noise (AWGN) varies from 8.2 dB for code rate of 16/18 with 1024-QAM on all subcarriers to 9.2 dB for code rate of 1/2 with QPSK on all subcarriers[4], allowing G.9960 to operate reliably and efficiently in wide range of channel characteristics and noise environments. Simulation results on the performance of the G.9960 FEC scheme for various code rates and number of decoding iterations in the presence of AWGN are reported in [28].

8.8.2.3   The Frame

A transmit frame (PHY frame) consists of a preamble, header and payload (Figure 8.13). The preamble is composed of sections S_1–S_3, each consists of N_S symbols. Symbols of section S_2

----

[4] Assuming a sum-product decoder performing 20 iterations with flooding schedule.

are inverted relatively to symbols of S_1, forming a reference point to detect the start of the received frame. Windowing is applied at the edges of each section for spectrum compatibility.

The header carries settings of all programmable parameters related to the payload, such as guard interval, bit loading and FEC parameters. The parameters of the header are unified per medium to ensure interoperability and selected to allow reliable detection of the header over noisy channels even without preliminary channel estimation. The payload includes one or more FEC codewords. Each codeword carries a segment of the transmitted data, a header identifying the carried segment and the CRC to detect errored codewords for selective retransmission.

8.8.2.4   MIMO

Recommendation ITU-T G.9963 specifies how to add to G.hn a Multiple Input Multiple Output (MIMO) capability. MIMO transceivers are able to transmit over three power line conductors (phase, neutral, and ground) in more than one Tx port and receive in more than one Rx port, thus providing an increased data rate and enhancing the connectivity (i.e. service coverage) of the home network. MIMO is addressed in detail in Section 5.7.

8.8.3   *Overview of the Data Link Layer of G.hn*

8.8.3.1   **Media Access Methods**

G.9960 defines *synchronized* media access, i.e. transmissions in the domain are coordinated by the DM and synchronized with the Media Access (MAC) cycle. The MAC cycle, in turn, can be synchronized with the mains – to cope with periodically time-varying behavior of channel response and noise caused by electrical devices and appliances plugged into the power line[5]. Each MAC cycle is divided into time intervals associated with transmission opportunities (TXOP) assigned by the DM for nodes in the domain. The DM assigns at least one TXOP to transmit the Media Access Plan (MAP) frame, which describes the boundaries of the TXOPs assigned for one or several following MAC cycles. The latter protects from MAP erasures caused mainly by impulse noise. Other TXOPs are assigned by the DM to nodes requesting to transmit application data (e.g. video services, data services, VoIP). All nodes in the domain synchronize with the MAC cycle, read and interpret the MAP and transmit only during the TXOPs assigned for them by the DM. Thus, collisions can be avoided for particular connections. The DM sets the order, type and duration of TXOPs based on requests from nodes and available bandwidth; the schedule can change from one MAC cycle to another due to variations in medium characteristics, in user application or when the number of nodes in the domain changes.

To address different applications, three types of TXOP are defined:

- *Contention-free TXOP* (CFTXOP) implements a pure time-division media access (TDMA): only one node can transmit during this TXOP – targets services with fixed bandwidth and strict QoS (e.g. video).

---

[5] The input impedance and injected noise of household electrical devices often depends on the instantaneous amplitude of the AC mains voltage which causes a periodically time-varying channel response and a periodically time-varying noise.

- *Shared TXOP with managed time slots* (STXOP) implements a managed carrier-sense media access with collision avoidance (CSMA/CA), similar to ITU-T G.9954 – beneficial for services with flexible bandwidth where QoS is an issue (e.g. VoIP, games, interactive video).
- *Contention based TXOP* (CBTXOP) is a shared TXOP, in which assigned nodes may contend for transmission based on the frame priority, similar to HomePlug AV [37] – for best effort services with several priority levels.

An STXOP is divided into a number of short time slots (TS). Each TS is assigned for a particular node to transmit a frame with a particular priority. If the node assigned for the TS has a frame of the assigned priority ready it transmits it, otherwise it skips the TS and passes the transmission opportunity to the node/priority assigned for the next TS. The node assigned to transmit in the next TS monitors the medium (by carrier sensing) and waits till no activity in the medium. Thus, despite that STXOP is shared between several nodes, no collision occurs if carrier sensing is sufficiently reliable.

Transmission during CBTXOP is arranged by contention periods. At the beginning of the contention period, each contending node indicates the priority of the frame it intends to send using priority signaling (PRS). The PRS selects nodes with frames of highest priority: only these nodes are allowed to contend, while all others back off to the next contention period. The probability of collision between the selected nodes is reduced by a random pick of the particular transmission slot inside the contention window. From the beginning of the window, all selected nodes monitor the medium (by carrier sensing). If the medium is inactive at the slot picked by the node, the node transmits its frame otherwise it backs off to the next contention period. The described principles of G.9960 media access are shown in Figure 8.14.

To facilitate virtual carrier sensing every frame indicates its duration in the frame header. Also, request-to-send (RTS) and clear-to-send (CTS) messages, similar to IEEE 802.11, are defined to reduce the time loss in case of collision and improve operation in presence of hidden nodes.

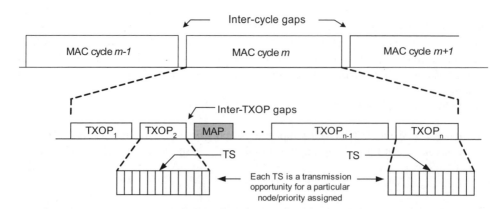

Figure 8.14   G.9960 media access (TXOP$_1$ is assigned as a CFTXOP, TXOP$_2$ and TXOP$_n$ are assigned as STXOP, and TXOP$_{n-1}$ is assigned as a CBTXOP).

8.8.3.2 Security

Since G.9960 is intended to operate over shared media, such as power line and coax, its threat model includes two kinds of threats: external and internal. In both cases, the goal is to protect against attackers with reasonably powerful computing resources, but with no access inside operating nodes.

External threat implies an attacker capable of eavesdropping on transmissions and sending frames within the network, but out of network access credentials. Internal threat is from a legitimate user of the network, who has an illegitimate interest in the communications of another user, or in access to specific network client. In case of hidden nodes, communications between two particular nodes may pass through a relay node, causing a 'man-in-the-middle' threat.

Concerning external threats, G.9960 defines an authentication procedure based on the Diffie-Hellman algorithm and the Counter with Cipher Block Chaining-Message Authentication Code algorithm (CCM) which uses AES-128. Against internal threats, typical for public installations, G.9960 defines pair-wise security, i.e. a unique encryption key is assigned to each pair of communicating nodes and is unknown to all other nodes. Pair-wise security maintains confidentiality between users within the network and builds another layer of protection against an intruder that has broken through the network admission control. The expected grade of security in G.9960 is the same or stronger than the one defined in the most recent specification for WLAN IEEE 802.11n.

References

1. Y.-J. Lin, H. A. Latchman, M. Lee, and S. Katar, A power line communication network infrastructure for the smart home, *IEEE Wireless Commun.*, 9(6), 104–111, Dec. 2002.
2. H. A. Latchman, K. H. Afkhamie, S. Katar, R. E. Newman, B. Mashburn, and L. Yonge, High speed multimedia home networking over powerline, in *Nat. Cable Telecommun. Assoc. Nat. Show Tech. Papers,* San Francisco, USA, Apr. 3–5, 2005, 9–22.
3. M. K. Lee, R. E. Newman, H. A. Latchman, S. Katar, and L. Yonge, HomePlug 1.0 powerline communication LANs – Protocol description and performance results, *Int. J. Commun. Syst., Special Issue: Powerline Commun. and Applic.*, 16(5), 447–473, May 2003.
4. Current Technologies, Sep. 2008. [Online]. Available: http://www.currenttechnologies.com
5. Y.-J. Lin, H. A. Latchman, S. Katar, and M. K. Lee, A comparative performance study of wireless and power line networks, *IEEE Commun. Mag.*, 41(4), 54–63, Apr. 2003.
6. M. K. Lee, H. A. Latchman, R. E. Newman, S. Katar, and L. Yonge, Field performance comparison of IEEE 802.11b and HomePlug 1.0, in *Proc. IEEE Conf. Local Comput. Netw.*, Tampa, USA, Nov. 6–8, 2002, 598–599.
7. E. Mikoczy, D. Sivchenko, B. Xu, and J. I. Moreno, IPTV services over IMS: Architecture and standardization, *IEEE Commun. Mag.*, 46(5), 128–135, May 2008.
8. B. Ji, A. Rao, M. Lee, H. A. Latchman, and S. Katar, Multimedia in home networking, in *Proc. Int. Conf. Cybern. Inform. Technol., Syst. Applic.*, vol. 1, Orlando, USA, Jul. 21–25, 2004, 397–404.
9. 802.11n: Next-generation wireless LAN technology, Broadcom Corporation, White paper, Apr. 2006. [Online]. Available: https://www.broadcom.com/collateral/wp/802_11n-WP100-R.pdf
10. Ultra-wideband (UWB) technology: Enabling high-speed wireless personal area networks, Intel Corporation, Ultra-Wideband (UWB) White Paper, 2004. [Online]. Available: http://www.intel.com/technology/comms/uwb/
11. A. Monk, S. Palm, A. Garrett, R. Lee, and R. Leacock, MoCA Protocols: What exactly is this MoCA thing? MoCA Alliance, Tech. Rep., Nov. 2007. [Online]. Available: http://www.mocalliance.org
12. HomePNA and IPTV, HomePNA Alliance, Website, Apr. 2007. [Online]. Available: http://www.homepna.org

13. J. S. Barnes, A physical multi-path model for powerline distribution network propagation, in *Proc. Int. Symp. Power Line Commun. Applic.*, Tokyo, Japan, Mar. 24–26, 1998, 76–89.
14. K. Dostert, Telecommunications over the power distribution grid – possibilities and limitations, in *Proc. Int. Symp. Power Line Commun. Applic.*, Essen, Germany, Apr. 2–4, 1997, 1–9.
15. H. Hrasnica, A. Haidine, and R. Lehnert, *Broadband Powerline Communications: Network Design*. John Wiley & Sons, 2004.
16. OPERA Technology White Paper, Open PLC European Research Alliance, Tech. Rep., Jul. 2007. [Online]. Available: http://www.ist-opera.org
17. HD PLC White Paper, HD-PLC Alliance, Tech. Rep., Nov. 2005. [Online]. Available: http://www.hd-plc.org
18. HomePlug AV White Paper, HomePlug Powerline Alliance, Tech. Rep., Aug. 2005. [Online]. Available: http://www.homeplug.org
19. K. H. Afkhamie, S. Katar, L. Yonge, and R. E. Newman, An overview of the upcoming HomePlug AV standard, in *Proc. IEEE Int. Symp. Power Line Commun. Applic.*, Vancouver, Canada, Apr. 6–8, 2005, 400–404.
20. IEEE 1901 Working Group. [Online]. Available: http://grouper.ieee.org/groups/1901
21. IEEE P1901, Standard for broadband over power line networks: Medium access control and physical layer specifications. [Online]. Available: http://grouper.ieee.org/groups/1901/index.html
22. S. Galli and O. Logvinov, Recent developments in the standardization of power line communications within the IEEE, *IEEE Commun. Mag.*, 46(7), 64–71, Jul. 2008.
23. Unified high-speed wire-line based home networking transceivers – system architecture and physical layer specification, ITU-T, Recommendation G.9960, 2011. [Online]. Available: https://www.itu.int/rec/T-REC-G.9960
24. Unified high-speed wire-line based home networking transceivers – data link layer specification, ITU-T, Recommendation G.9961, 2014. [Online]. Available: http://www.itu.int/rec/T-REC-G.9961
25. Unified high-speed wire-line based home networking transceivers – management specification, ITU-T, Recommendation G.9962, 2013.
26. Unified high-speed wire-line based home networking transceivers – multiple input/multiple output specification, ITU-T, Recommendation G.9963, 2011.
27. Unified high-speed wire-line based home networking transceivers – specification of spectrum related components, ITU-T, Recommendation G.9964, 2011.
28. V. Oksman and S. Galli, G.hn: The new ITU-T home networking standard, *IEEE Commun. Mag.*, 47(10), 138–145, Oct. 2009.
29. Coexistence mechanism for wireline home networking transceivers, ITU-T, Recommendation G.9972, 2010.
30. S. Galli, A. Kurobe, and M. Ohura, The Inter-PHY protocol (IPP): A simple co-existence protocol for shared media, in *Proc. IEEE Int. Symp. Power Line Commun. Applic.*, Dresden, Germany, Mar. 30–Apr. 1, 2009, 194–200.
31. S. Galli, H. Koga, and N. Kodama, Advanced signal processing for PLCs: Wavelet-OFDM, in *Proc. IEEE Int. Symp. Power Line Commun. Applic.*, Jeju Island, Korea, Apr. 2–4, 2008, 187–192.
32. H. C. Ferreira, L. Lampe, J. E. Newbury, and T. G. Swart, Eds., *Power Line Communications: Theory and Applications for Narrowband and Broadband Communications over Power Lines*, 1st ed. John Wiley & Sons, Hoboken, 2010.
33. D. Ayyagari and W.-C. Chan, A coordination and bandwidth sharing method for multiple interfering neighbor networks, in *Proc. IEEE Consum. Commun. Netw. Conf.*, Las Vegas, USA, Jan. 3–6, 2005, 206–210.
34. S. Katar, B. Mashburn, K. Afkhamie, H. Latchman, and R. Newman, Channel adaptation based on cyclostationary noise characteristics in PLC systems, in *Proc. IEEE Int. Symp. Power Line Commun. Applic.*, Orlando, USA, Mar. 26–29, 2006, 16–21.
35. S. Katar, L. Yonge, R. Newman, and H. Latchman, Efficient framing and ARQ for high-speed PLC systems, in *Proc. IEEE Int. Symp. Power Line Commun. Applic.*, Vancouver, Canada, Apr. 6–8, 2005, 27–31.
36. H. A. Latchman, S. Katar, L. W. Yonge, and S. Gavette, *HomePlug AV and IEEE 1901: A Handbook for PLC Designers and Users*. Wiley-IEEE Press, 2013.
37. HomePlug Power line Alliance. [Online]. Available: http://www.homeplug.org/

9

PLC for Smart Grid

I. Berganza, G. Bumiller, A. Dabak, R. Lehnert, A. Mengi, and A. Sendin

9.1 Introduction

Smart grid is still a relatively new concept for many utilities worldwide. While there is no single, generally agreed definition of smart grid [1]–[8], many utilities are currently engaged in the process of adapting their grids to integrate recent advances in electronics and information and communications technology (ICT), to provide improved energy supply quality based on remote monitoring and control of the different electric grid assets. Smart metering is the application with the greatest support both from the industry and utilities, which find in it an opportunity to build the foundation for a larger scope smart grid, while reaping some immediate benefits in terms of savings and commercial opportunities from real time access to customers' smart meters.

Power line communication (PLC) is a natural communications technology for smart grids, as it uses the existing power cables. For more than a hundred years, amplitude-modulation (AM) carrier based communications has been used by the power network operators to transfer status and alarm messages between power plants and substations. This technology operated on long-wave (LW) frequencies, e.g. in the range from 24 kHz to 500 kHz. It has been a long-haul system covering distances of several hundreds of kilometers. It has been widely used on high-voltage (HV) lines. The available bandwidth of a few kHz was sufficient for the above applications. It also has been used as operator internal telephony. These implementations may be considered as the first PLC systems.

When fiber based optical communications became available, operators started to install fiber links in underground cables and overhead using existing poles. This increased the bandwidth significantly. It is now standard on HV lines and, due to the high data rate, it allows to sell communications capacity to other operators.

In medium-voltage (MV) networks, fibers are rarely included in the power cabling. While at present, MV substations are connected to the communications network mainly via

Power Line Communications: Principles, Standards and Applications from Multimedia to Smart Grid, Second Edition.
Edited by Lutz Lampe, Andrea M. Tonello, and Theo G. Swart.
© 2016 John Wiley & Sons, Ltd. Published 2016 by John Wiley & Sons, Ltd.

digital subscriber lines (DSL), private pilot cables (copper pairs) or cellular radio techniques, communications to substations may be realized by PLC as will be presented in this chapter.

On low-voltage (LV) lines there are definitely no extra communication cables, e.g. fibers. Similar to telephony cabling, the 'last mile' in electricity accounts for the major investments because of the sheer number of connected customers. Hence, here PLC is an excellent candidate to bring customers into the smart grid. In addition, the communication terminals, e.g. electronic meters, of customers are quite often below ground level, which is not the best scenario for radio systems.

9.1.1 PLC Technology Classification

Reference [9] proposes a classification of PLC for smart grid, based on the frequency bands PLC systems have been using. This classification is useful as, for decades, higher and wider frequency bands have progressively been employed for PLC, as technology made their use practical. Frequency bands have also direct impact on signal reach and effective bandwidth, which ultimately define data rate and consequently the applications for the PLC technologies. Following [9], PLC systems are classified as follows.

- Ultra-Narrowband (UNB) PLC: Systems which operate from close to 300 Hz to 3 kHz (SLF and ULF bands). 'Ripple control' systems are an example of the use of these frequencies; most of them were designed for one-way communications. These systems convey very low data rates (usually less than 100 bps) over tens or even hundreds of kilometers.
- Narrowband (NB) PLC: Systems which work at frequencies between 3 kHz and about 500 kHz. These include regulated bands such as the CENELEC A to D-bands defined in EN 50065-1 (Europe, 3–148.5 kHz), the FCC provisions in clause 15.113 of Title 47 of the Code of Federal Regulations (USA, 9–490 kHz), the band specified in ARIB STD-T84 (Japan, 10–450 kHz) and Chinese band (3–500 kHz). These systems usually show a signal reach which, depending on the power grid, is in the order of hundreds of meters to several kilometers. NB PLC systems can be further divided into:
  - Low Data Rate (LDR) NB PLC, employing single carrier modulations for throughputs of hundreds of bps to few kbps, and
  - High Data Rate (HDR) NB PLC, employing multiple carrier modulations for throughputs up to hundreds of kbps.
- Broadband (BB) PLC (also known as Broadband over Power Line (BPL)): Systems which use frequency bands anywhere from 1.8 MHz to 250 MHz, covering distances from hundreds of meters to a few kilometers, and with data rates ranging from several Mbps to hundreds of Mbps.

9.1.2 Electricity Grid

The electric grid is the basic infrastructure supporting the different kinds of PLC. Continuing the preliminary discussion above, we provide some analysis on how the distribution grid is organized, to better understand the possible PLC applications in each segment of the grid. This is especially important when different PLC systems are deployed over the same grid and other technologies are additionally used to create a complete telecommunications network. Furthermore, different national and regional characteristics of the grids have an impact on the extent to which PLC can be considered for smart grid when compared to other telecommunications alternatives.

9.1.2.1 Grid Description

A power system usually comprises four distinct pieces: power generation plants, transmission lines, substations and the distribution grid. The topology and characteristics of electric networks present differences in the various parts of the world, even among neighboring countries. Furthermore, power system components have different features depending on the age of the infrastructure. We will mainly focus this discussion on distribution grids, comprising MV and LV segments.

The concept of smart grid is specific to the element of the power system it is applied to, since the needs are unique for each part of the network. PLC is usually understood as a communications access technology and thus it is used in the access segment of the electricity grid. This means the MV and LV segments of the grid, the closest to customers (see Figure 9.1). Nevertheless, there are also application scenarios for PLC over existing HV power lines, to avoid costly deployment of other telecommunications technologies (e.g. optical fiber).

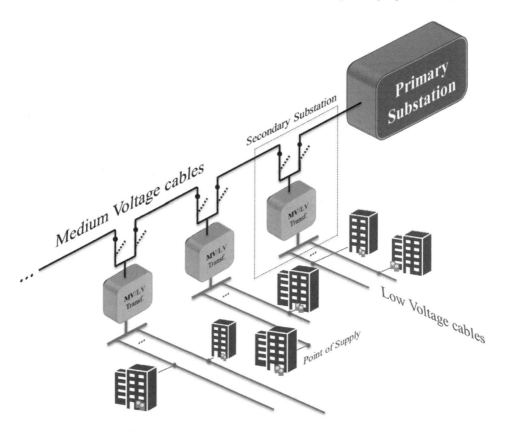

Figure 9.1 MV and LV electricity distribution grid.

The MV segment is the part of the grid between the HV lines and the MV/LV transformers (which can be found sheltered inside secondary substations or on poles). MV lines can be overhead or underground; both behave differently with respect to the performance of the PLC channel, since impedances are different and propagation is usually found to be better in overhead topologies. If overhead and underground installation is mixed (e.g. for crossing a motorway) a strong attenuation can be observed. This is caused by wave impedance mismatches between overhead and underground cables. Signal injection needs also to be adapted for each case since coupler installation depends on the type of cable and infrastructure. Lastly the grid architecture is different since overhead power lines follow a bus topology with mechanical switches in parts of the grid, whereas underground networks usually configure a point to point topology where the nodes are the MV to LV transformers. This is an interesting aspect which favors PLC deployments in underground MV distribution, as point to point connections are easier to establish, operate and maintain.

The LV segment is the part of the grid which reaches the final customers. The LV segment presents two main features:

- It is a ubiquitous and heterogeneous grid, and it reaches all final points with minimum investment. It presents huge diversity of configurations and different generations of assets coexist.
- Its characteristics as a PLC channel depend heavily on customer premises, where loads of all kinds are randomly connected and disconnected. This means that impedances are constantly changing throughout the grid, and multiple noise patterns appear randomly and unpredictably from unspecified sources. Hence, it is not unusual to deal with the whole LV systems as a set of separate subnetworks (each one usually depending on a single MV/LV transformer) each with its own characteristics and performance pattern.

There are at least some known factors which need to be considered to understand if/how PLC can optimally be used for communications:

- HV and MV segments: Voltage levels are important, since they are a critical factor to design the couplers.
- LV segment: Distances, total number of LV customers and customer density per subnetwork. These three variables are not totally independent (see [10]). If for a certain technology the distance between nodes is too large (so there is too much attenuation) or if the customer density is too low even with repetition of PLC signals (as explained in [11]), or if the total number of customers per subnetwork is not large enough to allow for a return on investment, PLC might not be a recommended solution.

When PLC is actually a viable technology for any of the HV, MV and/or LV segments, the variables above are the ones that each PLC implementation has to consider.

9.1.2.2 Grid Regional Differences

Although the structure of electric grids is different in different regions of the world, the functional principles are basically the same. Grids are different in aspects like voltage levels, extension and reach of the MV grid, and hence the nature and structure of the LV grid. A good example is the dissimilarity between European and North American networks. The European model uses MV grid voltage levels typically ranging from 10 kV to 24 kV, both in overhead and underground distribution. LV grids stemming from each secondary substation cover around 150 customers on average, and distances from LV bus bars to meters are rarely above 200 m. MV to LV transformers are often built in a 'star' configuration with 400 V among phases, and 230 V between phase and neutral. The North American model, in contrast, employs voltages usually between 4 kV and 34 kV in MV, and distances ranging from 15 km to 50 km. MV to LV transformers produce voltages of 120 V or 200 V depending on load type. Typical total LV cable lengths are close to 300 m, with an average 10 customers or less per transformer. PLC signals may be injected between phase and neutral (120 V).

The following paragraphs give additional details for the European model, since PLC system development are more popular in Europe (e.g. smart metering systems).

HV grid in Europe: HV usually refers to levels above 35 kV. It usually starts at 45 kV and 69 kV, and common transmission voltages above those include 110 kV, 132 kV, 150 kV, and 220 kV with transport levels normally above 300 kV (e.g. 400 kV). The topology especially at higher voltages is meshed/redundant, meaning that these are the most reliable voltage levels. HV grids cover the larger distances over tens or hundreds of kilometers, usually over sparsely populated areas and using overhead lines.

MV grid in Europe: MV topologies can be generally divided into three distinct groups.

- Radial topology: Radial lines are used to connect primary substations with transformers, and the transformers among them. These MV lines or 'feeders' may go to one or multiple transformers. Radial systems keep central control of all the substations. They show a tree-shaped configuration when they grow in complexity. Radial topologies are the easiest and most cost-effective to deploy, operate and maintain.
- Ring topology: This topology is fault tolerant, so it overcomes the main weakness of radial topologies (i.e. a fault at one element of the line, interrupts service –outage– in the rest of connected substations). A ring topology is an improved evolution of the radial topology connecting to other MV lines to create redundancy. The grid is operated as radial, but in the event of a feeder fault, certain elements are switched to reconfigure the grid such that the affection to customers is minimized. Faulty lines are most of the time automatically restored soon afterwards.
- Networked topology: This topology occurs when primary substations and MV/LV transformers are connected through multiple MV lines. Reconfiguration options in case of fault are thus multiple, and in the event of any failure more than one course of action is possible.

LV grid in Europe: LV grids usually show complex and heterogeneous topologies, more so than MV grids. Reasons for that are multiple: need to serve areas of very different sizes, high variance in the number of end customers, average

Table 9.1   Typical data for electricity grids in Europe (SS: Secondary Substation)

Parameter	High Density Residential Area	Low Density Residential Area
Type of SS	Underground or above ground inside a building	House-type or over a pole
Transformers per SS	2	1
Average number of customers per SS	250–320	100 (10–200)
LV feeders per transformer	6–8	6–8
Average length of the LV lines (m)	150	300 (100–800)
Type of LV line	Underground	Overhead
Customers per metering room	10–25	1–4

consumption of different customers, country- and utility-specific operating procedures, standards which may change during the years so different versions are to be found simultaneously in the grid, etc. A transformer typically provides service to several LV lines, with one or more sharing location at the premises. LV topology is typically radial, with multiple branches stemming from the main feeder and in turn from other branches. Networked grids also exist, along with ring or dual-feed configurations. LV lines are typically shorter than MV lines, and their main characteristics differ depending on the service area (see Table 9.1 from [12]).

9.1.3   Requirements

Any utility, which wants to connect customers to the smart grid needs a communications network which is reliable, does have the bandwidth and latency specified, and is secure against attacks and eavesdropping.

Data rate: Metering data consisting of a few bytes at the application level and transferred, e.g. every 15 minutes does not require a high data-rate link. The same holds for control commands to switch a remote photo-voltaic (PV) system. Recently, the requirements of encryption mechanisms to protect sensitive data have resulted in increased data rates. When PLC is used as the backbone for some areas of a smart grid, large volumes of Ethernet traffic are usually transported, and the requirements for a point to point link between substations range between hundreds of kbps up to around 40 Mbps.

Latency: Power grid stability requires that demand and power generation are continuously balanced. As power generated by renewable sources is intermittent and usually does not have buffer such as a battery or the spin momentum of a generator, fast control loops must be implemented. Currently reaction times in the range of a full wave at 50 Hz (or 60 Hz) are discussed. Accordingly, an end-to-end delay (one way) of 10 ms may have to be realized. In a control loop hierarchy, this requirement may be relaxed to a range of few seconds. Again, for the case of broadband PLC being used for the transport of smart grid services between substations, low latency requirements for VoIP and low rate video stream must be met (some hundreds of ms maximum) and it is especially important to control jitter (i.e. the variance of latency).

Privacy and security: Energy consumption in a household, especially the daily curve of consuming electrical energy belongs to private data which may be used only by the metering company and here only in an aggregated form. Control commands for demand side management (DSM) or for the distributed power generation must reach its destination without interference from eventual intruders. Therefore encryption techniques are needed to ensure security.

The same holds for the local energy production by e.g. a PV generator. Also other data such as remote control of appliances, are typically personal data. Therefore most of the communication between a home and the utilities' data processing center contains private information which has to be protected against e.g. eavesdropping and loss. So encryption techniques to ensure privacy are needed. Both require extra effort w.r.t. the usable (net) data rate and the grade of interconnection in the communications network graph. Usually state of the art encryption and authentication techniques are used by PLC systems. It also must be carefully considered, when designing a PLC system, which protocols will implement security and which will not (e.g. encryption at MAC level or at Application level).

Availability and reliability: Whenever (part of) the power grid fails, there must be mechanisms to restart the power generators and, in an ordered sequence, also the power consumers. This sequence may be run separately in power islands before these islands are re-connected. Alternatively, the power network restart procedure may be centrally controlled for a larger area. Anyway, the controlled restart, also called 'black start', needs a communication network, which is operational from the beginning of the procedure. Therefore the smart grid communication network must have its own power supply to support black-start situations. If the communication network is realized by e.g. leased lines provided by a public network operator, the communication network operator has to fulfill the above requirements by, e.g. installing an uninterrupted power supply (UPS). For decades, PLC systems operated with batteries have been well known to continue normal communications even in the event of grid faults on its own transmission power lines. However, it is difficult to predict when a line fault will impede PLC transmission and reception or not.

In the case of virtual power plants consisting of many distributed generators which produce non-stationary power from renewable sources (wind turbines, PVs), the backup power for the communication may be difficult to ensure. Especially when power consumers become 'prosumers' (producer and consumer) it may be very difficult to guarantee the power supply of the communications part of the network.

Extensive operational experience could be derived from different projects realized in Europe. For example, an industry relation group containing several utilities, chip manufacturers and system integrators defined and published the following additional requirements for NB PLC systems [13].

- For utility enterprises especially those in Germany with its market model of extreme complexity it is essential to find a PLC solution in which all devices of different manufacturers fully support interchangeability.
- To achieve this aim the manufacturers of semiconductor devices must offer to the device manufacturers as much as possible components which are compatible with each other in the PLC networks, like the Ethernet components in Ethernet networks.
- The PLC network must provide a transmission technology being completely transparent for other communication protocols. In particular, it must be possible to use the IP and the transport protocols located above this (UDP and TCP).
- The available payload bandwidth and the response times must be dimensioned sufficiently to support all currently discussed tasks in smart metering standard operation processes and all smart grid applications which are not security-relevant.
- The necessary operational availability and robustness have to be provided by the PLC network inherently, e.g. without any controlling intervention by an operator.
- The interchangeability requirement will have to be proven by adequate conformity tests and certificates.
- Behind a PLC end point single nodes in sub-networks need to be addressable (e.g. via wireless network for submetering). Therefore the addressing required generally by IP must be transmitted beyond the PLC end point.
- For an efficient operation of a PLC network it must provide monitoring information which allows evaluating the current availability and the analysis of operational disturbances.
- To allow an efficient on-site installing of PLC components the installation must be possible without parametrization with a minimum time needed like 'Plug and Play'.
- Bringing into service of installed PLC components must be possible time-decoupled of its installation. The PLC component must allow bringing it into communication service automatically without an employee on-site.

9.1.4   Applications

Utilities have a number of applications where a communication network is required:

- Smart metering
- Control of distributed power sources
- Demand side management (DSM)
- Control loop
- Company internal communications.

> **Smart metering:** Depending on country specific regulations, smart meters are already widely deployed. Meters are typically connected to the LV network. In Europe, mostly PLC is used as the communications technology as PLC can reach the basement of a building easier than a radio technology whereas in the Americas and Asia cellular radio or some special radio systems are preferred.
>
> **Control of distributed power sources:** Renewable energy sources (wind turbines, PVs) are feeding into every voltage level of the power network. Big wind turbines may be connected to the HV layer, smaller ones to the MV layer. Big PV systems

are connected to MV lines whereas the many private roof-top PV systems feed into the LV network. Communications at the HV layer today is provided by fibers deployed in the earth wire. Communications to the substations at the MV may either be implemented by fibers, by DSL or cellular radio provided by public telecom operators. Again PLC is a suitable technology to reach homes on the LV power grid.

Demand side management: DSM is an approach to avoid peak load situations by peak load clipping. That is, some consumers (appliances) are switched off. It may be used on every voltage level in the power network if there is an appropriate consumer. DSM makes sense for consumers with a thermal latency. For example, a refrigerator may be switched off for some minutes without having a temperature increase outside of its specifications. Similarly, any kind of heating or cooling appliances may be switched without disturbing its main functionality.

Control loop: Power production and demand must always be in balance. In the past, a few (big) power plants were used as generators to serve a large number (millions) of consumers. Today, there is a transition to (many) distributed generators feeding the consumers. This makes the control loop more complex. In the past a simple rotational speed controller at the electrical generator has been sufficient to keep the distribution network stable. It is obvious that a single control loop cannot be built with millions of sources and sinks. So the control architecture will most probably be hierarchical. Possibly there will be control loops at the HV, MV, and the LV layer.

Company internal communications: This requires a variety of technologies, depending on the specific use case. Utility internal telephone and data communication may run on the fiber-based HV network. This is a fixed network. If technicians need mobile communication, then private mobile radio (PMR) systems may be used. Also public cellular operators may provide a fixed-mobile integrated communication solution. Definitely the reliability of this communication network is very important as it is not only used for routine tasks but, especially, for handling emergency situations.

Today mainly smart metering and secondary control of PV power generation together with demand side management are the drivers to implement PLC technology in the LV layer. In the past decade there have been testbeds to study the behavior of PLC technology for these purposes, e.g. [14], [15]. For example, Table 9.2 summarizes the requirements for a SG communications testbed using PLC run at a regional utility in Germany in summer 2014. Besides meter reading with variable readout intervals, fast PV system control and DSM were aimed at. The tests were run with artificial (background) traffic to evaluate the performance limits of the communication system.

9.1.5   Outline

In the remainder of this chapter, we first present an overview of the more recently developed standards for PLC for smart grid communication in Section 9.2. This is followed by a review

Table 9.2   PLC network requirements list for a PLC testbed in Germany

Parameter	Value
Frequency band used	2 to 30 MHz
Max. PHY data rate	200 Mbps
Max. power consumpt. slave	3 watts (standby)
Max. power consumpt. master	5 watts (standby)
IP version	IPv6
Min. no. of slaves addressable	1000
Coupling	3 phase
Network transparency	Bridged (transparent to the user)
Redundancy for master	Yes
Redundancy for slave	No (optional)
Addressing	Static or dynamic
Repeater function	In every modem
Min. modem distance (comm. span)	30 m
Min. data rate (IP layer)	10 Mbps
Max. latency (single hop)	15 ms
Max. unavailability (network)	10E-4

of norms and standards that serve the regulation of PLC for smart grid in different parts of the world in Section 9.3. Then, Section 9.4 provides an in-depth discussion of the use of PLC to support various smart grid applications, which also includes deployment examples.

9.2   Standards

In this section, we present an overview over more recent HDR NB PLC standards developed specifically for smart grid communications. These can be considered a second generation of standards, following a suite of sometimes proprietary, mostly LDR NB PLC specifications that have been used by electric utilities for a long time. The first generation of PLC systems are narrowband and low data rate (few hundreds of bps to a few kbps) operating at frequencies below 500 kHz. Examples include ISO/IEC 14908-3 (LonWorks), ISO/IEC 14543-4-5 (KNX), CEA-600.31 (CEBus), X10, Insteon, IEC 61334-5-1/2/4 and also the above-mentioned ripple control systems, the AMR Turtle system and the two-way automatic communications system (TWACS), which use UNB PLC [9].

At around 2010, several organizations (industry alliances and standards developing organizations) started standardizing a new generation of PLC systems operating at frequency bands below 500 kHz. These HDR NB PLC systems use multicarrier modulation, specifically orthogonal frequency-division multiplexing (OFDM), similar to broadband PLC systems, which were consolidated in IEEE 1901 and ITU-T G.hn standards and are discussed in more detail in Chapter 8. Basically, four separate ITU-T recommendations for NB PLC were approved in late 2012 and published afterwards. Furthermore, the IEEE 1901.2 Working Group published the IEEE 1901.2 Standard in 2013. The standards are summarized in the following.

- ITU-T G.9901 [16] Narrowband OFDM Power Line Communication Transceivers-Power Spectral Density (PSD) Specification. Recommendation ITU-T G.9901 specifies the control parameters that determine spectral content, PSD mask requirements, a set of tools to support the reduction of the transmit PSD, the means to measure this PSD for transmission over power line wiring, as well as the allowable total transmit power into a specified termination impedance. It complements the system architecture, PHY and data link layer (DLL) specifications in Recommendations ITU-T G.9902 (G.hnem), ITU-T G.9903 (G3-PLC) and ITU-T G.9904 (PRIME).

- ITU-T G.9902 [17] Narrowband OFDM Power Line Communication Transceivers for ITU-T G.hnem Networks. This recommendation contains the PHY and the DLL specifications for the ITU-T G.9902 narrowband OFDM power line communication transceivers, operating over alternating current and direct current electric power lines over frequencies below 500 kHz.
- ITU-T G.9903 [18] Narrowband OFDM Power Line Communication Transceivers for G3-PLC Networks. This recommendation contains the PHY and the DLL specification for the G3-PLC narrowband OFDM power line communication transceivers for communications via alternating current and direct current electric power lines over frequencies below 500 kHz.
- ITU-T G.9904 [19] Narrowband OFDM Power Line Communication Transceivers for PRIME Networks. Recommendation ITU-T G.9904 contains the PHY and DLL specification for PRIME narrowband OFDM power line communication transceivers for communications via alternating current and direct current electric power lines over frequencies in the CENELEC A-band.
- IEEE 1901.2 [20] IEEE Standard for Low-Frequency (less than 500 kHz) Narrowband Power Line Communications for Smart Grid Applications. This standard contains the PHY/MAC layer, coexistence and EMC requirements for narrowband PLC via alternating current, direct current, and non-energized electric power lines using frequencies below 500 kHz.

These standards have similarities at the PHY layer but differ significantly in their approach to supporting higher layers. This will be discussed in more detail in the following sections. Additionally, the broadband PLC HomePlug GreenPHY specification will also be covered, since it was devised to support smart grid applications such as plug-in electric vehicles, smart home automation and smart meters within the customer's premises. We refer to Chapter 8 for details about broadband PLC standards, which were devised for high data-rate access and in-home communications and thus can also be part of solutions for smart grid communications as discussed in the previous section and in Section 9.4.

9.2.1 ITU-T G.9902 G.hnem Standard

As per scope of ITU-T G.9902 G.hnem [17], the standard was developed for the main smart grid applications, namely Advanced Metering Infrastructure (AMI), plug-in electrical vehicles (PEV), and various home energy management applications. It contains the specifications of PHY layer and DLL, supporting the default protocol IPv6. Other network layer protocols can also be supported using an appropriate convergence sublayer.

9.2.1.1 Physical Layer

The ITU-T G.9902 G.hnem control parameters determining spectral content, PSD mask requirements and related tools to support the reduction of the transmit PSD have been published in Recommendation ITU-T G.9901 [16]. The standard defines several bandplans. There are: CENELEC A-band (35.938 kHz to 90.625 kHz), CENELEC B-band (98.4375 kHz to 120.3125 kHz), CENELEC CD-bands (125 kHz to 143.75 kHz), FCC (34.375 kHz to 478.125 kHz), FCC-1 (154.6875 kHz to 487.5 kHz), FCC-2 (150 kHz to 478.125 kHz) and ARIB (154.7 kHz to 403.1 kHz).

The PHY transmitter is based on G3-PLC and consists of concatenated forward error correction (FEC) that makes use of Reed-Solomon (RS) and convolutional coding. The inner convolutional coding operates with a rate $1/2$ and constraint length 7. The outer RS encoder operates with input blocks up to 239 bytes. A two dimensional interleaving scheme is used to avoid the occurrence of burst errors which may be caused by time and frequency dependent noises. The ITU-T G.9902 supports coherent QAM modulations with 1, 2, 3 or 4 bits per carrier. The channel estimation is accomplished by the channel estimation symbols positioned in the PHY frame.

9.2.1.2   MAC Layer

The ITU-T G.9902 G.hnem topology is logically constructed in a domain structure, where each domain is a particular set of nodes associated to this domain. Each node is identified by its domain ID and node ID. One node in the domain is assigned as a domain master (DM). The DM controls operation of all other nodes and performs admission, resignation and other domain wide management operations. G.9902 devices of different domains communicate with each other via Inter-Domain Bridges (IDB). IDBs are simple data communications bridges that link multiple domains, enabling a node in one domain to pass data to a node in another domain. An ITU-T G.9902 node of a standard profile offers prioritized contention-based CSMA/CA medium access with four priorities. The three lower priorities are intended for user data frames, and the fourth priority is for frames carrying emergency signaling. As in the IEEE 1901.2 standard, ITU-T G.9902 has an agnostic structure of the MAC layer, which supports both layer 2 and layer 3 routing.

9.2.2   ITU G.9903 G3-PLC Standard

The G3-PLC specification was published in August 2009 by Maxim Integrated Products, Inc. (USA) as an open specification to meet the requirements of Electricité Réseau Distribution France (ERDF) in the smart grid applications: grid to utility meter applications, AMI, and other smart grid applications such as PEV, home automation and home area networking (HAN) communications scenarios. In 2011, under the leadership of ERDF, the G3-PLC Alliance was formed to promote and maintain the specification. The executive members of the alliance include twelve members, which are EDF, ERDF, Enexis, Maxim Integrated Products, STMicroelectronics, Texas Instruments, Cisco, Itron, Landis & Gyr, Nexans, Sagemcom and Trialog. In December 2012, the G3-PLC standard was published by the ITU as Recommendation ITU-T G.9903: "Narrowband orthogonal frequency division multiplexing power line communication transceivers for G3-PLC networks" and is available at [18]. In addition, G3-PLC control parameters determining spectral content, PSD mask requirements and related tools to support the reduction of the transmit PSD have also been published in Recommendation ITU-T G.9901 [16].

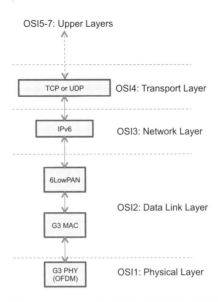

Figure 9.2   OSI reference model of ITU-T G.9903 G3-PLC standard.

The G3-PLC standard defines PHY, MAC and adaptation layers to enable IP-based data communication over the low and medium-voltage electrical grids. Figure 9.2 shows the communication layers and the scope of the G3-PLC standard based on the ITU-T G.9903 Recommendation.

One of the main characteristics of G3-PLC PHY layer is the additional features such as robust mode, adaptive tone mapping, and two-dimensional interleaving to cover scenarios with more severe noise impairments. G3-PLC also addressed LV to MV and MV to LV communication. Additionally, IEEE 802.15.4-based MAC layer is used above the OFDM based PHY layer. IPv6 packets are transmitted over the power line channel using the 6LoWPAN adaptation layer. The layer 2.5 mesh routing protocol is provided to determine the best path between the remote network nodes. Parameters of the ITU-T G.9903 G3-PLC standard are listed in Table 9.3.

9.2.2.1   Physical Layer

The G3-PLC standard applies OFDM based PHY layer to efficiently utilize the limited bandwidth channels. The standard may operate in the CENELEC A-band (35.938 kHz to 90.625 kHz), the FCC-1 band (154.6875 kHz to 487.5 kHz), the optional FCC-1a band (154.687 kHz to 262.5 kHz), the optional FCC-1.b band (304.687 kHz to 487.5 kHz) and the ARIB band (154.7 kHz to 403.1 kHz). Depending on the used frequency, the maximum number of carriers at the transmitter can be selected as NCARR = 128. This results in a minimum IFFT size of 256. Operating on the CENELEC Band, the OFDM sampling frequency is given as 0.4 MHz, [16]. Hence, the frequency spacing between OFDM carriers equal to 400 kHz/256 = 1.5625 kHz for CENELEC and 1.2 MHz/256 = 4.6875 kHz for the FCC band-plans, where 1.2 MHz is the sampling frequency operating on the FCC bands. The number

Table 9.3   Major technical features of the ITU-T G.9903 G3-PLC standard

Frequency bands	CENELEC A (35.938 kHz to 90.625 kHz), FCC-1 (154.6875 kHz to 487.5 kHz), FCC-1a (154.687 kHz to 262.5 kHz), FCC-1.b (304.687 kHz to 487.5 kHz) and ARIB (154.7 kHz to 403.1 kHz)
Coding/Modulation	OFDM using DBPSK, DQPSK or D8PSK Modulation, optionally BPSK, QPSK, 8-PSK and 16-QAM
Maximum Data Rate	Up to 300 kbps depending on the modulation and the frequency band
Data Link Layer	IEEE 802.15.4 MAC Frame Format/the adaptation sublayer based on IETF RFC 4944.
Channel Access	Carrier sense multiple access with collision avoidance (CSMA/CA) mechanism with a random back-off time.
Convergence Layer	IPv6 6LoWPAN
Network Topology	Mesh Network Routing based on LOADng
Network Formation	Mesh routing protocol
Repeating	Repeater mode available
Security	EAP-PSK, AES-128 key and CCM* encryption

of carriers and their locations in the bandplans are specified by the tone mask, which offers multiple possibilities depending on the application. An example is the tone mask index 0 in the FCC-above-CENELEC band for automotive applications, which defines 72 carriers in the frequency band between 154.6875 kHz and 487.5 kHz.

An overview of the PHY layer transmitter is shown in Figure 9.3. Forward error correction (FEC) makes use of Reed-Solomon (RS) and convolutional coding, which provide good performance against impulsive and burst-type errors, respectively. Three different modes are defined in the standard and illustrated in Table 9.4. For the low quality links, the Super-ROBO is mainly selected.

A two dimensional interleaving scheme is used to avoid the occurrence of burst errors which may be caused by time and frequency dependent noises. G3-PLC supports different modulation constellations, such as DBPSK, DQPSK, D8PSK and optional coherent modulation techniques. Based on the quality of the received signal, the receiver feeds back the type of modulation scheme to be applied at the transmitter. Hence, each carrier can be differentially or coherently modulated up to 4 coded bits per carrier. The standard can support a maximum PHY data rate of up to 300 kbps, see Table 9.5 for examples. Furthermore, the subcarriers with low SNR may be switched off to increase the reliability of the overall transmission.

9.2.2.2   MAC Layer

The G3-PLC network topology is constructed in a way that every node communicates with the network controller (or coordinator). G3-PLC supports meshed network structure. As a routing protocol, the Lightweight On-demand Ad hoc Distance-vector Routing Protocol — Next Generation (LOADng) [21] is provided at layer 2 (L2). The reasons for this choice are the need to handle looped low voltage grids, to exploit multiple routes to the network coordinator, low memory requirements and to avoid any intellectual property issues. In the case of L2 routing based on LOADng, each MAC packet is forwarded hop-by-hop to the

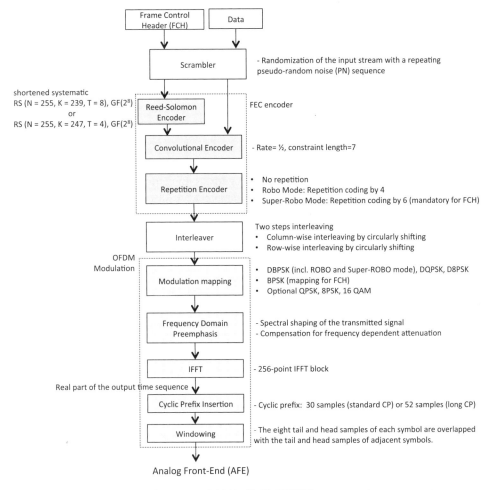

Figure 9.3   ITU-T G.9903 G3-PLC PHY layer transmitter.

destination. Thus, each node in the network does not necessarily reach the destination directly, but instead may use hopping to transfer its message. As illustrated in Figure 9.4, each node in the network is equipped with a neighbor table and a routing table. The neighbor table of a node contains the short MAC addresses of the neighbor nodes and the related PHY

Table 9.4   The modes of operation in ITU-T G.9903 G3-PLC PHY layer

	RS in $GF(2^8)$	Convolutional Encoder	Repetition Code
Normal Mode	$N = 255, K = 239, T = 8$	$R = 1/2, k = 7$	—
Robo Mode	$N = 255, K = 247, T = 4$	$R = 1/2, k = 7$	by 4
Super-Robo Mode	$N = 255, K = 247, T = 4$	$R = 1/2, k = 7$	by 6

Table 9.5   Maximum data rates of ITU-T G.9903 G3-PLC standard at PHY layer

Frequency Band	Robo (bps)	DBPSK (bps)	DQPSK (bps)	D8PSK (bps)	Max D8PSK (bps)
CENELEC A (36 kHz to 91 kHz)	4 500	14 640	29 285	43 928	46 044
FCC (150 kHz to 487.5 kHz)	21 000	62 287	124 575	186 683	234 321
FCC (10 kHz to 487.5 kHz)	38 000	75 152	150 304	225 457	298 224

parameters used for the communication. These PHY parameters may be modulation type, tone map, link quality indicator, etc. Since PLC is frequently asymmetrical, meaning that the quality of the channel is different depending on the communication direction, the PHY parameters are also dependent on the direction of the communication. The channel estimation based on tone map request/response is provided in the G3-PLC standard to measure the channel quality. The neighbor table is updated with the reception of either a packet from a neighbor node or a tone map response. Furthermore, the routing table of a node contains the short MAC addresses of the next node to be hopped. When a node receives a MAC packet, the destination node will be searched in its routing table. If the destination is already available in the table, the packet will be forwarded to the next hop of the route. If not, a route discovery mechanism will be initialized to find the best route to the destination. The LOADng protocol aims to find and maintain a bi-directional route to any destination in the network. If a connection is detected as unidirectional, it will be added to a Blacklisted Neighbor Set to avoid the selection of the same unidirectional link repeatedly.

Channel access in G3-PLC is accomplished by using the CSMA/CA mechanism with a random back-off time. It applies the basic idea of listen-before-talk. The CSMA/CA generates a random back-off time, an integer value corresponding to the number of time slots. If a node with a frame to transmit initially senses a busy channel, it waits until the channel becomes

Figure 9.4   Example of neighbor and routing tables for ITU-T G.9903 G3-PLC MAC layer.

idle. If the channel becomes idle, the node decrements its back-off timer until the channel becomes busy again or the timer reached zero. If the channel becomes busy before the timer reaches zero, then the node freezes its timer. When the timer is finally decremented to zero, the node transmits its frame. If two or more nodes decrement their time at the same time, a collision occurs, and each node has to regenerate a new back-off time. Prioritized access to the channel is also granted in the standard for the applications, in which an urgent message shall be delivered as soon as possible.

The G3-PLC MAC layer provides feedback to upper layers in the form of positive and negative acknowledgments (ACK or NACK) and also performs packet fragmentation and reassembly.

The network access control and authentication of the G3-PLC standard are provided in layer 2 with an AES symmetric encryption using a 128-bit shared secret, also known as pre-shared key or PSK. The keys are distributed via coordinator node. The authentication is granted based on the knowledge of the PSK at the other party. The confidentiality and integrity service is also provided in layer 2 by means of CCM* type of ciphering, which is a minor variation of the CCM (cypher counter mode) of encryption. The MAC frames are CCM encrypted and decrypted at every hop providing confidentiality and integrity.

9.2.2.3 Adaptation Layer

In the G3-PLC standard, the IPv6 over Low power Wireless Personal Area Network (6LoW-PAN) has been chosen to adapt IPv6, an internet network layer, into the power line communication. 6LoWPAN integrates the layer 2 routing, header compression, fragmentation and security and provides a network protocol usage to transport protocols such as TCP, UDP and ICMPv6.

9.2.2.4 Coexistence with other PLC Networks

Three mechanisms are provided in the G3-PLC standard to allow coexistence between ITU-T G.9903 devices and other PLC technologies operating in the same frequency range.

- Frequency separation: This mechanism avoids frequencies used by other networks by the use of ITU-T G.9903 bandplans and by limiting the out-of-band signal level.
- Frequency notching: This mechanism allows to notch out one or more subcarriers in order to avoid interferences for the other networks operating in the same frequency band.
- Preamble-based coexistence mechanism: This mechanism allows sharing the same channel with the other PLC technologies, which are using the same frequency band and supports this coexistence mechanism. This mechanism transmits a sequence of neutral coexistence preamble symbols at a specific frequency, or multiples of a specific frequency, depending on the bandplan. The coexistence preamble symbols use only carriers on the frequencies generated from the following equation: Frequency(Preamble) = 1.5625 kHz × 6n, where $n = 0$ through 54. Depending upon the bandplan over which the PLC devices are operating, the subset of the carrier indices corresponding to the frequencies within operating bandplan shall be used to generate the coexistence preamble symbol. For example, for transmissions using the FCC-above-CENELEC band, only carriers in the frequency range of this band shall be used.

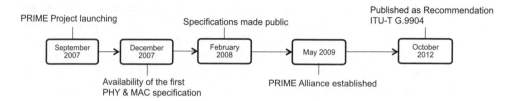

Figure 9.5   ITU-T G.9904 PRIME standard development timeline.

9.2.3   *ITU-T G.9904 PRIME Standard*

The PRIME (PoweRline Intelligent Metering Evolution) specification was developed by PRIME Alliance to provide an OFDM based narrowband PLC standard over the CENELEC A-band. Figure 9.5 illustrates the timeline of the PRIME Project. Formed in 2009 under the leadership of Spanish utility Iberdrola, the PRIME Alliance gathered many industry partners together with an initial objective to establish an open, public and non-proprietary PLC specification to support cost-effective smart metering functionalities in Spain. The founding members are Iberdrola, STMicroelectronics, Texas Instruments, Landis+Gyr, Itron, Current Group, Ziv Group and Advanced Digital Design. The main application area is the communication between the meters in the low-voltage power line networks and the base node, which is usually inside the data concentrator installed in the transforming station/secondary substation. At the transforming substation, the data concentrator communicates with the metering data management (MDM) system using other communication technologies. In fact, the PRIME specification today is being deployed in the field by other utilities such as Gas Natural Fenosa Spain, EDP Portugal and ENERGA Poland.

In October 2012, the PRIME standard was published by the ITU as Recommendation ITU-T G.9904: "Narrowband orthogonal frequency division multiplexing power line communication transceivers for PRIME networks" and is available from ITU's website [19]. The standard contains the specifications of PHY layer and DLL, including Convergence Layers for IEC61334-4-32, IPv4 and IPv6 profiles.

In addition, PRIME control parameters determining spectral content, PSD mask requirements and related tools to support the reduction of the transmit PSD have also been published in Recommendation ITU-T G.9901 [16].

Figure 9.6 shows the communication layers and the scope of the ITU-T G.9904 Recommendation. The ITU-T G.9904 PHY layer is based on OFDM data transmission only in CENELEC A-band and provides a maximum data rate of 128.6 kbps. It should be noted that a new release of PRIME specification PRIME version 1.4 was published in October 2014 [22], that includes changes at the PHY and MAC level bringing improvements such as increased robustness and throughput, band expansion, bandplan flexibility, whilst ensuring backwards compatibility. The specification details given in the next sections pertain to the ITU-T G.9904 recommendations [19] which was approved in October 2012.

The MAC layer defines a subnetwork as a group of service nodes managed by a base node and uses Carrier Sense Multiple Access (CSMA/CA) together with Time Division Multiplex (TDM) for the channel access. The convergence layer connects the upper layers into the PRIME MAC layer performing traffic mapping and data compression.

An overview of the ITU-T G.9904 PRIME Standard is listed in Table 9.6.

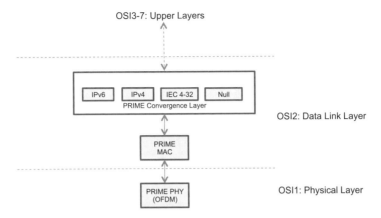

Figure 9.6   OSI reference model of ITU-T G.9904 PRIME standard.

9.2.3.1   Physical Layer

The PRIME PHY layer is based on OFDM and operates in the CENELEC A-band, which is defined from 3 kHz to 95 kHz. An overview of the PHY layer transmitter is shown in Figure 9.7. OFDM is used as a multicarrier transmission technique, which divides the available spectrum into many subcarriers, with the first subcarrier centered at 42 kHz and the last subcarrier centered at 89 kHz. The sampling frequency is chosen as $F_s = 250$ kHz with an FFT size of 512. Hence, this results in a frequency spacing between OFDM carriers equal to 488 Hz. The number of subcarriers used for data is 96 and an additional one pilot subcarrier. A convolutional encoder can optionally be activated to increase the reliability of the communication or deactivated to enable higher data rates. The optional convolutional encoder is followed by the mandatory scrambler, which randomizes the bit stream to reduce the peak values at the output of the IFFT. A block interleaver is applied afterwards to randomize the occurrence of bit errors prior to decoding. The standard considers the differential modulation schemes to detect the changes in the phase of the received signal. Three possible constellations are provided, namely DBPSK, DQPSK, and D8PSK. Thus, PHY data rates of up to 130 kbps PHY are reached, see Table 9.7. At the output of the IFFT, the OFDM symbol is cyclically extended by 48 samples to create the cyclic prefix.

Table 9.6   Major technical features of the ITU-T G.9904 PRIME Standard

Frequency bands	42–89 kHz in CENELEC A Band
Coding/Modulation	OFDM using DBPSK, DQPSK or D8PSK Modulation
Maximum Data Rate	Up to 128.6 kbps in CENELEC A-band
Channel Access	Carrier Sense Multiple Access (CSMA/CA) and Time Division Multiplex (TDM)
Convergence Layer	IEC-432, IPv4, IPv6, NULL
Network Topology	Tree structure
Network Formation	Connection oriented, can be logically seen as a tree structure. A base node manages the network resources and connections.
Security	128 bit AES encryption of data and its associated cyclic redundancy check (CRC).

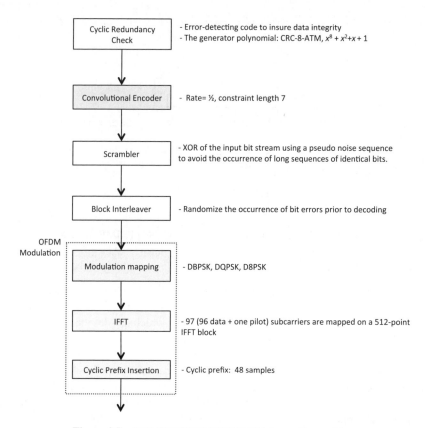

Figure 9.7   ITU-T G.9904 PRIME PHY Layer Transmitter.

9.2.3.2   MAC Layer

A PRIME network is composed of two device types, Base Nodes and Service Nodes, as illustrated in Figure 9.8.

- The Base Node is at the root of the tree and acts as master node that provides connectivity to the subnetwork. There is a single base node for each subnetwork identified by its universal 48 bit MAC address (the EUI-48; IEEE Std 802-2001 [23]). Since the base node may not

Table 9.7   Theoretical maximum data rates of ITU-T G.9904 PRIME standard at PHY layer

	DBPSK	DQPSK	D8PSK
Convolutional Code On	21.4 kbps	42.9 kbps	64.3 kbps
Convolutional Code Off	42.9 kbps	85.7 kbps	128.6 kbps

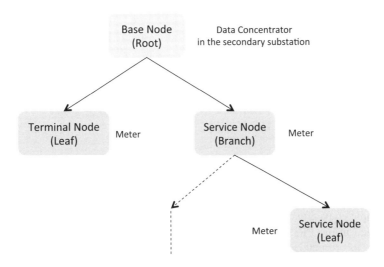

Figure 9.8   Example of an ITU-T G.9904 PRIME subnetwork.

have a direct communication with other nodes in the subnetwork, the data are forwarded via switching nodes.

- Any other node of the subnetwork is a Service Node. Service Nodes are dynamically configured by the Base Node either as leafs of the tree or branch points of the tree.
  - Since the nodes in the network do not necessarily have a direct connection with the master node, the switching nodes (also referred to as branch nodes) are configured by the Base Node to relay data to/from the devices in their domain from/to the base node. The forwarding or neighbor table in each node is maintained by beacon signals sent from switches.
  - Terminal or leaf nodes will not relay messages, and are available for the Base Node to promote them to switches if needed.

Each subnetwork is identified by its base node's 48-bit universal MAC address, which is assigned during the manufacturing process.

The MAC layer specified in PRIME defines a channel access method using both CSMA/CA and TDM. A MAC frame is a time unit of 618 ms continuously broadcasted, which is composed of starting beacon signals, one shared-contention period (SCP) and the optional contention free period (CFP), see Figure 9.9. The SCP defines the time slots, during which the base and service nodes should attempt to transmit using CSMA/CA. The CSMA/CA avoids collisions resulting from simultaneous attempts to access the channel. The CFP is granted by the base node for use by nodes requiring deterministic network access or guaranteed bandwidth.

The PRIME MAC provides an optional Automatic Repeat reQuest (ARQ) functionality to achieve a reliable data transmission.

Two profiles of security are provided in the PRIME standard. The first profile leaves security to the upper layers, whereas the second profile is based on 128 bit AES encryption.

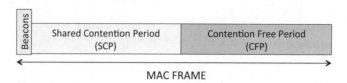

Figure 9.9   MAC frame format in the ITU-T G.9904 PRIME standard.

9.2.3.3   Convergence Layer

The Convergence Layer functions as an adaptation layer between the underlying MAC layer and the upper protocol layers. Its main task is to receive the traffic from upper supported layers and to encapsulate them into MAC Service Data Units (SDUs). The PRIME specification defines multiple convergence layers for transmitting. IPv4 and IPv6 based connections have been supported in the PRIME standard, whereas DLMS/COSEM applications can also be provided over its original LLC layer protocol IEC 61334-4-32, which is based on IEEE 802.2. The Null support in the convergence layer addresses the applications that do not need any special convergence capability, e.g. firmware upgrade frames.

9.2.4   The IEEE 1901.2 Standard

The IEEE 1901.2 Working Group was established in 2010 with the goal of developing a standard for low-frequency (< 500 kHz) narrowband PLC over indoor and outdoor electrical wiring [20]. The standard was approved by the IEEE Standards Association (IEEE-SA) in November 2013 and published in December 2013 as an active standard. Figure 9.10 illustrates the timeline of the process of developing IEEE 1901.2 Standard. The standard defines an FFT-OFDM-based PHY/MAC specification and the link layer security requirements. This standard addresses grid to utility meter, grid automation, electric vehicle to charging station, and within home area networking communications scenarios. Lighting and solar panel PLC are also potential uses of this communications standard. Additionally, the coexistence mechanism with the other classes of low-frequency narrowband and the broadband PLC devices and the electromagnetic compatibility (EMC) requirements are also defined in the standard.

Since the G3-PLC standard was used as the basis for the IEEE 1901.2 PLC standard, there are strong technical similarities between G3-PLC ITU-T G.9903 and IEEE 1901.2. The main characteristics of IEEE 1901.2 standard, however, is in the MAC technology which supports both mesh-under and route-over routing approaches and is capable of supporting a wider range of IPv6 functionalities. An overview of the IEEE 1901.2 standard is given in Table 9.8.

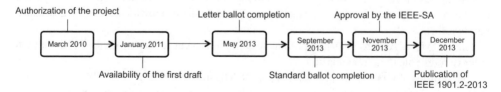

Figure 9.10   IEEE 1901.2 Working Group standard development timeline.

Table 9.8   Major technical features of the IEEE 1901.2 standard

Frequency bands	CENELEC A-band (35–91 kHz), CENELEC B-band (98–122 kHz), FCC-above-CENELEC (155–488 kHz), FCC (10–487.5 kHz) and ARIB (155–403 kHz)
Coding/Modulation	OFDM using DBPSK, DQPSK or D8PSK Modulation, optionally BPSK, QPSK, 8-PSK and 16-QAM
Maximum Data Rate	Up to 300 kbps depending on the modulation and the frequency band
Data Link Layer	IEEE 802.15.4 MAC Frame Format/the adaptation sublayer based on IETF RFC 4944.
Channel Access	Carrier sense multiple access with collision avoidance (CSMA/CA) mechanism with a random back-off time.
Convergence Layer	IPv6 6LoWPAN
Network Topology	Star, tree, or mesh
Routing	Either Layer 2 LOADng or Layer 3 RPL protocol
Repeating	Repeater mode available
Security	Based on 802.15.4, use AES-128 in CCM mode.

9.2.4.1   Frequency Band Usage and Coexistence

The IEEE 1901.2 defines operation in several frequency regions, referred to as bandplans. For each particular bandplan, the start and the stop center frequencies are shown in Figure 9.11.

9.2.4.2   Physical Layer

The IEEE 1901.2 physical layer is based on OFDM using advanced channel coding techniques. Depending on the used frequency, the minimum sampling frequency and the maximum number of carriers at the transmitter can be selected to be $F_s = 1.2$ MHz and NCARR = 128,

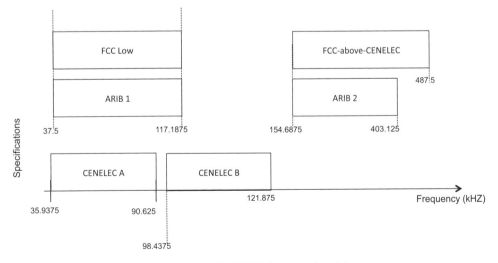

Figure 9.11   IEEE 1901.2 frequency bandplans.

respectively. This results in a minimum IFFT size of 256 with frequency spacing between OFDM carriers equal to $1.2\,\text{MHz}/256 = 4.6875$ kHz. Each carrier can be differentially or coherently modulated with up to 4 coded bits per carrier. The standard can support a maximum PHY data rate of up to 500 kbps. The number of carriers and their locations in the bandplans are specified by the tone mask, which offers multiple possibilities depending on the application. Since the IEEE 1901.2 PHY layer signal processing is similar to that for ITU-T G.9903 G3-PLC, we refer to Section 9.2.2.1 for more information on the technical specifications of the PHY layer.

9.2.4.3 MAC Layer

The IEEE 1901.2 MAC layer uses services and primitives from IEEE Std 802.15.4-2006. The network model can be simplified to two network devices as follows.

- Personal area network (PAN) coordinator, which is the coordinator of the network and can be seen as a master node.
- IEEE 1901.2 device, which contains an implementation of the IEEE 1901.2-2013 MAC and a physical interface to the physical medium.

The network can be built as tree, mesh or star. However, every network needs at least one IEEE 1901.2 device to work as the coordinator of the network. Frames of up to 1280 octets are the basic unit of data, providing different types, such as data, acknowledgment, beacon or MAC command frames. The physical medium is accessed through the CSMA/CA protocol, supporting:

- Unslotted version of the algorithm for the mandatory non-beacon PAN, in which the PAN does not regularly emit beacons.
- Slotted version of the algorithm for the optional beacon-enabled PAN.
- Prioritized access to the channel.

The main characteristics of IEEE 1901.2 standard, different from the ITU-T G.9903 G3-PLC, is the agnostic structure of the MAC layer which supports both mesh layer 2 and route-over routing layer 3 approaches. As a mesh routing protocol, LOADng is provided at layer 2, which is described in Section 9.2.2.2. In case of layer 3 routing, the Routing Protocol for Low Power and Lossy Networks (RPL) specification is supported. The RPL protocol is defined in the Routing Over Low Power and Lossy (ROLL) working group of The Internet Engineering Task Force (IETF) and provides an IPv6 based routing solution for IP smart object networks. When using RPL, each physical hop within the network is an IPv6 hop. More information on the RPL can be found in IETF RFC 6550 [24]. IEEE 1901.2 adopted 6LoWPAN to adapt IPv6 into the power line communication.

The network access control and authentication of the IEEE 1901.2 standard are provided in layer 2 with an AES symmetric encryption using a 128-bit shared secret, as described in Section 9.2.2.2.

The IEEE 1901.2 standard also contains coexistence mechanisms as discussed in Section 9.2.2.4.

9.2.5   HomePlug Green PHY Specification

The Green PHY (GP) specification was developed by the HomePlug alliance in 2010 as a subset of HomePlug AV to support smart grid applications such as plug-in electric vehicles, smart home automation and smart meters within the customer's premises. Since a high data rate broadband PLC connection is not needed for such applications, the HomePlug GP specification is designed to meet requirements such as low power consumption and cost and reliable communication. Furthermore, the HomePlug GP was adopted in ISO 15118 as a Europe-wide standard for communication between charging stations and electric vehicles.

HomePlug GP uses the same frequency band (2 MHz to 30 MHz), basic modulation scheme (OFDM), and FEC (Turbo Codes) as defined in HomePlug AV. However, the major difference between the two HomePlug standards lies in the peak PHY rate. HomePlug GP supports a peak PHY rate of 10 Mbps with a restriction of OFDM subcarrier modulation exclusively to QPSK and supporting only ROBO mode, therefore eliminating the need for adaptive bit loading and management of tone maps. These restrictions enable low-cost and low-power devices, which can also interoperate with HomePlug AV/IEEE 1901 devices.

9.3   Regulation

In this section we discuss regulations pertinent for PLC for smart grid in different parts of the world. We focus on the frequency band below 500 kHz and the USA, Europe and Japan. For the MHz range, it should be noted that the Code of Federal Regulations, Title 47, Part 15, from the U.S. Federal Communications Commission (FCC) devotes all of Subpart G to define very specific equipment authorization, marketing and administrative guidelines, radiated emission limits for MV and LV power lines, and notching requirements for Access (i.e. other than In-House) BPL between 1.705 MHz and 80 MHz. In the case of Europe, the main regulation to which broadband PLC equipment must comply is Directive 2004/108/EC on Electromagnetic Compatibility, to be replaced by Directive 2014/30/EU. We recommend that the content in this section is considered together with the information on EMC in Chapter 3.

9.3.1   USA

Narrowband PLC systems operating in the frequency range of 9–490 kHz as discussed in the previous section are governed by regulations in the Code of Federal Regulations, Title 47, Part 15, from the U.S. FCC [25]. A number of definitions in this Part 15 are interesting and relevant in this regard.

- An 'incidental radiator' is defined in FCC part 15.3(n) [25] as *a device that generates radio frequency energy during the course of its operation although the device is not intentionally designed to generate or emit radio frequency energy*. Examples of incidental radiators are dc motors, mechanical light switches, etc.
- An 'intentional radiator' is defined in FCC part 15.3(o) [25] as *a device that intentionally generates and emits radio frequency energy by radiation or induction*. Examples for intentional radiators are thus WLAN, Bluetooth, cell phone, Zigbee transceivers. These devices intentionally transmit RF energy through the antenna.

- An 'unintentional radiator' is defined in FCC part 15.3(z) [25] as *A device that intentionally radiates radio frequency energy for use within the device, or that sends radio frequency signals by conduction to associated equipment via connected wiring, but which is not intended to emit RF energy by radiation or induction.* Examples of unintentional radiators would be wired communication systems like Ethernet, DSL, cable modems, etc., as also PLC systems.
- Section 15.3 clause (t) of [25] specifies a 'power line carrier system' as *An unintentional radiator employed as a carrier current system used by an electric power utility entity on transmission lines for protective relaying, telemetry, etc. for general supervision of the power system. The system operates by the transmission of radio frequency energy by conduction over the electric power transmission lines of the system. The system does not include those electric lines which connect the distribution substation to the customer or house wiring.*
- Section 15.3 clause (f) of [25] defines a 'carrier current system' to be *A system, or part of a system, that transmits radio frequency energy by conduction over the electric power lines. A carrier current system can be designed such that the signals are received by conduction directly from connection to the electric power lines (unintentional radiator) or the signals are received over-the-air due to radiation of the radio frequency signals from the electric power lines (intentional radiator).*

Section 15.113 of FCC part 15 lays out that power utilities can operate power line carrier systems within the 9 kHz to 490 kHz band on an 'unprotected, non-interference basis'. It is further specified that this section does not permit operation on power lines connecting customers to distribution substations. Thus, for PLC use cases in the LV segment of the grid, the clauses regulating carrier current systems would apply.

Section 15.107 clause (c) of FCC part 15 specifies that carrier current systems are not subject to conducted limits. However the systems are subject to the following requirements:

(a) Clause (c)(1) of 15.107 in [25] specifies that all carrier current systems with fundamental emissions in 535–1705 kHz AM frequency range and intended to be received by AM receivers have no limit on conducted emissions. However all other carrier current systems have to meet a conducted emissions requirement of 1000 μV within the frequency band 535–1705 kHz, as measured using a 50 μH/50 Ω LISN.

    The latter condition would apply to the considered narrowband PLC systems. Thus, while FCC does not impose a conducted emission limit for narrowband PLC systems for in-band transmission, it does limit out-of-band conducted emissions. This in turn governs the in-band transmission limit, depending on the filtering that can be achieved in the narrowband PLC systems on the transmit side.

(b) Clause (c)(3) of 15.107 in [25] specifies that carrier current systems transmitting below 30 MHz are subject to radiated emissions in section 15.109(e). Section 15.109(e) specifies that carrier current systems operating in the 9 kHz to 30 MHz range shall comply with the emission limits of intentional radiators in section 15.209.

    Section 15.209 clause (a) gives the radiated limits as shown in Table 9.9.

    Section 15.209(d) [25] specifies that 'The emission limits shown in the above table are based on measurements employing a CISPR quasi-peak detector except for the frequency bands 9–90 kHz, 110–490 kHz and above 1000 MHz. Radiated emission limits in these three bands are based on measurements employing an average detector'.

Table 9.9 Applicable radiated emission limits from Section 15.209 in [25]

Frequency (MHz)	Field Strength (microvolts/meter)	Measurement Distance (meters)
0.009–0.490	2400/F(kHz)	300
0.490–1.705	24000/F(kHz)	30
1.705–30.0	30	30
30–88	100	3
88–216	150	3
216–960	200	3
Above 960	500	3

9.3.2 Europe

Directives from the European Commission (EC) define the EU regulatory structure, while official (*de jure*) standardization organizations such as ETSI, CEN and CENELEC describe the specific limits. As an EU rule, PLC equipment is classified as telecommunication equipment and is controlled in the member countries of EU under the framework of the EMC Directive 2004/108/EC. The legal structure and the relevant rules are laid out in the EMC Directive, where it differentiates between installations and consumer products. Both market access and interference complaint handling are covered by the EMC Directive. The primary goal of the EMC Directive is to ensure the unrestricted movement of apparatus, installations and networks within the EU, and to preserve a suitable EMC environment. The secondary goal on interference is complied with by guaranteeing that electromagnetic disturbances emitted by an apparatus does not interfere with the functioning of another apparatus as well as radio and telecommunication networks, related equipment and electricity distribution networks, but also by ensuring that the apparatus has a sufficient level of inherent immunity to permit it to operate as envisioned in the presence of electromagnetic disturbances. Therefore, emission and immunity requirements are an essential part of the protection requirements. When products comply with the EMC regulation, market access is approved. For this purpose, product specific compliant tests are necessary. Directive 2004/108/EC is a so called 'New Approach Directive' requiring manufacturers' self-declaration. According to this directive a manufacturer is permitted to evaluate the EMC of its products by:

(a) compliance with a European Harmonized Standard, which is EN 50065-1 for PLC in the spectrum from 3 kHz to 148.5 kHz, or
(b) carrying out an EMC assessment centered on manufacturers' own procedures and approaches.

The certification of these procedures and methods by an independent notified body (as defined by the EMC Directive) are normally included in option (b). These procedures and methods are usually the result of discussions in the relevant standardization and compliance platforms. Please also refer to Section 3.6 for further details.

Irrespective of which option has been selected, the manufacturer is compelled to provide a statement of conformity declaring the compliance of its products with the protection

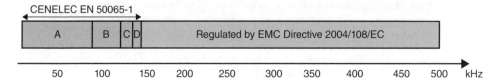

Figure 9.12 Regulation and standardization of the PLC in the frequency range up to 500 kHz.

requirements based on the EMC Directive. The 'Conformité Européenne' (CE) marking is proof of product compliance with all appropriate directives (e.g. safety, etc.).

Currently option (b) remains a valid choice, although one day a European harmonized standard should be available for the 150–500 kHz PLC band, see Figure 9.12. Should a European harmonized standard become available, a manufacturer will have to provide an explanation if there is a preference to maintain internal procedures and methods. In such a scenario, a potential argument could be the existence of an installed embedded base without interference problems at that time. The argumentation process is only needed for new products. Certified products deployed do not need to be recertified.

9.3.2.1   Limits for PLC in the European Market

Due to the lack of certainty and reproducibility of the radiated measurement from a power line, the measurement of conducted emissions is established in EMC product standards. Conducted emission measurement is made by connecting and operating the Equipment Under Test (EUT) at a defined network (see Chapter 3). The IEEE 1901.2 standard defines the limits of conducted disturbances of the PLC port of the EUT as listed in Table 9.10 for devices use in Europe.

The methods of measurement for PLC devices in the frequency range 3–148.5 kHz and in the frequency range 150–500 kHz are described in Sections 9.3.2.2 and 9.3.2.3, respectively.

9.3.2.2   Measurement Methods for PLC Devices Operating in the Frequency Range 3–148.5 kHz

Section 3.6.3.1.1 and Table 3.3 give the EMC regulation limits for PLC frequencies of 9–148.5 kHz. In this section, we explain the methods of measurement according to the EN 50065-1 standard in detail. Important features are:

- the determination of the bandwidth of the signal,
- the allowed maximum output voltage for this signal over the whole bandwidth,
- out of band emission requirements.

9.3.2.2.1   Determination of Signal Bandwidth

The output signal spectrum is determined by the use of a spectrum analyzer having a peak detector and a 100 Hz resolution bandwidth. The transmitter shall operate in such a way that the bandwidth and output signal magnitude have the greatest values permitted by the manufacturer's specification. The spectral width (B in Hz) is defined by the length of the interval where all the frequency lines are less than 20 dB below the maximum spectral

Table 9.10 Conducted disturbance limits for PLC usage and out-of-band per frequency range

Frequency Range	Limits During PLC Transmission	Out-of-band Disturbance Limits
3 kHz to 9 kHz	CENELEC EN 50065-1 2011, clauses 6.3.1.1; 6.3.2.1	CENELEC EN 50065-1 2011, clause 7.2.1
9 kHz to 95 kHz 'CENELEC A'	CENELEC EN 50065-1 2011, clauses 6.3.1.2; 6.3.2.2	CENELEC EN 50065-1 2011, clause 7.2.2
95 kHz to 125 kHz 'CENELEC B'	CENELEC EN 50065-1 2011, clauses 6.3.1.3 6.3.2.3	
125 kHz to 140 kHz 'CENELEC C'		
140 kHz to 148.5 kHz 'CENELEC D'		
150 kHz to 500 kHz	Class A: −22 dBm/Hz Quasipeak, −35 dBm/Hz Average, Class B (−35)–(−45) dBm/Hz Quasipeak, (−45)–(−55) dBm/Hz Average	CISPR 22: 2008, Tables 1 & 2
>500 kHz to 30 MHz 'BPL'	Out of scope	
30 MHz to 1000 MHz	Out of scope	CISPR 22: 2008, Tables 5 & 6 (radiated)

line (see Figure 9.13). Further, the signal shall be considered as a narrowband signal if its bandwidth is less than 5 kHz and as a wide-band signal if the bandwidth is equal to or greater than 5 kHz. For example, a PRIME signal transmitting in the 40–90 kHz band would be a wide-band signal.

9.3.2.2.2 *Allowed Maximum Output Voltage*
The measurement of the signal is done using a line impedance stabilization network (LISN) as in Figure 4 of EN 50065-1 [26], given in Figure 9.14, where the artificial mains network

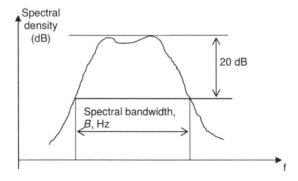

Figure 9.13 Measurement of spectral bandwidth.

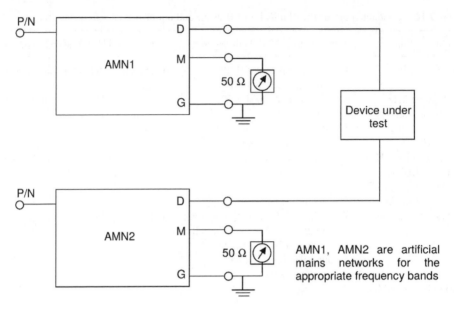

Figure 9.14 Line impedance stabilization network (LISN) for measurement of the maximum output voltage. From [26, Figure 4].

(AMN) in Figure 9.14 is specified in Figure 5 of EN 50065-1 and shown in Figure 9.15. Further as specified in Section 6.3.1.2 of [26], measurement is done as per CISPR 16-1:1993. As given in Figure 7 of [26] for frequencies from 9 kHz to 150 kHz a measurement bandwidth of 200 Hz will be used. For frequencies from 150 kHz to 30 MHz, a measurement bandwidth of 9 kHz will be used. The measurement given in (a) and (b) below is rms-average measurement. The allowed maximum output voltage is specified by clause 6.3.1.2 of the EN 50065-1 document as follows:

(a) Narrowband signal: The measured level shall not exceed 134 dBμV at 9 kHz decreasing linearly with the logarithm of frequency to 120 dBμV at 95 kHz.

(b) Wide-band signals: The measured level shall not exceed 134 dBμV.

In addition, when measured with a peak detector with 200 Hz bandwidth, no part of the spectrum of the signal shall exceed 120 dBμV.

9.3.2.2.3 Out-of-band Emissions

Section 7 of EN 50065-1 gives the requirements for the out-of-band (OOB) emissions. Note that the OOB emissions requirements hold good for outside the sub-bands (bands A, B, C, and D, see Table 3.3) in which the device is transmitting. The OOB emission limits are shown in Figure 9.16. Note the change in measurement bandwidth for the quasi-peak at 150 kHz to 9 kHz as against the measurement bandwidth of 200 Hz for frequencies below 150 kHz. For example, in the case of PRIME, which operates in the Band A of EN 50065-1, the OOB applies for frequencies above 95 kHz.

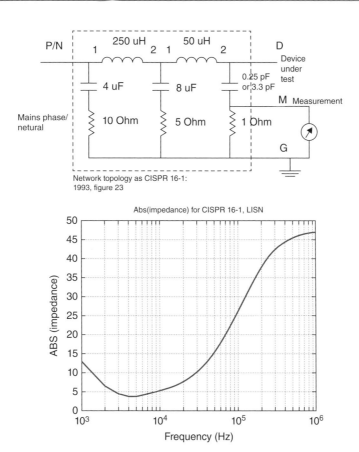

Figure 9.15 Circuitry for artificial mains network (AMN) for testing as per EN 50065-1 requirements and its impedance response as a function of the frequency (for a single branch). From [26, Figure 5].

9.3.2.3   Measurement Methods for PLC Devices Operating in the Frequency Range 150–500 kHz according to the IEEE 1901.2 Standard

The IEEE 1901.2 Standard [20] provides for four independent power level considerations for PLC access data transmission in the frequency range of 150 kHz to 500 kHz, see Table 9.10. For a PLC system using the 150 kHz to 500 kHz band for data transmission, the limits are measured as follows.

9.3.2.3.1   Out-of-band Conducted Disturbance Limits for 3 kHz to 148.5 kHz
The out-of-band disturbance limits are given in EN 50065-1 2011, clause 7.2.1, [26].

For the 3–9 kHz band, the PLC device is plugged in an AMN with the specified termination impedance 50 Ω ∥ (50 μH + 1.6 Ω). The measurement shall be made on each phase, and the resolution bandwidth of the measurement equipment should be set to 100 Hz. If the measured signal shows signal power near the limit line, the affected frequencies need to be observed for 15 seconds and the maximum level must be documented.

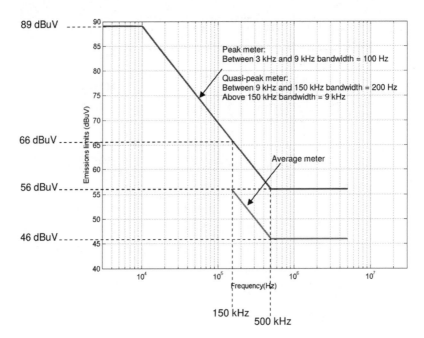

Figure 9.16   Out-of-band (OOB) emission limits in EN 50065-1. From [26, Figure 7].

For the 9–148.5 kHz band, the termination impedance is set to 50 Ω ‖ (50 μH + 5 Ω). The coupling unit has to provide the same functions as in the measurement of the frequency range 3–9 kHz, where the resolution bandwidth of the measurement equipment is set to 200 Hz. The frequencies of the power peaks near the limit have to be observed for 15 seconds and the maximum power has to be documented.

9.3.2.3.2   In-band Conducted Disturbance Limits for 150 kHz to 500 kHz

The in-band disturbance limits are given in the IEEE 1901.2 standard [20]. The measurement setup differs from the out-of-band disturbance measurements below 150 kHz. An Impedance Stabilization Network (ISN) is placed between the PLC device and the AMN. The ISN is described with the asymmetrical termination impedance 150 ± 20 Ω (EN 55022, [27]) and longitudinal conversion loss (LCL)-EUT side > 55 dB. The resolution bandwidth is set to 9 kHz. The frequencies near the limit have to be observed for 15 seconds, and the maximum power has to be documented.

9.3.2.3.3   Out-of-band Conducted Disturbance Limits for 500 kHz to 30 MHz

The out-of-band disturbance limits are given in EN 55022, [27]. The PLC device is plugged in an AMN with the specified termination impedance of 50 Ω ‖ 50 μH with a resolution bandwidth of 9 kHz. The frequencies near the limit have to be observed for 15 seconds and the maximum power has to be documented.

Figure 9.17   ARIB measurement setup for in-band measurement of the transmitted PLC signal.

9.3.3   Japan

The Association of Radio Industries and Businesses (ARIB) in Japan regulates and permits PLC transmission in the frequency band of 10–450 kHz. As with other regulations, there is a regulation for in-band transmission and a regulation for OOB emission. A combination of these governs the filtering requirements for PLC systems.

9.3.3.1   In-band Measurement Setup for ARIB

For the in-band signal transmission the measurement setup is shown in Figure 9.17.

The power measurement is done with a 10 kHz resolution bandwidth and sweeping over the transmit signal frequency range over a 10 Ω load. There are two different cases for the in-band transmission specification for ARIB:

- In case the signal transmission frequency falls in the 10–200 kHz band then the allowed transmission power is 30 mW/10 kHz RMS.
- In case the signal transmission frequency range exceeds 200 kHz then the allowed transmission power is 10 mW/10 kHz RMS.

9.3.3.2   Out-of-band Emission Requirement for ARIB

The measured OOB emissions are specified to be 56 dBμV in the frequency range of 450 kHz to 5 MHz and 60 dBμV in the range of 5 MHz to 30 MHz. The measurement done is quasi-peak with a resolution bandwidth of 9 kHz.

Let us consider the example of IEEE 1901.2 signal transmission using OFDM and transmitting up to 450 kHz. Assume the difference between the quasi-peak to rms of the OFDM signal to be ~7 dB. In this case, a 10 mW transmit power over a 10 Ω resistance implies a transmit voltage of 0.316 V. This implies an in-band quasi-level that would be measured by a CISPR measurement equipment in a resolution bandwidth of 9 kHz to be $20 \log_{10}(0.316 \text{ V}/10^{-6} \text{ }\mu\text{V}) + 7$ (rms to Quasi) $-$ 3 dB (positive/negative part of spectrum) $-$ 6 dB (CISPR measurement) = 108 dBμV. This implies an IEEE 1901.2 transmitter has to reject its in-band signal by $108 - 56 = 52$ dB. We note that this is taken as an example implementation for IEEE 1901.2 for the purposes of illustration. In the case of an individual manufacturer these rejection numbers may be different depending on the different trade-offs in their implementation.

9.4   Applications

Transmission and distribution networks are the focus of smart grid developments. Transmission usually includes HV lines and substations, an environment in which ICT already has a high degree of integration with the grid. Transmission assets are high cost investments and have usually been deployed concurrently with broadband communications (optical fiber, microwave radio, etc.) which along with local electronics allow for enhanced remote control and automation. Distribution has historically not been coupled with ICT. The degree of remote control and/or automation of MV grids varies largely among utilities, but it rarely amounts to a significant percentage of the assets. Thus, operation of MV remains mostly manual. In the case of LV grids, any level of automation is the exception. At most, some kind of (semi) automated, remote metering of LV assets (mainly meters) exists.

As already mentioned in the Introduction section, PLC is a technology with a long history and tradition in electricity companies. For decades there have been plenty of applications and different implementations used at utilities. The natural confluence of smart grid and PLC technologies is more recent. The place of PLC within the smart grid development has been discussed in [9], reckoning that PLC is *an excellent and mature technology that can support a wide variety of applications from the transmission side to the distribution side and also to and within the home*.

However, the applicability of PLC to the different services needed by utilities to configure an optimized and cost-effective solution depends mainly on the characteristics of the utility, the features and performance of each different PLC technology type, and the architecture to be deployed (whether broadband and/or narrowband PLC will be used, which segments of the grid will be covered with PLC (MV and/or LV), etc.).

9.4.1 PLC as a Telecommunication Backbone Technology

Primary and secondary substations are the main premises to be served in the modern concept of a smart grid. On the remote end of the grid, smart meters are also obvious service points. PLC may be used to transport aggregated data flows, expanding its use from a pure last-mile access technology, to a backbone transport technology in the access segment, mainly from the secondary substations (SSs) over MV power lines. This subsection covers the use of broadband PLC over MV power lines.

In the context of smart metering, SSs are the natural concentration or gateway points when PLC is used to reach meters through the LV grid. Even if meters communicate with wireless means (e.g. mesh radio, as in many US smart metering deployments using private communications [28]), substations could be used as locations to concentrate or perform gateway or routing functions over the data coming from the smart meters.

In the general context of smart grid the role of substations as communications nodes is even more evident, since data which needs to be exchanged with operation and control centers can use the power lines for communication purposes.

The role of PLC and its effective use as a backbone technology to communicate with substations for operational needs (e.g. distribution automation purposes), largely depends on several aspects that need to be considered in detail:

- Feasibility of the PLC signal coupling;
- Data-rate requirements;
- Communications resiliency;
- Network planning procedures.

9.4.1.1   Feasibility of Signal Coupling

In order to inject PLC signals at substations to be transmitted over MV power lines, the main challenge is to interface with line voltages in the order of kilovolts, from LV-powered PLC devices which usually work at MHz frequency range. The utility industry has extensive experience on this for over one hundred years.

Usually two kind of coupling methods are considered, see also Chapter 4.

- Capacitive couplers (see Figure 9.18): They are usually solid, resilient devices with known and predictable behavior. They are very flexible since they allow for installation in different setups, and allow for an effective coupling of broadband signals. A drain coil (which may physically exist as such, or it may actually be provided by the inductance of the primary matching/isolation transformer winding) makes for the device-side voltage to be almost zero at power frequency (50 or 60 Hz), so that the coupling capacitor is effectively connected to ground at power frequency.
- Inductive couplers: They have been shown to be a flexible, easy-to-install solution which allows for some PLC links to be established at relatively low cost. Their behavior however may vary along with line impedance, to the point of ceasing functioning when the breaker inside the switchgear in which it is installed is open. Because of that, usually smart grid applications (which need reliable functioning even in situations with open breakers and switches) employ capacitive couplers.

In the case of capacitive couplers, the PLC signal is usually injected directly between one phase and ground. This is usually a simpler process in older masonry and open-air (metal-clad and metal-enclosed) switchgears, in which the phase contacts are immediately accessible for the installer. Typical requirements for capacitive couplers are to present a 50 Ω BNC connector for the PLC device, to be sturdy and easy to install for long periods of unmaintained operation (e.g. 25 years), and to comply with utility and safety requirements such as external insulation (see Figure 9.19). Broadband capacitive couplers are usually assumed to be tested

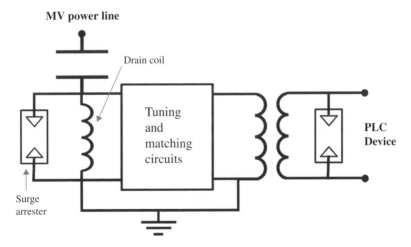

Figure 9.18   Capacitive coupler elements.

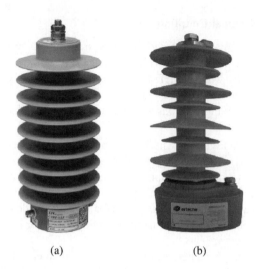

(a)   (b)

Figure 9.19 Capacitive couplers for masonry and open-air switchgear. Courtesy of ZIV (http://www.ziv.es) and Arteche (http://www.arteche.es).

against IEC 60358:1990 'Coupling capacitors and capacitor dividers' and IEC 60481:1974 'Coupling devices for power line carrier systems'. Coupling capacitors for broadband PLC are often optimized for HF (3–30 MHz), and the electrical capacitance ranges between 400 and 2000 pF (usually the larger the capacitance the more expensive the coupler and the better the performance). Surge arresters are usually required both for line-side and device side.

Nowadays capacitive couplers also need to be installed inside modern gas insulated switchgear (GIS) units. In these, the switchgear is isolated in a closed container filled with SF6 gas (sulphur hexafluoride), which safely contains arc discharges. Space availability is scarce and direct access to the MV cables limited. Thus the most usual method is to install the coupler directly into the standardized separable connector (e.g. according to EN 50181:2010 in Europe) which fixes the bushing to the cable (see Figure 9.20).

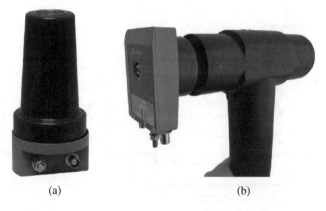

(a)   (b)

Figure 9.20 Capacitive couplers for Gas Insulated Switchgear. Courtesy of ZIV (http://www.ziv.es) and Arteche (http://www.arteche.es).

Requirements are similar for these couplers: 50 Ω BNC connector for the PLC device, to stand large periods of continuous operation and to comply with utility and safety requirements. Since the space is constrained, coupling capacitors are usually below 500 pF.

Inductive couplers must be carefully considered before usage for critical operation PLC communications, since they are prone to be affected by impedance variations when the MV grid changes (e.g. a breaker opens). They usually are clamped around one MV phase cable, grounded on their ground terminal and connected to a PLC device through a 50 Ω BNC connector.

It has to be understood that between different countries and utilities there may be large differences in the MV distribution grids. Age of assets varies greatly and deployment philosophies and technologies have historically been different across countries and even utilities in the same country. Thus, it is common that specific PLC coupling solutions have to be tailored to the type of MV assets each utility has installed, which usually includes underground and overhead networks, one phase and three phase cables with associated switchgear.

One interesting attempt at normalizing requirements for PLC couplers is IEEE 1675-2008 'Standard for Broadband over Power Line Hardware', specifically addressing the US market up to 38 kV. Among others it specifies altitude and temperature service conditions, grounding, external insulation, safety, voltage and electrical requirements. It defines type and/or routine tests for coupler insertion loss, partial discharge, 60 second withstand, lighting and flashover. Finally it describes mechanical strength tests and gives guidelines on product marking and installation procedures.

9.4.1.2 Data-rate Requirements

A telecommunications backbone application in a utility needs to carry grid related services.

The throughput required for smart metering scenarios is variable, and depends heavily on the number of smart meters per SS, the type of data being transferred, the application and communication layers being used, and the required response time (e.g. frequency of the meter readings). The summary in [29] includes a number of typical use cases, and the results are revealing. The number of different operations that can be performed in a normal smart metering system is higher than one would expect in a traditional AMR system. Furthermore, the number of smart meters per SS and the response time have a similar impact on application-level throughput. If the number of smart meters in an SS is doubled, the throughput needs to be doubled. If the response time is 1 hour instead of 15 minutes, the throughput is reduced by 4 times. The typical values for the demanding use cases introduced in [29] show an on average less than 3 kbps at the application level per smart meter, considering TCP/IP traffic in 15 minutes response-time scenarios. For an SS with 100 smart meters, the total throughput would be of 300 kbps.

The most common other smart grid application is remote control. Remote terminal units (RTUs) are present at the SSs, and communicate with central SCADA systems with protocols that of late flow over TCP/IP. Some common examples of these protocols are IEC 60870-5-104 [30] and DNP3 TCP/IP [31]. Although the throughput demanded by these protocols heavily depends on the parametrization made at system level to cover the specific requirements of each utility, some common estimations can be made. Reference [32] states that application-level data-rate requirements will be in the 9.6 kbps to 100 kbps range. Common use of generally available 2G services for remote control offer the conclusion that an average throughput of 10 kbps to 20 kbps is in the range of what commonly configured SCADA's need. On the low throughput range end, an average of less than 1 kbps can typically be measured in real IEC 60870-5-104 dedicated channels, with peaks rarely exceeding 2 kbps per SS.

Considering the aggregate throughput for adjacent SSs with communications needs, the PLC technology to be considered should be broadband. If the number of PLC-connected SSs is limited, NB PLC solutions could be used. As already pointed out earlier above, we mostly focus on broadband PLC (i.e. BPL) for MV grids.

9.4.1.3 Communications Resiliency

Resiliency of communications is tightly linked with the capability to operate the grid precisely when it is needed. Resiliency is not usually a technology issue, but a network planning and operation concern. An example of this are public networks. Public networks (especially radio networks, e.g. 2G and 3G) are designed and operated to offer services to residential customers, and as such, offer a certain degree of availability. This availability, provided considering the different propagation considerations and channel disturbances, together with the network (devices) and the planning and maintenance procedures providing network resiliency, configure the final service.

Thus if we consider that once a network is deployed, the end-to-end availability is fixed due to natural constraints, we need to concentrate on providing network protection and redundancy schemes, to keep it in those availability levels. This is what we mean by telecommunications resiliency. When it comes to PLC, power line channels are affected by well-known natural effects (noise being the most important one), that have an impact on availability. Thus with a network using PLC as the communication technology to point-to-point link network devices, we need to configure schemes able to provide resiliency. PLC technologies are mostly focused on OSI layer 1 and layer 2 procedures, and as such, upper layer mechanisms should be used to configure this resiliency. The problem of network resiliency in PLC channels is related to the networking (TCP/IP) technologies commonly used on top of the broadband PLC technologies, and to the different channel qualities that may be found in PLC communications. When the PLC channel lack of availability is measured as a degradation of throughput and/or latency, and not as a traditional abrupt loss-of-signal message [33], telecommunications resiliency must be based on the TCP/IP layer on top of the PLC links used to communicate to the electricity related premises, using east or west-bound links out of them, depending on which one is considered available.

9.4.1.4 Network Planning Procedures

There are several aspects that need to be considered when planning broadband PLC over MV power lines.

The first is the structure of the power lines, which in communication network terms means to discover if the connectivity structure is bus-like or point-to-point. This heavily depends on whether overhead lines or underground cables are used.

The second aspect is related to signal attenuation. The capability to reach certain distances over different MV cable types has been a concern since the early days of PLC. On the one hand, the maximum transmitted signal level is limited by regulation, and reach is limited by the cables (materials, configuration, disposition, etc.) and grid elements in the propagation path. The details will vary depending on the utility, but for some of them, excessive distances in power lines will force the introduction of network repetition points or even the impossibility to reach the far end. On the other hand, overreach may be a concern in point-to-point configurations, as the MV transformers allow part of the PLC signal to be coupled from one MV line to the other, and thus the overreach is translated into interference in the other end.

The third aspect is related to the MAC layer capabilities. MAC layer characteristics depend on the broadband PLC standard or solution chosen. An industry solution, the OPERA technology [12], has been demonstrated in the field in a large scale since 2008, and can be used as a reference. OPERA-based systems have been reported working in telecommunications backbone applications with thousands of links for smart grid applications [34], and that has offered a set of grid and technology-associated planning rules. With TDMA-based OPERA technology, network planning is a process that can be implemented configuring TDMA domains, each working in a different frequency band (FDMA) to control the interfering possibilities of the different TDMA domains. Thus different TDMA groups of SSs cover the deployment area interleaving the different frequency domains (frequency bands).

9.4.1.5   A Real Deployment

It is difficult to find reliable information showing the real situation and use of broadband PLC as a telecommunication backbone technology for smart grid. The largest publicly reported deployment is in Iberdrola's grid in Spain. Other broadband PLC usages as a telecommunication backbone have been reported in several pilots, as those reported by Power Plus Communications (PPC) in Poland [35] and the INTEGRIS Research and Development project [36].

The Iberdrola smart grid deployment bases part of its telecommunications network transport on 3490 MV broadband PLC links (as reported in [34]; [37] reports in December 2014 a total of 8000; and as of July 2015 a total of 11 000+ were working in the field). This deployment continues to grow to eventually cover 10.8 million meters. The architecture of the smart grid deployment is based on a segmented approach that covers the MV grid first as far as the SSs, and from there to the smart meters (on the LV grid using the PRIME HDR NB PLC). In the network segment communicating to the SSs, broadband PLC is used to provide an extended layer 2 local area network. Thus the SSs are connected in broadband PLC groups ('cells', a group of generally connected SSs using the same frequency band; see the set of lines of the same color represented in Figure 9.21), each of which is connected to the central system via a non-PLC link. This architecture leverages broadband PLC to make smart grid communications future-proof, in such a way that whenever the smart grid communications need to extract the maximum throughput available from the broadband PLC deployment (smart meters do not need a high bandwidth, but smart grid solutions such as the ones described in Section 9.4.4 will need it), the backbone connectivity can be improved to offer higher bandwidth to the group of SSs, without needing to deploy new communications infrastructure in each of the SSs.

This broadband PLC deployment is multi-vendor, multi-service, and it is controlled through a network management system to operate and maintain this telecommunications network

Figure 9.21 Broadband PLC (BPL) links connecting SSs (Iberdrola's grid in Spain).

remotely. Reference [34] highlights that 56% of the links performed between 20 and 40 Mbps. The latency of these links (which are connected to ADSL backbones to reach central systems) was in the range of 66 to 200 ms (it is important to note that latency in a pure ADSL segment varies from 62 to 112 ms; typically, the latency for the transmission inside broadband PLC links is in the range of 10 to 20 ms per hop).

9.4.1.5.1 Planning Methodology for MV Grids
One of the challenges of BPL deployment on MV links is the ability to make a proper PLC communications planning that develops a network that can perform within the targeted performance objectives. The planning rules that are needed to carry out this PLC planning and to be applicable to FDMA/TDMA systems (such as the ones reported in the previous section), need to be established considering:

- Frequency isolation, so that two subsystems working in the same frequency band cannot interfere with each other.
- Time resource sharing for the case that the physical media realizes a bus topology.

For both aspects, technology performance needs to be evaluated working on the existing media. Thus, each of the frequency bands needs to be assessed with respect to distance reach and power level, in each of the available media (i.e. MV cables). With this data, maximum reach can be derived, and minimum separation distance is calculated. The assessment needs to be performed in two steps: the first step will take into consideration theoretically available information [12], and the second step will be the validation of the theoretical assumption in the different scenarios.

For instance, the study in [34] concluded that the classification of MV cables according to PLC performance could be reduced to two broad groups: oil-filled paper insulated cables (referred to as 'old' type cables), and cables employing XLPE, EPR or PVC insulation ('new' type cables, which allow for higher-distance PLC links). While the former group exhibits higher attenuation, the latter can achieve higher distances. Neither group is 'good' or 'bad' in terms of performance, as if distance is large it is so both for reach and for interference in nearby SSs. Eventually this information is to be combined with the frequency band used, as lower bands tend to propagate better than higher bands ([34] uses 2 to 18 MHz frequency range, divided in two bands, from 2 to 7 MHz, and from 8 to 18 MHz). Even more, this last statement needs to be supported by the channel responses of broadband PLC couplers.

Once the matrix of values mentioned above has been fixed, the values need to be considered with another group of parameters to control global performance (throughput and latency), within the limits of the possibilities of the technology and its implementation in the product under use and the economic aspects. These parameters include maximum reachable distances, guard distances to keep interference below allowed levels, and the number of SSs connected in the same time and frequency domain to control the latency and the available data rate.

The network planning processes produce different possible configurations, each of them with different advantages and disadvantages. Different figures of merit can be defined, depending on the specific interest of the utility to adapt to each specific scenario.

9.4.1.5.2   High Availability Solutions to Improve Communications Resiliency

The communication topology created with broadband PLC deployments, based on MV power lines and SSs, may typically be either a tree or a linear topology. As the typical network deployment configuration assumes the existence of isolated groups of PLC-connected SSs (with a variable size), the tree or linear (as in Figure 9.22) structure is a topology repeated multiple times (one per each broadband PLC cell). Thus, the most general assumption is that ring topologies cannot be produced with broadband PLC, and the existence of multiple backbones is to be granted if we want to protect communications in the case that a single backbone connection fails. The broadband PLC network must manage to keep connectivity (OSI layer 1 and layer 2), to enable layer 3 connected elements to progress through one of the available existing backbones.

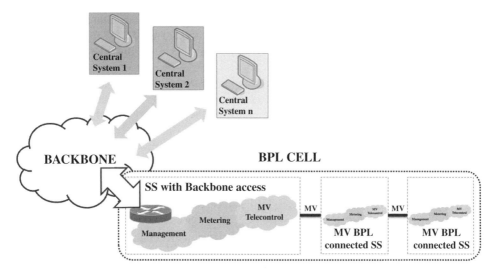

Figure 9.22   MV broadband PLC group connected to central systems.

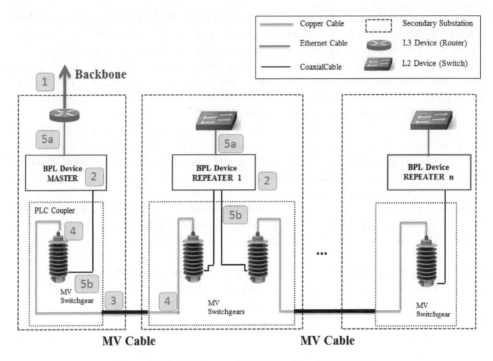

Figure 9.23   Fault types in MV broadband PLC (BPL).

Figure 9.23 shows the points of failure for a broadband PLC connection chain. Reference [34] analyzes redundancy for OPERA-based PLC technology. PLC coupler units need to be installed at each of the ends of the broadband PLC link to inject PLC signals into the power lines. These elements are connected in such a way that communication is possible in any of the positions of the switches of the SSs. If an SS is connected to two other SSs, typically it will have two couplers, one pointing in the direction of each of the connected SSs.

OPERA-based PLC networks need a device acting as the master of the cluster. This device is to be located preferably at the SS with access to the backbone to avoid extra possibilities of backbone disconnection. The different failure possibilities shown in Figure 9.23 are described below, and must be considered in the high availability scenarios:

- Backbone Fault (point 1 in Figure 9.23): When the main backbone access is not available, the PLC cell cannot work.
- PLC Device Fault (either the hardware or the power supply, point 2 in Figure 9.23): If the master fails, the entire PLC cell is unavailable. If a repeater fails, the cell splits into two, and the two half-cells must manage to work simultaneously not to lose any part of the cell.
- PLC Device Link Fault (point 3 in Figure 9.23): This failure may be caused by a high attenuation in the PLC link (MV cable cut or degradation), unexpected noise or low SNR, PLC interference, or saturation in short PLC links.

- Coupler Fault (point 4 in Figure 9.23) or the coupler is not working properly (e.g. bad installation): This situation will possibly cause faults or instability of the link.
- Connection Fault (point 5 in Figure 9.23): The Ethernet cable or the connector itself between switch/router and PLC device (point 5a in Figure 9.23), or the coaxial cable or the BNC connector between PLC device and coupler unit (point 5b in Figure 9.23) may be broken.

9.4.2 PLC in Protective Relaying

In the utility industry, system protection is one of the most important, well-funded and researched disciplines. Its basic target is to detect problems with power system components, and then usually isolate these components from the rest of the system. The set of elements used in system protection includes protective relays, associated communication systems, voltage and current sensing devices, and DC control circuitry. The components which traditionally system protection applies to are generators, transformers, transmission (and distribution, more recently) lines, buses and capacitor banks. These elements are mainly present in HV power lines, and increasingly in MV lines.

9.4.2.1 Pilot Relaying

Specifically when line protection is considered, protective relays at the end of transmission lines which communicate with each other are able to clear faults faster, increase fault sensitivity, and restore power quicker. Historically the term 'pilot' meant that between the ends of the transmission line there was an interconnecting channel of some sort over which information (the pilot signal) could be conveyed. Pilot relaying refers to the communication network implemented on the line to transmit and receive high-speed 'trip or do not trip' commands to/from the remote end of the line, where usually a substation exists. Pilot relays can communicate signals through various channels: telephone line service delivered by a public telecommunications company, fiber optic or metallic (pilot wire) cable installed along the length of the line, microwave links between substations and, of course, PLC over the lines themselves.

Traditionally there have been a number of protection schemes associated with pilot relaying. The most common are:

- Blocking (DCB) or permissive (POTT) pilot relaying schemes,
- Line current differential, which is used normally in shorter lines and requires enhanced latency and performance,
- Direct transfer trip (DTT), technically not a pilot scheme but a remote trip mechanism which requires a pilot channel for the protection of a certain system component (not necessarily a transmission line).

One of the more important electricity utility power lines' related telecommunication services is the one used to protect the power lines themselves. From the telecommunications perspective, the service to provide protective pilot relaying for transmission and distribution lines is known as teleprotection. The communication services for teleprotection in power lines have traditionally been closely related to the protective systems themselves, and at the same time very demanding in terms of performance.

A good example of such performance is the ENA technical specification TS 48-6-7 [38] for communication services for teleprotection systems in the UK. Although among the interfaces admitted for this service PLC is not mentioned, the requirements are a good example of the necessary performance.

The highest tolerance for these communications in terms of propagation delay (end-to-end delay) is 30 ms, and 6 ms is the lowest. The same value applies for differential delay (difference between the go and return path propagation delay). As for the admitted bit error rate (BER), the technical specification defines a parameter called long term BER that should be less than 10^{-8}.

9.4.2.2 Test Deployments

Although PLC started to be used by utilities originally for voice telephony (carrier-current systems), its use for pilot relaying was initiated in the 1930s and became well established over the 1940s and 1950s. Conventional analog PLC on HV lines (pilot channel-based PLC used to provide simple commands) has been used by most utilities for protective purposes for decades.

Broadband PLC has only recently been considered for protective relaying (HDR NB PLC could also be applicable, although no real applications have been reported yet) and may become an attractive alternative for system protection in the future. Supported by the U.S. Department of Energy's National Energy Technology Laboratory (NETL) [39] and carried out by utility AEP along with vendor GridEdge Networks [40], a proof of concept demonstration took place in 2006 over a 0.5 mile, 46 kV line in West Virginia (USA), which succeeded in using broadband PLC signals in the 2 MHz to 35 MHz band for a POTT pilot relaying scheme, which traditionally had not been advised for PLC teleprotection. Furthermore several 69 kV lines in West Virginia and Ohio were trialled afterwards, in which also DCB and Line Current Differential pilot schemes were implemented. Average throughputs in excess of 10 Mbps and typical latencies of some 5 ms were measured within the FCC emission limits for broadband PLC devices. 138 kV lines have also been tested.

It is only logical that HV transmission lines actually represent a relatively good channel for MHz-range broadband PLC. Cables are manufactured to the highest standards, showing constant conductor section and uniform characteristics, while there are less discontinuities or taps than in lower voltages. Signals have less attenuation or noise level in HV lines, and additionally it is easier to deploy noise mitigation techniques such as differential (phase to phase) coupling.

Relying on this excellent performance characteristics, [41] projects the use of this technology not only for protection circuits but for wide band applications such as remote control (SCADA), station video-surveillance and others.

With a different approach (kHz, instead of MHz band) [42] introduces a new use of this PLC in FCC band for distributed generation control applications. Utility independent power producers need to connect to the utility's grid and this requires to extend the protection from the substation to the distributed generation sites, to prevent islanding conditions with anti-islanding PLC-based protection systems.

9.4.3   PLC in Smart Metering

The transition from remote metering to smart metering is blurred by the different technologies used to provide the connectivity to the remotely accessed meters. PLC has been a dominant technology in the connectivity of the different generations of smart meters in the last decades in many of the grids all over the world (and especially in Europe). Both UNB and NB PLC systems have populated the smart metering systems, with a special focus in Europe. From the ripple control systems to the first limited-band NB PLC systems, and then to the HDR NB PLC systems, PLC communications have been prevalent in the access to the communications-capable-electronic meters [43]. Broadband technologies have also been used for these purposes, but to a very limited extent in some cases (see [44]) and with limited success in others (see Korean case of ISO/IEC 12139-1:2009 [45]).

When PLC is used for smart metering purposes, it has traditionally been assumed that a data concentrator should exist. The data concentrator is the distributed intelligence element that, located in the SS, is able to access the smart meters with a certain autonomous intelligence, and that it is able to temporally store the meter readings until the meter data management (MDM) system requests them. This data concentrator has been both a prevalent concept and a likely consequence of the limitations of both high bandwidth backbone connectivity (to be provided in the data concentrators) and the low throughput of some of the LDR NB PLC systems. However, nowadays there is an increased interest in the end-to-end management of smart meters. The existence of broadband telecommunication access in many of the SSs, together with the use of MV broadband PLC, has made a new concept in the form of a communications gateway for the SS appear, in such a way that this element would be providing a direct connectivity between the MDM systems and the smart meters through a series of communications technologies, PLC being the one covering at least the last mile to the meter. Thus, seamless connectivity between the systems and the smart meters can be achieved, and this gateway would be providing the telecommunications media adaptation without any application layer intelligence in the SS.

The evolution of the throughput of the narrowband systems also deserves some attention. LDR NB PLC systems have traditionally offered a few kbps to the system and it has been usual to create ad-hoc application layers capable of improving end-to-end user experience. Of late, this situation has been solved by HDR NB PLC systems, that can afford to absorb the inefficiencies of some standard higher application layers, and the consequences of interoperability at meter level. Their inherent broader-band characteristics have opened a new dimension for smart metering, projecting it into smart grids.

Smart metering is not directly related to in-home environments. However, many of the value added services that could be delivered through smart metering infrastructure could be maximized if the smart meter control could be linked to in-home appliances [46]. Thus active demand management programs could be developed, in such a way that system price orders could be delivered to customers, and these customers could react to them modifying consumption patterns directly delivered to connected appliances through pre-existing programs. An example of this type of proposal can be seen in [47], where a mixed architecture of HDR NB PLC in the SS to the smart meter segment and BPL in the in-home access is proposed. Focusing on the HDR NB PLC domain, some technologies [48] already support the extension of PLC communications directly to the in-home segments through a connectivity service known as Direct Connection. However, not all systems offer a good performance in large in-home/building environments as [49] suggests.

9.4.3.1 PLC Deployments

This section covers the most representative massive PLC deployments reported in Europe, with a special focus on NB PLC systems. Europe has been the region of the world where these systems have mostly been adopted for smart metering deployments, due to the clear definition of CENELEC A-band for utility operational purposes [26].

The PLC band availability, and the emergence of PLC systems (even though some of them might have been proprietary), created the conditions for different utilities to find their business cases to begin deploying AMI solutions, that are in many cases transitioning to smart metering systems with the increased adoption of HDR NB PLC technologies. Of late, with the interest of the European Union in the energy efficiency, many of these smart metering deployments are being mandated, and even if some countries have decided not to use PLC as the main communications media [50] the majority of the utilities are either deploying NB PLC systems or engaged with assessments of the most suitable PLC technology for their coming deployments. The Openmeter project [51] provided a starting point for these assessments and technology evolution, as a consequence of mandates M/441 [52] and M/490 [1].

Enel was one of the pioneers of the massive PLC deployments for smart metering purposes. The Telegestore project (remote management system) has been in place since 2001 [53]. Telegestore system allows the remote reading of the consumption of its customers and the remote management of contractual operations. Enel has installed more than 30 million smart meters in Italy [54]. Endesa in Spain, a subsidiary of Enel, is also proceeding with the deployment of Meters And More [55] technology, based on Enel's solution in Italy (3.5 of their 13 million meters [56]). Meters And More technology can be considered LDR NB PLC technology, with a BPSK modulation allowing a bit-rate of 4800 bps.

Vattenfal engaged in a smart meter deployment of 600 000 units in Sweden [57], with ANSI/EIA – 709 CENELEC A-band technology (LonWorks, by Echelon). It was complete by 2008 [58]. The PLC technology used has evolved and been referenced by ETSI [59] and promoted through the OSGP Alliance [60], and offers a few kbps with a BPSK modulation in CENELEC A-band.

ERDF made a deployment of 'Linky' technology in 300 000 smart meters between 2010 and 2011 [61]. Linky PLC technology [62] functional specification is based on IEC 61334-5-1 PHY and MAC technology. Thus, S-FSK (spread-frequency shift keying) modulation is used with a modulation rate of 2400 bps.

Enexis, a utility in the Netherlands with 2.6 million electricity meters, was proactively pushing the evolution of ITU-T G.9903 (G3-PLC) technology. Enexis was preparing [63] the deployment of its G3-PLC based smart meter deployment, as it was stated in [64]. Their approach was to combine the deployment of 2G-based smart meters with PLC meters, and the architecture proposal makes use of the so-called communication gateways in the SSs, together with the IP nature of their smart meters. Their objective was to have the MDM systems access the smart meters with the same architecture, independently of the last mile telecommunications solution. However, the reported performance data of the first 900 G3-PLC smart meters deployed in the field in 2013 is short [64]. The latest news in [65] introduces some uncertainty on the future of G3-PLC technology within the utility.

Iberdrola is a Spanish utility with 10.8 million electricity meters. Spain has a legal mandate that forces all meters to be substituted by smart meters by the end of the year 2018. Iberdrola is deploying smart meters using the ITU-T G.9904 (PRIME) standard. The experience of

Iberdrola in the first part of its massive roll-out is well covered by [10], [11], [66], and the latest results for their LV smart meter deployment can be found in [34]. The deployment is based in the data concentrator smart meter approach, but offers the novelty of the pre-production stress tests [10], that offer the possibility of understanding and solving any deployment problem. Reference [67] is worth mentioning, as in that paper a definition of availability can be found, so that comparisons among different PLC systems can be performed. The latest news on this deployment can be found in [68] and reports a total of 4.2 million ITU-T G.9904 PRIME smart meters in the field.

ENERGA OPERATOR, a utility in Poland with around 2.8 million customers, is in the process of deploying PRIME technology. In [69] they announced the deployment of 310 000 PRIME smart meters in part of their service area (25% of Polish territory).

Other utilities are already deploying smart meters. According to the figures of Meter On project in June 2013 [70], Gas Natural Fenosa (Spain) had deployed 0.5 of their 4 million meters and Hidrocantabrico (Spain) 0.15 of their 0.7 million meters. These companies have chosen ITU-T G.9904 PRIME smart meters.

Overall, we note that LDR NB PLC are the mostly deployed systems for smart metering. However, HDR systems are being deployed more massively to the extent that their feasibility has been demonstrated, and that their performance exceeds that of their predecessors.

9.4.4 PLC for Low Voltage Grid Control in Smart Grid Applications

An immediate further step in the evolution of the smart metering systems is their evolution and integration into smart grid systems. Smart grids where smart metering is integrated extend the control capabilities present in other voltage levels of the grid (HV, MV) to the LV grid. LV grid assets need to be controlled in the same way other grid segments are, and thus the full control of the grid will be achieved.

However, it is common knowledge [71] that LV grids are not as well documented in utility records as the MV and HV grid components. The basic cause for this situation is that operational databases are not static and change from one grid intervention to the other. Thus, LV grid knowledge and control are seriously handicapped by this lack of reliable information, and LV grid operation and maintenance are not performed efficiently. System average interruption duration index (SAIDI) and the system average interruption frequency index (SAIFI) improvement, common electricity grid quality index, are seriously affected by this inefficiency.

LV grid management should be based on a minimum number of considerations:

- A dynamic smart meter connectivity management is needed, to avoid the need to manually update the connectivity and location of smart meters ([72] shows a typical manual operation instrument to check connectivity in LV grid). A secondary gain of this capability will be the possibility to balance the load of the transformers through the proper connection of smart meters.
- Grid outages need to be identified and located in real time in LV grid segments.
- Tampering attempts need to be identified and narrowed in the involved transformers.
- Remote control of LV grid needs to be available to increase the reliability of the grid in its last segment closest to the customer.

9.4.4.1   Advantages and Examples Using PLC for Smart Grid Operation

Connectivity: One of the first direct advantages of using PLC communications to access smart meters is that in a natural way, the smart meter connected to a feeder is receiving the signal coming from the transformer to which it is electrically connected. This is an obvious issue, that helps to identify the MV/LV transformer connected to the different smart meters. The advantage of this feature is clear if we consider that even if all utilities have drawings and records of LV feeder connectivity, this data is often not accurate or not updated. This mismatch has been quantified in [73] mentioning that the utility records can be between 15% and 20% inaccurate. In fact, this reference mentions that Duke Energy (USA utility) and Enel (European utility) have already benefited from the deployment of PLC technology in this respect, contrary to what happens if the smart meter is based on non-PLC communications (e.g. radio).

Transformer connectivity is important, not only because the utility can understand which customers are to get disconnected when a transformer fails, but to properly balance the load of the transformers, when customers are connected typically through either three or single phase. In the absence of this information, utility procedures restrict themselves to try to connect single-phase customers randomly to each transformer, feeder or phase. However, this approach is not optimal to balance the load of the transformer, and the connectivity of customers (meters) to transformer, feeder and phase is needed.

Reference [74] reports the example of Taipower (Taiwan Power Company), which needed to provide connectivity data to its assets management system. This data was easier to obtain for overhead lines, but not readily available for the underground distribution system. The first non-customer friendly way of getting this information was based on the interruption of customer power service, disconnecting the fuse of each distribution transformer to locate the customers connected, identifying those with power outage. The alternative and more sophisticated procedure was based on PLC technology. A PLC-based transceiver was developed, involving a signal modulation which included an individual transformer coding. This signal was injected in the secondary side of the transformer, and from there traveled as far as the PLC enabled smart meters, to retrieve the transformer code.

References [67] and [75] demonstrate the evolution of connectivity assessment including the feeder identification. In densely populated LV grids, it is easy to find hundreds of meters connected to the same transformer. The way these meters are organized to connect to the transformer is through different LV feeders. As a consequence, the next level of the granularity is the identification of the LV feeder connectivity. The proposal presented in [67], [75] is based on the PRIME HDR NB PLC standard, but could possibly be extended to any CENELEC A-band PLC system with a regular amount of control messages exchanged over the PLC media and a hierarchical PLC connection structure. The proposal uses a combined assessment of the power level of the smart meter-received messages at the transformer in the different LV feeders (with one inductive coupler per LV feeder), and the PLC communication logical topology. The algorithms used provide an increased connectivity accuracy, that improves as the PLC network becomes stable and manages to stay connected.

Non-technical losses: Non-technical losses are those that cannot be attributed to technical aspects, but to other aspects such as administrative considerations (unknown connections not registered due to administrative errors) and illegal connections (tampering). This last aspect has been key in the transition towards smart metering in classical AMR deployment examples (e.g. Enel [76]), as it has helped to justify the investments needed. The tampering level has been estimated by some distribution network operator [76] with an average of one case in 50% of the MV to LV SSs. One way to detect tampering is based on the duplication of measurements. Two metering devices per point of supply (and one of them non-reachable by the user) are used, so that the consumption at the transformer side can be compared with the addition of the consumptions of the individual smart meters [77].

Outage monitoring: Outage control in traditional utilities with no control or monitoring of the LV grid has traditionally been based on customer reporting and on-line complaints. When an area of a city was disconnected in an LV segment of the grid, the only way a utility noticed is by customer reaction and on-site information. However, with the advent of smart metering systems, the disconnection of part of the grid is directly translated in the non-availability of the installed smart meters. If the utility is able to manage the data coming from its smart meters, and the status of the smart meters is reported in real time, outage information is available.

Reference [78] elaborates on the algorithm to detect the unavailable smart meters together with some elements placed in specifically located elements of the network. Polling of the smart meters is used to identify the smart meters affected by the power outage. Modern HDR NB PLC systems can probably avoid application layer polling if they make use of the PLC control layer messages, that run beneath the applications layer and keep the PLC communications layer stable and controlled. The work in [43], [67] specifically addresses the need to correlate geographical information of the smart meters' location, with the communication aspects of the smart meters.

Remote control: The connection of remotely controllable elements in the LV grid is a reality (e.g. photovoltaic plants). Some of these elements are LV elements to control the LV grid itself. An example use case is the LV grid switch, to mechanically connect LV feeders to different SSs. Reference [79] presents a feasibility study on the use of PRIME to allow the remote control of such elements. IP convergence layer would allow the smooth integration of any modern telecontrol application protocol. The MAC data frame would make use of the contention frame period time reservation, to allow this remote control services to work with the controlled quality of service that the remote control of critical services requires.

9.5 Conclusions

PLC technology is an important enabler to the vision of Smart Grid, at all grid power levels. HV power line protective relay systems are being revisited applying PLC systems ready to cope with their strict and demanding requirements. Broadband and narrowband PLC solutions are being deployed in MV and LV grids, which are integral parts of smart grid applications reaching to customers and meters. Smart metering is without doubt the current main focus for smart grid implementation and the application of PLC. Although broadband PLC systems are attractive for smart meter access, their application has mainly been demonstrated at large scale at the MV level, as a telecommunications backbone solution to transport the smart metering data collected at SSs. At the same time, from the SSs to the meters (LV grid) several HDR NB PLC technologies have emerged and been standardized, and some of them have been deployed in the field at a massive (millions of units) scale. The large field deployments have demonstrated the maturity of PLC, both in the narrowband and broadband domain. But smart grid is more than smart metering. The use of PLC systems in smart grids that are more than smart metering is guaranteed, as the nature of the PLC (currently used in e.g. smart metering deployments) offers some characteristics, such as the grid connectivity relationship (especially in the LV grid, as Section 9.4.4.1 describes), that other telecommunication technologies cannot offer. By means of a proper consideration of these aspects, many applications can be developed to provide an enhanced LV control that will make LV grid segment control equivalent to the control that exists in MV and HV segments today.

References

1. M/490 standardization mandate to European Standardisation Organisations (ESOs) to support European smart grid deployment, European Commission, Directorate-General for Energy, Brussels, Belgium, Smart Grid Mandate, 2011. [Online]. Available: http://ec.europa.eu/energy/gas_electricity/smartgrids/doc/2011_03_01_mandatem490_en.pdf

2. Jeju smart grid project, 2009, Jeju Smart Grid Test-Bed: Jeju, Korea. [Online]. Available: http://smartgrid.jeju.go.kr/eng/

3. The Climate Group on behalf of the Global eSustainability Initiative (GeSI), SMART 2020: Enabling the low carbon economy in the information age, 2008, Brussels, Belgium. [Online]. Available: http://www.theclimategroup.org/_assets/files/Smart2020Report.pdf

4. R. Adam and W. Wintersteller, From distribution to contribution. commercializing the smart grid, 2008, Booz & Company: New York, NY, USA. [Online]. Available: http://www.booz.com/media/uploads/From_Distribution_to_Contribution.pdf

5. J. Miller, The smart grid – an emerging option. [Online]. Available: http://www.netl.doe.gov/smartgrid/referenceshelf/presentations/IRPS-Miller.pdf

6. European Commission, European smart grids technology platform: Vision and strategy for Europe's electricity networks of the future, 2006, Brussels, Belgium. [Online]. Available: http://ec.europa.eu/research/energy/pdf/smartgrids_en.pdf

7. Electric Power Research Institute (EPRI), IntelliGridSM: Smart power for the 21st century, 2005, product ID: 1012094, Palo Alto, CA, USA.

8. United States Department of Energy (DOE), GRID 2030 – A national vision for electricitys second 100 years, 2003, Office of Electric Transmission and Distribution: Washington, DC, USA. [Online]. Available: http://energy.gov/sites/prod/files/oeprod/DocumentsandMedia/Electric_Vision_Document.pdf

9. S. Galli, A. Scaglione, and Z. Wang, For the grid and through the grid: The role of power line communications in the smart grid, *Proc. IEEE*, 99(6), 998–1027, Jun. 2011.

10. A. Sendin, I. Berganza, A. Arzuaga, A. Pulkkinen, and I. H. Kim, Performance results from 100,000+ PRIME smart meters deployment in Spain, in *Proc. IEEE Int. Conf. Smart Grid Commun.*, Tainan, Taiwan, Nov. 5–8, 2012, 145–150.

11. A. Arzuaga, I. Berganza, A. Sendin, M. Sharma, and B. Varadarajan, PRIME interoperability tests and results from field, in *Proc. IEEE Int. Conf. Smart Grid Commun.*, Gaithersburg, USA, Oct. 4–6, 2010, 126–130.

12. OPERA, Open PLC European research alliance for new generation PLC integrated network. [Online]. Available: http://www.ist-world.org/ProjectDetails.aspx?ProjectId=cac045d4ca6740c796b80906299b14f3&Source DatabaseId=7cff9226e582440894200b751bab883f

13. ISO/IEC, DKE AK 0.141 PLC of K461: National Requirements for narrowband PLC solutions, 2010. [Online]. Available: https://www.dke.de/de/Service/Installationstechnik/Documents/Nationale Anforderungen an Schmalband-PLC.pdf

14. V. B. Pham, V. A. Nguyen, and L. P. Do, A communication system for smart grids using powerline communications (PLC) technology – field trials and initial measurement results, in *Proc. Int. Conf. Commun. Electron.,* Hue, Vietnam, Aug. 1–3, 2012.

15. A. Haidine, A. Portnoy, S. Mudriivskyi, and R. Lehnert, DLC+VIT4IP project: High-speed NB-PLC for smart grid communication – design of field trial, in *Proc. IEEE Int. Symp. Power Line Commun. Applic.,* Beijing, China, Mar. 27–30, 2012, 88–93.

16. Narrowband orthogonal frequency division multiplexing power line communication transceivers – power spectral density specification, ITU-T, Recommendation G.9901, Nov. 2012. [Online]. Available: http://www.itu.int/rec/T-REC-G.9901-201211-I/en

17. Narrowband orthogonal frequency division multiplexing power line communication transceivers for ITU-T G.hnem networks, ITU-T, Recommendation G.9902, Oct. 2012. [Online]. Available: http://www.itu.int/rec/T-REC-G.9902

18. Narrowband orthogonal frequency division multiplexing power line communication transceivers for G3-PLC networks, ITU-T, Recommendation G.9903, May 2013. [Online]. Available: http://www.itu.int/rec/T-REC-G.9903

19. Narrowband orthogonal frequency division multiplexing power line communication transceivers for PRIME networks, ITU-T, Recommendation G.9904, Oct. 2012. [Online]. Available: http://www.itu.int/rec/T-REC-G.9904-201210-I/en

20. IEEE 1901.2-2013 for low-frequency (less than 500 kHz) narrowband power line communications for smart grid applications, IEEE Standards Association, Active Standard IEEE 1901.2-2013, 2013. [Online]. Available: http://standards.ieee.org/findstds/standard/1901.2-2013.html

21. T. Clausen, A. C. de Verdiere, J. Yi, A. Niktash, Y. Igarashi, H. Satoh, U. Herberg, C. Lavenu, T. Lys, and J. Dean, The lightweight on-demand ad hoc distance-vector routing protocol - next generation (LOADng). [Online]. Available: https://tools.ietf.org/html/draft-clausen-lln-loadng-12

22. PRIME Specification revision v1.4, Specification for powerline intelligent metering evolution, Oct. 2014. [Online]. Available: http://www.prime-alliance.org/wp-content/uploads/2014/10/PRIME-Spec_v1.4-20141031.pdf

23. IEEE standard for local and metropolitan area networks. overview and architecture, IEEE Standards Association, Standard 802-2001 (R2007), 2007.

24. RPL: IPv6 routing protocol for low-power and lossy networks, Internet Engineering Task Force, Tech. Rep. IETF RFC 6550, Mar. 2012.

25. US Federal Communications Commission (FCC), Title 47 of the Code of Federal Regulations, 47 CFR /S15, Sep. 19, 2005, part 15, http://www.fcc.gov/encyclopedia/rules-regulations-title-47

26. Signalling on low-voltage electrical installations in the frequency range 3 kHz to 148.5 kHz, Part 1: General requirements, frequency bands and electromagnetic disturbances, European Committee for Electrotechnical Standardization (CENELEC), Brussels, Belgium, Standard EN 50065-1:2001+A1:2010, 2001.

27. Information technology equipment - Radio disturbance characteristics - Limits and methods of measurement, European Committee for Electrotechnical Standardization (CENELEC), Standard EN 55022:2010, 2010.

28. B. Lichtensteiger, B. Bjelajac, C. Müller, and C. Wietfeld, RF mesh systems for smart metering: System architecture and performance, in *Proc. IEEE Int. Conf. Smart Grid Commun.,* Gaithersburg, USA, Oct. 4–6, 2010, 379–384.

29. High-level smart meter data traffic analysis (For: ENA), Engage Consulting Limited, Document Ref ENA-CR008-001-1.4, May 2010. [Online]. Available: http://www.energynetworks.org/modx/assets/files/electricity/futures/smart_meters/ENA-CR008-001-1 4_Data Traffic Analysis_.pdf

30. Telecontrol equipment and systems – Part 5-104: Transmission protocols – Network access for IEC 60870-5-101 using standard transport profiles, International Electrotechnical Comission, Geneva Standard IEC 60870-5-104 ed.2.0, Jun. 2006.

31. Distributed network protocol website. [Online]. Available: http:/www.dnp.org

32. P. L. Fuhr, W. Manges, and T. Kurugant, Smart grid communications bandwidth requirements. An overview, Feb. 2011, extreme Measurement Communications Center Oak Ridge National Laboratory. [Online]. Available: http://trustworthywireless.ornl.gov/pdfs/Smart-Grid-Communications-Overview-Bandwidth-2011.pdf

33. Loss of signal (LOS), alarm indication signal (AIS) and remote defect indication (RDI) defect detection and clearance criteria for PDH signals, ITU-T, Recommendation G.775, Oct. 1998. [Online]. Available: https://www.itu.int/rec/dologin_pub.asp?lang=e&id=T-REC-G.775-199810-I!!PDF-E&type=items

34. A. Sendin, J. Simon, I. Urrutia, and I. Berganza, PLC deployment and architecture for smart grid applications in Iberdrola, in *Proc. IEEE Int. Symp. Power Line Commun. Applic.*, Glasgow, Scotland, Mar. 30–Apr. 2, 2014, 173–178.

35. J. Koźbiał and T. Wolski, Medium voltage BPL installations: case study from Poland. Smart metering central & eastern Europe 2011. [Online]. Available: http://www.ppc-ag.de/files/2011_smart_metering_cee_ppc_mikronika.pdf

36. Integris: new infrastructure for smart grids, Mar. 2013. [Online]. Available: http://www.enel.com/en-GB/media/news/integris-new-infrastructure-for-smart-grids/p/090027d981f0c977

37. bmp Telecommunications Consultants, bmp TC Broadband PLC Atlas, Dec. 2014. [Online]. Available: http://www.bmp-tc.com/download/WBPLAtlas 2015 Orderform & Description.pdf

38. Communications services for teleprotection systems, Energy Networks Association, ENA Technical Specification 48-6-7 Issue 2 2013, Dec. 2013.

39. B. Renz, Broadband over power lines could accelerate the transmission smart grid, May 2010, dOE/NETL (National Energy Technology Laboratory), DOE/NETL-2010/1418. [Online]. Available: http://www.netl.doe.gov/FileLibrary/research/energyefficiency/smartgrid/articles/06-02-2010_Broadband-Over-Power-Lines.pdf

40. N. Sadan, Transitioning from copper networks with B-PLC, *UTC Journal*, 27–29, 2013, 4th Quarter.

41. N. Sadan, M. Majka, and B. Renz, Advanced P&C applications using broadband power line carrier (B-PLC), in *DistribuTECH Conf. and Exhibition*, San Antonio, USA, Jan. 24–26, 2012.

42. N. Sadan, Distributed generation transfer trip protection using power line carrier technology, in *Utilities Telecom Council Region 1&2 Presentations*, Atlantic City, USA, Sep. 11–13, 2013.

43. A. Sendin, I. Peña, and P. Angueira, Strategies for power line communications smart metering network deployment, *Energies*, 7(4), 2377–2420, Apr. 2014.

44. Q. Liu, B. Zhao, Y. Wang, and J. Hu, Experience of AMR systems based on BPL in China, in *Proc. IEEE Int. Symp. Power Line Commun. Applic.*, Dresden, Germany, Mar. 29–Apr. 1, 2009, 280–284.

45. ISO/IEC, Information technology - Telecommunications and information exchange between systems - Powerline communication (PLC) - High speed PLC medium access control (MAC) and physical layer (PHY) - Telecontrol equipment and systems - Part 1: General requirements, Jul. 2009, ISO/IEC 12139-1.

46. F. Lobo-Llata, A. Cabello, F. Carmona, J. C. Moreno, and D. Mora, Home automation easing active demand side management, in *CIRED Workshop*, Lyon, France, Jun. 7–8, 2010.

47. GAD Project, GAD PROJECT: Active and efficient electric consumption management. [Online]. Available: http://gad.ite.es/index_en.html

48. PRIME Alliance website. [Online]. Available: http://www.prime-alliance.org/

49. L. Di Bert, S. D'Alessandro, and A. M. Tonello, Enhancements of G3-PLC technology for smart-home/building applications, *J. Elect. Comput. Eng.*, vol. 2013, 2013, article ID 746763, doi: 10.1155/2013/746763.

50. OFGEM, Transition to smart meters. [Online]. Available: https://www.ofgem.gov.uk/electricity/retail-market/metering/transition-smart-meters

51. OPEN Meter Project. [Online]. Available: http://www.openmeter.com

52. European Commission, Enterprise and Industry Directorate General, Standardisation mandate to CEN, CENELEC and ETSI in the field of measuring instruments for the development of an open architecture for utility meters involving communication protocols enabling interoperability, Brussels, Belgium, 2009, m/441 EN. [Online]. Available: http://www.etsi.org/images/files/ECMandates/m441EN.pdf

53. Enel, Telegestore - Italy. [Online]. Available: http://www.enel.com/en-GB/innovation/smart_grids/smart_metering/telegestore/

54. ——, Enel's smart meter is the world's benchmark. [Online]. Available: http://www.enel.com/en-GB/media/news/enels-smart-meter-is-the-world-s-benchmark/p/090027d981a1b2f2

55. Meters and more Association. [Online]. Available: http://www.metersandmore.com/

56. R. Denda, The Meter-ON project – key findings, in *Sustainable Energy Week*, Brussels, Belgium, Jun. 24–28, 2013.

57. Telvent, Smart metering solution in Sweden. [Online]. Available: http://www.echelon.com/partners/partner-programs/partner_highlight/telvent/Vattenfall_en_nd.pdf

58. J. Söderbom, Smart meter roll out experiences from Vattenfall. [Online]. Available: http://esmig.eu/sites/default/files/presentation_by_johan_soederbom.pdf

59. PowerLine Telecommunications (PLT); BPSK narrow band power line channel for smart metering applications [CEN EN 14908-3:2006, modified], European Telecommunications Standards Institute, Tech. Specification ETSI TS 103 908 v1.1.1 (2011-10), 2011. [Online]. Available: http://www.etsi.org/deliver/etsi_ts/103900_103999/103908/01.01.01_60/ts_103908v010101p.pdf

60. OSGP Alliance, The open smart grid protocol. [Online]. Available: http://www.osgp.org

61. Le compteur communicant Linky dERDF: Une expérimentation réussie (in French), Électricité Réseau Distribution France, Tech. Rep., Jul. 2011. [Online]. Available: http://www.erdf.fr/medias/dossiers_presse/DP_ERDF_010711_1.pdf

62. Linky PLC profile functional specifications, Électricité Réseau Distribution France: Metering Department, Tech. Rep. ERDF-CPT-Linky-SPEC-FONC-CPL, version: V1.0, 2009. [Online]. Available: http://www.erdf.fr/medias/Linky/ERDF-CPT-Linky-SPEC-FONC-CPL.pdf

63. Enexis, Annual report 2013, 2013. [Online]. Available: https://www.enexis.nl/Documents/investor-relations/enexis-annual-report-2013.pdf

64. ——, G3-PLC at Enexis: Description of G3-PLC technology and pilot results at Enexis, Amsterdam, Netherlands, Oct. 2013, European Utility Week. [Online]. Available: http://www.g3-plc.com/sites/default/files/document/20131015 G3-PLC at Enexis European Utility Week October 15th version 1.0....pdf

65. Berg Insight AB, Smart Metering in Europe 11th Edition, Dec. 2014. [Online]. Available: http://www.reportlinker.com/p02522763-summary/Smart-Metering-in-Europe-11th-Edition.html

66. I. Berganza, A. Sendin, A. Arzuaga, M. Sharma, and B. Varadarajan, PRIME on-field deployment first summary of results and discussion, in *Proc. IEEE Int. Conf. Smart Grid Commun.*, Brussels, Belgium, Oct. 17–20, 2011, 297–302.

67. A. Sendin, I. Berganza, A. Arzuaga, X. Osorio, I. Urrutia, and P. Angueira, Enhanced operation of electricity distribution grids through smart metering PLC network monitoring, analysis and grid conditioning, *Energies*, 6(1), 539–556, Jan. 2013.

68. Metering International, Press Release February 2015, Smart meters Europe: Iberdrola rolls out over 4m, Linky's Brittany pilot, Dec. 2014. [Online]. Available: http://www.metering.com/smart-meters-europe-iberdrola-rolls-out-over-4m-linkys-brittany-pilot/

69. ENERGA, Energa-Operator carries on with building up the smart metering system, Feb. 2013. [Online]. Available: http://media.energa.pl/en/pr/233518/energa-operator-carries-on-with-building-up-the-smart-metering-system

70. Meter-ON project website. [Online]. Available: http://www.meter-on.eu/

71. D. Pollock, Boost power grid resilience – exploring communications for real-time network visibility, *Electricity Today – Transmisssion & Distribution*, 27(3), 8–12, Apr. 2014.

72. Ariadna Instruments. [Online]. Available: http://www.ariadna-inst.com/v2/pub/en/

73. Echelon, Automatic topology mapping with PLC. [Online]. Available: http://www.echelon.com/technology/power-line/topology.htm

74. C.-S. Chen, T.-T. Ku, and C.-H. Lin, Design of PLC-based identifier to support transformer load management in Taipower, *IEEE Trans. Ind. Applic.*, 46(3), 1072–1077, May–Jun. 2010.

75. L. Marron, X. Osorio, A. Llano, A. Arzuaga, and A. Sendin, Low voltage feeder identification for smart grids with standard narrowband PLC smart meters, in *Proc. IEEE Int. Symp. Power Line Commun. Applic.*, Johannesburg, South Africa, Mar. 24–27, 2013, 120–125.

76. P. Kadurek, J. Blom, J. F. G. Cobben, and W. L. Kling, Theft detection and smart metering practices and expectations in the Netherlands, in *Proc. IEEE PES Innovative Smart Grid Technol. Conf. Europe*, Gothenburg, Sweden, Oct. 11–13, 2010, 1–6. [Online]. Available: http://www.cricte2004.eletrica.ufpr.br/anais/IEE_ISGT_2010/2048141.pdf

77. I. H. Cavdar, A solution to remote detection of illegal electricity usage via power line communications, in *Proc. IEEE Power Eng. Soc. General Meeting*, vol. 1, Denver, USA, Jun. 6–10, 2004, 896–900.

78. H. Kuang, B. Wang, and X. He, Application of AMR based on powerline communication in outage management system, in *Proc. Asia-Pacific Power and Energy Eng. Conf.*, Chengdu, China, Mar. 28–31, 2010, 1–4.

79. A. Sendin, I. Urrutia, M. Garai, T. Arzuaga, and N. Uribe, Narrowband PLC for LV smart grid services, beyond smart metering, in *Proc. IEEE Int. Symp. Power Line Commun. Applic.*, Glasgow, Scotland, Mar. 30–Apr. 2, 2014, 168–172.

10

PLC for Vehicles

F. Nouvel and L. Lampe

10.1 Introduction

Most of the often discussed use cases for power line communications (PLC) are connected to communication over AC power networks in homes or in the distribution domain. However, there is another fairly diverse application domain for PLC, which is its use in the autarkic power line networks of vehicles, which are also dominantly DC. The basic motivation for PLC is again the possible reuse of an existing infrastructure and the resulting benefits in terms of ease of deployment. In the case of vehicles two additional factors speak in favor of PLC, which are weight and space. The use of PLC can reduce weight and space of the wiring harness required for both power supply and data communications, which has become a more and more important issue due to the fast proliferation of electronics in vehicles [1].

In this chapter, we elaborate on the use of PLC for vehicles, with a focus on PLC in cars. First, we continue our discussion about advantages of PLC in this application domain in Section 10.2. This is followed by an overview of studies on PLC in different vehicles, from cars to trains, and their key results in Section 10.3. Section 10.4 discusses the main challenges associated with using and implementing PLC for intra-vehicle communications. Part of this section is closely related to the characterization of the PLC channel in vehicles presented in Section 2.9. Then, Section 10.5 presents results for an actual PLC implementation in a car. We close this chapter by a discussion on a recently suggested alternative communication infrastructure, and how PLC could be part of a converged network solution of intra-vehicle (or intra-car) communications in Section 10.6.

10.2 Advantages of PLC

PLC is always then an advantageous method of communication when communication between points connected to a power grid is considered. Many such points exist in vehicles. The reason for this is the massive deployment of electronics in many types of modern vehicles. For

Power Line Communications: Principles, Standards and Applications from Multimedia to Smart Grid, Second Edition.
Edited by Lutz Lampe, Andrea M. Tonello, and Theo G. Swart.
© 2016 John Wiley & Sons, Ltd. Published 2016 by John Wiley & Sons, Ltd.

example, referring to automobiles, we have witnessed an exponential increase in the number of electronic systems since the 1970s, with up to 70 electronic control units (ECUs) exchanging 2500 signals for luxury cars of the year 2004 [2]. A 2007 report by Strategy Analytics estimated that, some 2 billion nodes per year will be connected in car electronic networks by 2014 [3]. This trend leads to considerable pressure on automotive communication networking to accommodate the intra-vehicle information flow. Using extra wires for communication complicates the wiring harness causing an increase in production and installation costs and extra weight and space. The reuse of electricity wires for data communications through PLC however avoids the additional wires and, in addition to cost, weight and space benefits, PLC also enables easy retrofitting of electronic components.

The potential benefits from PLC are even more pronounced in electric vehicles (EVs), in particular electric cars. Compared to conventional internal combustion engine (ICE) vehicles, additional communication for power and battery management is required, supporting regular operation but also complex diagnostics and maintenance. All these communication tasks can be accommodated through PLC, avoiding extra wires and thus, by reducing size and weight of the wiring harness, leading to further improvements in vehicle energy efficiency.

Furthermore, PLC can be used as an independent and redundant communication link, without any wiring overhead. This will reduce failure risks and support 'X by wire' concepts, in which electrical systems replace mechanical and hydraulic ones, with the result of better mileage per Joule. Referring again to EVs, and in particular, plug-in EVs, PLC also enables communication between vehicle and the charging infrastructure over the power plug [4].

10.3   Studies of PLC for Vehicles

In this section, we will review the body of work that has studied the use of PLC in different vehicular environments, namely in automobiles, aircrafts and space ships, ships and trains.

10.3.1   Automotive PLC

As stated in [2], *the use of networks for communications between the ECUs of a vehicle in production cars dates from the beginning of the 1990s.* The reason for introducing communication networks, or buses, was that the use of point-to-point communication links was not scalable with the increasing number of electronic components. As already mentioned above, the demand for efficient networking, including requirements on providing the physical medium (i.e. wires) for communication, has only increased with modern cars being high-technology mechatronical systems. This is the main incentive for considering automotive PLC.

10.3.1.1   Network Classification

The communication networks in cars serve a large variety of purposes, which can roughly be divided into low-speed and high-speed and time-triggered and event-triggered communications. As explained in more detail in [2], the Society for Automotive Engineers (SAE) defines four classes of automotive communication networks based on transmission speed and applications. Class A and B networks for low data-rate not time-critical communication use low-speed event triggered protocols. High-speed real-time communication in Class C and

Class D networks relies on high-speed event and time-triggered protocols, respectively. The latter are used for multimedia data, such as e.g. streaming of audio and/or video signals from music or video players or from cameras monitoring the vehicle perimeter, and safety critical applications, such as, e.g. X-by-wire systems, which impose high requirements on availability of resources and reliability of communications.

One of the first studies of automotive PLC is presented in [5]. It considers the use of spread spectrum technology to accomplish robust and multi-user transmission. Another early work is the dissertation [6], which develops a coded single-carrier PLC system with adaptive equalization and simulates its performance in a model based on measured channel and noise realizations in a passenger car. Data rates of 2 Mbps and 4 Mbps are demonstrated and carrier frequencies of 8 MHz and 12 MHz are identified as favorable. A number of more recent works have considered the use of multicarrier modulation for vehicular PLC. More specifically, technology originally developed for in-home PLC, namely HomePlug 1.0, HomePlug AV, and HD-PLC, was tested in automotive environments. These studies include [7]–[10]. The results are promising in that data rates of up to 10 Mbps could be achieved using an approximately 30 MHz bandwidth. The achievable rates are dependent on the topology and the transmit power spectral density (PSD), which according to [9] should be in the range of -60 dBm/Hz and -80 dBm/Hz to satisfy electromagnetic disturbance limits. Since the medium access control (MAC) protocol used in HomePlug 1.0 is carrier sense multiple access with collision avoidance (CSMA/CA), it would be suitable for Class C networks, while the hybrid time-division multiple access (TDMA) and CSMA/CA of HomePlug AV and HD-PLC could support Class D networks. The considered frequency band is extended up to 100 MHz in [11]. For a set of measured channels and noise power spectral densities representative for different vehicle operation conditions, data rates above 100 Mbps are shown to be achievable. This would support standards for audio video bridging such as the IEEE 802.1 AVB standard.

10.3.1.2   CAN/LIN over PLC

While the above-mentioned works aim at generally high data-rate transmission mostly using established PLC technologies, PLC physical layer designs that mimic the bit-wise arbitration signaling method of the controller area network (CAN) system are studied in [12]. CAN is a highly popular in-vehicle network for data communication between ECUs and real-time control [2]. Bit-wise contention detection and resolution for 'DC-bus' systems is discussed and analyzed in [13]. Commercial solutions based on the proprietary DC-BUS technology for CAN messages of up to 250 kbps are provided by Yamar Electronics Ltd. Their PLC based product family (see [14]), with maximal data rates of 1.3 Mbps and carrier frequencies in the 1.75 to 13 MHz range, also includes other protocols, in particular the popular local interconnect network (LIN), which is a low data-rate serial communication system used for Class-A networks. A LIN protocol over PLC is also presented in [15]. PLC uses binary phase-shift keying (BPSK) and binary amplitude-shift keying (BASK) for master-to-slave and slave-to-master transmission, the reason for which is not entirely clear though. Carrier frequencies of 100 kHz and 2 MHz have been considered, and requirements with regard to coupling and applicable voltage levels are discussed. The benefits of using multiple frequency bands, due to a diversity effect, are demonstrated in [16], considering measured channel transfer functions and PLC with coded 3-ary frequency-shift keying (FSK).

A very different approach to automotive PLC is advocated in [17]. It is suggested to modify the DC wiring harness to provide a more benign medium for PLC. The concept relies on paired wires and the integration of additional impedances for impedance matching.

10.3.1.3 Electric Vehicles

More recently, PLC in electric vehicles (EVs) has been studied. In [18], some commentary is given on communication requirements for EVs and the use of PLC. Channel measurements and modeling approaches are reported in [19]–[22]. Experiments in [21] using commercial modems from Yamar for PLC in an EV demonstrate reliable communication with data rates of about 1 Mbps. Measurement based capacity evaluations for another EV in [22] show 70-th percentiles of data rates at 240 kbps, 140 kbps, 60 kbps, and 140 Mbps for PLC in the Cenelec A, B, C bands and the 2 to 28 MHz band, respectively.

10.3.1.4 Vehicle-to-infrastructure PLC

Finally, PLC has also been considered for communication between the EV and the charging infrastructure. The IEC 61851 standard defines two charging modes, of which Modes 3 and 4 require a control pilot signal [23]. In [24] it is suggested to transmit this pilot via a common-mode PLC signal. The cables used for charging through dedicated EV charging infrastructure, so-called electric vehicle service equipment (EVSE), include a pilot wire though. Then PLC can be used to send data for vehicle identification and billing, grid load optimization, etc. [23] over the AC or DC power cables [25], [26]. Alternatively, PLC signals can be sent over the pilot wire and ground link. This is preferable for the channel characteristics are not affected by AC/DC loads and possible regulatory constraints for PLC do not apply. In fact, since the control wire is not a power line, we cannot really speak of PLC anymore. Both narrowband and broadband PLC solutions have been adopted for the physical layer of communication between EV and EVSE. The SAE J2931/2,3,4 standards define FSK PLC, G3-PLC, and HomePlug Green PHY (HPGP), respectively. The ISO/IEC 15118-3 standard adopted the broadband HPGP as the mandatory PHY/MAC layer technology and the narrowband G3-PLC is specified as an optional mode [27].

10.3.2 PLC in Aircraft and Spacecrafts

The possibility of reducing the mass, volume, and complexity and assembly costs of the wiring harness make PLC also attractive for use in aircrafts and spacecrafts [28], [29]. Even more so than in automotives, on-board electronics is a dominant element in air and space vehicles, and thus the potential for PLC to enable communication for automation and control purposes. On the negative side, there are significant concerns about electromagnetic interference from PLC signals transmitted over unshielded single-wire networks [30]. Depending on the wiring structure, compliance with EMC limits is limiting the applicability of PLC.

With the goal of minimizing the wiring of flight control systems in commercial aircrafts, [31] presents a new system architecture whose key element is to place the control electronics closer to individual actuators, which in turn reduces the long wire runs in large aircrafts. It is further suggested that the data bus between control electronics and flight computers can be transmitted over the two power wires for DC supply. Altogether, the reduction in wiring weight for a Boeing 777 was quantified as approximately 900 pounds, much of it is due to the new architecture rather than PLC though. Using PLC for data communication between flight control and remote electronics is also discussed in [32]. The authors go one step further and replace multiple point-to-point links between control and remote electronics by a shared link. Different multiple access options for the PLC link are discussed, and it is argued that about 3000 m of wires can be removed with a net weight saving of about 17 kg.

While PLC in the above works operated on two power wires, [28] presents experimental results for PLC in a military aircraft that uses a single-wire power bus structure with the aircraft's chassis as return path. Measurements of channel transfer functions in the 10 MHz to 100 MHz band show a highly frequency-selective channel, and OFDM with frequency-division duplexing is suggested as a PLC modulation format to meet data-rate and delay requirements. Since the power lines are not shielded and single-wire, the issue of electromagnetic interference is considered a main concern.

From 2008 to 2012 the research project Transmissions in Aircraft on Unique Path wirEs (TAUPE) [33] funded by the European Community investigated solutions to simplify the (communication) architectures in commercial aircrafts. In particular, to decrease the number and weight of wires in modern aircrafts, merging the power and communication networks using (i) PLC and (ii) Power over Data (PoD) technologies were studied. The reference applications considered were the Cabin Lighting System (CLS) and the Cockpit Display System (CDS), and the wiring architectures were based on the Airbus 380.

A number of publications present the results from the TAUPE project. The feasibility of PLC in aircrafts is discussed [34]. In particular, a simulation model for PLC in the CLS is developed based on multi-conductor transmission line principles. Wire-to-ground (i.e. common mode (CM)) and wire-to-wire (i.e. differential mode (DM)) transmission are compared in terms of insertion gain and data throughput. For the latter, the signal and noise power spectral densities are adjusted according to the RTCA/DO-160 standard for environmental test of avionics hardware. Based on these assumptions, it is found that DM PLC provides data rates between 18 Mbps and 62 Mbps using the 1.8 MHz to 30 MHz band. Due to the low signal power, CM PLC on the other hand is not able to reach all network nodes. To deal with this problem, [30] suggests to replace the single-wire with a double-wire structure, which would allow for DM PLC and reduce electromagnetic emissions and cross-talk compared to CM PLC. This so-called bifiliar approach would use wire pairs with a smaller cross-section than the replaced wires, to minimize the weight increase. Measurement results demonstrated compliance of PLC with RTCA/DO-160 requirements, while achieving sufficient throughput and latency. Further documentation of the results from the TAUPE project are available at [33].

Another simulation study, similar to [34] but for the transport aircraft, was performed in [35]. Considering PLC in the 1 MHz to 30 MHz frequency range with bit-loaded OFDM, feasibility of PLC in this aircraft is demonstrated. However, compliance with emission limits is not discussed.

Spacecrafts are clearly more customized vehicles produced in small numbers, and considering the high demands on weight, data rate, and reliability, fibre optics would seem to be well suited as communication medium. Nevertheless, it is argued in [36] that PLC could be a reasonable alternative or, perhaps more likely, provide an independent back-up network. Considering the wiring structure in a NASA Space Shuttle and using the multi-conductor transmission line method to obtain channel transfer functions, the authors predict throughputs of several tens of Mbps with PLC. The European Space Agency (ESA) has also considered PLC on spacecrafts, including satellites and launch vehicles [29]. As outlined in [29], PLC would be a good candidate to replace a large number of discrete command lines by transmitting on/off switching and other low level commands over the power line. The feasibility and EMC compliance of PLC for such links, in particular a point-to-point link over twisted-pair wire, is studied in [37]. DM and CM signaling are compared in terms of channel transfer function, radiated emissions, and noise immunity. This study includes experimental tests with a narrowband PLC system developed for automotive applications by Yamar Electronics Ltd. (see Section 10.3.1). It is shown that DM PLC is a workable solution for application in spacecrafts.

10.3.3   PLC in Ships

Ships are an interesting application scenario for PLC. Due to the large run length of cables, in e.g. cruise ships, the potential advantages of PLC in terms of reducing complexity and cost of the wiring infrastructure are substantial. Similarly, retrofitting of older ships with a data communication infrastructure, e.g. LAN in cabins, could be done relatively non-invasively. Furthermore, wireless solutions face signal penetration challenges due to metal structures in ships.

A first study on PLC in ships was done in [38]. More specifically, the signal-to-noise ratio for PLC signals was measured to determine the possibility of PLC and favorable frequency bands. While the measured link is not clearly specified, the result plots in [38] demonstrate the dependence of the PLC link quality on the operation of equipment on board of the ship. Experimental results for cabin-to-cabin and in the wheel house of a 'typical cruise ship' in [39] show UDP throughputs of hundreds of kbps with HomePlug 1.0 compatible PLC modems. A model based approach for PLC in (navy) ships is presented in [40]. The power distribution path between a propulsion and control system was modeled through multi-conductor transmission line principles, and achievable data rates were determined in the form of channel capacities. For reasonable transmit powers of between 10 mW and 100 mW, data rates of tens of Mbps are reported, indicating the feasibility of PLC. However, the noise environment was assumed AWGN with a relatively arbitrary noise level. A similar approach was chosen in [41], testing PLC for a switchboard-to-switchboard communication link on a cruise ship. Since the ship was not powered, the effect of the power signal and loads on PLC could not be included. The frequency range of up to 8 MHz was found suitable for PLC. The measurement-based data-rate analysis, again for a cruise ship, in [42] makes somewhat more promising predictions than the previous works. The links between an MV/LV substation on a lower deck and a distribution board on an upper deck as well as between a distribution board and a room service panel were considered. The measurements provided information about the channel transfer function, from which, again under the assumption of AWGN, achievable data rates of hundreds of Mbps were predicted. The extension of the 2 MHz to 30 MHz band to up to 50 MHz showed a significant gain in data rate. Furthermore, it is shown that MIMO transmission sending two independent signals over the ship's three-wire system almost doubles the achievable rate.

An interesting result published in [43] indicates that common-mode coupling at the transmitter and differential-mode reception might be preferable over differential-mode transmission in terms of signal transfer for cargo ships. This is done against the background that braided cables with the metal wire braid grounded to the ship's hull are widely used for cargo ships, and thus radiation is not problematic.

10.3.4   PLC in Transit Systems

Another application domain for PLC in vehicles are transit systems. In a broader sense, this includes vehicle-to-infrastructure communications such as between the subway trains and control centers. Such a scenario, more specifically narrowband PLC over a 750 VDC traction network, is investigated in [44]. Using a measurement study, channel transfer function and noise are characterized, and an OFDM system solution is proposed to enable reliable low data-rate transmission. Specific aspects of this environment for PLC are the noise time variance, since impulse disturbances occur together with the DC current, and the Doppler effect, due to the relative motion between train and infrastructure.

The use of PLC within a train is discussed in [45]. This is a preliminary work pointing out the usability of standardized cables that run through the entire train for PLC.

10.4   Challenges for PLC

While PLC is a seemingly excellent solution for communications in vehicles with reduced weight and cost of wiring harnesses [1], its wide introduction and commercial success has yet to be seen. It is actually not trivial to establish the exact harness simplifications due to PLC. A study on the harness of a Ford Focus vehicle determined that the total length of the communication wires is about 245 m, weighing about 2 kg [46]. Considering the overall weights of vehicles, passengers, and loads, the possible weight reduction due to PLC is not very impressive. The associated savings in manufacturing costs due to reduced raw material and simpler harness assembly can be quite notable though. On the other hand, a PLC node requires a more complicated coupling interface to the DC lines. These economical considerations are somewhat moving targets, depending on the evolution of communications applications in future vehicles but also on regulatory influences especially on automotive manufacturers and even commodity prices. Furthermore, possible complementary or alternative solutions such as a gigabit Ethernet backbone solution as considered in IEEE P802.3bp [47] will play a role.

In this section, we summarize the main technical challenges faced by PLC, for which we focus our discussion on PLC in cars.

10.4.1   Characteristics of the Power Line Channel

The general characteristics of the power line channel in vehicles are discussed in Section 2.9. In terms of signal transfer, the most pronounced characteristics are as follows.

- The channel transfer function is frequency selective due to signal reflections at points of impedance mismatch, leading to multi-path propagation.
- The network impedance is not well defined as it is affected by the impedances of loads connected to the DC power line. Furthermore, the impedance is highly varying with frequency, but also changing in time, as devices continuously change their operation points.
- Since all/most loads are connected to the same battery, the PLC channel is a broadcast channel without well defined boundaries. In fact, the high-frequency communication signals may even propagate through open relays [20].
- The wiring topology differs between many car manufacturers and vehicle types.

The frequency-selective nature of the channel causes the blocking (notching) of some frequency bands, which can be critical for very narrowband PLC. In particular, due to the model, link, and time dependent location of these notches, narrowband solutions require adaptive frequency selection mechanisms. Alternatively, relatively broadband communication, such as multicarrier (typically in the form of orthogonal frequency division multiplexing (OFDM)) solutions mentioned in Section 10.3.1 should be used to cope with the frequency selectivity. The broadcasting nature of the channel means that all/most links share the same communication medium. Multiple access mechanisms, like using different frequency bands for different applications, need to be able to cope with the shared access while at the same time meeting the quality-of-service requirements. One way to enable spatial reuse is the deployment of filters to communication-wise separate parts of the wiring harness. Of course, this compromises the concepts of PLC being a reuse technique, but it can be developed further towards conditioning the harness to be more amenable for communication [17].

The model, link, and time varying network impedance have consequences for signal coupling. To illustrate this point, Figures 10.1 and 10.2 show the access impedance ranges measured in an ICE and hybrid electric car as reported in [48] and [20], respectively. The Smith chart results are normalized to a reference impedance of $Z_0 = 50\,\Omega$. The union of the colored regions represents the complete range of access impedance values measured for frequencies between 100 kHz and 100 MHz. When ignoring some isolated large peaks the regions for impedance values shrink to the lighter shaded areas. The smallest shaded (left-most) area in each figure covers the impedance values when reducing the frequency ranges considered for PLC to 100 kHz to 40 MHz and to 30 MHz to 40 MHz, respectively. The large range of possible impedance is obvious from the figures. Reducing the considered frequency band does significantly shrink the impedance range. However, impedance matching, preferably adaptively so as to adjust itself to the specific link, is required. Due to the frequency selective nature of the impedance (see also Section 2.9, Figure 2.117), wideband matching is difficult to achieve. However, matching over one or multiple, say, 1 MHz-wide band or approximate broadband matching to improve signal transfer should be feasible [49].

10.4.2 Noise and Interference

Since PLC shares the communication medium with the devices connected to the wiring harness for power supply, it is exposed to the unintended high-frequency signals induced by those devices into the power lines. These transient signals are regulated through international and national standards to guarantee immunity of devices. For example, the DIN 40839 defines five test pulses to model transient signals experienced on DC supply lines in cars (cf. [6, Appendix A]). From the viewpoint of PLC systems, such transients are seen as impulsive noise. Measurement studies in e.g. [6], [50], [51] have shown the presence of such impulse noise in the frequency range of PLC. Pulse amplitudes and durations of several volts and microseconds have been measured, respectively, as well as bursts of impulses, in particular during acceleration and braking. Furthermore, [52] shows that sparks produced by the spark plugs of the engine cylinders generate noise impulses of hundreds of millivolts measured at a cigarette lighter outlet. These impulses occur periodically, with the pulse frequency being determined by the revolution cycle of the engine. Components used in the electric vehicles also cause periodic noise components. Switching DC/DC converters and the square waveform current driving the electric engine are identified as sources for such interference for narrowband

NORMALIZED IMPEDANCE AND ADMITTANCE COORDINATES

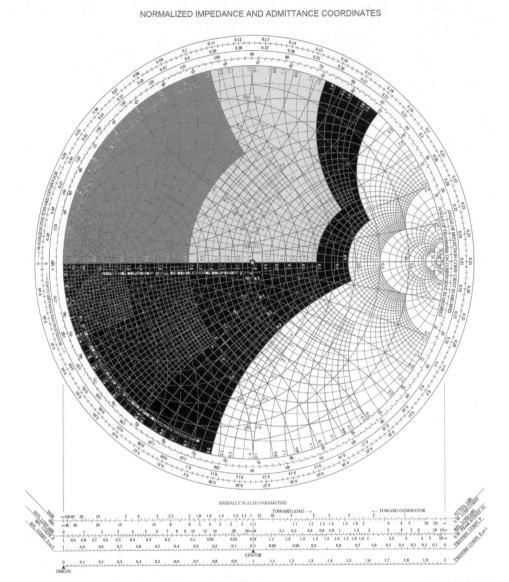

Figure 10.1 Smith chart illustration of access impedance range according to measurement results for an ICE vehicle. All shaded (or colored) areas: complete range of measured results for 100 kHz to 100 MHz. Smaller shaded (green and yellow, top and left) areas: reduced range when excluding some isolated large peak variations. Smallest shaded (green, top left) area: reduced range for frequency range from 100 kHz to 40 MHz. [49] © 2012 IEEE.

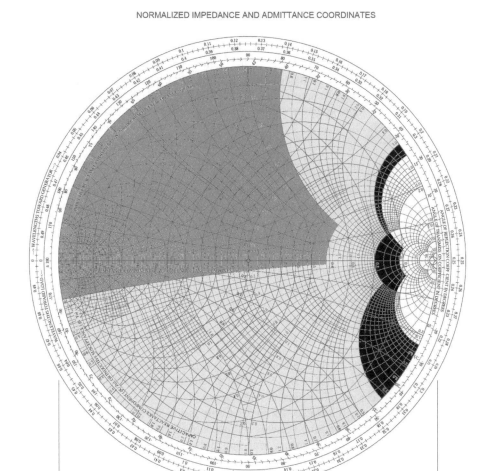

Figure 10.2 Smith chart illustration of access impedance ranges according to measurement results for a hybrid electric vehicle. All shaded (or colored) areas: complete range of measured results for 100 kHz to 100 MHz. Smaller shaded (green and yellow, left) areas: reduced range when excluding some isolated large peak variations. Smallest shaded (green, left-most) area: reduced range for frequency range of 30 MHz to 40 MHz. [20] © 2012 IEEE.

and broadband PLC in [22]. Another source of interference for PLC are radio systems, such as AM and FM radio, which couple into the power lines of vehicles.

The presence of impulsive and narrowband noise is not unique to in-vehicle PLC systems, see Section 2.7. To cope with such noisy environments, two principled mechanisms can be applied. First, noise suppression at the receiver side is possible. This is likely effective against narrowband interference, for which multicarrier systems disregard the strongly inter-fered signal at several carriers. The lost data can be compensated through the application of error-correction coding. Noise impulses of the magnitudes and durations mentioned above can hardly be dealt with through suppression though. For this type of interference, retransmission of interfered data packets, either on request or proactively using packet repetitions when trans-mitting data. The latter would be appropriate for time-critical data. Both mechanisms require that in-vehicle PLC systems integrate sufficient redundancy to ensure reliable transmission over the noisy channel.

10.4.3   Electromagnetic Compatibility (EMC)

As discussed in Chapter 3, EMC of PLC concerns the immunity of PLC systems to electro-magnetic signals and the emission of electromagnetic signals due to the operation of PLC. The former has partly been addressed in the previous section, considering noise and interference. In this section, we briefly discuss regulations for emissions from PLC systems in vehicles.

The EMC regulations for PLC systems summarized in Chapter 3 typically apply to PLC over AC power lines. Furthermore, it is assumed that the power lines are part of the common electricity grid. For PLC in vehicles, we often have DC lines, and the grid is operating autonomously, powered by batteries or local power generators. Depending on the nature of the vehicle, there are international, national, and manufacturer specific norms that apply.

CISPR 12 and 25 and ISO 11451/11452 are the relevant international EMC standards for vehicles, which makes them a first point of reference for vehicular PLC systems. In the automotive domain, further EMC requirements are specified by car manufacturers. Example specifications are Ford's EMC-CS-2009 [53] or the requirements for LIN, CAN, and FlexRay interfaces by German car manufacturers, cf. [15, Figure 7].

For aeronautic systems, the Radio Technical Commission for Aeronautics (RTCA) has developed 'Environmental Conditions and Test Procedures for Airborne Equipment' (RTCA DO-160), which includes radio-frequency emission and susceptibility tests and limits. Other national standards are applied for electromagnetic compatibility, such as for example the U.S. military standard MIL-STD-461, the GAM-T-13 (FR 1982) standard in France and the DEF-STAN 59-41 Part 3 (UK 1995) standard in the UK. The conducted emission limits of DO-160 are specified in terms of the spectral density of the common mode current. They are 20 dBμA/kHz and 40 dBμA/kHz in the 0.15 to 30 MHz range for power lines and inter-connecting cables, respectively. These values would be equivalent to a PSD of, respectively, -83 dBm/Hz and -63 dBm/Hz at 50 Ω. However, if differential mode signaling can be used, transmission with -50 dBm/Hz has been found to meet the common-mode limit [54]. In the case of the MIL-STD 461, the maximum conducted emission level is 60 dBμV/10 kHz at 50 Ω for signals up to 10 MHz. The measurement setup includes a 20 dB attenuation, so that this would correspond to -67 dBm/Hz. In this case, both common and differential mode signals are measured. In [55, Figure I.4.9], the author presents a comparison between all the EMC standards. We can observe that the military standard GAM-T-13 Terre is the most restrictive. Furthermore, as mentioned in [37], additional mission and/or payload specific limits may be applied by space agencies.

For ships, [39] considers the electromagnetic emission requirements specified in IEC 60533 'Electrical and electronic installations in ships – Electromagnetic compatibility' and reports that HomePlug 1.0 compatible PLC modems are in compliance with this norm.

The mentioned standards should be used to determine permissible transmission levels for PLC system. As discussed in Chapter 3, this is a non-trivial task as the conversion of conducted (desired) to radiated (undesired) PLC signals depends on many factors, in particular power line topology and signal coupling. For PLC in vehicles, one of the decisive factors is whether wire-to-wire or wire-to-ground coupling is applied. Some studies mentioned in Section 10.3 have investigated EMC compliance of PLC in vehicles, but it is difficult to derive generally applicable constraints on, for example, the PSD of PLC signals.

10.4.4 Real Time Constraints

As already mentioned, many communication applications in vehicles, in particular control application, impose strict real-time constraints. Automotive networks like high-speed CAN or FlexRay have been designed for real-time delivery of messages. However, they do not support high data-rate applications.

Against this background, there has been a push to adopt Ethernet for converged vehicular backbone networks [56], [57]. Ethernet would naturally support high data rates, but modifications have to be made to provide latency guarantees. Studies for Ethernet (over dedicated twisted pair wires) for vehicular networks have suggested a switched Ethernet architecture to alleviate the collision problem that occurs for a shared Ethernet bus [57]. Furthermore, IEEE 802.1Q message prioritization have been applied to meet delay constraints. In doing so, [58] shows simulation results for switched Ethernet networks, which meet end-to-end delay requirements of 10 ms–100 ms. Furthermore, to also satisfy maximal-delay requirements of 100 μs or below for critical control messages, the Ethernet maximum transmission unit (MTU) is reduced.

In the aircraft domain, the Avionics Full Duplex Switched Ethernet (AFDX) protocol has recently been developed to meet the stringent real-time constraints [59]. It is also based on Ethernet, but with modifications to allow for a deterministic timing behavior and bounded latency and jitter. In addition to this, the network is also redundant to achieve fault tolerance.

Broadband PLC transmission can be used as the physical layer for Ethernet. Since broadband PLC employs multicarrier modulation, parameters such as symbol-length will need to be adjusted to deal with the aforementioned real-time constraints. While such modifications are doable, a difficulty however is the broadcast nature of the PLC channel. Being a reuse technology, no switch-like devices can be deployed, and thus the switched-Ethernet architecture is not possible.

10.5 An Experimental Implementation

In this section, we present results for an actual implementation of in-vehicle PLC, namely for PLC in a passenger car. We first introduce the experimental setup and then discuss the measured transfer function and data throughput results. This section is based on the Ph.D. thesis work by P. Tanguy, and we refer to [60], [61] for further details.

10.5.1 Vehicle PLC Testbed

The PLC experiments have been conducted in a Peugeot 407 SW, which is an ICE car.

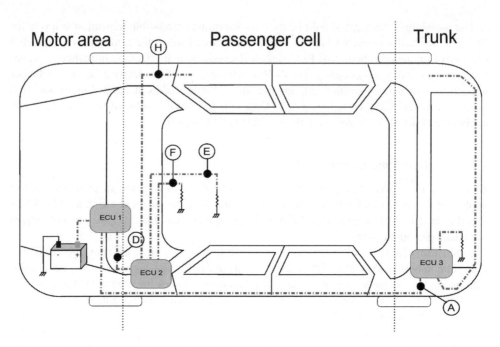

Figure 10.3 Illustration of the experimental setup. The five PLC measurement points are indicated by the upper-case letters.

Five PLC measurement points (A, D, E, F and H) have been considered as illustrated in Figure 10.3. These points are spread widely over the entire vehicle. Point A is a DC power supply in the trunk near the headlights. Points E and F are a cigarette lighter and another DC power supply. Finally, point H is situated on the right side of the vehicle, directly linked with the fuse box, and point D is near the fuse box on the left side of the car.

To include the effect of electric loads on PLC transmission, several states of the car have been considered. These are:

(a) The engine is turned off (SC1).
(b) The engine is turned on, but the car is not moving (SC2).
(c) The engine is turned on, the car is not moving, and electric devices like lights, hazard lights, radio, windscreen wipers, and power windows are used (SC3).
(d) The car is in motion and devices as under (c) are used (SC4).

For PLC transmission, broadband PLC modems using the HomePlug AV (HPAV) and HD-PLC industry standard, respectively, have been used. These were the Devolo dLAN200AV modem for HPAV and the PLC Panasonic BL-PA510KT modem for HD-PLC. These modems are usually deployed for indoor communication. For the operation in the car, the coupling to the DC lines has been modified. Furthermore, the modems do not use the (absent) AC mains cycle for synchronization. The modems have been connected to laptops sending and receiving TCP and UDP traffic and measuring the throughput.

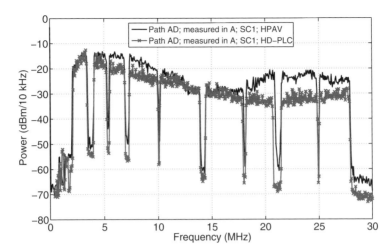

Figure 10.4 Power spectral density in dBm/10 kHz measured with a spectrum analyzer (RBW = 10 kHz, MAX-hold function) for HD-PLC and HPAV signals. Measured at point A for Scenario 1.

10.5.2 Results and Discussion

Figure 10.4 shows a comparison between the power spectral density (PSD) of the HD-PLC and HPAV signals on the power line. For this result, a spectrum analyzer has been used at point A during a PLC session between points A and D. The spectrum analyzer has a resolution bandwidth (RBW) of 10 kHz and the MAX Hold function was used. The PSD specified in the HD-PLC and HPAV standards is −50 dBm/Hz for a 100 Ω termination impedance. This corresponds to a power of −10 dBm in the 10 kHz RBW. The measured PSDs in Figure 10.4 illustrate the coupling loss due to impedance mismatch in the 2 to 28 MHz transmission band. We note that the deep notches are part of the HPAV/HD-PLC PSD mask, which are applied to avoid interference with wireless communication systems.

Figures 10.5 and 10.6 show the measured insertion gain $|S_{21}|$ for different paths[1] and the three use cases of Scenario 1, 2 and 3. In particular, Figure 10.5 shows the results for paths between the rear and the front of the vehicle, and Figure 10.6 shows the results for paths in the front area. We observe the strong frequency selectivity of the channel, indicating many reflection points along the paths. There are significant differences between the insertion gains for different links and for the same link in different vehicle states. This demonstrates the effect of loads connected to the harness on the PLC signal transfer.

Next, Figure 10.7 shows measured throughputs for different paths and Scenario 1. We observe that throughputs achieved with HD-PLC are slightly higher than those for HPAV for all paths in this scenario while the PSD is lower. This is due to the fact that the wavelet OFDM in HD-PLC does not require a guard interval as applied for OFDM in HPAV. The throughputs are in excess of 35 Mbps, which compares well to the 10 Mbps of the FlexRay protocol. Furthermore, the throughputs are relatively similar for different paths, with the highest throughput for the path HD, which experiences the lowest average attenuation.

---

[1] The link between points X and Y is referred to as path XY.

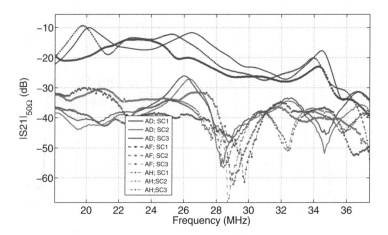

Figure 10.5 Insertion gain for different paths in and states of the vehicle. Paths are between front and rear parts of the car.

Finally, Figures 10.8 and 10.9 show the measured throughputs for different paths and Scenarios 1 to 3 for HD-PLC and HPAV, respectively. Like for the results for Scenario 1 the achieved throughputs are higher than for the FlexRay protocol. The HPAV throughputs are somewhat less path dependent than those for HD-PLC. Again, the highest throughputs are achieved for the path HD. While there are some variations in throughput for different paths, there are little differences because of changes in the vehicle state described by Scenarios 2, 3 and 4. On the other hand, the throughput drops notably when changing from Scenario 1 to the other scenarios (cf. Figure 10.7). This indicates that the powering of electric loads in the vehicle has a strong effect on the PLC link qualities. Part of this difference can be explained by the insertion gain. In Figures 10.5 and 10.6, we can see that the insertion gains for Scenarios

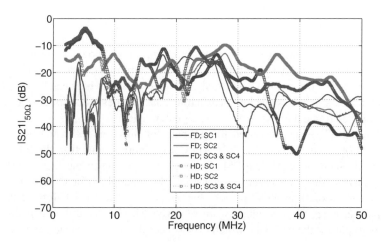

Figure 10.6 Insertion gain for different paths in and states of the vehicle. Paths are in the front part of the car.

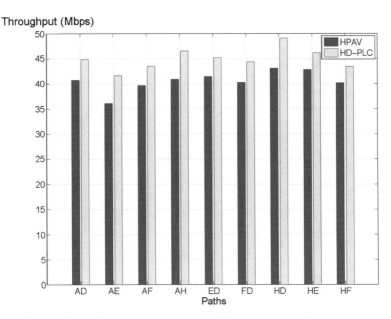

Figure 10.7 Measured throughputs for different paths and Scenario 1.

2, 3, and 4 are consistently worse than for Scenario 1. The only exception is the path HD, for which also throughput is only slightly degraded. Furthermore, also the noise environment usually worsens when the ignition has been turned on.

The throughput results suggest that point-to-point high data-rate communication is possible with PLC, using existing broadband PLC standards. In addition to applying modifications for

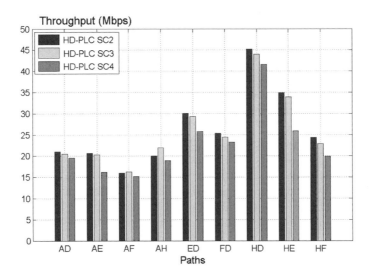

Figure 10.8 HD-PLC throughputs for different paths and Scenarios 2, 3 and 4.

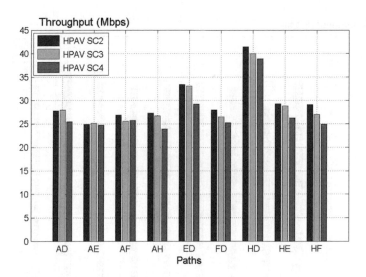

Figure 10.9 HPAV throughputs for different paths and Scenarios 2, 3 and 4.

also supporting applications with stringent real-time constraints, the fact that the power line medium needs to be shared between different links has to be considered when developing PLC-based solutions for vehicles.

10.6   Alternative to and Integration of PLC

As already mentioned in Section 10.4.4, the use of Ethernet over dedicated wires is being considered for communications in vehicles. Typically, Ethernet is used as a high-bandwidth network complementary to existing in-vehicle network technologies such as CAN, LIN, or FlexRay. But as discussed in Section 10.4.4, Ethernet is also envisaged as a converged network for (almost) all data communication applications in vehicles.

In addition to the support of high data rates, another advantage of using Ethernet is that it is a globally recognized standard that supports IP-based networking. This facilitates the adaptation of existing IP-based applications from the consumer and business markets for automotive use and also the development of the vision of connected vehicles through vehicle-to-vehicle and vehicle-to-infrastructure communication. In November 2011, the One-Pair Ether-Net (OPEN) Alliance Special Interest Group (SIG) was formed to 'encourage wide scale adoption of Ethernet-based networks as the standard in automotive networking applications' [62]. The use of single unshielded twisted pair (UTP) cables is considered for low cost and weight of cables.

Ethernet over UTP is clearly an alternative to PLC, with the usual advantages (dedicated communication network) and disadvantages (installation of cables required) compared to PLC. The integration of PLC as a hybrid Ethernet over UTP and PLC is an interesting opportunity though. It would avoid the problems of a completely shared medium (for PLC) and the need to provide a dedicated infrastructure for every connection (for UTP). A switched Ethernet network is possible, using a combination of UTP for the backbone and PLC for branching.

This line of thought can be developed further considering the emergence of Power over Ethernet (PoE). PoE utilizes standard Ethernet cables to transmit both power and data on the same wires to connected devices. The IEEE 802.3at PoE standard provides up to 25.5 W of power, and industry standards go up to about 50 W. However, Category 5 or higher cables are used to achieve these ratings. For PoE using UTP, lower power levels would apply.

Nevertheless, the integration of PoE and PLC is an interesting 'one wire' approach to deliver power and data to electronic components. PLC would be needed when the power level of PoE is too low and/or UTP cables are not desired.

References

1. P. A. Janse van Rensburg and H. C. Ferreira, Automotive power-line communications: Favourable topology for future automotive electronic trends, in *Proc. Int. Symp. Power Line Commun. Applic.*, Kyoto, Japan, Mar. 26–28, 2003, 103–108.
2. N. Navet, Y. Song, F. Simonot-Lion, and C. Wilwert, Trends in automotive communication systems, *Proc. IEEE*, 93(6), 1204–1223, Jun. 2005.
3. Strategy Analytics Automotive Electronics report, Automotive multiplexing protocols: Cost/performance driving new protocol adoption, Nov. 2007. [Online]. Available: http://www.businesswire.com/news/home/200711150061 21/en/STRATEGY-ANALYTICS-Automotive-Electronics-Network-Market-Stretch
4. C. C. Chan, The state of the art of electric, hybrid, and fuel cell vehicles, *Proc. IEEE*, 95(4), 704–718, Apr. 2007.
5. F. Nouvel, G. El Zein, and J. Citerne, Code division multiple access for an automotive area network over power-lines, in *Proc. IEEE Veh. Technol. Conf.*, vol. 1, Stockholm, Sweden, Jun. 8–10, 1994, 525–529.
6. A. Schiffer, Design and evaluation of a powerline communication system in an automotive vehicle, Ph.D. dissertation, Technical University of Munich, Germany, 2001.
7. W. Gouret, F. Nouvel, and G. El Zein, Additional network using automotive powerline communication, in *Proc. Int. Conf. ITS Telecommun.*, Chengdu, China, Jun. 21–23, 2006, 1087–1089.
8. F. Nouvel and P. Maziéro, X-by-wire and intra-car communications: power line and/or wireless solutions, in *Proc. Int. Conf. ITS Telecommun.*, Phuket, Thailand, Oct. 22–24, 2008, 443–448.
9. V. Degardin, M. Lienard, P. Degauque, and P. Laly, Performances of the HomePlug PHY layer in the context of in-vehicle powerline communications, in *Proc. IEEE Int. Symp. Power Line Commun. Applic.*, Pisa, Italy, Mar. 26–28, 2007, 93–97.
10. P. Tanguy, F. Nouvel, and P. Maziéro, Power line communication standards for in-vehicle networks, in *Proc. Int. Conf. ITS Telecommun.*, Lille, France, Oct. 20–22, 2009, 533–537.
11. J. J. Sánchez-Martínez, A. B. Vallejo-Mora, J. A. Cortés, F. J. Cañete, and L. Díez, Performance analysis of OFDM modulation on in-vehicle channels in the frequency band up to 100 MHz, in *Proc. Int. Conf. Broadband and Biomedical Commun.*, Málaga, Spain, Dec. 15–17, 2010.
12. T. Enders and J. Schirmer, Automotive powerline communications – a new physical layer for CAN, in *Proc. Int. CAN Conf.*, Munich, Germany, Oct. 14–16, 2003.
13. O. Amrani and A. Rubin, Contention detection and resolution for multiple-access power-line communications, *IEEE Trans. Veh. Technol.*, 56(6), 3879–3887, Nov. 2007.
14. DC-BUS Products, Accessed May 2013. [Online]. Available: http://www.yamar.com/products.php
15. S. De Caro, A. Testa, and R. Letor, A power line communication approach for body electronics modules, in *Proc. European Power Electron. Applic.*, Barcelona, Spain, Sep. 8–10, 2009, 1–10.
16. M. Mohammadi, L. Lampe, M. Lok, S. Mirabbasi, M. Mirvakili, R. Rosales, and P. Van Veen, Measurement study and transmission for in-vehicle power line communication, in *Proc. IEEE Int. Symp. Power Line Commun. Applic.*, Dresden, Germany, Mar. 29–Apr. 1, 2009, 73–78.
17. T. Huck, J. Schirmer, T. Hogenmuller, and K. Dostert, Tutorial about the implementation of a vehicular high speed communication system, in *Proc. IEEE Int. Symp. Power Line Commun. Applic.*, Vancouver, Canada, Apr. 6–8, 2005, 162–166.
18. E. Bassi, F. Benzi, L. Almeida, and T. Nolte, Powerline communication in electric vehicles, in *Proc. IEEE Int. Elect. Mach. and Drives Conf.*, Miami, USA, May 3–6, 2009, 1749–1753.
19. S. Barmada, M. Raugi, M. Tucci, and T. Zheng, Power line communication in a full electric vehicle: Measurements, modelling and analysis, in *Proc. IEEE Int. Symp. Power Line Commun. Applic.*, Rio de Janeiro, Brazil, Mar. 28–31, 2010, 331–336.

20. N. Taherinejad, R. Rosales, L. Lampe, and S. Mirabbasi, Channel characterization for power line communication in a hybrid electric vehicle, in *Proc. IEEE Int. Symp. Power Line Commun. Applic.*, Beijing, China, Mar. 27–30, 2012, 328–333.

21. S. Barmada, M. Raugi, M. Tucci, Y. Maryanka, and O. Amrani, PLC systems for electric vehicles and smart grid applications, in *Proc. IEEE Int. Symp. Power Line Commun. Applic.*, Johannesburg, South Africa, Mar. 24–27, 2013, 23–28.

22. M. Antoniali, M. De Piante, and A. M. Tonello, PLC noise and channel characterization in a compact electrical car, in *Proc. IEEE Int. Symp. Power Line Commun. Applic.*, Johannesburg, South Africa, Mar. 24–27, 2013, 29–34.

23. P. van den Bossche, N. Omar, and J. van Mierlo, Trends and development status of IEC global electric vehicle standards, *J. Asian Electr. Veh.*, 8(2), 1409–1414, Dec. 2010.

24. C. Bleijs, Low-cost charging systems with full communication capability, in *Proc. Int. Battery, Hybrid and Fuel Cell Electric Veh. Symp.*, Stavanger, Norway, May 13–16, 2009, 1–9.

25. D. Shaver, TI helps developers design affordable, robust and high-performance communications between plug-in electric vehicles (PEVs) and electric vehicle supply equipment (EVSE), white paper, Apr. 2012. [Online]. Available: http://www.ti.com/lit/wp/slyy031/slyy031.pdf

26. C.-U. Park, J.-J. Lee, S.-K. Oh, J.-M. Bae, and J.-K. Seo, Study and field test of power line communication for an electric-vehicle charging system, in *Proc. IEEE Int. Symp. Power Line Commun. Applic.*, Beijing, China, Mar. 27–30, 2012, 344–349.

27. Road vehicles – Vehicle to grid communication interface – Part 3: Physical and data link layer requirements, International Organization for Standardization, ISO/DIS 15118, 2012.

28. C. H. Jones, Communications over aircraft power lines, in *Proc. IEEE Int. Symp. Power Line Commun. Applic.*, Orlando, USA, Mar. 26–29, 2006, 149–154.

29. J. Wolf, Power line communication (PLC) in space – Current status and outlook, in *Proc. ESA Workshop on Aerospace EMC*, Venice, Italy, May 21–23, 2012, 1–6.

30. S. Dominiak, S. Serbu, S. Schneele, F. Nuscheler, and T. Mayer, The application of commercial power line communications technology for avionics systems, in *Proc. IEEE/AIAA Digital Avionics Syst. Conf.*, Williamsburg, USA, Oct. 14–18, 2012, 7E1–1–7E1–14.

31. E. L. Godo, Flight control system with remote electronics, in *Proc. IEEE/AIAA Digital Avionics Syst. Conf.*, vol. 2, Irvine, USA, Oct. 27–31, 2002, 13B1–1–13B1–7.

32. J. O'Brien and A. Kulshreshtha, Distributed and remote control of flight control actuation using power line communications, in *Proc. IEEE/AIAA Digital Avionics Syst. Conf.*, St. Paul, USA, Oct. 26–30, 2008, 1.D.4–1–1.D.4–12.

33. Transmissions in Aircraft on Unique Path wirEs (TAUPE). [Online]. Available: http://www.TAUPE-Project.eu

34. V. Degardin, I. Junqua, M. Lienard, P. Degauque, and S. Bertuol, Theoretical approach to the feasibility of power-line communication in aircrafts, *IEEE Trans. Veh. Technol.*, 62(3), 1362–1366, Mar. 2013.

35. M. D'Amore, K. Gigliotti, M. Ricci, and M. S. Sarto, Feasibility of broadband power line communication aboard an aircraft, in *Proc. Int. Symp. Electromagn. Compat. (EMC Europe)*, Hamburg, Germany, Sep. 8–12, 2008, 1–6.

36. S. Galli, T. Banwell, and D. Waring, Power line based LAN on board the NASA space shuttle, in *Proc. IEEE Veh. Technol. Conf.*, vol. 2, Milan, Italy, May 17–19, 2004, 970–974.

37. F. Grassi, S. A. Pignari, and J. Wolf, Channel characterization and EMC assessment of a PLC system for spacecraft DC differential power buses, *IEEE Trans. Electromagn. Compat.*, 53(3), 664–675, Aug. 2011.

38. J. Yazdani, M. Scott, and B. Honary, Point to point multi-media transmission for marine application, in *Proc. IEEE Int. Symp. Power Line Commun. Applic.*, Athens, Greece, Mar. 27–29, 2002, 171–175.

39. E. Liu, Y. Gao, G. Samdani, O. Mukhtar, and T. Korhonen, Powerline communication over special systems, in *Proc. IEEE Int. Symp. Power Line Commun. Applic.*, Vancouver, Canada, Apr. 6–8, 2005, 167–171.

40. A. Akinnikawe and K. L. Butler-Purry, Investigation of broadband over power line channel capacity of shipboard power system cables for ship communication networks, in *Proc. IEEE Power & Energy Soc. General Meeting*, Calgary, Canada, Jul. 26–30, 2009, 1–9.

41. S. Barmada, L. Bellanti, M. Raugi, and M. Tucci, Analysis of power-line communication channels in ships, *IEEE Trans. Veh. Technol.*, 59(7), 3161–3170, Sep. 2010.

42. M. Antoniali, A. M. Tonello, M. Lenardon, and A. Qualizza, Measurements and analysis of PLC channels in a cruise ship, in *Proc. IEEE Int. Symp. Power Line Commun. Applic.*, Udine, Italy, Apr. 3–6, 2011, 102–107.

43. S. Tsuzuki, M. Yoshida, Y. Yamada, H. Kawasaki, K. Murai, K. Matsuyama, and M. Suzuki, Characteristics of power-line channels in cargo ships, in *Proc. IEEE Int. Symp. Power Line Commun. Applic.*, Pisa, Italy, Mar. 26–28, 2007, 324–329.

44. P. Karols, K. Dostert, G. Griepentrog, and S. Huettinger, Mass transit power traction networks as communication channels, *IEEE J. Sel. Areas Commun.*, 24(7), 1339–1350, Jul. 2006.

45. S. Barmada, A. Gaggelli, A. Musolino, R. Rizzo, M. Raugi, and M. Tucci, Design of a PLC system onboard trains: Selection and analysis of the PLC channel, in *Proc. IEEE Int. Symp. Power Line Commun. Applic.*, Jeju Island, Korea, Apr. 2–4, 2008, 13–17.

46. N. Taherinejad, Estimation of length and weight of communication wires in a typical car, University of British Columbia, Tech. Rep., 2013.

47. K. Pretz, Fewer wires, lighter cars, The Institute (IEEE). [Online]. Available: http://theinstitute.ieee.org/benefits/standards/fewer-wires-lighter-cars

48. N. Taherinejad, R. Rosales, S. Mirabbasi, and L. Lampe, A study on access impedance for vehicular power line communications, in *Proc. IEEE Int. Symp. Power Line Commun. Applic.*, Udine, Italy, Apr. 3–6, 2011, 440–445.

49. ——, On the design of impedance matching circuits for vehicular power line communication systems, in *Proc. IEEE Int. Symp. Power Line Commun. Applic.*, Beijing, China, Mar. 27–30, 2012, 322–327.

50. V. Degardin, P. Laly, M. Lienard, and P. Degauque, Impulsive noise on in-vehicle power lines: Characterization and impact on communication performance, in *Proc. IEEE Int. Symp. Power Line Commun. Applic.*, Orlando, USA, Mar. 27–29, 2006, 222–226.

51. V. Degardin, M. Lienard, P. Degauque, E. Simon, and P. Laly, Impulsive noise characterization of in-vehicle power line, *IEEE Trans. Energy Convers.*, 50(4), 861–868, Nov. 2008.

52. A. B. Vallejo-Mora, J. J. Sánchez-Martínez, F. J. Cañete, J. A. Cortés, and L. Díez, Characterization and evaluation of in-vehicle power line channels, in *Proc. IEEE Global Telecom. Conf.*, Miami, USA, Dec. 6–10, 2010, 1–5.

53. Component EMC Specifications EMC-CS-2009, Ford Motor Company, Tech. Specifications. [Online]. Available: http://www.fordemc.com/docs/requirements.htm

54. S. Dominiak, H. Widmer, M. Bittner, and U. Dersch, A bifilar approach to power and data transmission over common wires in aircraft, in *Proc. IEEE/AIAA Digital Avionics Syst. Conf.*, Seattle, USA, Oct. 16–20, 2011, 7B4–1–7B4–13.

55. M. Beltramini, Contribution à l'optimisation de l'ensemble convertisseur / filtres de sortie vis à vis des contraintes CEM avion (in French), Ph.D. dissertation, Institut National Polytechnique de Toulouse, Toulouse, France, 2011.

56. Y. Kim and M. Nakamura, Automotive Ethernet network requirements, IEEE 802.1 AVB Task Force Meeting, Mar. 2011.

57. K. Matheus, Ethernet in cars: an idea whose time has come, *Automotive Eng. Int. Online*, Jun. 2012. [Online]. Available: http://www.sae.org/mags/aei/11142

58. Y. Lee and K. Park, Meeting the real-time constraints with standard Ethernet in an in-vehicle network, in *Proc. IEEE Intell. Veh. Symp.*, Gold Coast, Australia, Jun. 23–26, 2013, 1313–1318.

59. ARINC Specification 664: Aircraft Data Networks, part 7 – avionics full duplex switched ethernet (AFDX) network, Aeronautical Radio, Inc., Specification, 2004.

60. P. Tanguy, Etude et optimisation d'une communication par courant porteur á haut débit pour l'automobile (in French), Ph.D. dissertation, Institut National des Sciences Appliquées de Rennes, France, 2012.

61. F. Nouvel, P. Tanguy, S. Pillement, and H. M. Pham, Experiments of in-vehicle power line communications, in *Advances in Vehicular Networking Technologies*, M. Almeida, Ed. InTech, 2011, ch. 14, 255–278.

62. Official website of Open Alliance Special Interest Group, 2011. [Online]. Available: http://www.OPENsig.org

11

Conclusions

L. Lampe, A. M. Tonello, and T. G. Swart

Power line communications (PLC) is a topic with many facets. From a communication engineering perspective there are notable similarities to wireless communications. Being a reuse technique, the signal propagation and noise environment is harsh and usually not under the control of the communication system designer. But being a wireline medium, it also shares commonalities to wireline communication, most notably a certain determinism of signal propagation as well as a high penetration in, e.g., indoor environments. PLC is unique among the communication technologies in that the communication signal is superposed to a relatively much higher voltage signal, which in turn requires protection of the communication equipment. This as well as the understanding of the characteristics of the communication medium call for power engineering expertise when designing, analyzing and deploying PLC systems. From an application and business perspective, PLC has experienced a long journey from its original use by electric utility companies to industry and home automation to Internet access and multimedia communication and recent renewed interest in modern PLC solutions for smart grid applications. The latter has seen significant innovations, which we expect to continue especially in the context of the unique 'through the grid' property of PLC.

The objective of this second edition of *Power Line Communications* is to provide an updated comprehensive coverage of the topic.

Channel characterization, covered in great detail in Chapter 2, is certainly at the heart of understanding and designing systems for communications over power lines. We have presented a number of different approaches for power line channel characterization used in the past. These can broadly be categorized into phenomenological (or top-down) and physical (or bottom-up) approaches. Models for channel frequency responses resulting from the phenomenological approaches show some similarities to models used in wireless communications. The bottom-up approaches have strong commonalities with those used for other wireline communication techniques. The noise scenario is quite unique in its richness and perhaps hostility for communications. The significant channel characterization efforts, which have accelerated over the past 15 years especially for broadband PLC, have led to a fairly solid understanding

Power Line Communications: Principles, Standards and Applications from Multimedia to Smart Grid, Second Edition.
Edited by Lutz Lampe, Andrea M. Tonello, and Theo G. Swart.
© 2016 John Wiley & Sons, Ltd. Published 2016 by John Wiley & Sons, Ltd.

and accepted modeling paradigms for signal propagation. Substantial progress has also been made on noise modeling. Still, further research efforts are required towards the consolidation of standardized channel models, applicable to the different grid domains and types of PLC systems.

One of the big challenges for the wide acceptance of PLC lies in issues surrounding electromagnetic compatibility (EMC), which have been discussed in Chapter 3. Due to the nature of power line installations, which are not built to prevent radiation, the potential of broadband PLC systems causing harmful interference to radio services has been a strongly contentious issue ever since. These concerns are mostly related to broadband PLC, and recent regulatory efforts especially in Europe will help to provide certainty about emission limits. Similarly, the development of cognitive transmission concepts provides PLC systems with the capability to adjust to varying EMC requirements. Future research will need to address challenges arising from the emission of non-PLC electronic devices especially in frequency bands used by narrowband PLC.

As mentioned above, communication and power systems operate at very different voltage and power levels. Therefore, they cannot merely be interconnected. Chapter 4 presented principles of coupling, including coupling circuits and their analysis and approaches for low-voltage, medium-voltage and high-voltage coupling. Low-loss, small-form factor and adaptive coupling solutions that provide sufficient protection to both humans and the modems connected to the coupler will remain an important topic of research and development.

In Chapter 5 we presented a fairly comprehensive discussion of modulation, coding and detection techniques suitable for both narrowband and broadband PLC systems. These include the more classical single-carrier transmission methods and the relatively newer multicarrier transmission formats, which have been adopted for all recent high data-rate PLC specifications. In addition, current and voltage modulation techniques were introduced as practical and low-cost solutions which are quite unique to PLC. The presented ultra-wideband impulsive transmission methods are another intriguing direction for PLC. Multiple-input multiple-output communications as discussed in this chapter have already been included in PLC standards, which is indicative of the integration of the latest communication-theoretic innovations into practical PLC systems. We expect this trend to continue, and further research to contribute to more channel and interference adaptive transmission methods.

The PLC channel is inherently a broadcast channel with no well defined boundaries. Chapter 6 addressed the allocation of resources among a set of communication nodes, possible organized in networks of multiple cells. Here again similarities to wireless communication, especially cellular wireless, are exploited for the proper design of resource allocation methods. In addition, the evolution of classical signal repetition to modern relaying or more generally collaborative communication was discussed in this chapter. We believe that further innovation in medium access protocols is needed especially in the context of throughput efficient transmission over impulse-noise (or burst-noise) channels, which render existing protocols for random access with frequent (per-packet) acknowledgments inefficient.

Since PLC is applicable whenever there is an electrification infrastructure, there are numerous application domains for this technology. Our coverage in Chapters 7 to 10 included home and industry automation, in which PLC has long played an important role, multimedia PLC systems, for which broadband PLC has become an established solution and alternative and extension for wireless systems, smart grid communication, for which PLC has been the original and in a sense 'organic' choice, and finally vehicular communication, which is one of the examples for PLC over direct-current lines. Much of the interest in and consolidation of modern broadband and high data-rate narrowband PLC systems can be seen from the list of industry and international PLC standards that have been developed and approved during the past 10 years, especially for the multimedia and smart grid application scenarios. We expect those standards to evolve further considering, for example, updates to regulations and improved transmission and access techniques.

In conclusion, PLC is a mature technology that has experienced significant innovation especially during the past 20 plus years. It serves a broad range of applications, not unlike wireless communications. Innovations of the technology should address reliability and efficiency of communication also in the face of changes to the transmission environment due to proliferation of new types of electronic loads.

Index

Power Line Communications: Principles, Standards and Applications from Multimedia to Smart Grid, Second Edition.
Edited by Lutz Lampe, Andrea M. Tonello, and Theo G. Swart.
© 2016 John Wiley & Sons, Ltd. Published 2016 by John Wiley & Sons, Ltd.